Asymptotic Expansions of Integrals

Asymptotic Expansions of Integrals

Norman Bleistein
Colorado School of Mines

Richard A. Handelsman
Glencoe, Illinois

Dover Publications, Inc., New York

Published in Canada by General Publishing Company, Ltd., 30 Lesmill Road, Don Mills, Toronto, Ontario.

Published in the United Kingdom by Constable and Company, Ltd., 10 Orange Street, London WC2H 7EG.

This Dover edition, first published in 1986, is an unabridged and corrected republication of the work first published by Holt, Rinehart and Winston, New York, 1975. A new Preface has been written for this edition by the authors.

Manufactured in the United States of America

Dover Publications, Inc., 31 East 2nd Street, Mineola, N.Y. 11501

Library of Congress Cataloging-in-Publication Data

Bleistein, Norman.

Asymptotic expansions of integrals.

Reprint. Originally published: New York: Holt, Rinehart and Winston. 1975. With new pref.

Bibliography: p.

Includes index.

1. Integrals. 2. Asymptotic expansions. I. Handelsman, Richard A. II. Title.

QA311.B58 1986 515′.43 85-29317

ISBN 0-486-65082-0

To our wives, Sandy and Cheryl, for their loving support;
to our children, Steven, Abby, Gary, and Robin;
and in memory of our teacher, Bob Lewis.

Preface to the Dover Edition

In the eleven years since the preface to the first edition was written, the research that I (N.B.) have carried out has repeatedly confirmed the power and utility of asymptotic analysis, in general, and asymptotic expansions of integrals, in particular. I have also found that my presentations of this material in the classroom have followed very closely the presentations of this book, some of which were first prepared as part of a set of lecture notes, ten years before the first edition was printed!

The first edition has retained a loyal following, which has been extremely gratifying to both of us. This has been true despite an embarrassingly large number of typographical and other errors, which admittedly detracted from the overall value of the original edition. The present edition has been prepared by directly correcting the original pages and then photographing the corrected pages. This method minimizes the probability of introducing new errors, while maximizing the likelihood of eliminating the old errors. Thus, we hope that this Dover edition will be welcome both to owners of the original edition and to new readers.

In regard to corrections, we would like to express a special debt of gratitude to John Boersma, whose careful reading and extensive list of corrections have provided the bulk of outside information for this edition.

The bittersweet process of producing a book such as this remains an important part of our professional and personal growth as applied mathematicians. We would not have missed that experience for anything!

N. Bleistein
R.A. Handelsman

Preface to the First Edition

Asymptotic analysis is that branch of mathematics devoted to the study of the behavior of functions in particular limits of interest. This book is concerned with the theory and technique of asymptotic expansions of functions defined by integrals. Although the subject matter might appear narrow at first glance, in actuality its scope is quite large, and it is particularly relevant to applied mathematics. Indeed, the solutions to a large class of applied problems can, by means of integral transforms, be represented by definite integrals. Exact numerical values are often difficult to obtain from such representations, in which event one must resort to some method of approximation.

We might also mention that many of the techniques to be developed here have counterparts in other areas of asymptotic analysis. Thus it is expected that once these techniques are learned, they will prove useful in the other areas as well.

In the past, asymptotic analysis was considered more an art than a discipline. This was mostly due to the fact that researchers in the field were from a wide variety of scientific areas and their methods were developed for specific problems or for a narrow group of problems. As a result, asymptotic techniques appear to be rather ad hoc in nature. One of the goals of our book is to establish that there are certain underlying principles in the asymptotic analysis of integrals which enable one to attack problems in a systematic way.

It is hoped that this book will be used by a variety of people. Basically we feel that the student, the user, and the researcher will all find the book of value. Because the needs of each of these groups are obviously different, we have attempted to start each chapter with either elementary motivational material or an informal development of the subject matter. While the initiated user or researcher might find this material useful for reference purposes, it is expected that neither will find it necessary to read these sections in detail. In fact, such readers will likely find the material in Chapters 1–3 familiar and will, in most instances, want to concentrate on Chapters 4–9.

To aid the person who merely wishes to use certain results and is not particularly interested in the theory from which these results were obtained, we have prepared an extensive index. This index lists and locates all of the special functions whose asymptotic expansions have been derived in the text and also locates the formulas associated with the major techniques of asymptotic analysis.

The book is primarily oriented as a text for students of applied mathematics. One quickly learns that asymptotic analysis lies at the heart of applied mathematics and that it is crucial for the student to have a solid foundation in the subject. Although this book covers only a portion of the entire field, we feel that a course based on its contents will go a long way in developing the desired foundations.

The prerequisites for such a course are minimal. Indeed, we feel that any student having a good background in advanced calculus, differential equations, and complex variables can adequately handle the contents.

Actually the book contains more material than can be covered in a single quarter or even in a single semester course. In fact, we believe that a full year would be required to cover the entire text. Nevertheless, meaningful one-quarter and one-semester courses can be readily designed. A one-quarter course might consist of the material in Chapters 1, 2, and 3 and sections 4.1, 5.1, 6.1, and 7.1–7.3. This covers the so-called standard methods that is, integration by parts, Watson's lemma, LaPlace's method, stationary phase, and steepest descents. For a one-semester course one might add to the above the remaining portions of Chapters 4 and 7. Chapter 4 is devoted to the Mellin transform method, and the latter sections of Chapter 7 are concerned with less elementary aspects of the method of steepest descents.

The amount of material included in a one-year course is dependent on the thoroughness with which each topic is treated. We might mention however, that some topics, especially those in the chapter on uniform methods, are not intended for use in an introductory course; rather, they have been included for the sake of completeness.

A great deal of effort has been devoted to preparing the exercises for each chapter. As in so many courses of an applied nature, we feel that the material can best be learned by doing as many problems as possible. Whenever feasible, the answers to complicated exercises are given by simply asking the student to "show that...". We might also mention that new results are to be derived in the exercises in several instances.

Some remarks should be made concerning the selection of material. The subject matter is itself sufficiently vast that it would be impossible to treat it completely in a single volume of reasonable size. Thus we had to be selective in preparing our work. In the final analysis, we chose those topics which we felt were most useful and which most lent themselves to a coherent and systematic presentation. Of course, our selection is biased by our own experience, and we recognize that other authors might quite honestly come up with different ones.

One aspect of asymptotic analysis has been deliberately omitted from our treatment, namely, that of obtaining precise numerical estimates or errors. In other words, we have in all cases contented ourselves with establishing the asymptotic nature of our results by means of traditional order estimates and have not attempted to develop methods for attaching numerical values to errors.

We wish to say a few words concerning other works in the area of asymptotic analysis of integrals and our system of referencing in general. There are several existing books, as well as a multitude of research papers, that are worthy of mention. We have found it impossible, however, to make specific reference to all of these and have therefore been selective in our referencing. Basically we have chosen those references which we feel are most useful from the point of view of the student.

We have used a two-decimal-point numbering system for equations, theorems, lemmas, examples, and figures. Thus Figure 7.2.3. refers to the third figure in the second section of Chapter 7. We deviate from this practice for the case of "one of a kind." Thus Theorem 4.2 refers to the *only* theorem in section 2 of Chapter 4. The second section of Chapter 4 is itself referenced as section 4.2.

Finally we wish to thank many people who have aided us in the preparation of this work. In particular, we acknowledge the helpful discussions with and the encouragment from Professors J.B. Keller, D. Ludwig, and the late R.M. Lewis. We thank Professors V. Barcilon and W.E. Olmstead for reading the manuscript and offering their constructive criticisms. We also thank E. Adams, R. Mager, and D. Strawther for testing the text and exercises and K. Russell for an excellent job of typing. Finally, we would like to take this opportunity to express our gratitude to those agencies who have over the years supported our research efforts, many of which have been incorporated into the text. These agencies are the Air Force Cambridge Research Laboratories, the Air Force Office of Scientific Research, the National Science Foundation, and the Office of Naval Research.

September, 1974
Denver, Colorado
Chicago, Illinois

N. Bleistein
R.A. Handelsman

Contents

CHAPTER 1. **Fundamental Concepts**

1.1.	*Introduction*	1
1.2.	*Order Relations*	6
1.3.	*Asymptotic Power Series Expansions*	9
1.4.	*Asymptotic Sequences and Asymptotic Expansions of Poincaré Type*	14
1.5.	*Auxiliary Asymptotic Sequences*	19
1.6.	*Complex Variables and the Stokes Phenomenon*	21
1.7.	*Operations with Asymptotic Expansions of Poincaré Type*	25
1.8.	*Exercises*	33
	References	40

CHAPTER 2. **Asymptotic Expansions of Integrals: Preliminary Discussion**

2.1.	*Introduction*	41
2.2.	*The Gamma and Incomplete Gamma Functions*	42
2.3.	*Integrals Arising in Probability Theory*	44
2.4.	*Laplace Transform*	46
2.5.	*Generalized Laplace Transform*	50
2.6.	*Wave Propagation in Dispersive Media*	53
2.7.	*The Kirchhoff Method in Acoustical Scattering*	55
2.8.	*Fourier Series*	60
2.9.	*Exercises*	62
	References	68

CHAPTER 3. **Integration by Parts**

3.1.	*General Results*	69
3.2.	*A Class of Integral Transforms*	76
3.3.	*Identification and Isolation of Critical Points*	84
3.4.	*An Extension of the Integration by Parts Procedure*	89
3.5.	*Exercises*	92
	References	101

CHAPTER 4. *h*-Transforms with Kernels of Monotonic Argument

4.1. *Laplace Transforms and Watson's Lemma* 102
4.2. *Results on Mellin Transforms* 106
4.3. *Analytic Continuation of Mellin Transforms* 110
4.4. *Asymptotic Expansions for Real* λ 117
4.5. *Asymptotic Expansions for Real* λ *: Continuation* 125
4.6. *Asymptotic Expansions for Small Real* λ 130
4.7. *Asymptotic Expansions for Complex* λ 136
4.8. *Electrostatics* 146
4.9. *Heat Conduction in a Nonlinearly Radiating Solid* 152
4.10. *Fractional Integrals and Integral Equations of Abel Type* 155
4.11. *Renewal Processes* 159
4.12. *Exercises* 165
References 178

CHAPTER 5. *h*-Transforms with Kernels of Nonmonotonic Argument

5.1. *Laplace's Method* 180
5.2. *Kernels of Exponential Type* 187
5.3. *Kernels of Exponential Type: Continuation* 195
5.4. *Kernels of Algebraic Type* 199
5.5. *Expansions for Small* λ 206
5.6. *Exercises* 209
References 218

CHAPTER 6. *h*-Transforms with Oscillatory Kernels

6.1. *Fourier Integrals and the Method of Stationary Phase* 219
6.2. *Further Results on Mellin Transforms* 224
6.3. *Kernels of Oscillatory Type* 229
6.4. *Oscillatory Kernels: Continuation* 238
6.5. *Exercises* 245
References 251

CHAPTER 7. The Method of Steepest Descents

7.1. *Preliminary Results* 252
7.2. *The Method of Steepest Descents* 262
7.3. *The Airy Function for Complex Argument* 281
7.4. *The Gamma Function for Complex Argument* 286
7.5. *The Klein-Gordon Equation* 292
7.6. *The Central Limit Theorem for Identically Distributed Random Variables* 298

7.7. *Exercises* 302
References 319

CHAPTER 8. **Asymptotic Expansions of Multiple Integrals**

8.1. *Introduction* 321
8.2. *Asymptotic Expansion of Double Integrals of Laplace Type* 322
8.3. *Higher-Dimensional Integrals of Laplace Type* 331
8.4. *Multiple Integrals of Fourier Type* 340
8.5. *Parametric Expansions* 354
8.6. *Exercises* 359
References 366

CHAPTER 9. **Uniform Asymptotic Expansions**

9.1. *Introduction* 367
9.2. *Asymptotic Expansions of Integrals with Two Nearby Saddle Points* 369
9.3. *Underlying Principles* 379
9.4. *Saddle Point near an Amplitude Critical Point* 380
9.5. *A Class of Integrals That Arise in the Analysis of Precursors* 387
9.6. *Double Integrals of Fourier Type* 393
9.7. *Exercises* 401
References 411

Appendix 413
General References 417
Index 421

Asymptotic Expansions of Integrals

1 Fundamental Concepts

1.1. Introduction

Simply stated, asymptotic analysis is that branch of mathematics devoted to the study of the behavior of functions at and near given points in their domains of definition. Suppose then that $f(z)$ is a function of the complex variable z. Suppose further that we wish to study f near the point $z = z_0$. If f is analytic at $z = z_0$, then the desired behavior can be determined by studying its Taylor series expansion about $z = z_0$.

Now suppose that $z = z_0$ is a singularity of f. If it is either a pole or a branch point, then again the analysis can be reduced to the investigation of convergent series expansions. However, if $z = z_0$ is an essential singularity of f, then no such reduction is possible and the analysis is far more complicated. Partly for this reason, we shall find that most often our investigations will involve the study of functions near their points of essential singularity.

In this chapter we shall consider, for the benefit of those readers unfamiliar with asymptotic analysis, some of the more fundamental concepts of the subject. It is not our aim to be exhaustive in this regard, but rather to present enough introductory material so that the techniques to be presented in future chapters can be well understood. Therefore, the present chapter shall consist of several definitions and theorems to place the subject on a firm mathematical foundation and also, perhaps more importantly, several heuristic discussions designed to give the reader an intuitive feel for what asymptotic analysis is all about.

Let us begin by considering a particular example. We define

$$I(x) = xe^x \int_x^\infty \frac{e^{-t}}{t}\, dt. \tag{1.1.1}$$

Here x is real and nonnegative. The integral in (1.1.1) is often referred to as the *exponential integral* and is denoted by $E_1(x)$. The reason for the factor xe^x will be clear from the discussion below.

Suppose it is desired to approximate $I(x)$ at certain values of x. More precisely, suppose that an estimate of I correct to three significant figures is required. Our first inclination might be to seek a series representation of I and then use the appropriate partial sums to obtain the desired approximations. Since to expand a function in a series about a given point requires a knowledge of the value of the function at that point, it should be clear that selecting any finite nonzero value of x about which to expand I would not be helpful. However, by applying L'Hospital's rule in (1.1.1) we find that

$$\lim_{x \to 0+} I = 0, \qquad \lim_{x \to \infty} I = 1. \tag{1.1.2}$$

To expand I in a series about $x = 0$ is by no means a simple matter. The reason for this is essentially due to the fact that $E_1(x)$ has a logarithmic singularity at the origin and hence $I(x)$ does not have a Taylor series expansion about $x = 0$. Nevertheless an expansion can be obtained. We leave its derivation to a later chapter and merely quote the result here:

$$E_1(x) = \int_x^\infty \frac{e^{-t}}{t}\, dt = -\log x - \gamma + \sum_{n=1}^\infty \frac{(-1)^{n+1} x^n}{n \cdot n!}. \tag{1.1.3}$$

Here γ is the so-called *Euler-Mascheroni* constant and is defined by

$$\gamma = \lim_{m \to \infty} \left[\sum_{n=1}^m \frac{1}{n} - \log m \right] = 0.5772157\ldots. \tag{1.1.4}$$

Thus, upon combining (1.1.1) and (1.1.3) we obtain

$$I(x) = xe^x \left[-\log x - \gamma + \sum_{n=1}^\infty \frac{(-1)^{n+1} x^n}{n \cdot n!} \right] \tag{1.1.5}$$

We note that the series in (1.1.5) converges for all x and that (1.1.5) is a series (although not a power series) representation of I about $x = 0$. When we examine (1.1.5) we quickly discover that, for "moderate" values of x, the convergence is painfully slow. Indeed, for $x = 10$, upward of 40 terms must be retained to achieve an estimate of $I(10)$ accurate to three significant figures. Moreover, as x gets larger, the situation worsens. Upon reflection, we realize that (1.1.5) is an expansion of I about $x = 0$ and hence we should only expect to obtain accurate estimates using relatively few terms for x "small."

Thus, in trying to estimate I for $x \geq 10$ say, it is natural to seek an expansion about $x=\infty$. Such an expansion is readily obtained by repeatedly integrating by parts in (1.1.1). Indeed, after N integrations by parts, we obtain

$$I(x)=\sum_{n=0}^{N-1}\frac{(-1)^n\,n!}{x^n}+(-1)^N\,N!\,xe^x\int_x^\infty\frac{e^{-t}}{t^{N+1}}\,dt \tag{1.1.6}$$

which is an exact expression.

At first glance we are pleased with (1.1.6) because we have represented $I(x)$ by a series whose terms involve inverse powers of x. In fact, we are tempted to let N go to infinity in (1.1.6) and set

$$I(x)=\sum_{n=0}^{\infty}\frac{(-1)^n\,n!}{x^n}. \tag{1.1.7}$$

Our pleasure is short-lived, however, when we realize that the series (1.1.7) diverges for *all* x. Indeed,

$$\left|\frac{(n+1)\text{st term}}{n\text{th term}}\right|=\frac{n}{x} \tag{1.1.8}$$

which, for every fixed x, increases without bound as $n\to\infty$.

Having learned long ago that divergent series are "bad" while convergent series are "good," we are inclined to discard these last results altogether. Admittedly, we must discard (1.1.7) because it is a meaningless statement, but, as we shall now show, to discard (1.1.6) would be decidedly premature.

Let us set

$$S_N(x)=\sum_{n=0}^{N-1}\frac{(-1)^n\,n!}{x^n}, \tag{1.1.9}$$

$$\mathscr{E}(x,N)=(-1)^N\,N!\,xe^x\int_x^\infty\frac{e^{-t}}{t^{N+1}}\,dt. \tag{1.1.10}$$

Thus $S_N(x)$ is the Nth partial sum of the divergent series (1.1.7) and $\mathscr{E}(x,N)$ is the error made in approximating $I(x)$ by $S_N(x)$. We now make the observation that, because x is positive, $\mathscr{E}(x,N)$ is positive when N is even and negative when N is odd. This implies that

$$\begin{aligned} S_N(x)\le I(x)\le S_{N+1}(x), &\qquad N \text{ even},\\ S_{N+1}(x)\le I(x)\le S_N(x), &\qquad N \text{ odd}. \end{aligned} \tag{1.1.11}$$

Hence, for any x, the actual value of $I(x)$ must lie between two successive partial sums of the divergent series (1.1.7).

As (1.1.11) shows, a succession of upper and lower bounds for $I(x)$ can be obtained by evaluating $S_N(x)$, $N=1,2,\ldots$. We have not as yet, however, determined how good any of these bounds are. One point is clear; for fixed x, the best approximation of $I(x)$ by $S_N(x)$ is achieved for that integer N which minimizes $|\mathscr{E}(x,N)|$. Furthermore, this optimum value of N, call it $\hat{N}(x)$,[1] must be finite because $\lim_{N\to\infty}|\mathscr{E}(x,N)|=\infty$ due to the divergence of (1.1.7). Also, because $\mathscr{E}(x,N)$ alternates in sign with N, $|\mathscr{E}(x,N)|$ is less than $N!/x^N$, the absolute value of the first term omitted in (1.1.7) when approximating $I(x)$ by $S_N(x)$. Thus, for fixed x, we might expect that the estimate $S_N(x)$ improves with N so long as the absolute value of the ratio of successive terms remains less than or equal to 1. Hence, from (1.1.8) we are led to predict that

$$\hat{N}(x)=[x]=\text{Greatest integer less than or equal to } x. \tag{1.1.12}$$

Finally, we observe that for fixed N, $|\mathscr{E}(x,N)|$ is a monotonically decreasing function of x and $\lim_{x\to\infty}|\mathscr{E}(x,N)|=0$. Thus, for any fixed N, $S_N(x)$ becomes a better approximation of I as x increases, but the error is zero only in the limit $x=\infty$.

Many of the results obtained above are summarized in Table 1.1. In this table the numbers in the column headed $I(x)$ have been obtained by numerical means and can be taken as exact to the indicated accuracy.

It is of interest to compare the results for $x=1$, $x=10$, and $x=100$. We see from the table that the best approximation of $I(1)$ is given by $S_1(1)$ as predicted by (1.1.12). $S_1(1)$, however, is not "close" to $I(1)$, the percentage error being approximately 66 percent. The best approximation of $I(10)$ is indeed afforded by $S_{10}(10)$ which we see is correct to three significant figures. It will be recalled that 40 terms of the convergent expansion (1.1.5) are required to obtain equivalent accuracy. Finally, for $x=100$, we see that $S_4(100)$ is already correct to five significant figures and, presumably, the next 96 or so partial sums yield still better approximations of $I(100)$.

We can conclude, therefore, that it would have been a mistake to discard the expansion about $x=\infty$ obtained via integration by parts, because a great deal of information about $I(x)$ has been garnered from (1.1.6). We might ask what the feature of the divergent series (1.1.7) is that makes it so useful in approximating I for x, say, greater than 10 and, in the same vein, why the utility of the convergent series (1.1.5) is rather limited in this region. The answer simply stated is that for $x\geq 10$, the convergent series "initially" diverges from the true value of $I(x)$ while the divergent series "initially" converges toward this value. Hence, we can obtain a reasonable approximation to I by taking relatively few terms of the divergent series whereas many more terms of the convergent series are needed to achieve the same degree of accuracy.

We might now conclude that, when x is "large," we should always use the partial sums $S_N(x)$ to approximate $I(x)$ rather than the partial sums of (1.1.5). Why this is not quite so brings out the inherent disadvantage of divergent series. In using the partial sums of (1.1.5) to approximate $I(x)$, we can make the error as small as we please, no matter how large x is, by simply taking sufficiently many terms. In using the partial sums $S_N(x)$ to approximate $I(x)$, the smallest

[1] $\hat{N}(x)$ need not be unique.

Table 1.1

x	$I(x)$	S_1	S_2	S_3	S_4	S_5	S_6	S_7	S_8	S_9	S_{10}
1	0.59634	1	0	2.0000	−4.0000						
2	0.72266	1	0.5000	1.0000	0.2500	1.7500					
3	0.78625	1	0.667	0.8999	0.6667	0.9626	0.4688				
5	0.85212	1	0.8000	0.8800	0.8352	0.8736	0.8352	0.8820			
10	0.91563	1	0.9000	0.9200	0.9140	0.9164	0.9152	0.91592	0.91542	0.91581	0.91544
100	0.99019	1	0.9900	0.9902	0.99019						

error obtainable is dictated by the value of x. Moreover, this minimum error is never zero unless $x=\infty$. Thus, if we require $I(10)$ to four significant figures, then we cannot use any of the partial sums $S_N(10)$ because, at best, they afford three significant figure accuracy.

As a practical matter, in any problem where $I(x)$ is to be approximated, some a priori upper bound, say $\mathscr{E}_0$, would be placed on the tolerable error. We know that for fixed N, $|\mathscr{E}(x,N)|$ decreases monotonically to zero as x increases to infinity. Therefore, no matter how small $\mathscr{E}_0$ is, we can find an x_0 such that

$$|\mathscr{E}(x_0,\hat{N}(x_0))| \leq \mathscr{E}_0. \tag{1.1.13}$$

Then, for all $x \geq x_0$, the partial sum $S_{\hat{N}(x_0)}(x)$ yields an estimate of $I(x)$ accurate to within the tolerable error. Actually as x increases from x_0, the number of terms required to achieve the desired accuracy decreases from $\hat{N}(x_0)$. Indeed we have in the extreme $I(\infty)=S_1(\infty)$.

With the above example as motivation, we shall, in the sections to follow, develop a theory that will enable us to systematically exploit the advantageous features of divergent series.

1.2. Order Relations

Throughout this book, because we shall be concerned with studying the behavior of functions as their arguments approach some limiting value, we shall find indispensable the use of the so-called order or "O" relations defined below.

In this section x represents a real or complex variable. Let us suppose that $f(x)$ and $g(x)$ are two functions of x defined and continuous in a domain R and that x_0 lies in $\bar{R}$, the closure of R.

DEFINITION 1.2.1. LARGE "O" SYMBOL. Suppose that, as $x \to x_0$ through values in R, there exists a constant k, that is, a quantity independent of x, and a neighborhood N_0 of x_0 such that

$$|f(x)| \leq k|g(x)| \tag{1.2.1}$$

for all x in $N_0 \cap R$. Then we say that, as $x \to x_0$, $f(x)$ is large "O" of $g(x)$ and write symbolically

$$f(x)=O(g(x)), \qquad x \to x_0 \text{ in } R. \tag{1.2.2}$$

Similarly, we introduce the following.

DEFINITION 1.2.2. SMALL "o" SYMBOL. Suppose that, for any $\varepsilon > 0$, there exists a neighborhood N_ε of x_0 such that

$$|f(x)| \leq \varepsilon |g(x)| \tag{1.2.3}$$

for all x in $N_\varepsilon \cap R$. Then we say that, as $x \to x_0$, $f(x)$ is small "O" of $g(x)$ and write symbolically

$$f(x) = o(g(x)), \qquad x \to x_0 \text{ in } R. \tag{1.2.4}$$

Thus, so long as $g(x)$ is not zero in R,[2] $f = O(g(x))$ as $x \to x_0$ if the ratio f/g remains bounded as $x \to x_0$ in R and $f = o(g(x))$ as $x \to x_0$ if the limit of this ratio is zero as $x \to x_0$ in R. We note that if $f = o(g(x))$ as $x \to x_0$, then necessarily $f = O(g(x))$ in this limit. The converse, however, need not be true.

EXAMPLE 1.2.1. Let $R = (0,1), f(x) = \sqrt{x}$, and $g(x) = \sin \sqrt{x}$. Then we have by L'Hospital's rule

$$\lim_{x \to 0+} \left[\frac{\sqrt{x}}{\sin \sqrt{x}} \right] = 1 \tag{1.2.5}$$

so that

$$\sin \sqrt{x} = O(\sqrt{x}) \qquad \text{and} \qquad \sqrt{x} = O(\sin \sqrt{x}), \tag{1.2.6}$$

as $x \to 0+$.

EXAMPLE 1.2.2. Let $R = (-\infty, \infty)$ and

$$f(x) = \begin{cases} 0, & -\infty < x \leq 0, \\ e^{-1/x}, & 0 < x < \infty. \end{cases} \tag{1.2.7}$$

If $g(x) = x^m$ where m is any complex number, then

$$\lim_{x \to 0+} \left[\frac{f(x)}{g(x)} \right] = \lim_{\xi \to \infty} \left[\frac{\xi^m}{e^\xi} \right] = 0 \tag{1.2.8}$$

and clearly

$$\lim_{x \to 0-} \left[\frac{f(x)}{g(x)} \right] = 0. \tag{1.2.9}$$

Hence, $f(x) = o(x^m)$ as $x \to 0$ for all complex numbers m.

EXAMPLE 1.2.3. Let x be a complex variable and let R be the sector $0 < |x| < \infty$, $|\arg(x)| \leq \pi/2 - \delta$. If δ is positive, then, as is easily verified,

$$e^{-x} = o(x^m) \tag{1.2.10}$$

[2] Except possibly at x_0.

as $|x| \to \infty$ in R for all complex numbers m. On the other hand, if δ is negative, then (1.2.10) does not hold for any complex number m as $|x| \to \infty$ in R due to the rapid growth of e^{-x} in the left half-plane $\text{Re}(x) < 0$.

At first glance it might appear that if, as $x \to x_0, f(x) = O(g(x))$ but $f \neq o(g(x))$, then $g(x) = O(f(x))$. That this is not necessarily so can be seen from the following counterexample.

EXAMPLE 1.2.4. Let $R = (0,1)$, $f(x) = x \sin(x^{-1})$, $g(x) = x$, and $x_0 = 0$. In $\bar{R}$, because

$$\left|\frac{f(x)}{g(x)}\right| = |\sin(x^{-1})| \leq 1 \tag{1.2.11}$$

we have $f = O(g)$ as $x \to 0+$. We also have that

$$\lim_{x \to 0+} \left[\frac{f}{g}\right] = \lim_{x \to 0+} [\sin(x^{-1})] \tag{1.2.12}$$

is undefined and hence $f \neq o(g)$ as $x \to 0+$. Finally,

$$\left|\frac{g}{f}\right| = \left|\frac{1}{\sin(x^{-1})}\right| \tag{1.2.13}$$

is not bounded as $x \to 0+$, so that $g \neq O(f)$ in this limit.

As the above discussion shows, the order relations allow us to compare the behavior of functions in some prescribed limit. It is often useful, when considering many functions, to have one function or set of functions that serve as a scale on which comparisons are made. This idea will be exploited in later sections of this chapter. Here, however, we wish to use it to introduce the concept of the order of vanishing of a function at a point.

If we wish to study $f(x)$ as $x \to x_0$, then the simplest comparison functions are the (not necessarily integer) powers of $(x - x_0)$. Suppose that, as $x \to x_0$,

$$f(x) = O((x - x_0)^{\Delta}) \tag{1.2.14}$$

for some real number $\Delta = \Delta_0$. Then clearly (1.2.14) holds for all $\Delta \leq \Delta_0$. If

$$\delta = \sup\{\Delta | f(x) = O((x - x_0)^{\Delta})\}, \tag{1.2.15}$$

then f is said to vanish at $x = x_0$ to order δ.[3] Of course δ may be negative in which event f becomes unbounded as $x \to x_0$. This growth, however, can be interpreted as a negative decay.

It may or may not be true that

$$f(x) = O((x - x_0)^{\delta}), \qquad x \to x_0. \tag{1.2.16}$$

Indeed, let $f = \log x$, $R = (0,1)$, and $x_0 = 0$. We have $\log x = 0(x^{\Delta})$, as

[3] More precisely, δ is the order of vanishing of f at $x = x_0$ with respect to the powers of $(x - x_0)$.

$x \to 0+$, for all $\Delta < 0$ and $\log x \neq O(x^\Delta)$ for $\Delta > 0$. Thus, $\delta = 0$, but $\log x \neq O(1)$ as $x \to 0+$.

From Example 1.2.1 we have that $\sin \sqrt{x}$ vanishes to order $\frac{1}{2}$ as $x \to 0+$; while from Example 1.2.3 we have that as $x \to \infty$ in the right half complex x plane, e^{-x} vanishes to infinite order or, equivalently, faster than any power of x^{-1}.

There are many useful formulas involving combinations of order relations whose validity follows directly from the basic definitions. Below we list some of the more important ones. In each of these, the limit $x \to x_0$ in R is to be understood.

(1) $O(O(f)) = O(f)$.
(2) $O(o(f)) = o(O(f)) = o(o(f)) = o(f)$.
(3) $O(fg) = O(f)O(g)$.
(4) $O(f)\, o(g) = o(fg)$.
(5) $O(f) + O(f) = O(f)$.
(6) $o(f) + o(f) = o(f)$.
(7) $o(f) + O(f) = O(f)$.

To illustrate how the above formulas are to be interpreted we shall express (1) in expanded form. For example, (1) states that if $g = O(h)$ and $h = O(f)$, as $x \to x_0$, then $g = O(f)$ as $x \to x_0$. We can therefore conclude that if $g = O(h)$ as $x \to x_0$, then the order of vanishing of g at $x = x_0$ is at least as large as that of h.

To conclude this section we wish to point out that there are several operations permissible with order relations. An important result is that an order relation can be integrated with respect to the independent variable. Indeed, suppose that R is an interval on the real line and $f = O(g)$ as $x \to x_0$ in R. Then

$$\int_x^{x_0} f(t)\, dt = O\left(\int_x^{x_0} |g(t)|\, dt\right), \qquad x \to x_0 \text{ in } R. \tag{1.2.17}$$

We leave the proof of (1.2.17) to the exercises.

In general, order relations cannot be differentiated. That is, if $f = O(g)$ as $x \to x_0$, then it is *not* true in general that $f' = O(g')$ as $x \to x_0$.

1.3. Asymptotic Power Series Expansions

Let us briefly reconsider the function

$$I(x) = xe^x \int_x^\infty \frac{e^{-t}}{t}\, dt \tag{1.3.1}$$

studied in Section 1.1 and interpret the results obtained there in terms of the order relations of Section 1.2. We have from (1.1.6)

$$I(x) - \sum_{n=0}^{m-1} \frac{(-1)^n\, n!}{x^n} = \mathscr{E}(x,m)$$

$$= (-1)^m\, m!\, xe^x \int_x^\infty \frac{e^{-t}}{t^{m+1}}\, dt, \qquad m = 1, 2, \ldots, \tag{1.3.2}$$

which relates I, the mth partial sum of the divergent series (1.1.7) and the truncation error $\mathscr{E}(x,m)$.

We now make the claim that $\mathscr{E}(x,m) = O(x^{-m})$ as $x \to \infty$. Indeed, this follows by L'Hospital's rule which yields

$$\lim_{x\to\infty} \{x^m\, \mathscr{E}(x;m)\} = (-1)^m\, m! \lim_{x\to\infty} \left\{x^{m+1} \int_x^\infty \frac{e^{-t}}{t^{m+1}}\, dt / e^{-x}\right\} \tag{1.3.3}$$

$$= (-1)^m\, m!.$$

Thus, as previously noted, the error made in approximating $I(x)$, as $x \to \infty$, by the first m terms of (1.1.6) is of the order of[4] the first omitted term.

It is readily seen that the preceding result can be written in the following equivalent ways:

$$\lim_{x\to\infty} \left\{x^m \left[I(x) - \sum_{n=0}^{m} \frac{(-1)^n\, n!}{x^n}\right]\right\} = 0, \qquad m = 0, 1, 2, \ldots. \tag{1.3.4}$$

$$I(x) = \sum_{n=0}^{m} \frac{(-1)^n\, n!}{x^n} + O(x^{-m-1}), \qquad x \to \infty, \quad m = 0, 1, 2, \ldots. \tag{1.3.5}$$

It is the property expressed by (1.3.4) and (1.3.5) that makes the divergent series (1.1.7) useful in approximating I as $x \to \infty$. We now introduce the concept of asymptotic power series based on just this property.

For the present we assume that x is a real variable whose domain is R and that x_0 is a finite point in $\bar{R}$, the closure of R. We then have the following.

DEFINITION 1.3.1. Let $f(x)$ be defined and continuous on R. The formal[5] power series $\Sigma_{n=0}^{\infty} a_n(x - x_0)^n$ is said to be an asymptotic power series expansion of f, as $x \to x_0$ in R, if the conditions

$$\lim_{x\to x_0} \left\{(x - x_0)^{-m} \left[f - \sum_{n=0}^{m} a_n\, (x - x_0)^n\right]\right\} = 0, \qquad m = 0, 1, 2, \ldots \tag{1.3.6}$$

are satisfied.

It is readily seen that conditions (1.3.6) are equivalent to

$$f(x) = \sum_{n=0}^{m} a_n\, (x - x_0)^n + O(x - x_0)^{m+1}, \qquad x \to x_0 \text{ in } R, \qquad m = 0, 1, 2, \ldots. \tag{1.3.7}$$

[4] The phrase "of the order of" always refers to the large rather than the small "o" symbol.

[5] By a formal series we mean any infinite series where no assumption is made concerning its convergence.

We note that neither (1.3.6) nor (1.3.7) implies the convergence of the formal power series $\Sigma_{n=0}^{\infty} a_n (x - x_0)^n$. They are simply statements concerning the behavior, as $x \to x_0$, of the error made in approximating f by the partial sums of this series. Thus, in general, we cannot set f equal to the series and hence we introduce the notation

$$f(x) \sim \sum_{n=0}^{\infty} a_n (x - x_0)^n, \qquad x \to x_0 \tag{1.3.8}$$

to imply that conditions (1.3.6) hold and, in particular, to allow for the possible divergence of the right-hand side.

DEFINITION 1.3.2. Assume that (1.3.6) is satisfied for $m = 0, \ldots, N - 1$, but not for $m = N$. Then we say that $\Sigma_{n=0}^{N-1} a_n(x - x_0)^n$ is an asymptotic power series of f as $x \to x_0$ to N terms and write

$$f(x) \sim \sum_{n=0}^{N-1} a_n (x - x_0)^n, \qquad x \to x_0. \tag{1.3.9}$$

Under the assumptions made we can now only conclude that

$$f(x) = \sum_{n=0}^{N-1} a_n (x - x_0)^n + o(x - x_0)^{N-1}, \qquad x \to x_0. \tag{1.3.10}$$

It is a simple matter to adapt our definition of asymptotic power series to the important case where $x_0 = \infty$. Indeed, we have the following.

DEFINITION 1.3.3. The formal series $\Sigma_{n=0}^{\infty} a_n x^{-n}$ is said to be an asymptotic power series expansion of f as $x \to \infty$ if the following equivalent sets of conditions are satisfied:

$$\lim_{x \to \infty} \left\{ x^m \left[f - \sum_{n=0}^{m} a_n x^{-n} \right] \right\} = 0, \qquad m = 0, 1, 2, \ldots, \tag{1.3.11}$$

$$f(x) = \sum_{n=0}^{m} a_n x^{-n} + O(x^{-m-1}), \qquad x \to \infty, \qquad m = 0, 1, 2, \ldots. \tag{1.3.12}$$

If (1.3.11) and (1.3.12) hold, then we write

$$f(x) \sim \sum_{n=0}^{\infty} a_n x^{-n}, \qquad x \to \infty. \tag{1.3.13}$$

Thus, the analysis of Section 1.1 coupled with (1.3.3) shows that

$$xe^x \int_x^{\infty} \frac{e^{-t}}{t}\, dt \sim \sum_{n=0}^{\infty} \frac{(-1)^n n!}{x^n}, \qquad x \to \infty. \tag{1.3.14}$$

As we have noted, the asymptotic conditions (1.3.6) imply neither the convergence nor the divergence of the formal series $\Sigma_{n=0}^{\infty} a_n (x-x_0)^n$. If the series converges to $f(x)$ throughout some neighborhood of $x=x_0$, then f is a real analytic function in this neighborhood and can be studied by using the powerful theorems associated with such functions. The more interesting case arises when the formal series actually diverges in R, except of course at $x=x_0$. In that event $f(x)$ must have some sort of singularity at $x=x_0$ in the sense that $x=x_0$ must be a point of nonanalyticity for f.

We remind the reader that, if the asymptotic power series diverges, then for any value of $x \neq x_0$, the optimum number of terms of the series to be used in approximating f must be finite and the corresponding optimum error is not zero. In our work throughout this book we shall not dwell on obtaining the optimum number of terms to be retained and shall, whenever possible, determine an infinite asymptotic expansion.

At first glance, the concept of an asymptotic expansion may appear rather foreign to many readers. Upon reflection, however, everyone who has studied the calculus will realize that he has already been exposed to such expansions. Indeed, although it is rarely stated in such terms, Taylor's theorem with remainder is a theorem in asymptotic analysis. For any function $f(x)$ having N continuous derivatives at $x=x_0$, Taylor's theorem with remainder states

$$\begin{aligned} f(x) &= \sum_{n=0}^{m} \frac{f^{(n)}(x_0)}{n!}(x-x_0)^n \qquad (1.3.15)\\ &\quad + \frac{1}{(m+1)!}(x-x_0)^{m+1} f^{(m+1)}(x_0+\theta_m[x-x_0]), \qquad 0<\theta_m<1,\\ &= \sum_{n=0}^{m} \frac{f^{(n)}(x_0)}{n!}(x-x_0)^n + O(x-x_0)^{m+1}, \qquad x \to x_0,\\ &\qquad\qquad m = 0, 1, \ldots, N-1. \end{aligned}$$

Then it immediately follows that

$$f(x) \sim \sum_{n=0}^{N-1} \frac{f^{(n)}(x_0)(x-x_0)^n}{n!}, \qquad x \to x_0 \qquad (1.3.16)$$

is an asymptotic power series expansion of f to N terms as $x \to x_0$.

Let us now suppose that f is infinitely differentiable at $x=x_0$. We then say that f belongs to the class $C^\infty(x_0)$. We can then let N go to infinity in (1.3.16) to obtain the infinite asymptotic power series expansion

$$f(x) \sim \sum_{n=0}^{\infty} \frac{f^{(n)}(x_0)(x-x_0)^n}{n!}, \qquad x \to x_0. \qquad (1.3.17)$$

Note that we have not set $f(x)$ equal to its Taylor series, for nothing we have assumed implies the convergence of this series. In other words, f need not be analytic at $x=x_0$.

To pursue this matter a bit further, let us define the function space $\tilde{C}^{\infty}(x_0)$ which includes those functions infinitely differentiable at $x = x_0$ but *not* analytic there. Such functions arise often in asymptotic analysis. Unfortunately, they are easier to define than to construct. If $f \in \tilde{C}^{\infty}(x_0)$, then $f^{(n)}(x_0)$ is finite for each n but the Taylor series

$$\sum_{n=0}^{\infty} \frac{f^{(n)}(x_0)}{n!}(x - x_0)^n$$

does not converge to f throughout any neighborhood of $x = x_0$. This can happen in two ways. The Taylor series may actually converge throughout some neighborhood of $x = x_0$ but not to $f(x)$. Alternatively, the series might diverge for all $x \neq x_0$. We shall illustrate both of these instances in the following examples.

EXAMPLE 1.3.1. Consider

$$f(x) = \begin{cases} e^{-1/x}, & x > 0, \\ 0, & x \leq 0. \end{cases} \tag{1.3.18}$$

Here $f(x)$ and all of its derivatives are continuous and equal to zero at $x = 0$. (See Example 1.2.2.) Although f is infinitely differentiable at $x = 0$, it is not analytic there because its Taylor series sums to zero for all x, a result which disagrees with f for x positive. However, because f is infinitely differentiable at $x = 0$ we do have the asymptotic power series expansion

$$f(x) \sim 0, \qquad x \to 0. \tag{1.3.19}$$

EXAMPLE 1.3.2. Let us now consider

$$f(x) = \sum_{m=0}^{\infty} e^{-m} \cos m^2 x; \qquad -1 \leq x \leq 1. \tag{1.3.20}$$

We first note that the infinite series is uniformly convergent for $x \in [-1,1]$ as are all of the series obtained by successively differentiating (1.3.20) term-by-term. Therefore, we can conclude that $f(x)$ and all of its derivatives are continuous in $[-1,1]$ and furthermore the derivatives of f can be obtained by successive term-by-term differentiation of (1.3.20).

At the origin only the derivatives of even order are nonzero and we have

$$f^{(2n)}(0) = (-1)^n \sum_{m=0}^{\infty} e^{-m}(m^2)^{2n}, \qquad n = 0, 1, 2, \ldots. \tag{1.3.21}$$

Thus, because f is infinitely differentiable at the origin, we can immediately write

$$f(x) \sim \sum_{n=0}^{\infty} \frac{(-1)^n x^{2n}}{(2n)!} \left(\sum_{m=0}^{\infty} e^{-m} m^{4n} \right), \qquad x \to 0. \tag{1.3.22}$$

Consider now the absolute value of the term of degree $2n$, $n > 0$, in (1.3.22). We have

$$\frac{|x^{2n}|}{(2n)!} \left(\sum_{m=0}^{\infty} e^{-m} m^{4n} \right) \geq \left(\frac{|x|}{2n} \right)^{2n} \left(\sum_{m=0}^{\infty} e^{-m} m^{4n} \right) \geq \left(\frac{|x| \, m^2}{2n} \right)^{2n} e^{-m} \tag{1.3.23}$$

which holds for $m = 0, 1, 2, \ldots,$ because one term of a sum of positive numbers is certainly less than the sum. Furthermore, for $x \neq 0$, strict inequality holds in (1.3.23). If we set $m = 2n$, then (1.3.23) becomes

$$\frac{|x|^{2n}}{(2n)!} \left(\sum_{m=0}^{\infty} e^{-m} m^{4n} \right) > \left(\frac{2n \, |x|}{e} \right)^{2n}, \qquad x \neq 0. \tag{1.3.24}$$

Because the right-hand side of (1.3.24) exceeds 1 for all n greater than $e/2|x|$, the series in (1.3.22) diverges for all nonzero x. Therefore, f is infinitely differentiable at $x = 0$ but is not analytic there.

In the following two sections we shall introduce generalizations of asymptotic power series. We wish to comment here, however, that the terms of any asymptotic expansion can depend on one or more parameters. *If the asymptotic conditions are satisfied independent of the parameters as these parameters range over some domain in parameter space, then we say that the asymptotic expansion holds uniformly in the parameter(s) on this domain.*

1.4. Asymptotic Sequences and Asymptotic Expansions of Poincaré Type

The asymptotic power series expansion of a function $f(x)$, as $x \to x_0$, was defined in Section 1.3. As (1.3.6) shows, it involves the sequence of functions $\{(x - x_0)^n\}$, $n = 0, 1, \ldots$. Its utility stems from the fact that the difference between f and the Nth partial sum of the expansion is $O(x - x_0)^N$ as $x \to x_0$. That is to say, the truncation error is of the order of the first term omitted or, equivalently, vanishes, as $x \to x_0$, to a higher order than any of the terms retained.

Upon reflection we realize that the validity of the above remarks is a consequence of the simple relation

$$(x - x_0)^{m+1} = o((x - x_0)^m), \qquad x \to x_0, \tag{1.4.1}$$

which actually holds for all complex numbers m. There are, of course, many sequences of functions which have the property that all ratios of successive terms vanish as $x \to x_0$. For example, the sequence $\{\sin^n(x - x_0)\}$, $n = 0, 1, \ldots$

has this property. It is reasonable to expect that a meaningful generalization of asymptotic power series can be achieved in terms of such sequences. The remainder of this section shall be devoted to the development of this generalization.

Let us assume that R is an open interval on the real line, that x is a real variable, and that $x_0 \varepsilon \bar{R}$. Motivated by (1.4.1), we now give the following.

DEFINITION 1.4.1. The sequence of functions $\{\phi_n(x)\}$, $n = 0, 1, 2, \ldots$ is called an *asymptotic sequence* as $x \to x_0$ in R if, for every n, $\phi_n(x)$ is defined and continuous in R and

$$\phi_{n+1}(x) = o(\phi_n(x)), \qquad x \to x_0. \tag{1.4.2}$$

It follows from (1.4.1) that $\{(x - x_0)^n\}$ is an asymptotic sequence as $x \to x_0$. The list below yields further examples in the limits indicated.

(1) $\{(x - x_0)^{\gamma_n}\}$, $x \to x_0$, $\mathrm{Re}(\gamma_{n+1}) > \mathrm{Re}(\gamma_n)$,
(2) $\{x^{-\gamma_n}\}$, $x \to \infty$, $\mathrm{Re}(\gamma_{n+1}) > \mathrm{Re}(\gamma_n)$,
(3) $\{[g(x)]^n\}$, $x \to x_0$, $g(x_0) = 0$,
(4) $\{g(x)\,\phi_n(x)\}$, $x \to x_0$.

In (4) $\{\phi_n\}$ is an asymptotic sequence as $x \to x_0$ while in (3) and (4) $g(x)$ is continuous and not identically zero in any neighborhood of $x = x_0$.

Our generalization of asymptotic power series is given in the following.

DEFINITION 1.4.2. Let $f(x)$ be continuous in R and let $\{\phi_n(x)\}$ be an asymptotic sequence as $x \to x_0$ in R. Then the formal series $\Sigma_{n=0}^{\infty} a_n \phi_n(x)$ is said to be an infinite asymptotic expansion of $f(x)$, as $x \to x_0$, with respect to $\{\phi_n\}$ if the equivalent sets of conditions

$$f(x) = \sum_{n=0}^{m} a_n \phi_n + O(\phi_{m+1}), \qquad x \to x_0, \qquad m = 0, 1, 2, \ldots \tag{1.4.3}$$

$$\lim_{x \to x_0} \left[\frac{f - \sum_{n=0}^{m} a_n \phi_n}{\phi_m} \right] = 0, \qquad m = 0, 1, 2, \ldots \tag{1.4.4}$$

are satisfied. Moreover, the asymptotic expansion is said to be of *Poincaré type*.

As in the case of asymptotic power series, this definition neither implies nor precludes the convergence of the formal series. Hence, whenever (1.4.3) holds, we write symbolically

$$f(x) \sim \sum_{n=0}^{\infty} a_n \phi_n, \qquad x \to x_0, \tag{1.4.5}$$

to allow for the possible divergence of the right-hand side. We wish to emphasize that the usefulness of (1.4.5) is a consequence of (1.4.3) which states that, upon truncating the formal series, the resulting error vanishes, as $x \to x_0$, at a well-defined rate.

DEFINITION 1.4.3. If (1.4.3) holds only for $m = 0, 1, \ldots, N-1$, then

$$f(x) \sim \sum_{n=0}^{N-1} a_n \phi_n, \qquad x \to x_0 \tag{1.4.6}$$

is said to be an asymptotic expansion of f to N terms with respect to the asymptotic sequence $\{\phi_n\}$. We note that, in the case of finite expansions, (1.4.3) and (1.4.4) are not quite equivalent. Indeed, if (1.4.4) holds for $m = 0, 1, \ldots, N-1$, then we can only conclude that

$$f(x) = \sum_{n=0}^{N-1} a_n \phi_n + o(\phi_{N-1}). \tag{1.4.7}$$

Now suppose that we are given a function f and an asymptotic sequence $\{\phi_n\}$ as $x \to x_0$. It follows directly from (1.4.4) that *if* an infinite asymptotic expansion of f exists with respect to this sequence, then the coefficients a_n of the expansion are defined recursively by

$$a_n = \lim_{x \to x_0} \left[\frac{f - \sum_{j=0}^{n-1} a_j \phi_j}{\phi_n} \right], \qquad n = 0, 1, 2, \ldots . \tag{1.4.8}$$

If, however, the limits in (1.4.8) exist and are finite for $n = 0, 1, \ldots, N-1$, but a_N is either undefined or infinite, then we can only conclude that the finite expansion (1.4.6) holds.

We shall presently illustrate the above remarks by an example but first we establish the uniqueness of asymptotic expansions of Poincaré type in the following.

THEOREM 1.4. Let

$$f(x) \sim \sum_{n=0}^{N-1} a_n \phi_n, \qquad x \to x_0, \tag{1.4.9}$$

be an asymptotic expansion to N terms of f with respect to the asymptotic sequence $\{\phi_n\}$. Then the coefficients a_n are uniquely determined.

PROOF. Assume there exists a second expansion given by

$$f(x) \sim \sum_{n=0}^{N-1} b_n \phi_n(x), \qquad x \to x_0. \tag{1.4.10}$$

The coefficients b_n must satisfy

$$b_n = \lim_{x \to x_0} \left[\frac{f - \sum_{j=0}^{n-1} b_j \phi_j}{\phi_n} \right], \qquad n = 0, 1, \ldots, N-1. \tag{1.4.11}$$

It follows from (1.4.8) and (1.4.11) that

$$a_0 = b_0 = \lim_{x \to x_0} \left[\frac{f}{\phi_0} \right]. \tag{1.4.12}$$

Upon assuming that $a_n = b_n$ for $n = 0, 1, 2, \ldots, i < N-1$, we immediately obtain

$$a_{i+1} - b_{i+1} = \lim_{x \to x_0} \left[\frac{f - \sum_{j=0}^{i} a_j \phi_j - f + \sum_{j=0}^{i} b_j \phi_j}{\phi_{i+1}} \right] = 0. \tag{1.4.13}$$

It therefore follows by induction that $a_n = b_n$, for $n = 0, 1, 2, \ldots, N-1$, and the theorem is proved.

EXAMPLE 1.4. Let us again consider the function

$$I(x) = xe^x \int_x^\infty \frac{e^{-t}}{t}\, dt \tag{1.4.14}$$

introduced in Section 1.1. In that section we obtained, via integration by parts, an asymptotic expansion of I, as $x \to \infty$, with respect to the asymptotic sequence $\{x^{-n}\}$. This expansion can be directly obtained from (1.4.8). Indeed, upon applying L'Hospital's rule we find that

$$a_0 = \lim_{x \to \infty} I(x) = \lim_{x \to \infty} \left[\frac{x \int_x^\infty t^{-1} e^{-t}\, dt}{e^{-x}} \right] = \lim_{x \to \infty} \left[1 - \frac{1}{x} \right] = 1, \tag{1.4.15}$$

$$a_1 = \lim_{x \to \infty} \left[\frac{x^2 \int_x^\infty t^{-1} e^{-t}\, dt - xe^{-x}}{e^{-x}} \right] = \lim_{x \to \infty} \left[1 - \frac{2x \int_x^\infty t^{-1} e^{-t}\, dt}{e^{-x}} \right]$$

$$= -1. \tag{1.4.16}$$

The higher coefficients can also be determined in this manner, but the labor involved increases quite rapidly. In addition, the above procedure suffers from a defect in that its application requires a good deal of information about the answer sought before we begin. In particular, although we are free to select any asymptotic sequence we desire, we will not get very far unless the sequence chosen reflects very closely the actual behavior of I as $x \to \infty$.

To illustrate these remarks, suppose that we attempt to find an asymptotic expansion of (1.4.14), as $x \to \infty$, with respect to the asymptotic sequence $\{x^{-2n}\}$, $n = 0, 1, 2, \ldots$. We have

$$a_0 = \lim_{x \to \infty} I(x) = 1 \tag{1.4.17}$$

as before, but

$$a_1 = \lim_{x \to \infty} \left[\frac{x^3 \int_x^\infty t^{-1} e^{-t}\, dt - x^2 e^{-x}}{e^{-x}} \right] = \infty. \tag{1.4.18}$$

Hence, the best we can say is that $I \sim 1$ is an asymptotic expansion of I, to one term, with respect to the asymptotic sequence $\{x^{-2n}\}$.

Thus, we can conclude that unless we have some a priori knowledge about the correct asymptotic sequence in a given problem, it is unlikely that a meaningful asymptotic expansion can be obtained by using (1.4.8) directly.

Let us now consider some additional facts concerning asymptotic expansions of Poincaré type. We first point out that a given function $f(x)$ may have, as $x \to x_0$, an asymptotic expansion with respect to each of the several asymtotic sequences. Indeed, let

$$f(x) = \sqrt{1 - \sin x}. \tag{1.4.19}$$

Upon expanding the square root we obtain

$$f(x) \sim 1 - \frac{1}{2}\sin x - \frac{1}{2^2 \cdot 2!}\sin^2 x - \frac{1 \cdot 3}{2^3 \cdot 3!}\sin^3 x - \cdots, \qquad x \to 0. \tag{1.4.20}$$

Here the asymptotic sequence is $\{\sin^n x\}$, $n = 0, 1, \ldots$. Alternatively,

$$f(x) = \frac{\cos x}{\sqrt{1 + \sin x}} \sim \sum_{n=0}^{\infty} \frac{(-1)^n x^{2n}}{(2n)!\sqrt{1 + \sin x}}, \quad x \to 0 \tag{1.4.21}$$

is an asymptotic expansion of f with respect to the asymptotic sequence $\{x^{2n}/\sqrt{1 + \sin x}\}$, $n = 0, 1, \ldots$. That both (1.4.20) and (1.4.21) are actually convergent expansions is of no consequence here.

We now point out that two distinct functions can have the same asymptotic expansion. To illustrate this assume that

$$f(x) \sim \sum_{n=0}^{\infty} a_n x^{-n}, \qquad x \to \infty. \tag{1.4.22}$$

Because

$$\lim_{x \to \infty} (x^n e^{-x}) = 0, \qquad n = 0, 1, 2, \ldots, \tag{1.4.23}$$

we also have

$$f(x)+e^{-x}\sim\sum_{n=0}^{\infty}a_n x^{-n},\qquad x\to\infty. \tag{1.4.24}$$

Suppose then we consider the set of functions defined on a domain R and having the same asymptotic expansion as $x\to x_0$. These functions are said to be *asymptotically equivalent* with respect to the underlying asymptotic sequence. That is to say, a given asymptotic sequence $\{\phi_n\}$ establishes an equivalence relation between functions defined on R and having an asymptotic expansion of Poincaré type with respect to that sequence. In fact, we have that, as $x\to x_0$, f and g are asymptotically equivalent with respect to $\{\phi_n\}$ if

$$h(x)=f(x)-g(x)=o(\phi_n),\qquad x\to x_0$$

for $n=0, 1, 2, \dots$.

Intuitively, f and g are asymptotically equivalent with respect to $\{\phi_n\}$ if their difference h is asymptotically zero on the scale induced by this asymptotic sequence, that is, if, in terms of $\{\phi_n\}$, h cannot be distinguished from zero as $x\to x_0$.

It is instructive to view a given asymptotic sequence as defining an underlying scale of measurement. In fact, this notion suggests still a further generalization of our concept of an asymptotic expansion. This generalization shall be discussed in the following section and shall prove to be essential in our subsequent work.

1.5. Auxiliary Asymptotic Sequences

In the preceding section we considered asymptotic expansions of Poincaré type. In such expansions the terms of the underlying asymptotic sequence appear explicitly. If we look upon an asymptotic sequence as defining a scale on which to measure functions near a particular point, then a further generalization of the concept of an asymptotic expansion is suggested. In this generalization, we allow the asymptotic sequence to play an auxiliary role and we do not insist that its terms appear explicitly in the expansion.

The basic idea is formulated in the following.

DEFINITION 1.5. Let $\{\phi_n\}$ be an asymptotic sequence as $x\to x_0$. Then the formal series $\Sigma_{n=0}^{\infty} f_n(x)$ is said to be an asymptotic expansion of a given function $f(x)$ with respect to the *auxiliary asymptotic sequence* $\{\phi_n\}$ if

$$f(x)=\sum_{n=0}^{N} f_n(x)+o(\phi_N(x)),\qquad x\to x_0,\quad N=0, 1, 2, \dots . \tag{(1.5.1)}$$

We express this symbolically by

$$f \sim \sum_{n=0}^{\infty} f_n(x), \qquad x \to x_0, \quad \{\phi_n\}, \tag{1.5.2}$$

and, when no confusion can arise, we shall omit any explicit reference to the asymptotic sequence.

The main advantage of this generalization lies in the fact that the functions $f_n(x)$ need not themselves form an asymptotic sequence and hence our concept of an asymptotic expansion is significantly broadened. We note that if $\{f_n\}$ is not an asymptotic sequence as $x \to x_0$, then (1.5.2) is not an asymptotic expansion of Poincaré type.

EXAMPLE 1.5. Let $x_0 = \infty$ and consider the sequence $\{f_n\}$ where

$$f_n = \frac{\cos nx}{x^n}, \qquad n = 0, 1, 2, \dots . \tag{1.5.3}$$

We observe that because

$$\lim_{x \to \infty} \left[\frac{\cos (n+1)x}{x \cos nx}\right] = \lim_{x \to \infty} \left[\frac{f_{n+1}(x)}{f_n(x)}\right]$$

is undefined, $\{f_n\}$ is not an asymptotic sequence as $x \to \infty$. The difficulty is due to the fact that in every neighborhood of ∞ there are infinitely many points at which $\cos nx$ vanishes while $\cos (n+1)x$ does not.

We do have, however,

$$\left|\frac{\cos nx}{x^n}\right| \le |x^{-n}|, \qquad n = 0, 1, 2, \dots \tag{1.5.4}$$

and therefore it is reasonable to expect that we might obtain a result such as

$$f(x) = \sum_{m=0}^{N} \frac{a_n \cos nx}{x^n} + o(x^{-N}), \qquad x \to \infty, \qquad N = 0, 1, 2, \dots \tag{1.5.5}$$

for some function $f(x)$. If so, then (1.5.5) yields an asymptotic expansion of f in the sense of Definition 1.5. The auxiliary asymptotic sequence here, of course, is $\{x^{-n}\}$.

For expansions of the type (1.5.1), the auxiliary asymptotic sequence is by no means unique. Indeed, suppose that $\{\psi_n\}$ is another asymptotic sequence as $x \to x_0$ with the property that

$$\phi_n(x) = O(\psi_n(x)), \qquad x \to x_0, \qquad n = 0, 1, 2, \dots . \tag{1.5.6}$$

It then follows from (1.5.1), (1.5.6), and relation (2) on page 9 that

$$f(x) = \sum_{n=0}^{N} f_n(x) + o(\psi_N), \qquad x \to x_0, \qquad N = 0, 1, 2, \dots . \tag{1.5.7}$$

Hence, $f \sim \Sigma_{n=0}^{\infty} f_n(x)$ is an asymptotic expansion of f with respect to the auxiliary asymptotic sequence $\{\psi_n\}$ as well.

Some care must be taken in selecting the auxiliary asymptotic sequence if meaningful estimates of the truncation errors

$$\mathscr{E}(x,N) = f(x) - \sum_{n=0}^{N} f_n(x), \qquad N = 0, 1, 2, \ldots \tag{1.5.8}$$

are to be obtained. Indeed, suppose that (1.5.1) holds but that

$$f_m = o(\phi_m) \tag{1.5.9}$$

for some nonnegative integer m. We then claim that the auxiliary asymptotic sequence $\{\phi_n\}$ is too coarse or crude a scale on which to measure $\mathscr{E}(x,m-1)$ as $x \to x_0$. This is established by noting that if m is a positive integer, then

$$\lim_{x \to x_0} \left| \frac{\mathscr{E}(x,m-1)}{\phi_m} \right| \leq \lim_{x \to x_0} \left[\left| \frac{f - \sum_{n=0}^{m} f_n}{\phi_m} \right| + \left| \frac{f_m}{\phi_m} \right| \right] = 0 \tag{1.5.10}$$

while, if $m = 0$, then

$$\lim_{x \to x_0} \left| \frac{f}{\phi_0} \right| \leq \lim_{x \to x_0} \left[\left| \frac{f - f_0}{\phi_0} \right| + \left| \frac{f_0}{\phi_0} \right| \right] = 0. \tag{1.5.11}$$

To ensure that the auxiliary asymptotic sequence $\{\phi_n\}$ is *useful* for estimating the truncation errors $\mathscr{E}(x,N)$ we must insist that, in addition to (1.5.1), the conditions

$$f_n = 0(\phi_n), \qquad f_n \neq o(\phi_n), \qquad x \to x_0, \quad n = 0, 1, 2, \ldots \tag{1.5.12}$$

are satisfied.

We wish to emphasize that, in almost all problems, we first determine, by some means, the formal series Σf_n *which is expected to yield a good approximation of* f *in the limit under consideration.* Whether it does or not is independent of the particular auxiliary asymptotic sequence chosen as the underlying scale of measurement. *The selection of the auxiliary asymptotic sequence is an a posteriori step in the analysis.* It is introduced solely to enable us to obtain meaningful estimates for the truncation errors $\mathscr{E}(x,N)$, $N = 0, 1, 2, \ldots$.

1.6. Complex Variables and the Stokes Phenomenon

In the previous two sections we considered the asymptotic expansion of a function $f(x)$, as $x \to x_0$, in the case where x is a real variable. We now want to extend our results by allowing the independent variable to be complex. Fortunately, only minor modifications of the definitions given in Sections 1.4

and 1.5 are needed to achieve the desired generalization. There are, however, some interesting and informative aspects of the complex case that do not arise when the argument of f is restricted to be real.

To begin, let z be a complex variable whose domain is R and suppose that z_0 is a point in $\bar{R}$. Suppose, further, that $f(z)$ is defined on R and that we wish to study its behavior as $z \to z_0$ in R.

Because R is a two-dimensional region, in order to define any asymptotic expansion of f as $z \to z_0$ we must now specify the permissible directions of approach to z_0. With this purpose in mind, let us introduce the open sectorial region $S_{\alpha\beta}$ which is assumed to lie in R and which is defined by

$$\begin{aligned} S_{\alpha\beta} &= \{z: 0 < |z - z_0|; \quad \alpha < \arg(z - z_0) < \beta\},^{6} \quad z_0 \text{ finite}, \\ S_{\alpha\beta} &= \{z: |z| > \rho; \quad \alpha < \arg z < \beta\}, \quad z_0, \text{ the point at } \infty. \end{aligned} \tag{1.6.1}$$

We extend our concept of an asymptotic sequence to the complex case in the following.

DEFINITION 1.6.1. The sequence $\{\phi_n(z)\}$, $n = 0, 1, 2, \ldots$, is said to be an asymptotic sequence as $z \to z_0$ in $S_{\alpha\beta}$, if each $\phi_n(z)$ is continuous in R and if the conditions

$$\phi_{n+1}(z) = o(\phi_n(z)), \qquad n = 0, 1, 2, \ldots \tag{1.6.2}$$

hold uniformly as $z \to z_0$ in $S_{\alpha\beta}$.

To illustrate how we generalize the previously made definitions to the complex case, we consider asymptotic expansions of Poincaré type in the following.

DEFINITION 1.6.2. Let $\{\phi_n\}$ be an asymptotic sequence as $z \to z_0$ in $S_{\alpha\beta}$. The formal series $\sum_{n=0}^{\infty} a_n \phi_n$ is said to be an asymptotic expansion of f, as $z \to z_0$ in $S_{\alpha\beta}$, with respect to the asymptotic sequence $\{\phi_n\}$ if the conditions

$$f(z) = \sum_{n=0}^{m} a_n \phi_n + o(\phi_m), \qquad m = 0, 1, 2, \ldots \tag{1.6.3}$$

hold uniformly as $z \to z_0$ in $S_{\alpha\beta}$.

If (1.6.3) holds, then we write

$$f \sim \sum_{n=0}^{\infty} a_n \phi_n, \qquad z \to z_0 \text{ in } S_{\alpha\beta}. \tag{1.6.4}$$

We note that Definition 1.6.2 differs from the corresponding one given in Section 1.4 only in that now a sector of validity is specified. Because all of the definitions and results of Sections 1.4 and 1.5 can be extended via the modification, we shall not state these extensions here.

[6] If $z - z_0 = re^{i\theta}$, then $\arg(z - z_0) = \theta$.

Let us now suppose that $f(z)$ is an analytic and single-valued function in $S_{\alpha\beta}$ and suppose further that, as $z \to z_0$ in $S_{\alpha\beta}$, $f(z)$ has an asymptotic expansion with respect to the asymptotic sequence $\{\phi_n\}$ given by (1.6.4). Here we assume that each function ϕ_n is analytic and single-valued in $S_{\alpha\beta}$. It may occur that both $f(z)$ and its asymptotic expansion can be analytically continued outside of $S_{\alpha\beta}$, but that as certain rays through $z = z_0$ are crossed, the analytic continuation of the asymptotic expansion ceases to be the asymptotic expansion of the analytic continuation of f. This occurrence is known as the *Stokes phenomenon* and the rays across which it occurs are called *Stokes lines.*

The Stokes phenomenon is perhaps not so easy to understand for many readers. The following questions naturally arise: When should we expect it to occur? If it occurs, then how do we determine the locations of the Stokes lines? Finally, what is the significance of the Stokes phenomenon? Because we feel that the phenomenon is an important aspect of asymptotic analysis we shall attempt to at least partially answer these questions.

To begin, let us suppose that f is not single-valued in any 2π neighborhood of $z = z_0$ but that each ϕ_n in (1.6.4) is. It is then reasonable to expect that (1.6.4) will become invalid when a branch cut for $f(z)$ is crossed. In other words, a branch cut for $f(z)$ emanating from $z = z_0$ will be a Stokes line for (1.6.4). Alternatively, if f is single-valued in a 2π neighborhood of $z = z_0$, but at least one of the functions ϕ_n is not, then a branch cut for any $\phi_n(z)$ emanating from $z = z_0$ will be a Stokes line. Such occurrences of the Stokes phenomenon are easily understood and in these cases the Stokes lines are easily located. However, as we shall now see there are further possibilities.

Let us consider the function $f(z) = e^{-1/z}$. This function is analytic everywhere except at the origin where it has an isolated essential singularity. As $z \to 0$ in the region $\text{Re}(z) > 0$, $f(z)$ has an asymptotic expansion with respect to the asymptotic sequence $\{z^n\}$ given by

$$e^{-1/z} \sim 0. \tag{1.6.5}$$

The analytic continuation of $f(z)$ into the left half-plane is obvious as is the continuation of the asymptotic expansion (1.6.5). We also note that both f and the asymptotic expansion are single-valued in the entire plane. It is clear, however, that for $\text{Re}(z) \le 0$, $e^{-1/z}$ is not asymptotic to zero as $z \to 0$. In fact, $e^{-1/z}$ has no asymptotic power series expansion as $z \to 0$ in this region at all. We must conclude, therefore, that the positive and negative imaginary axes are Stokes lines for the asymptotic expansion (1.6.5).

From this last discussion we can begin to understand the basic cause for the occurrence of the Stokes phenomenon. In obtaining (1.6.5) we sought to study the behavior as $z \to 0$ of a fairly complicated function in terms of the simplest of all asymptotic sequences, namely $\{z^n\}$. We have found that this sequence is inadequate to describe the behavior of $e^{-1/z}$ as $z \to 0$ outside of the region $\text{Re}(z) > 0$. This then suggests that, in general, *the occurrence of the Stokes phenomenon is a manifestation of the inadequacy of a particular asymptotic sequence to describe the behavior of the function under consideration outside of some sector. This is the main point of this section.*

It appears then that the sector of validity for a given asymptotic expansion depends solely on the asymptotic sequence chosen and that to maximize the sector of validity for a particular f we have to select the appropriate sequence. Of course, this idea can be carried to an extreme by selecting an asymptotic sequence having f as its first element. We would then arrive at the rather uninteresting result $f(z) \sim f(z)$. In this event, however, the Stokes phenomenon does not occur. Fortunately, a compromise can often be achieved as is illustrated in the following.

EXAMPLE 1.6. Let

$$f(z) = e^{-1/z}\, q(z). \tag{1.6.6}$$

Assume that, as $z \to 0$ in the sector $S_{\alpha\beta}$, $\alpha < -\pi/2$, $\beta > \pi/2$, q has an asymptotic power series expansion given by

$$q(z) \sim \sum_{n=0}^{\infty} a_n z^n. \tag{1.6.7}$$

It is readily seen that, as $z \to 0$ in $\mathrm{Re}(z) > 0$,

$$f(z) \sim 0 \tag{1.6.8}$$

with respect to the asymptotic sequence $\{z^n\}$. Moreover, as $z \to 0$ in the sectors $S_{\alpha,-\pi/2}$ and $S_{\pi/2,\beta}$, $f(z)$ has no asymptotic power series expansion at all.

With respect to the asymptotic sequence $\{e^{-1/z} z^n\}$ we have

$$f(z) \sim e^{-1/z} \sum_{n=0}^{\infty} a_n z^n \tag{1.6.9}$$

which is valid as $z \to 0$ in $S_{\alpha\beta}$. Thus, not only does (1.6.9) appear to be more informative than (1.6.8) in the region $\mathrm{Re}(z) > 0$, it has, in addition, a larger domain of validity.

We have perhaps given the impression that the Stokes phenomenon is something bad and to be avoided whenever possible. Although it is true that its occurrence serves to limit the domain of validity of a given asymptotic expansion, nevertheless its very occurrence does yield useful information. Indeed it places us on alert that a significant change in the asymptotic behavior of the function under study will arise when a Stokes line is crossed. Thus, in the above example, the fact that the imaginary axes are Stokes lines for (1.6.8) indicates that the asymptotic behavior of (1.6.6) as $z \to 0$ undergoes a significant change when these lines are crossed. This information is not contained in (1.6.9) and can only be recovered by some further analysis of the terms in the asymptotic sequence $\{e^{-1/z} z^n\}$.

We now wish to investigate further under what circumstances we might expect the Stokes phenomenon to occur. In particular, let us consider asymptotic power series expansions. Suppose first that $f(z)$ is analytic and single-valued in a 2π neighborhood of $z = z_0$. Then $f(z)$ has a convergent power series expansion about $z = z_0$ valid for all values of $\arg(z - z_0)$. In other words, the Stokes phenomenon does not occur.

Conversely, suppose that $f(z)$ is analytic and single-valued in the punctured disc $0 < |z - z_0| < \rho$. Suppose, further, that $f(z)$ has an asymptotic power series expansion about $z = z_0$ valid for all values of $\arg(z - z_0)$, that is, an expansion for which no Stokes phenomenon occurs. Then we claim that $f(z)$ is analytic at $z = z_0$ and the asymptotic power series actually converges. To see this we note that the existence of the asymptotic power series for all values of $\arg(z - z_0)$ implies that $f(z)$ is bounded in a 2π neighborhood of $z = z_0$. Hence, by Riemann's lemma, f has a removable singularity at $z = z_0$. This in turn implies that f has a convergent Taylor series expansion about $z = z_0$ which, by uniqueness, must coincide with the asymptotic power series.

The above discussion shows that the Stokes phenomenon will occur for an asymptotic power series expansion about $z = z_0$ only when the function under consideration has a singularity at $z = z_0$. Moreover, we find that if $z = z_0$ is an isolated singularity, then it must be an essential singularity.

To conclude this section we wish to quote a theorem due to Carleman which shows that an analytic function can have a divergent power series about a point on the boundary of its domain of analyticity.

THEOREM 1.6. Assume that, as $z \to 0$ in the sector $S_{\alpha\beta}$, with $\alpha > -\pi$ $\beta < \pi$, $f(z)$ has an asymptotic power series expansion given by

$$f(z) \sim \sum_{n=0}^{\infty} a_n z^n. \tag{1.6.10}$$

Then there exists a function $g(z)$ analytic in $S_{\alpha\beta}$ and such that, as $z \to 0$ in $S_{\alpha\beta}$,

$$(f - g) \sim 0 \tag{1.6.11}$$

with respect to the asymptotic sequence $\{z^n\}$.

1.7. Operations with Asymptotic Expansions of Poincaré Type

We shall now consider several operations that can be performed on asymptotic expansions of Poincaré type. No attempt to be exhaustive will be made and indeed, we shall limit our considerations to those operations that are most likely to prove useful in future chapters.

Throughout this section x is a complex variable unless otherwise stated and all asymptotic expansions are assumed to hold as $x \to x_0$ in the sector

$S_{\alpha\beta} = \{x : 0 < |x - x_0| < \rho, \alpha < \arg(x - x_0) < \beta\}$. We first prove the following.

THEOREM 1.7.1. Let $\{\phi_n(x)\}$, $n = 1, 2, \ldots$, be an asymptotic sequence as $x \to x_0$ in $S_{\alpha\beta}$ and let

$$f(x) \sim \sum_{n=1}^{N} a_n \phi_n(x), \qquad g(x) \sim \sum_{n=1}^{N} b_n \phi_n(x) \tag{1.7.1}$$

be, respectively, asymptotic expansions to N terms with respect to $\{\phi_n\}$ of the given functions $f(x)$ and $g(x)$. If γ and μ are arbitrary complex constants, then an asymptotic expansion to N terms of $h = \gamma f + \mu g$ with respect to $\{\phi_n\}$ is given by

$$h(x) \sim \sum_{n=1}^{N} (\gamma a_n + \mu b_n) \phi_n(x). \tag{1.7.2}$$

PROOF. The proof follows directly from Definition 1.6.2.

If the expansions in (1.7.1) are valid for arbitrarily large N, then so is (1.7.2). Also, Theorem 1.7.1 can under certain conditions be extended to infinite linear combinations of asymptotic expansions. The extension to finite linear combinations is immediate.

A generalization of Theorem 1.7.1 is obtained in the following.

THEOREM 1.7.2. Let $\{\phi_n\}$, $n = 1, 2, \ldots$ and $\{\psi_m\}$, $m = 1, 2, \ldots$ be asymptotic sequences as $x \to x_0$ in $S_{\alpha\beta}$. Suppose that for a given positive integer N there exists a positive integer M such that $\psi_M = O(\phi_N)$ as $x \to x_0$ in $S_{\alpha\beta}$. Suppose, further, that

$$\psi_m = \sum_{n=1}^{N} c_{mn} \phi_n + o(\phi_N), \qquad m = 1, 2, \ldots, M.^{7} \tag{1.7.3}$$

Finally, let f and g be given functions having the finite asymptotic expansions $f \sim \Sigma_{n=1}^{N} a_n \phi_n$ and $g \sim \Sigma_{m=1}^{M} b_m \psi_m$, respectively. Then, for any complex constants γ and μ,

$$\gamma f + \mu g \sim \sum_{n=1}^{N} \left[\gamma a_n + \mu\left(\sum_{m=1}^{M} b_m c_{mn}\right)\right] \phi_n. \tag{1.7.4}$$

PROOF. Here again the proof follows directly from the definitions of Section 1.4.

We can conclude from Theorems 1.7.1 and 1.7.2 that addition and subtraction of asymptotic expansions of Poincaré type are readily justifiable procedures. Unfortunately the same cannot be said for the operations of multiplication

[7] Because $\{\psi_m\}$ is an asymptotic sequence certain of the constants c_{mn} must be zero.

and division. In fact, suppose that $\{\phi_n\}$ and $\{\psi_m\}$ are asymptotic sequences and that $f \sim \Sigma a_n \phi_n$ while $g \sim \Sigma b_m \psi_m$. The formal multiplication of the two asymptotic expansions does not, in general, yield an asymptotic expansion of fg because, in general, neither can the set $\{\phi_n \psi_m\}$, $n, m = 1, 2, \ldots$ be arranged so as to form an asymptotic sequence, nor can each member of the set be expanded with respect to some common asymptotic sequence.

There *are* conditions which, when satisfied, are sufficient to justify the multiplication of two asymptotic expansions. Indeed we have the following.

THEOREM 1.7.3. Let $\{\phi_n\}$, $n = 1, 2, \ldots$, $\{\psi_m\}$, $m = 1, 2, \ldots$, and $\{\theta_k\}$, $k = 1, 2, \ldots$ be asymptotic sequences as $x \to x_0$ in $S_{\alpha\beta}$. Assume that for a given positive integer K there exist positive integers $N(K)$, $M(K)$ such that the relations

$$\phi_{N(K)} \psi_1 = O(\theta_K), \tag{1.7.5}$$

$$\phi_1 \psi_{M(K)} = O(\theta_K), \tag{1.7.6}$$

$$\phi_n \psi_m = \sum_{k=1}^{K} c_{nmk} \theta_k + o(\theta_K), \qquad 1 \le n \le N(K), \qquad 1 \le m \le M(K) \tag{1.7.7}$$

hold. Let f and g be given functions having the respective asymptotic expansions

$$f = \sum_{n=1}^{N(K)} a_n \phi_n + o(\phi_{N(K)}), \qquad g = \sum_{m=1}^{M(K)} b_m \psi_m + o(\psi_{M(K)}). \tag{1.7.8}$$

Then

$$fg = \sum_{k=1}^{K} c_k \theta_k + o(\theta_K), \tag{1.7.9}$$

where

$$c_k = \sum_{n=1}^{N(K)} \sum_{m=1}^{M(K)} c_{nmk} a_n b_m, \qquad k = 1, 2, \ldots, K. \tag{1.7.10}$$

PROOF. We first assume that $N(K)$, $M(K)$, and K are finite and write

$$\begin{aligned} fg &= \left[\sum_{n=1}^{N(K)} a_n \phi_n + o(\phi_{N(K)})\right] \left[\sum_{m=1}^{M(K)} b_m \psi_m + o(\psi_{M(K)})\right] \\ &= \sum_{n=1}^{N(K)} \sum_{m=1}^{M(K)} a_n b_m \phi_n \psi_m + o(\phi_N \psi_1) + o(\phi_1 \psi_M). \end{aligned} \tag{1.7.11}$$

Now upon inserting relations (1.7.7) into (1.7.11) and collecting terms we obtain

$$fg = \sum_{k=1}^{K} \sum_{n=1}^{N(K)} \sum_{m=1}^{M(K)} a_n b_m c_{nmk} \theta_k + o(\theta_K) + o(\phi_N \psi_1) + o(\phi_1 \psi_M). \tag{1.7.12}$$

Finally by using (1.7.5), (1.7.6), and (1.7.10) in (1.7.12) we obtain (1.7.9).

If any of the integers $N(K)$, $M(K)$, and K is infinite, then the theorem remains valid so long as the constants c_k are well defined, that is, so long as each of the infinite series (1.7.10) converge.

Theorem 1.7.3 is a rather general result concerning the multiplication of two asymptotic expansions. In many problems the sequences $\{\phi_n\}$, $\{\psi_m\}$, and $\{\theta_k\}$ are the same, that is,

$$\phi_n \phi_m = \sum_{k=1}^{N} c_{nmk} \phi_k + o(\phi_N), \qquad 1 \le n \le N, \qquad 1 \le m \le N. \tag{1.7.13}$$

If, in addition, $\phi_1 = O(1)$ as $x \to x_0$ in $S_{\alpha\beta}$, then $\{\phi_n\}$ is said to be a *multiplicative asymptotic sequence*. We note, for example, that $\{x^n\}$, $n = 0, 1, 2, \ldots$, is a multiplicative asymptotic sequence as $x \to 0$. If two functions f and g each have an asymptotic expansion with respect to a given multiplicative asymptotic sequence, then clearly Theorem 1.7.3 applies. We also have the following.

THEOREM 1.7.4. Let $\{\phi_n\}$, $n = 1, 2, \ldots$ be a multiplicative asymptotic sequence and let $\phi_1 = o(1)$ as $x \to x_0$. Let N be a given positive integer and assume that there exists a finite positive integer M such that $(\phi_1)^M = O(\phi_N)$. If

$$g(z) = \sum_{m=1}^{M} c_m z^m + o(z^M), \qquad z \to 0 \tag{1.7.14}$$

and if

$$z = z(x) = \sum_{n=1}^{N} a_n \phi_n + o(\phi_N), \tag{1.7.15}$$

then

$$f(x) = g(z(x)) = \sum_{n=1}^{N} A_n \phi_n + o(\phi_N). \tag{1.7.16}$$

Here the coefficients A_n are obtained by formally substituting (1.7.15) into (1.7.14) and collecting terms by making use of (1.7.13).

PROOF. Because M is finite, we have by hypothesis that

$$z^m = \sum_{n=1}^{N} b_{mn} \phi_n + o(\phi_N), \qquad m = 1, 2, \ldots, M \tag{1.7.17}$$

and that $z^M = O(\phi_N)$. The theorem then follows upon inserting (1.7.17) into (1.7.14) and collecting terms.

COROLLARY. Let $\{\phi_n\}$ be a multiplicative asymptotic sequence satisfying the hypotheses of Theorem 1.7.4. Let

$$h(x) = c + \sum_{n=1}^{N} a_n \phi_n + o(\phi_N), \qquad c \neq 0, \tag{1.7.18}$$

$$\omega(x) = d + \sum_{n=1}^{N} b_n \phi_n + o(\phi_N). \tag{1.7.19}$$

Then $v = \omega/h$ has an asymptotic expansion to N terms with respect to $\{\phi_n\}$.

PROOF. To establish the corollary we only need show that $f = h^{-1}$ has an asymptotic expansion of the form (1.7.18) because the result would then follow by Theorem 1.7.3. If we set $z = h - c$, then we have

$$z(x) = \sum_{n=1}^{N} a_n \phi_n + o(\phi_N) \tag{1.7.20}$$

and

$$f(x) = h^{-1} = \frac{1}{z+c} = \frac{1}{c} \sum_{m=0}^{M} (-1) \left(\frac{z}{c}\right)^m + o(z^M). \tag{1.7.21}$$

Upon applying Theorem 1.7.4 we obtain the desired result concerning f which completes the proof.

The final operations we shall consider are integration and differentiation of asymptotic expansions. Let us first suppose that $f(x,v)$ is a function of the complex variable x and the real parameter v. We suppose that as $x \to x_0$ in $S_{\alpha\beta}$

$$f(x,v) \sim \sum_{n=1}^{N} a_n(v)\, \phi_n(x) \tag{1.7.22}$$

is an asymptotic expansion of f to N terms which holds uniformly in v for $v_0 \leq v \leq v_1$. We now prove the following.

THEOREM 1.7.5. Let $s(v)$ be any integrable function of v such that the integrals

$$b_n = \int_{v_0}^{v_1} s(v)\, a_n(v)\, dv, \qquad n = 1, \ldots, N \tag{1.7.23}$$

exist and such that for all x in $S_{\alpha\beta}$ the integral

$$g(x) = \int_{v_0}^{v_1} s(v) f(x,v)\, dv \tag{1.7.24}$$

exists. Then, as $x \to x_0$ in $S_{\alpha\beta}$

$$g(x) \sim \sum_{n=1}^{N} b_n \phi_n(x), \tag{1.7.25}$$

that is, the asymptotic expansion (1.7.22) can be integrated term-by-term with respect to the parameter v.

PROOF. We have

$$g(x) = \int_{v_0}^{v_1} s(v)\left[\sum_{n=1}^{N} a_n(v)\,\phi_n(x) + \Phi_N(x,v)\right] dv \tag{1.7.26}$$

where $\Phi_N(x, v) = o(\phi_N(x))$ uniformly in v for $v \in [v_0, v_1]$. It then follows that

$$\int_{v_0}^{v_1} \Phi_N(x,v)\, s(v)\, dv = o(\phi_N(x)), \qquad x \to x_0 \text{ in } S_{\alpha\beta}, \tag{1.7.27}$$

and hence

$$g(x) = \sum_{n=1}^{N} b_n\,\phi_n + o(\phi_N), \qquad x \to x_0 \text{ in } S_{\alpha\beta}. \tag{1.7.28}$$

This completes the proof.

Of perhaps more interest is the question of whether or not an asymptotic expansion can be integrated with respect to the independent variable term-by-term. We first establish the following.

LEMMA 1.7.1. Let x be a real variable and let R be the interval (a,x_0). If $\{\phi_n\}$ is an asymptotic sequence of integrable functions as $x \to x_0$ in R, then the sequence $\{\Phi_n\}$ where

$$\Phi_n(x) = \int_x^{x_0} |\phi_n(t)|\, dt, \qquad n = 1, 2, \ldots \tag{1.7.29}$$

is also an asymptotic sequence as $x \to x_0$ in R.

PROOF. We have by (1.7.29) that

$$\Phi_{n+1}(x) = \int_x^{x_0} |\phi_{n+1}(t)|\, dt = o\left[\int_x^{x_0} |\phi_n(t)|\, dt\right] = o(\Phi_n(x)), \quad n = 1, 2, \ldots \tag{1.7.30}$$

as $x \to x_0$ in R. This completes the proof.

We now prove the following.

THEOREM 1.7.6. Let x be a real variable, $R = (a_0,x_0)$, and $\{\phi_n\}$ an asymptotic sequence of positive functions as $x \to x_0$ in R. Suppose that the integrals (1.7.29) exist for $x \in R$. If f is a given function such that

$$f \sim \sum_{n=1}^{N} a_n\,\phi_n(x), \qquad x \to x_0 \text{ in } R \tag{1.7.31}$$

and if

$$g(x) = \int_x^{x_0} f(t)\, dt \tag{1.7.32}$$

exists for $x \,\varepsilon\, (b, x_0)$, $a \leq b < x_0$, then

$$g(x) \sim \sum_{n=1}^{N} a_n \Phi_n(x), \qquad x \to x_0 \tag{1.7.33}$$

is an asymptotic expansion of g to N terms with respect to $\{\Phi_n\}$.

PROOF. We have

$$f = \sum_{n=1}^{N} a_n \phi_n + \psi_N \tag{1.7.34}$$

where $\psi_N(x) = o(\phi_N)$ as $x \to x_0$. Upon integrating both sides of (1.7.34) from x to x_0 we obtain

$$\begin{aligned} g &= \int_x^{x_0} \sum_{n=1}^{N} a_n \phi_n \, dx + \int_x^{x_0} \psi_N \, dx \\ &= \sum_{n=1}^{N} a_n \Phi_n + o\left(\int_x^{x_0} |\phi_N| \, dx\right) \\ &= \sum_{n=1}^{N} a_n \Phi_n + o(\Phi_N) \end{aligned} \tag{1.7.35}$$

which completes the proof.

We remark that although Lemma 1.7.1 and Theorem 1.7.6 deal only with functions of a real variable, they can without difficulty be extended to the case where x is a complex variable and where (1.7.29) and (1.7.32) are line integrals in the complex x plane.

Just as with convergent series, term-by-term differentiation of an asymptotic expansion is more difficult to justify than term-by-term integration. Indeed, an asymptotic expansion of the derivative of a function cannot, in general, be obtained by differentiating its asymptotic expansion term-by-term. That this is so is due mainly to the fact that differentiation of an asymptotic sequence does not, in general, yield an asymptotic sequence. To illustrate this, let $x_0 = 0$ and $\phi_n = x^n [\cos(x^{-n+1}) + \alpha]$, $\alpha > 1$, $n = 1, 2, \ldots$. We have that $\{\phi_n\}$ is an asymptotic sequence as $x \to 0$. However, because

$$\phi_n' = nx^{n-1} [\cos(x^{-n+1}) + \alpha] + (n-1) \sin(x^{-n+1})$$

we have that $\{\phi_n'\}$ is *not* an asymptotic sequence as $x \to 0$.

It may also occur that $f(x)$ has an asymptotic expansion with respect to $\{\phi_n\}$ but its derivative does not have an asymptotic expansion with respect to $\{\phi_n'\}$ even when $\{\phi_n'\}$ is an asymptotic sequence in the limit under consideration. To see this consider $f(x) = e^{-x} \sin(e^x)$. As $x \to \infty$ we have that $f(x) \sim 0$

with respect to the asymptotic sequence $\{x^{-n}\}$, $n=0, 1, 2, \ldots$. But $f'(x)=\cos(e^x)-e^{-x}\sin(e^x)$ has no asymptotic expansion with respect to this sequence.

We can establish conditions sufficient to justify term-by-term differentiation of asymptotic series. Indeed, we have the following.

THEOREM 1.7.7. Let x be a real variable, $R=(a,x_0)$, and $\{\phi_n\}$, $n=1, 2, \ldots$, be an asymptotic sequence as $x \to x_0$ in R. Suppose that $\phi_1 = o(1)$ as $x \to x_0$ in R and that each ϕ_n is differentiable in (a,x_0). Suppose, further, that $\{\psi_n\}$ with $\psi_n = \phi_n'$ is an asymptotic sequence as $x \to x_0$ in R. Finally, let N be a fixed positive integer and assume that ψ_N is positive in the interval (b,x_0) for some $b \geq a$.

If f is a given differentiable function in R such that

$$f(x)-f(x_0) \sim \sum_{n=1}^{N} a_n\,\phi_n(x), \qquad x \to x_0 \tag{1.7.36}$$

and if $f'(x)$ has an asymptotic expansion to N terms with respect to $\{\psi_n\}$, then

$$f'(x) \sim \sum_{n=1}^{N} a_n\,\psi_n. \tag{1.7.37}$$

PROOF. By hypothesis there exist constants $b_1, \ldots, b_N$ such that

$$f'(x) = \sum_{n=1}^{N} b_n\,\psi_n + w(x) \tag{1.7.38}$$

where $w(x) = o(\psi_N)$ as $x \to x_0$. Furthermore,

$$f(x)-f(x_0) = -\int_x^{x_0} f'(t)\,dt = -\sum_{n=1}^{N} b_n \int_x^{x_0} \psi_n(t)\,dt - \int_x^{x_0} w(t)\,dt. \tag{1.7.39}$$

If we take $b < x < x_0$ in (1.7.39) and use (1.2.17), then we obtain

$$f(x)-f(x_0) = \sum_{n=1}^{N} b_n\,\phi_n(x) + o(\phi_N(x)), \qquad x \to x_0. \tag{1.7.40}$$

Here we have used the fact that $\phi_n(x_0)=0$ for all n because, by assumption, $\phi_1(x)=o(1)$, $x \to x_0$. Upon comparing (1.7.36) and (1.7.40) and applying the uniqueness theorem (Theorem 1.4), we find that $a_n = b_n$, $n=1, 2, \ldots, N$ and the theorem is proved.

We might point out that there are two key assumptions made in the statement of Theorem 1.7.7. The first is that $\{\phi_n'\}$ is an asymptotic sequence as $x \to x_0$ and the second is that f' has an asymptotic expansion with respect to this sequence. The first condition is readily verified and, for example, is satisfied for the important case $\phi_n = (x-x_0)^n$, $n=1, 2, \ldots$. The second condition, however, is not so easily verified and in fact the verification itself often involves the direct determination of the asymptotic expansion of f'. It may occur that we have good reasons, either mathematical or physical, to suspect that f' has an asymptotic expansion of the desired form and can proceed on this basis. Furthermore, additional information about f, such as the fact that f satisfies a differential equation of a given class, can often be used to directly verify the second condition.

There is one case where the second condition can be eliminated entirely and we offer without proof the relevant theorem.

THEOREM 1.7.8. Let x be a complex variable and let $f(x)$ be a given function analytic in the sector

$$S_{\alpha\beta} = \{x: 0 < |x - x_0| < \rho, \qquad \alpha < \arg(x - x_0) < \beta\}.$$

Suppose that

$$f(x) \sim \sum_{n=0}^{N-1} a_n (x - x_0)^n \tag{1.7.41}$$

is an asymptotic power series expansion of f to N terms which holds as $x \to x_0$ in $S_{\alpha\beta}$. Then

$$f'(x) \sim \sum_{n=1}^{N-1} n\, a_n (x - x_0)^{n-1} \tag{1.7.42}$$

is an asymptotic power series expansion of f' to $N - 1$ terms. It holds, however, in a smaller sector about $x = x_0$. Indeed (1.7.42) is valid as $x \to x_0$ in $\alpha - \delta < \arg(z - z_0) < \beta - \delta$ for any $\delta > 0$.

1.8. Exercises

1.1. (a) Show that the function $I(x)$ defined by (1.1.1) can be written in the form

$$I(x) = x \int_0^\infty \frac{e^{-t}}{t + x}\, dt. \tag{1.8.1}$$

(b) Use the exact expression

$$\frac{x}{x+t} = \sum_{n=0}^{N} \left(\frac{-t}{x}\right)^n + \left(\frac{-t}{x}\right)^{N+1} \Big/ \left(1 + \frac{t}{x}\right)$$

in (1.8.1) to obtain

$$I(x) = S_N(x) + \mathscr{E}(x, N) \tag{1.8.2}$$

with $S_N(x)$ given by (1.1.9).

(c) Discuss the error integral $\mathscr{E}(x, n)$ in this representation.

1.2. (a) Show that (1.8.1) can be rewritten in the form

$$I(x) = x\int_0^\infty \frac{e^{-xt}}{1+t}\,dt. \tag{1.8.3}$$

(b) Integrate by parts $N+1$ times in (1.8.3) to obtain an expression of the form (1.8.2). Show that the approximation $S_N(x)$ obtained here agrees with that obtained in Exercise 1.1.

(c) Discuss the remainder integral in the form it arises in the present exercise.

1.3. (a) Again consider $I(x)$ defined by (1.1.1) and in particular the error $\mathscr{E}(x,N)$ made in approximating I by $S_N(x)$. Use either (1.1.10) or the expressions for $\mathscr{E}$ obtained in Exercises 1.1 and 1.2 to show directly that

$$|\mathscr{E}(x,N)| < \frac{N!}{x^N}. \tag{1.8.4}$$

(b) Use (1.8.4) and a table of values for $n!$ to obtain the following upper bounds on the minimum error.

x	$\mathscr{E}(x,[x])$
5	4×10^{-2}
10	3.7×10^{-4}
50	3×10^{-21}
100	9.4×10^{-43}

Note that the results for $x=5$ and $x=10$ given in Table 1.1 indicate that the actual minimum errors are approximately one-half of the bounds listed above.

1.4. Repeat the analysis carried out in Section 1.1 for $I(x)$ for the integral

$$ci(x) = -\int_x^\infty \frac{\cos t}{t}\,dt. \tag{1.8.5}$$

In particular:

(a) Integrate by parts repeatedly to obtain approximations of $ci(x)$ in terms of the partial sums (of a divergent series). Also obtain an explicit expression for the error integral that arises after $N+1$ integrations by parts.

(b) Show that $\hat{N}$, the optimum number of terms of the divergent series to use in approximating $ci(x)$, that is, the number yielding the smallest absolute error, is given by

$$\hat{N} = [x].$$

(c) Given the exact numerical values below, construct a table analogous to Table 1.1 for the function $ci(x)$.

x	$ci(x)$
1	0.33740
2	0.42298

4	-0.14098
5	-0.14205
10	-0.04545
100	-0.00513

1.5. Repeat Exercise 1.4 for the function

$$si(x) = -\int_x^\infty \frac{\sin t}{t}\,dt. \tag{1.8.6}$$

Here the table of part (c) is to be replaced by

x	$si(x)$
1	-0.62471
2	0.03462
4	0.18741
5	-0.02086
10	$+0.08755$
100	0.00858

1.6. Consider the function

$$F(x) = xe^{x^2}\,\mathrm{erfc}(x) = \frac{2}{\sqrt{\pi}}\,xe^{x^2}\int_x^\infty e^{-t^2}\,dt. \tag{1.8.7}$$

(a) Introduce the change of variable of integration $\tau = t^2$ to obtain

$$\int_x^\infty e^{-t^2}\,dt = \frac{1}{2}\int_{x^2}^\infty \frac{e^{-\tau}}{\sqrt{\tau}}\,d\tau. \tag{1.8.8}$$

Integrate by parts $N+1$ times in (1.8.8) and show that

$$F(x) = \frac{1}{\sqrt{\pi}}\left\{1 + \sum_{m=1}^{N}\frac{(-1)^m 1\cdot 3\cdot\cdots\cdot(2m-1)}{(2x^2)^m} + \frac{(-1)^{N+1}(1\cdot 3\cdot\ \cdots\ \cdot(2N+1))}{2^N}\,xe^{x^2}\int_x^\infty \frac{e^{-t^2}}{t^{2N+2}}\,dt\right\}. \tag{1.8.9}$$

(b) Show that the infinite series

$$\sum_{m=0}^{\infty}\frac{(-1)^m(1\cdot 3\cdot\cdots\ \cdot(2m+1))}{(2x^2)^m} \tag{1.8.10}$$

diverges for all x and that the optimum number of terms to use here is $\hat{N} = [2x^2] + 1$.

(c) Show how to derive (1.8.9) directly from (1.8.7) by using the fact that

$$te^{-t^2} = -\frac{1}{2}\frac{d}{dt}e^{-t^2}.$$

(d) Develop a table analogous to Table 1.1 for $F(x)$ using the exact values

x	$F(x)$
1	0.42758
2	0.51079
5	0.55352
10	0.56140

1.7. Show that if $f(x) = O(g(x))$ and $g(x) = O(f(x))$ as $x \to x_0$, then neither is $f(x) = o(g(x))$ nor is $g(x) = o(f(x))$ as $x \to x_0$.

1.8. Show that if $f(x) = O((x - x_0)^{\delta_0})$ as $x \to x_0$, then $f(x) = O((x - x_0)^{\delta})$ for all $\delta \le \delta_0$ as $x \to x_0$.

1.9. Give an example in which $f(x) = o(g(x))$ as $x \to x_0$ and yet $f(x)$ and $g(x)$ both vanish to the same order, at $x = x_0$, with respect to the powers of $(x - x_0)$.

1.10. Prove the statements (1) to (7) at the end of Section 1.2.

1.11. Verify the following statements:

(a) $\sin x^{1/3} = O(x^{1/3}), \quad x \to 0+.$
(b) $\sin x = O(x \cos x), \quad x \to 0.$
(c) $x \cos x = O(\sin x), \quad x \to 0.$
(d) $e^{-x} = O(1), \quad x \to 0.$
(e) $n! = o(n^n), \quad n \to \infty.$
(f) $\log x = o(x^\varepsilon), \quad x \to \infty, \quad \varepsilon > 0.$
(g) $(\log x)^k = o(x^\varepsilon), \quad x \to \infty, \quad \varepsilon > 0.$
(h) $\log(\log[\log x]) = o(\log(\log x)), \quad x \to 0+.$
(i) $e^{ix} = O(1), \quad x \to x_0, \quad \text{any } x_0.$
(j) $si(x) = O(x^{-1}), \quad x \to \infty.$

1.12. Verify that (1.8.10) is an asymptotic power series for the function $F(x)$ defined by (1.8.7).

1.13. Suppose that $f(z)$ has a Laurent expansion for $|z| > a$ and has at worst a pole of order k at $z = \infty$. Show that $z^k f(z)$ has an asymptotic power series as $|z| \to \infty$.

1.14. Verify that $\{\phi_n(x)\}$ is an asymptotic sequence in the indicated limit when

(a) $\phi_n = x^{-r_n} e^{ix}, \quad \text{Re}(r_0) < \text{Re}(r_1) < \cdots, \quad x \to \infty.$
(b) $\phi_n = x^{r_n} \sin \alpha x, \quad \text{any } \alpha \ne 0, \quad \text{Re}(r_0) < \text{Re}(r_1) < \cdots, \quad x \to 0.$
(c) $\phi_n = e^{-r_n x}, \quad \text{Re}(r_0) < \text{Re}(r_1) < \cdots, \quad x \to \infty.$
(d) $\phi_n = e^{r_n x}, \quad \text{Re}(r_0) < \text{Re}(r_1) < \cdots, \quad x \to -\infty.$

(e) $\phi_n = \left(F(x) - \frac{1}{\sqrt{\pi}}\right)^n$, F given by (1.8.7), $x \to \infty$.

(f) $\phi_n = \left(\frac{F(x)}{x}\right)^n$, F given by (1.8.7), $x \to \infty$.

1.15. In each of the following cases explain why $\{\phi_n\}$ is *not* an asymptotic sequence in the indicated limit.

(a) $\phi_n = x^{r_n} e^{ix}$, $\mathrm{Re}(r_0) < \mathrm{Re}(r_1) < \cdots$, $x \to \infty$.

(b) $\phi_n = x^{-r_n} \cos nx$, $\mathrm{Re}(r_0) < \mathrm{Re}(r_1) < \cdots$, $x \to +\infty$.

(c) $\phi_n = \begin{cases} x^{-n} \cos x, & n \text{ even}, \\ x^{-n} \sin x, & n \text{ odd}, \end{cases}$ $x \to \infty$.

1.16. Suppose that

$$f(x) \sim \sum_{n=0}^{\infty} a_n x^{r_n}, \qquad x \to 0.$$

Here all the a_n's are nonzero and $\mathrm{Re}(r_0) < \mathrm{Re}(r_1) < \cdots$. Suppose further that

$$f(x) \sim \sum_{n=0}^{\infty} b_n x^{s_n}, \qquad x \to 0.$$

Here all the b_n's are nonzero and $\mathrm{Re}(s_0) < \mathrm{Re}(s_1) < \cdots$. Show that $a_n = b_n$ and $r_n = s_n$ for all n.

1.17. The Bessel function $J_0(2x)$ has the following (convergent) power series about $x = 0$:

$$J_0(2x) = \sum_{n=0}^{\infty} \frac{(-1)^n x^{2n}}{(n!)^2}. \tag{1.8.11}$$

When possible, find at least two nontrivial terms in the asymptotic expansion of $J_0(2x)$, as $x \to 0$ with respect to the asymptotic sequence $\{\psi_n(x)\}$, $n = 0, 1, 2, \ldots$, defined by

(a) $\psi_n = (\sin x)^n$.

(b) $\psi_n = (\sin x)^{2n}$.

(c) $\psi_n = (\sin x)^{3n}$.

1.18. For each of the sequences $\{f_n(x)\}$ defined below, find an asymptotic sequence $\{\psi_n(x)\}$ which is such that, for $n = 1, 2, \ldots$, $f_n(x) = O(\psi_n(x))$ but $f_n(x) \neq o(\psi_n(x))$ in the indicated limit.

(a) $f_n = \dfrac{\sin nx}{x^{n/2}}$, $x \to \infty$.

(b) $f_n = \dfrac{\alpha si(x) + \beta ci(x)}{x^n}$, $x \to \infty$, α, β constants.

(c) $f_n = e^{-r_n x} \cos nx$, $\mathrm{Re}(r_0) < \mathrm{Re}(r_1) < \ldots$, $x \to \infty$.

1.19. (a) Consider the expansion derived in Exercise 1.4. Identify a useful auxiliary asymptotic sequence and show that the derived result is an asymptotic expansion with respect to this sequence.

(b) Repeat part (a) for the expansion derived in Exercise 1.5.

1.20. Suppose that $(z - z_0)^k f(z)$ has an asymptotic power series as $z \to z_0$, valid for all values of $\arg(z - z_0)$. Prove that $f(z)$ has at worst a pole of order k at $z = z_0$.

1.21. Let

$$I(z) = z \int_0^\infty \frac{e^{-t}}{t + z}\, dt, \qquad |\arg(z)| < \pi. \tag{1.8.12}$$

(a) Derive an asymptotic expansion of $I(z)$ as $|z| \to \infty$ in the given sector.

(b) Show that

$$I(z) = -2\pi i z e^z + z \int_0^\infty \frac{e^{-t}}{t + z}\, dt, \qquad |\arg(z) - 2\pi| < \pi. \tag{1.8.13}$$

(c) Show that the result derived in part (a) cannot be valid for $\arg(z) \geq 3\pi/2$ and hence the ray $\arg(z) = 3\pi/2$ is a Stokes line for the expansion.

(d) Show that the result in part (a) is valid for $\pi < \arg(z) < 3\pi/2$.

(e) Show that $\arg(z) = 3\pi/2$ is a Stokes line for the asymptotic expansion of $I(z)$ with respect to the asymptotic sequence $\{z^{-n}\}$ but is not a Stokes line for the asymptotic expansion of $I(z) + 2\pi i z e^z$ with respect to this sequence.

1.22. (a) Show that $\{\psi_n(z)\} = \{z^{-n} e^{|\mathrm{Re}(z)|}\}$, $n = 0, 1, 2, \ldots$ is an asymptotic sequence as $|z| \to \infty$.

(b) Let $ci(z)$ be defined by (1.8.5) with x replaced by z and the path of integration being $\mathrm{Im}(z) = \text{const}$. Show that $ci(z)$ has an asymptotic expansion, for $|z| \to \infty$ in the sector defined by $|\arg(z)| < \pi$, with respect to the auxiliary sequence $\{\psi_n(z)\}$ of part (a).

(c) Show that

$$ci(z) = 2\pi i \cos z - \int_z^\infty \frac{\cos t}{t}\, dt, \qquad |\arg(z) - 2\pi| < \pi, \tag{1.8.14}$$

and hence that $ci(z)$ has an asymptotic expansion with respect to $\{\psi_n\}$ in this sector. Show, however, that this expansion differs from the one in part (b).

1.23. (a) Suppose that $\{\phi_n(x)\}$ is an asymptotic sequence as $x \to x_0$ and that each ϕ_n is integrable. Prove that $\{\Phi_n\}$ with

$$\Phi_n = \int_{x_0}^x |\phi_n(t)|\, dt$$

is an asymptotic sequence as $x \to x_0$.

(b) Show that if $f(x)=O(g(x))$, as $x\to x_0$, with f and g integrable functions, then

$$\int_{x_0}^{x} f(t)dt = O\left(\int_{x_0}^{x} |g(t)|\,dt\right), \qquad x\to x_0.$$

1.24. Consider the sequences $\{\phi_n(x)\}$, $\{\Phi_n(x)\}$, and $\{\psi_n(x)\}$ as $x\to 0+$. Here, $n\geq 1$ and

$$\phi_n(x)=x^n\cos\frac{1}{x}, \qquad \Phi_n(x)=\int_0^x \phi_n(t)dt, \qquad \psi_n(x)=\int_0^x |\phi_n(t)|\,dt.$$

Show directly that $\{\phi_n(x)\}$ and $\{\psi_n(x)\}$ are asymptotic sequences but that $\{\Phi_n\}$ is not. [*Hint*: Examine the zeros of $\Phi_n(x)$ and $\Phi_{n+1}(x)$.]

1.25. Suppose that

$$f(x)=e^{-x^{-2}}\cos e^{x^{-2}}.$$

Find an asymptotic power series for $f(x)$ as $x\to 0$. Differentiate the series and show that it is *not* an asymptotic power series for $f'(x)$.

1.26. Consider the integral

$$I(\lambda)=\int_0^\lambda e^{t^2}\,dt.$$

(a) Integrate e^{z^2} counterclockwise around the rectangle in the complex z plane having vertices $z=0$, $z=\lambda$, $z=\lambda+iR$, and $z=iR$. Let $R\to\infty$ and show that

$$I(\lambda)=\frac{i\sqrt{\pi}}{2}-i\int_0^\infty e^{-(y-i\lambda)^2}\,dy. \tag{1.8.15}$$

(b) Note that

$$-2(y-i\lambda)e^{-(y-i\lambda)^2}=\frac{d}{dy}\left(e^{-(y-i\lambda)^2}\right). \tag{1.8.16}$$

Use (1.8.16) to derive an asymptotic expansion of the integral in (1.8.15) with respect to the asymptotic sequence $\{e^{\lambda^2}/\lambda^n\}$.

(c) Note that each term in the asymptotic expansion derived in part (b) is real as is $I(\lambda)$ itself. How then can the presence of the purely imaginary term $i\sqrt{\pi}/2$ be explained?

REFERENCES

T. CARLEMAN. *Les functions quasi-analytiques*. Paris, 1926.
A proof of Theorem 1.6 can be found in this book. Carleman refers to functions having asymptotic power series expansions valid only in sectors as quasi-analytic.

E. T. COPSON. *Asymptotic Expansions*. University Press, Cambridge, 1965.
The asymptotic expansion of the exponential integral is considered in great detail.

P. DAVIS. Uniqueness theory for asymptotic expansions in general regions. *Pac. J. Math.* **7**, 1, 849–859, 1957.
In this paper the uniqueness of asymptotic power series is discussed.

A. ERDÉLYI. *Asymptotic Expansions*. Dover, New York, 1956.
The basic ideas of asymptotic expansions are discussed in this book.

A. ERDÉLYI. General asymptotic expansions of Laplace integrals. *Arch. Rat. Mech. Anal.* **7**, 1, 1–20, 1961.
Section 2 of this paper discusses fundamental concepts and considers asymptotic expansions of Poincaré type.

A. ERDÉLYI and M. WYMAN. Asymptotic evaluation of integrals. *Arch. Rat. Mech. Anal.* **14**, 3, 217–260, 1963.
Section 2 of this paper discusses asymptotic expansions of Poincaré type and generalized asymptotic expansions with respect to auxiliary asymptotic sequences.

S. KAPLUN. *Fluid Mechanics and Singular Perturbations*. P. Lagerstrom, L. N. Howard, C.-S. Liu, Eds. Academic Press, New York, 1967.
An alternative development of the basic concepts underlying asymptotic expansions is presented.

H. POINCARÉ. Sur les intégrales irrégulières des équations linéaires. *Acta Mathematica* **VIII,** 295–344, 1896.
This is a fundamental paper in which asymptotic power series are defined as in Section 1.3.

G. G. STOKES. On the discontinuity of arbitrary constants which appear in divergent developments. *Trans. Camb. Phil. Soc.* **X,** 106–128, 1864. (This paper also appears in *Math. and Phys. Papers,* **IV,** 77–109, 1904.)
In this paper the occurrence of what is now called the Stokes phenomenon is discussed within the context of asymptotic expansions of Airy functions.

G. N. WATSON. *Theory of Bessel Functions*. University Press, Cambridge, (paperback), 1966.
The Stokes phenomenon is discussed in § 7.2.2. An historical discussion is given on page 202.

2 Asymptotic Expansions of Integrals: Preliminary Discussion

2.1. Introduction

In the chapters to follow, our main objective shall be to determine asymptotic expansions for functions defined by definite integrals. For the most part we will be concerned with one-dimensional integrals although Chapter 8 deals exclusively with a certain class of multidimensional integrals.

In one dimension, the integrals we shall study have the general form

$$I(\lambda) = \int_C H(z;\lambda)\, dz. \tag{2.1.1}$$

Here C is a given contour in the complex z plane. A typical problem will be to approximate $I(\lambda)$ for λ in some neighborhood of a prescribed point λ_0. Stated in this generality, that is, with no further restrictions placed on H and C, this problem cannot be solved by analytical methods.

It often occurs, however, that conditions can be placed on H which guarantee that the major contributions to the asymptotic expansion of I, as $\lambda \to \lambda_0$, are determined by the behavior of H in small neighborhoods of certain isolated points in the z plane called "critical points." Indeed, the "asymptotic evaluation of integrals" is almost entirely concerned with the development and application of techniques for the determination of the contributions from such critical points.

The particular techniques that we shall discuss in this book have been selected because of their widespread utility in the analysis of *applied problems.*

In the remainder of this chapter we shall consider several examples to illustrate how integrals of the form (2.1.1) arise in rather diverse areas of application. We expect that not only will these examples serve to motivate the discussions in future chapters, but also they will indicate the wide scope of the subject matter itself.

2.2. The Gamma and Incomplete Gamma Functions

There is a class of functions, the so-called special functions, that are of great importance in applied analysis. What makes them special is the fact that out of the multitude of functions, these have been singled out for detailed investigation; in other words, they have been well studied. Of course the reason they have been well studied is that they arise quite often in applications.

Of the existing special functions, the *gamma function*, denoted by $\Gamma(z)$, is one of the most important. This is due to the fact that it not only arises in the analysis of physical problems but also in such diverse mathematical areas as number theory, approximation theory, and probability theory just to name a few.

$\Gamma(z)$ is a complex-valued function with the property that, for n a nonnegative integer, $\Gamma(n+1) = n!$ and hence it serves to interpolate the factorial function. Of special interest to us are its integral representations which, in fact, can be used to define it. One of these is given by

$$\Gamma(z+1) = \int_0^\infty t^z e^{-t}\, dt = \int_0^\infty e^{-t+z\log t}\, dt \tag{2.2.1}$$

which is valid for $\operatorname{Re}(z) > -1$. We shall eventually be concerned with the asymptotic behavior of $\Gamma(z+1)$ as $|z| \to \infty$. In order to obtain an integral representation more suitable for determining this behavior we introduce

$$\lambda = |z|, \qquad z = \lambda\, e^{i\theta} \tag{2.2.2}$$

and set $t = \lambda s$ in (2.2.1) which then becomes

$$\Gamma(z+1) = I(\lambda;\theta) = \lambda^{z+1} \int_0^\infty \exp\{\lambda\, f(s;\theta)\}\, ds, \qquad \operatorname{Re}(z) = \lambda\cos\theta > -1. \tag{2.2.3}$$

Here

$$f(s;\theta) = e^{i\theta}\log s - s \tag{2.2.4}$$

and the multiplicative factor λ^{z+1} is defined by

$$\lambda^{z+1} = \exp\{(\lambda e^{i\theta} + 1)\log\lambda\}. \tag{2.2.5}$$

By using (2.2.3) our problem now is to determine an asymptotic expansion of $I(\lambda;\theta)$ as $\lambda \to \infty$ with $\theta = \arg(z)$ viewed as a parameter. The utility of (2.2.3) is lessened by the restriction $\operatorname{Re}(z) > -1$. To avoid this, we introduce a slightly different integral representation due to Hankel and given by

$$\Gamma(z+1) = \frac{e^{-\pi i z}}{2i\sin\pi z} \int_C t^z e^{-t}\, dt. \tag{2.2.6}$$

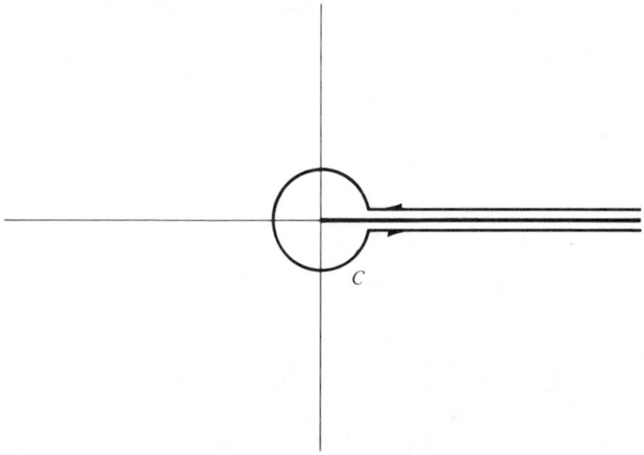

Figure 2.2 Contour of Integration for Representation of Gamma Function

Here C is a contour which originates at $+\infty$, runs in toward the origin just above the real axis, circles the origin once counterclockwise, and then returns to $+\infty$ just below the real axis. (See Figure 2.2.) This representation is valid for all complex z except integer values. At the nonnegative integers it has removable singularities, while at the negative integers it has simple poles, which reflect the actual behavior of $\Gamma(z+1)$.

Again suppose that we are interested in studying $\Gamma(z+1)$ as $|z|=\lambda\to\infty$ Upon introducing $t=\lambda s$ in (2.2.6) we obtain

$$\Gamma(z+1)=\frac{e^{-\pi i z}}{2i\sin\pi z}\lambda^{z+1}\int_C \exp\{\lambda f(s;\theta)\}\,ds. \tag{2.2.7}$$

Here f is given by (2.2.4) and we have used Cauchy's integral theorem to replace the new "stretched" contour by the original contour C.

The function

$$\Gamma(\nu,x)=\int_x^\infty e^{-t}t^{\nu-1}\,dt \tag{2.2.8}$$

is seen to be closely related to the gamma function and, indeed, is often referred to as an *incomplete gamma function* for obvious reasons. It also arises often in applications and, in particular, $\Gamma(0,x)=E_1(x)$, which is the exponential integral considered in Section 1.1, plays a key role in the analysis of the diffusion of light in a one-dimensional "milky" or "foggy" medium. (This latter problem is sometimes called the Milne problem.)

$\Gamma(\nu,x)$ is, of course, a function of the two variables x and ν. For ν fixed, the analysis of its behavior as $x\to\infty$ is straightforward and indeed has been carried out for the case $\nu=0$ in Section 1.1. It often occurs, however, that we are interested in the behavior of $\Gamma(\nu,x)$ as *both* ν and x get large. It then proves convenient to set

$$t = sx, \qquad \nu = \alpha x \tag{2.2.9}$$

in (2.2.8) which yields

$$\Gamma(\nu,x) = x^{\nu} \int_1^{\infty} s^{-1} \exp\{x\, f(s;\alpha)\}\, ds. \tag{2.2.10}$$

Here

$$f(s;\alpha) = -s + \alpha \log s. \tag{2.2.11}$$

As we shall find, the asymptotic behavior of $\Gamma(z)$ will be repeatedly considered throughout this book. Not only will we develop methods to determine this behavior, but also this information will prove useful in the asymptotic analysis of other functions.

2.3. Integrals Arising in Probability Theory

Let us consider an experiment whose possible outcomes are idealized as points in a set S called the *sample space* of the experiment. Let the function X be a mapping of S onto the extended real line. Such a function is called a *random variable* in probability theory. In order for the concept of a random variable to be of use, we must associate with its values a measure of probability. This is accomplished via what is known as the *probability distribution function* $F(t)$ of X which is defined by

$$F(t) = \text{Prob}\{X \le t\}. \tag{2.3.1}$$

We shall assume here that $F(t)$ is a given function and that it is sufficiently smooth for what follows. We note that $F(t)$ must be a nonnegative, monotonically nondecreasing function and such that

$$\lim_{t\to-\infty} F(t) = 0, \qquad \lim_{t\to\infty} F(t) = 1. \tag{2.3.2}$$

It is also convenient to introduce the associated *probability density function* $f(t)$ defined by

$$f(t) = \frac{dF}{dt}. \tag{2.3.3}$$

Indeed, it is often the case that we are given $f(t)$ rather than $F(t)$. It is important to note that f is nonnegative. Furthermore, because

$$F(t) = \int_{-\infty}^{t} f(\xi)\, d\xi \tag{2.3.4}$$

we have, by (2.3.2),

$$\int_{-\infty}^{\infty} f(\xi)\, d\xi = 1. \tag{2.3.5}$$

Of course, there are infinitely many random variables that can be considered for any given experiment. Of special importance is the notion of *independence*. Indeed, let X_1 and X_2 be two random variables defined on the same sample space. Let $F_1(t)$ and $F_2(t)$ denote their respective distribution functions. We now introduce $G(t)$, the *joint probability distribution function* of X_1 and X_2, defined by

$$G(t) = \text{Prob}\{X_1 \leq t;\, X_2 \leq t\}. \tag{2.3.6}$$

Then X_1 and X_2 are said to be *independent random variables* if and only if

$$G(t) = F_1(t)\, F_2(t). \tag{2.3.7}$$

Intuitively, we can say that X_1 and X_2 are independent when, for all t, the probability that $X_1 \leq t$ is completely unaffected by the value of X_2 and vice versa.

Suppose now that X_1 and X_2 are independent random variables with distribution functions $F_1(t)$, $F_2(t)$ and corresponding density functions $f_1(t)$, $f_2(t)$. Suppose further that X is positively distributed, that is,

$$f_1(t) = 0, \qquad -\infty < t < 0.$$

Let us consider a new random variable $X_3 = X_1\, X_2$ and attempt to determine its distribution function $F_3(t)$. As is readily shown

$$F_3(t) = \text{Prob}\{X_3 = X_1\, X_2 \leq t\} = \iint_D f_1(\xi_1)\, f_2(\xi_2)\, d\xi_1\, d\xi_2. \tag{2.3.8}$$

Here D is the domain defined by $\xi_1\, \xi_2 \leq t$, $\xi_1 \geq 0$. Upon writing (2.3.8) as an iterated integral we have

$$F_3(t) = \int_0^{\infty} f_1(\xi_1)\, [\int_{-\infty}^{t/\xi_1} f_2(\xi_2)\, d\xi_2]\, d\xi_1 = \int_0^{\infty} f_1(\xi_1)\, F_2\left(\frac{t}{\xi_1}\right) d\xi_1. \tag{2.3.9}$$

A problem of interest is to study the behavior of $F_3(t)$ as $t \to \infty$. It then proves convenient to write (2.3.9) as

$$F_3(t) = t\int_0^{\infty} f_1(t\xi)\, F_2\left(\frac{1}{\xi}\right) d\xi. \tag{2.3.10}$$

Of course, we know that $\lim_{t\to\infty} F_3(t) = 1$. However, it is desirable to determine just how F_3 approaches 1 in this limit.

Another useful concept in probability theory is that of the *characteristic function* of a random variable. It is simply the Fourier transform of the corresponding density function. Thus, if X is a random variable with density $f(\xi)$, then

$$\phi(\alpha) = \int_{-\infty}^{\infty} f(\xi)\, e^{i\alpha\xi}\, d\xi \tag{2.3.11}$$

is its characteristic function.

Now suppose that we have a sequence of mutually independent random variables $X_1, X_2, \ldots, X_n, \ldots,$ each having the same distribution function $F(t)$ and hence the same density function $f(t)$. Such a sequence is said to consist of identically distributed random variables. An important problem in probability theory is to consider the new random variable

$$Y_N = X_1 + X_2 + \cdots + X_N \tag{2.3.12}$$

and study the behavior of its density function $f_N(t)$ as $N \to \infty$.

It can be shown that

$$f_N(t) = \int_{-\infty}^{\infty}\int_{-\infty}^{\infty}\cdots\int_{-\infty}^{\infty} f(t-\xi_1) f(\xi_1 - \xi_2)\cdots f(\xi_{N-2} - \xi_{N-1})\, f(\xi_{N-1})\, d\xi_1 \ldots d\xi_{N-1}. \tag{2.3.13}$$

It then follows by the convolution theorem for Fourier transforms that

$$\phi_N(\alpha) = \int_{-\infty}^{\infty} f_N(t)\, e^{it\alpha}\, dt = [\phi(\alpha)]^N, \tag{2.3.14}$$

where $\phi(\alpha)$ is given by (2.3.11). Upon applying the Fourier inversion formula we obtain

$$f_N(t) = \frac{1}{2\pi}\int_{-\infty}^{\infty} [\phi(\alpha)]^N\, e^{-i\alpha t}\, d\alpha. \tag{2.3.15}$$

Because we are interested in the behavior of $f_N(t)$ as $N \to \infty$, it will prove convenient in our later analysis to set $\beta = t/N$ and write

$$f_N(\beta N) = \frac{1}{2\pi}\int_{-\infty}^{\infty} \exp[N\psi(\alpha;\beta)]\, d\alpha, \tag{2.3.16}$$

where

$$\psi(\alpha;\beta) = \log \phi(\alpha) - i\alpha\beta. \tag{2.3.17}$$

The asymptotic behavior of f_N as $N \to \infty$ will be studied in Section 7.6. There we shall show that, under suitable restrictions on $f(t)$, the density function f_N approaches the density function of a special distribution called the normal distribution. This result is well known in probability theory and, indeed, is a simple version of the fundamental *central limit theorem.*

2.4. Laplace Transform

The ordinary differential equation

$$a\frac{d^2x}{dt^2} + b\frac{dx}{dt} + cx = f(t) \tag{2.4.1}$$

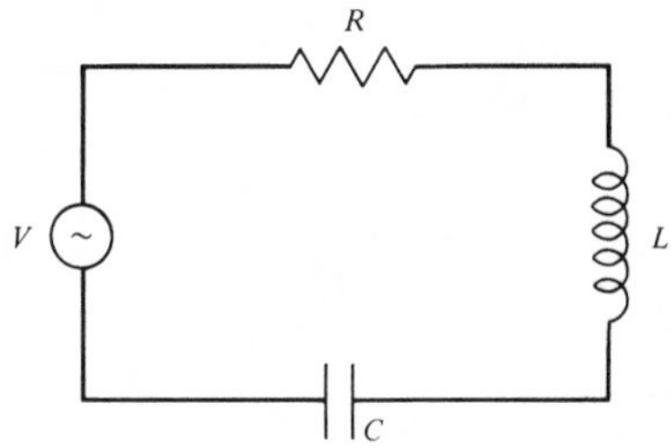

Figure 2.4.1. *RLC* Network

with a, b, and c given constants and f a prescribed function of t, can be used to model many physical phenomena. For example, if x represents the current flowing in the simple *RLC* circuit depicted in Figure 2.4.1, then in (2.4.1) a is the inductance of the coil L, b is the resistance of the resistor R, c is the admittance [(capacitance)$^{-1}$] of the capacitor C, and $f(t)$ is the derivative of the applied potential $V(t)$.

Alternatively, if x represents the displacement from equilibrium of the mass in the damped spring-mass system depicted schematically in Figure 2.4.2, then a is the mass m, b is the coefficient of damping, c is the spring constant, and $f(t)$ represents an externally applied force. We note that, in both of these problems, t represents time.

Let us now consider (2.4.1) and assume that x satisfies the homogeneous initial conditions $x(0) = 0$, $\dot{x}(0) = 0$. Moreover, in order to illustrate certain

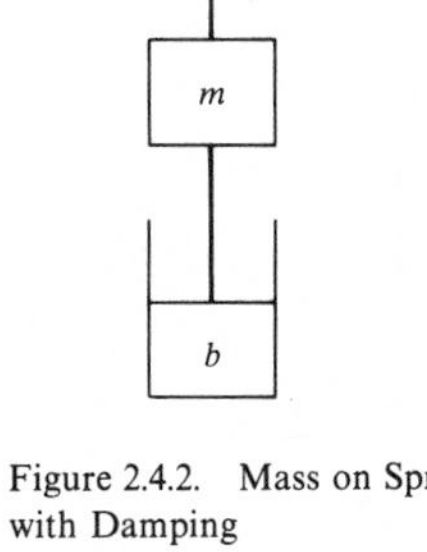

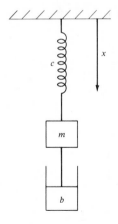

Figure 2.4.2. Mass on Spring with Damping

points, we further assume that the forcing function f is given by

$$f(t) = \gamma\, t^{1/2}\, e^{-\alpha t}, \qquad \alpha > 0. \tag{2.4.2}$$

Here γ is a constant whose dimensions are those of the left-hand side of (2.4.1) $\times$ (time)$^{-1/2}$, while α is a constant whose dimension is (time)$^{-1}$. We note that α^{-1} is often referred to as the *relaxation time* of $f(t)$.

By using the standard Laplace transform method, we can express the solution to (2.4.1) as a contour integral:

$$x(t) = \frac{\gamma}{4i\sqrt{\pi}} \int_\Gamma (s+\alpha)^{-3/2}\,(as^2 + bs + c)^{-1}\, e^{st}\, ds. \tag{2.4.3}$$

Here Γ is any infinite vertical contour in the complex s plane that lies to the right of every singularity of the integrand.[1]

Our next objective is to obtain a more suitable integral representation than (2.4.3) for analysis by the asymptotic methods to be developed in future chapters. To accomplish this, we first assume that the roots of $as^2 + bs + c = 0$ do not lie on the branch cut for $(s+\alpha)^{-3/2}$. Then, upon applying Cauchy's integral

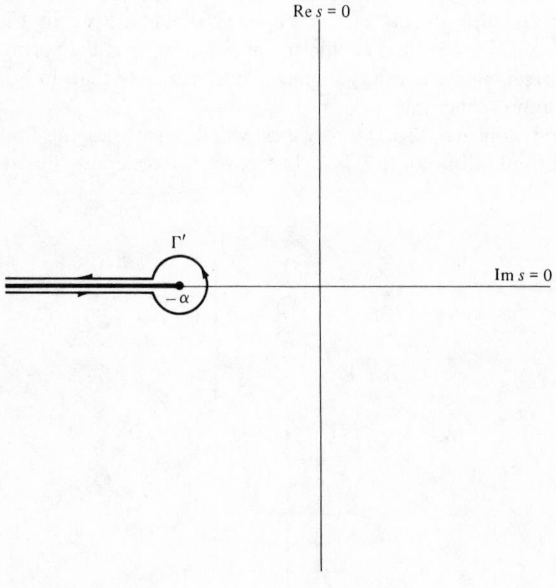

Figure 2.4.3. Contour of Integration for (2.4.4)

[1] We take the branch cut for $(s+\alpha)^{-3/2}$ to extend from $-\alpha$ to $-\infty$ along the real s axis and define $(s+\alpha)^{-3/2}$ to be positive for $s > -\alpha$.

theorem, we find that $x(t)$ is given by the residue contributions corresponding to these roots plus the integral

$$I(t) = \frac{\gamma}{4i\sqrt{\pi}} \int_{\Gamma'} (s+\alpha)^{-3/2} (as^2 + bs + c)^{-1} e^{st} \, ds. \tag{2.4.4}$$

Here Γ' is the loop contour depicted in Figure 2.4.3.

If we integrate by parts once in (2.4.4) to replace the factor $(s+\alpha)^{-3/2}$ by $(s+\alpha)^{-1/2}$, then we can shrink the circular part of Γ' onto the branch point $s = -\alpha$ with no additional contribution to $I(t)$. Let us assume that this has been done and set

$$s = \begin{cases} -\alpha + \rho e^{i\pi}, & \text{above the branch cut,} \\ -\alpha + \rho e^{-i\pi}, & \text{below the branch cut.} \end{cases}$$

Then we finally obtain

$$I(t) = \frac{\gamma \, e^{-\alpha t}}{\sqrt{\pi}} \int_0^\infty \frac{g(-\alpha - \rho)}{\rho^{1/2}} \, e^{-\rho t} \, d\rho, \tag{2.4.5}$$

where

$$g(\xi) = -\left\{ \frac{t}{a\xi^2 + b\xi + c} - \frac{2a\xi + b}{(a\xi^2 + b\xi + c)^2} \right\}. \tag{2.4.6}$$

We might want to study the behavior of $x(t)$ for both "large" and "small" times. This, of course, requires that we investigate $I(t)$ and the residue contributions in these limits. Because the residue terms are easily obtained and analyzed, we need only concern ourselves with the integral $I(t)$.

The question arises, however, just what is meant by either large or small time? Upon reflection, we realize that, as they stand, such expressions are meaningless. The problem is one of dimensionality. Indeed, we cannot talk of the largeness or smallness of a dimensional quantity, rather only of its size relative to some unit of measurement. In a given problem, this unit would presumably be selected to reflect typical values attained by the quantity in question.

In asymptotic analysis, we often study functions whose arguments are dimensional quantities. To avoid the difficulties just discussed, dimensionless variables must first be introduced. This is usually accomplished through changes of variables which are most often simple stretchings. If the scales induced by these stretchings are appropriate for the problem at hand, that is, if they are defined by "characteristic values" of the dimensional quantities involved, then we can proceed with the asymptotic analysis in terms of the constructed dimensionless variables.

To illustrate the procedure, we return to the integral $I(t)$ defined by (2.4.5). Here the relevant dimensional variable is time. There are several quantities that could serve as the characteristic time in this problem. Indeed, a/b, b/c, and α^{-1} are all reasonable possibilities. For the purpose of exposition, we arbitrarily choose the latter, namely the relaxation time of the forcing function $f(t)$. Thus, we introduce in (2.4.5)

$$\lambda = \alpha t, \qquad \rho = \alpha\sigma. \tag{2.4.7}$$

Here λ is clearly dimensionless and because the exponent ρt must be dimensionless, then σ is also.

In terms of λ and σ, (2.4.5) becomes

$$I = \tilde{I}(\lambda) = \frac{\alpha^{1/2}\,\gamma\, e^{-\lambda}}{\sqrt{\pi}} \int_0^\infty \frac{g_0(\sigma)\, e^{-\lambda\sigma}}{\sigma^{1/2}}\, d\sigma, \tag{2.4.8}$$

where

$$g_0(\sigma) = g(-\alpha(1+\sigma)). \tag{2.4.9}$$

We note that to study $\tilde{I}$ for either large or small λ is a meaningful problem. It must be emphasized, however, that the results obtained from any such analysis must be interpreted in terms of the time unit α^{-1}. For example, the behavior of $\tilde{I}(\lambda)$ for large λ yields the behavior of $I(t)$ for times large compared to the relaxation time of the forcing function.

2.5. Generalized Laplace Transform

In the previous section we considered an ordinary differential equation with constant coefficients and obtained an integral representation of its solution via the standard Laplace transform method. Here we shall consider a generalization of that method applicable to cases where the differential equation under investigation has coefficients which are polynomials in the independent variable. As we shall see, the method is most useful when the degrees of these polynomials are all less than the order of the differential equation itself, but can be adapted to other cases as well. (See Exercises 2.7 and 2.8.)

The main reason for introducing this generalized Laplace transform method is to obtain integral representations for many of the special functions of mathematical physics. Indeed, we shall describe the method by applying it to obtain integral representations for the solutions to Airy's differential equation

$$\frac{d^2 f}{dz^2} - zf = 0. \tag{2.5.1}$$

This equation arises in diverse areas of applied analysis.

Suppose now that we seek solutions to (2.5.1) of the form

$$f(z) = \int_C F(s)\, e^{sz}\, ds \tag{2.5.2}$$

assuming, of course, that such representations exist. In (2.5.2) the complex contour C and the function F are to be determined. If we formally substitute (2.5.2) in (2.5.1), then we obtain

$$\int_C (s^2 - z)\, F(s)\, e^{sz}\, ds = 0. \tag{2.5.3}$$

The presence of z in the factor $(s^2 - z)$ of the integrand is undesirable, but it can be removed by one integration by parts. Indeed, we then obtain

$$-F(s)\, e^{sz}\Big|_C + \int_C \left[s^2 F(s) + \frac{dF}{ds}\right] e^{sz}\, ds = 0. \tag{2.5.4}$$

Here, the first term is to be interpreted as the difference of the values of $-F(s)\, e^{sz}$ at the endpoints of the contour C (or the difference of limiting values when these endpoints are at infinity).

In order to satisfy (2.5.4), we first require that $F(s)$ satisfy the differential equation

$$\frac{dF}{ds} + s^2 F = 0 \tag{2.5.5}$$

in which event the integral vanishes identically. Once $F(s)$ is determined, the next step is to select C so that the endpoint contributions are zero as well.

Because (2.5.5) is an ordinary differential equation of first order it is easily solved. Indeed we find that to within a multiplicative constant

$$F = e^{-s^3/3}. \tag{2.5.6}$$

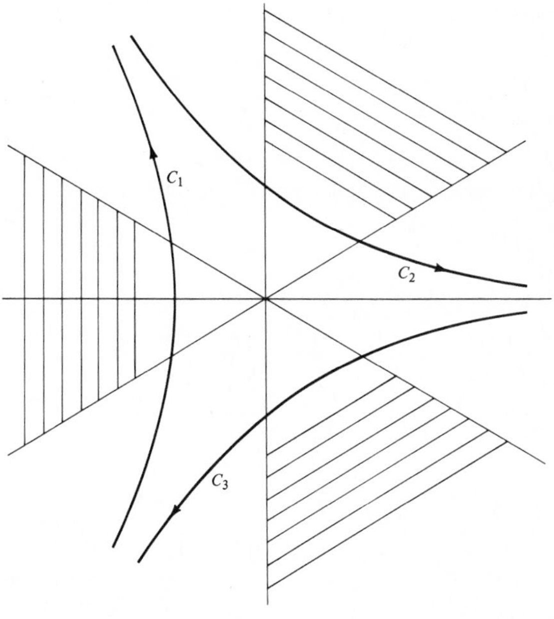

Figure 2.5. Contours for Integral Representations of Airy Functions

Hence, we must choose C so that

$$G(s;z) = F(s)\, e^{sz} = e^{-s^3/3 + sz} \tag{2.5.7}$$

vanishes at its endpoints. To accomplish this we consider the three sectors defined by

$$\frac{-\pi}{6} < \arg(s) < \frac{\pi}{6}, \quad \frac{\pi}{2} < \arg(s) < \frac{5\pi}{6}, \quad \frac{-5\pi}{6} < \arg(s) < \frac{-\pi}{2}. \tag{2.5.8}$$

We note that, as $|s| \to \infty$ within these sectors, $\mathrm{Re}(s^3) \to +\infty$. Thus, we can conclude that, for all complex values of z, we may choose C to be any infinite contour which starts at infinity in one of the sectors (2.5.8) and ends at infinity in either of the other two.

In this manner we obtain three distinct nontrivial solutions to (2.5.1) given by

$$f_n(z) = \frac{1}{2\pi i} \int_{C_n} e^{-s^3/3 + sz}\, ds, \qquad n = 1, 2, 3. \tag{2.5.9}$$

In (2.5.9) the contours C_1, C_2, and C_3 are as depicted in Figure 2.5. Any two of the functions $f_n(z)$, $n = 1, 2, 3$ can be shown to be linearly independent solutions to (2.5.1), while their sum is immediately seen to be zero by Cauchy's integral theorem. The solution $f_1(z)$ is usually denoted by $\mathrm{Ai}(z)$ and is called the Airy function of the first kind. Furthermore, for z real, say $z = x$, the contour C_1 can be deformed onto the imaginary axis with the result

$$\mathrm{Ai}(x) = \frac{1}{2\pi i} \int_{C_1} e^{sx - s^3/3}\, ds = 2\frac{1}{\pi} \int_{-\infty}^{\infty} \cos\left(\frac{\tau^3}{3} + \tau x\right) d\tau. \tag{2.5.10}$$

The remaining solutions $f_2(z)$ and $f_3(z)$ can both be written in terms of $\mathrm{Ai}(z)$. This is accomplished by introducing a change of variable in (2.5.9) which transforms C_2 (C_3) into C_1. Indeed, we then find that

$$\begin{aligned} f_2(z) &= e^{-2\pi i/3}\ \mathrm{Ai}(ze^{-2\pi i/3}) \\ f_3(z) &= e^{2\pi i/3}\ \mathrm{Ai}(ze^{2\pi i/3}). \end{aligned} \tag{2.5.11}$$

Finally, we mention that the function

$$\mathrm{Bi}(z) = i[f_2(z) - f_3(z)] \tag{2.5.12}$$

is called the Airy function of the second kind and is often used together with $\mathrm{Ai}(z)$ to form a set of two linearly independent solutions to (2.5.1).

Before proceeding, let us consider the main feature of the method. In the present case, the unwanted polynomial coefficient z in the integrand of (2.5.3) was eliminated via a single integration by parts, thereby introducing the first derivative of the transform $F(s)$. If polynomials of higher order than one had been present, then they would have been eliminated by successive integrations by parts. This process, however, introduces derivatives of $F(s)$ of higher order than the first. In fact, the degree of the polynomial coefficient of highest order in the original differential equation equals the order of the differential equation we have to solve to determine $F(s)$. Therefore, we can expect this technique to yield a simplification only when that degree is less than the order of the original differential equation. There are cases, however, where after a simple change of dependent variable this condition is satisfied by the new equation whereas it is not satisfied by the original equation.

As an example, let us consider Weber's differential equation

$$\frac{d^2w}{dz^2} + \left(r + \frac{1}{2} - \frac{z^2}{4}\right)w = 0, \qquad r = \text{const.} \tag{2.5.13}$$

This equation is of second order while its polynomial coefficient is of second degree. A direct application of the generalized Laplace transform would result in a differential equation for $F(s)$ that is essentially the same as (2.5.13) and hence no easier to solve. On the other hand, suppose that in (2.5.13) we set

$$w = e^{-z^2/4}\, u. \tag{2.5.14}$$

Then, after a simple calculation we find that u satisfies the differential equation

$$u'' - zu' + ru = 0 \tag{2.5.15}$$

which, in turn, satisfies the required condition. We could now proceed to apply the method of this section to determine the solutions to (2.5.15) and hence through (2.5.14) the solutions to (2.5.13). We choose, however, to leave this analysis to the exercises.

Let us now return to our discussion of Airy's equation and, in particular, to the integral representations of its solutions. In many problems it is important to know the asymptotic behavior of these solutions for large $|z|$. To prepare for such an analysis, it is convenient to set

$$\lambda = |z|^{3/2}, \qquad \theta = \arg(z) \tag{2.5.16}$$

and to then introduce the change of variable

$$s = \lambda^{1/3}\, t \tag{2.5.17}$$

in the integrals (2.5.9). In this manner we obtain

$$f_n(z) = \frac{\lambda^{1/3}}{2\pi i} \int_{C_n} \exp\left\{\lambda\left(t\, e^{i\theta} - \frac{t^3}{3}\right)\right\} dt, \qquad n = 1, 2, 3. \tag{2.5.18}$$

Here we have continued to denote the contours of integration by C_n because only a simple stretching was performed. The actual asymptotic analysis of the Airy function Ai(z) will be carried out in Chapters 4, 6, and 7. See also Exercise 4.3.

2.6. Wave Propagation in Dispersive Media

Let us consider the function $u(\mathbf{x},t)$ defined as the solution to the initial-value problem

$$c^2 \Delta u - u_{tt} - b^2 u = 0, \tag{2.6.1}$$

$$u(\mathbf{x},0) = f(\mathbf{x}), \qquad u_t(\mathbf{x},0) = g(\mathbf{x}), \qquad u(\mathbf{x},t) \equiv 0, \qquad t < 0. \tag{2.6.2}$$

Here $\mathbf{x} = (x_1, x_2, \ldots, x_n)$ is a position vector in an n–dimensional coordinate system, Δ is the Laplacian operator in $\mathbf{x}$ space, and t represents time. Although in the discussion that follows n can be any positive integer, in physical problems $n = 1, 2$, or 3.

The partial differential equation (2.6.1) is called the *Klein-Gordon equation.* It is of great interest in mathematical physics, being the simplest of the energy-preserving dispersive hyperbolic equations, and serves as a useful mathematical model for several physical phenomena. In particular, the solution u can be used to describe the propagation of electromagnetic waves in certain plasmas. In that event, c is the speed of light in a vacuum and b is the plasma frequency having therefore the dimension of $(\text{time})^{-1}$. Alternatively, we can look upon u as the vertical displacement of a vibrating string or membrane under the influence of a restoring force which is proportional to this displacement. Here again c is a speed and moreover $c^2 = \tau/\rho$ with τ the tension in the string and ρ the linear mass density. Because $b^2 u$ must have the dimensions of acceleration, we find, as before, that b has the dimension of $(\text{time})^{-1}$.

The solution to (2.6.1)–(2.6.2) is most readily obtained via the Fourier transform method applied in several dimensions. We shall merely quote the result here and leave the details of its derivation to the exercises. We find that

$$u(\mathbf{x},t) = \sum_{\pm} \int A_{\pm}(\mathbf{k}) \exp\{i[\mathbf{k}\cdot\mathbf{x} \mp \omega(k)\, t]\}\, d\mathbf{k}, \qquad \mathbf{k} = (k_1, k_2, \ldots, k_n). \tag{2.6.3}$$

Here

$$\omega(k) = \sqrt{c^2 k^2 + b^2}, \qquad k^2 = \mathbf{k}\cdot\mathbf{k} \tag{2.6.4}$$

and

$$A_{\pm}(\mathbf{k}) = \frac{1}{2(2\pi)^n} \int [f(\mathbf{x}) \mp (i\omega)^{-1} g(\mathbf{x})] \exp\{-i\mathbf{k}\cdot\mathbf{x}\}\, d\mathbf{x}. \tag{2.6.5}$$

In (2.6.5) the domain of integration is all of real $\mathbf{x}$ space. In (2.6.3), however, the domain of integration is somewhat harder to define. For the present, suffice it to say that it is infinite in extent and is to be chosen so that the condition $u \equiv 0$ for $t < 0$ is satisfied. We might mention that (2.6.4) is called the *dispersion relation* for the problem. The quantities $\mathbf{k}$ and ω are seen to have the dimensions of $(\text{length})^{-1}$ and $(\text{time})^{-1}$, respectively.

Because the arguments of u are dimensional quantities, we must introduce dimensionless variables before investigating its behavior in any asymptotic limit. With this purpose in mind we introduce

$$\begin{aligned} \lambda &= bt, \qquad \boldsymbol{\kappa} = b^{-1} c\, \mathbf{k}, \qquad \boldsymbol{\theta} = \frac{\mathbf{x}}{ct}, \\ \nu(\kappa) &= \sqrt{\kappa^2 + 1}, \qquad \kappa^2 = \boldsymbol{\kappa}\cdot\boldsymbol{\kappa}, \end{aligned} \tag{2.6.6}$$

which we note are all dimensionless quantities. We wish to point out that we have selected b^{-1} as the "characteristic time" of the problem. We now rewrite (2.6.3) as

$$U(\lambda;\boldsymbol{\theta}) = u(\mathbf{x},t) = \sum_{\pm} \int B_{\pm}(\boldsymbol{\kappa}) \exp\{i\lambda\,[\boldsymbol{\kappa}\cdot\boldsymbol{\theta} - \nu(\kappa)]\}\, d\boldsymbol{\kappa}. \tag{2.6.7}$$

Here

$$B_{\pm}(\boldsymbol{\kappa}) = (b\,c^{-1})^n A_{\pm}(b\,c^{-1}\,\boldsymbol{\kappa}). \tag{2.6.8}$$

As we shall see in future chapters, this form is particularly appropriate for the study of $U(\lambda,\boldsymbol{\theta})$ as $\lambda \to \infty$, that is, for the study of $u(\mathbf{x},t)$ for times large compared to b^{-1}. In the electromagnetic problem mentioned above, a typical value of b is $6 \times 10^{10}\,(\text{sec})^{-1}$, so that on a time scale with b^{-1} as unit, a microsecond must be considered a large time.

A large dimensionless parameter can be introduced into our problem in other ways. Indeed, let us suppose that the initial data, that is, the functions $f(\mathbf{x})$ and $g(\mathbf{x})$ in (2.6.2), are nonzero only for $|\mathbf{x}| \le r < \infty$. (We then say that the initial data has compact support; the support of a function $h(\mathbf{x})$ being the closure of the domain in $\mathbf{x}$ space where h is nonzero.) By introducing in (2.6.3)

$$\boldsymbol{\kappa} = r\mathbf{k}, \qquad \boldsymbol{\theta} = \frac{\mathbf{x}}{|\mathbf{x}|}, \qquad |\mathbf{x}| = (\mathbf{x}\cdot\mathbf{x})^{1/2} \tag{2.6.9}$$

and an appropriate nondimensionalization of t, we would then obtain an integral representation of u similar to (2.6.7) except now

$$\lambda = \frac{|\mathbf{x}|}{r}. \tag{2.6.10}$$

Thus, a study of the behavior of the transformed integral, as $\lambda \to \infty$, would yield information about the behavior of $u(\mathbf{x},t)$ at points whose distances from the origin $|\mathbf{x}| = 0$ are large compared to the size of the support of the initial data. We might mention that such results constitute what is called the *far-field approximation.*

We have introduced the Klein-Gordon equation not only because it is relevant to several interesting physical phenomena, but also because it serves as a motivation for the asymptotic analysis of the multiple integrals discussed in Chapter 8.

2.7. The Kirchhoff Method in Acoustical Scattering

In this section we shall describe the Kirchhoff method as applied to the problem of scattering of an acoustical wave by a closed convex body. Despite certain analytical inconsistencies in its derivation, the method is in wide use. We shall not attempt to explain or justify these inconsistencies here, but shall rather be content with showing how the application of the method leads to integrals which can be studied by the asymptotic techniques to be developed in this book. In doing this, we are taking the point of view that, when treating a physical problem, the worth of any analytical method is measured ultimately by how well the results predicted by it agree with those obtained by experiment. The Kirchhoff method has been shown to yield good agreement with experimental results in certain limits to be indicated below. We wish to point out that our presentation is based on an unpublished note by R. M. Lewis.

Suppose that an acoustic wave U_i, called the *primary field*, is incident on a finite, closed, convex body whose surface we denote by S. Upon incidence, U_i is scattered (reflected and diffracted) giving rise to a new wave called the *scattered field* and denoted by U_s. For simplicity, we shall assume from the outset that both U_i and U_s have harmonic time dependences and write

$$\begin{aligned} U_i &= u_i(\mathbf{x})\, e^{-i\omega t}, \\ U_s &= u_s(\mathbf{x})\, e^{-i\omega t}. \end{aligned} \tag{2.7.1}$$

Here $\mathbf{x} = (x_1, x_2, x_3)$ is the spatial coordinate vector, t is time, and ω is the frequency of the two waves.

Linear acoustical theory tells us that there exists a *total field* U which is the superposition of the primary and scattered fields, that is,

$$U = u(\mathbf{x})\, e^{-i\omega t}, \qquad u(\mathbf{x}) = u_i + u_s. \tag{2.7.2}$$

Moreover, $u(\mathbf{x})$, $u_i(\mathbf{x})$, and $u_s(\mathbf{x})$ must all satisfy the Helmholtz or reduced wave equation

$$(\Delta + k^2)\, u = 0 \tag{2.7.3}$$

in the region exterior to S. Here the quantity k is proportional to the frequency of the time harmonic wave u_i.

Given $u_i(\mathbf{x})$, our problem is to determine $u_s(\mathbf{x})$ such that (2.7.3) and one of the following two boundary conditions [denoted by $(\pm)$] are satisfied:

$$\left.\begin{aligned} &(+) \qquad u(\mathbf{x}) = 0, \qquad \text{that is,} \qquad u_s(\mathbf{x}) = -u_i(\mathbf{x}), \\ &(-) \qquad \frac{\partial u(\mathbf{x})}{\partial n} = 0, \qquad \text{that is,} \qquad \frac{\partial u_s(\mathbf{x})}{\partial n} = \frac{-\partial u_i(\mathbf{x})}{\partial n} \end{aligned}\right\}, \quad \mathbf{x} \in S. \tag{2.7.4}$$

In the second of these conditions, $\partial/\partial n = \mathbf{n}\cdot\nabla$ with $\mathbf{n}$ the unit outward normal to S. In addition, we must require that, in both cases,

$$\lim_{r\to\infty} u_s = 0, \qquad \lim_{r\to\infty} r\left(\frac{\partial u_s}{\partial r} - iku_s\right) = 0. \tag{2.7.5}$$

Here r represents the radial coordinate of a spherical coordinate system whose origin lies in the region interior to S and these conditions must be satisfied uniformly in the remaining two angular variables.

A few remarks concerning the various conditions given above are perhaps in order. In (2.7.4), the (+) boundary condition corresponds to what is called an "acoustically soft" or perfectly absorbing surface, while the (−) boundary condition corresponds to an "acoustically hard," "acoustically rigid," or perfectly reflecting surface. Of course, we could consider a combination of the two which would result in a boundary condition of the form

$$\alpha u + \beta \frac{\partial u}{\partial n} = 0, \qquad \mathbf{x} \varepsilon S, \qquad \alpha, \beta = \text{const.} \tag{2.7.6}$$

We shall, however, restrict our considerations to (2.7.4). The first of conditions (2.7.5) guarantees that u_s will have certain required regularity properties at infinity, while the second, called the *radiation condition*, ensures that at large distances from S the scattered field represents an *outgoing* travelling wave.

To solve our problem, we first introduce the free-space Green's function $G(\mathbf{x},\xi)$ for the Helmholtz operator $\Delta + k^2$ given by

$$G(\mathbf{x},\xi) = \frac{\exp\{ik\,|\mathbf{x}-\xi|\}}{4\pi\,|\mathbf{x}-\xi|}, \qquad |\mathbf{x}| = (\mathbf{x}\cdot\mathbf{x})^{1/2}. \tag{2.7.7}$$

Then, upon applying Green's theorem and using the known properties of G, we find that at points exterior to S, u_s is given exactly by

$$u_s(\mathbf{x}) = \int_S \left\{u(\xi)\frac{\partial G(\mathbf{x},\xi)}{\partial n} - G(\mathbf{x},\xi)\frac{\partial u(\xi)}{\partial n}\right\} dA. \tag{2.7.8}$$

Depending on which of the boundary conditions (2.7.4) is used, one of the two factors in the above integrand vanishes while the other is unknown. The Kirchhoff approximation consists essentially of a priori assuming that *both* u and $\partial u/\partial n$ are known on S no matter which of conditions (2.7.4) is used. We shall now proceed to describe the approximation in some detail. In doing so we shall introduce terminology suggested by that used when considering problems in optics.

Let us consider the wave fronts or equiphase surfaces of the incident field.[2] The orthogonal trajectories to these surfaces are curves which we shall call *rays*. These rays impinge on a portion L of the surface S called the *lit region*. That portion D of S not touched by incident rays is called the *dark region*. Finally, we denote by C the curve on S separating D from L. See Figure 2.7.1.

The key step in the Kirchhoff approximation is the replacement of the exact boundary conditions (2.7.4) by the approximate conditions

$$-u_s = \pm u_i, \qquad \frac{\partial u_s}{\partial n} = \pm \frac{\partial u_i}{\partial n} \quad \text{on } L, \tag{2.7.9}$$

[2] The actual construction of the phase ϕ of u_i, its level surfaces and corresponding orthogonal trajectories, involves what is called the geometrical optics approximation. Because u_i (x) must satisfy (2.7.3), it must be of the form $e^{ik\phi}v$ for k "large." This requirement serves to determine ϕ. We note that, in some sense, we are considering waves of high frequency.

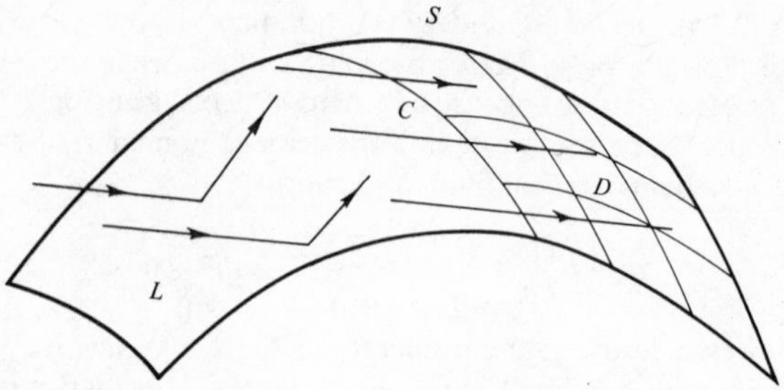

Figure 2.7.1. Lit Side and Dark Side of a Scattering Surface Separated by a Boundary Curve C

$$u = u_i + u_s = 0, \qquad \frac{\partial u}{\partial n} = \frac{\partial u_i}{\partial n} + \frac{\partial u_s}{\partial n} = 0 \text{ on } D. \tag{2.7.10}$$

In (2.7.9), the $(\pm)$ factors correspond to the $(\pm)$ boundary conditions in (2.7.4).

In order to motivate this replacement we assume for the moment that u_i is a plane wave propagating in the direction of the unit vector $\boldsymbol{\mu}_+$, that is,

$$u_i = e^{ik\boldsymbol{\mu}_+ \cdot \boldsymbol{x}} \tag{2.7.11}$$

and that S is a planar surface. In this event, both (2.7.9) and (2.7.10) are exact. In any other case they are inexact and are even incompatible in the sense that the assumed values of u_s are not compatible with the assumed values of $\partial u_s/\partial n$. Suppose, however, that the incident field is such that S is "nearly" planar over one wavelength. Heuristically, this means that S "appears" to be a planar surface so far as u_i is concerned. We can express this analytically quite simply in terms of ρ_1 and ρ_2, the principal radii of curvature of S. In fact, we need only require that k be such that

$$k\rho_j \gg 1, \qquad j = 1, 2 \tag{2.7.12}$$

at each point of S. If these conditions are satisfied, then we might expect that the error introduced by using (2.7.9) and (2.7.10) in place of (2.7.4) will not be great. Unfortunately, even when (2.7.12) holds, (2.7.9) and (2.7.10) are good approximations to the true boundary conditions on only part of S.

It can be shown, however, that there exists a region $\tilde{D}$ contained in D on which both u and $\partial u/\partial n$ are very "small." Moreover, we find that $D - \tilde{D}$ consists essentially of a "band" on S one of whose boundary curves is C, the other being a closed curve $\tilde{C}$ contained in D.[3] If we denote by d the maximum distance from C to $\tilde{C}$ on S, then it can be shown that the surface area of the band will be small compared to that of D whenever

$$(kd)^{1/3} \gg 1. \tag{2.7.13}$$

[3] See Figure 2.7.2.

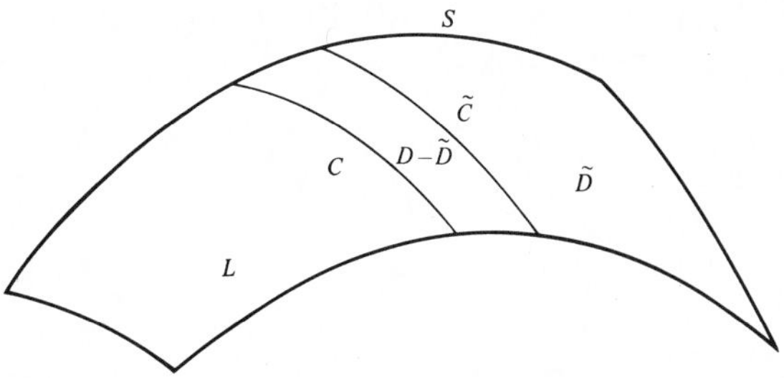

Figure 2.7.2. The Region D

Thus, if in addition to (2.7.12), (2.7.13) is satisfied, then the use of (2.7.10) in (2.7.8) should not seriously affect the result. We might also mention that for certain surfaces S, there will exist a "bright region" in D, but not in the band, throughout which (2.7.10) is a poor approximation. Here again this area is small so that the error introduced in (2.7.8) is negligible.

In the lit region there is another band bounded by C and a closed curve in L such that (2.7.9) is a good approximation except in this band. Again we find that if k satisfies (2.7.13) where d is now the maximum width of this new band, then the surface area of the band is small compared to that of L and we can use (2.7.9) in (2.7.8) with little error.

Suppose then we apply (2.7.9) and (2.7.10) in (2.7.8). We then obtain

$$u_s(\mathbf{x}) \approx \int_L \left\{ u_i(\boldsymbol{\xi})\,(1 \mp 1)\,\frac{\partial G(\mathbf{x},\boldsymbol{\xi})}{\partial n} - G(\mathbf{x},\boldsymbol{\xi})\frac{\partial u_i(\boldsymbol{\xi})}{\partial n}\,(1 \pm 1) \right\} dA \tag{2.7.14}$$

which presumably yields a good approximation to u_s when (2.7.12) and (2.7.13) are satisfied. We now propose to further simplify our expression for u_s by assuming that u_i is a plane wave of the form (2.7.11) and that the observation point $\mathbf{x}$ is "far" from the scattering body. To make this latter assumption more precise, we first suppose that the origin of our coordinate system is interior to S. Then we suppose that

$$\left|\frac{\xi}{\mathbf{x}}\right| \ll 1, \qquad \xi \in S. \tag{2.7.15}$$

When (2.7.15) is satisfied we can simplify the expressions for $G(\mathbf{x},\boldsymbol{\xi})$ and $(\partial G/\partial n)(\mathbf{x},\xi)$. Indeed, we have

$$|\mathbf{x} - \xi| \approx r - \boldsymbol{\mu}_- \cdot \xi.$$

Here $\boldsymbol{\mu}_- = \mathbf{x}/r$ and $r = |\mathbf{x}|$ so that $\boldsymbol{\mu}_-$ is a unit vector in the direction of $\mathbf{x}$. Then, after straightforward calculations, we find that

$$G(\mathbf{x},\boldsymbol{\xi}) \approx \frac{\exp\{ik\,(r - \boldsymbol{\mu}_- \cdot \xi)\}}{4\pi r},$$

$$\frac{\partial G}{\partial n}(\mathbf{x},\boldsymbol{\xi}) \approx -ik\,\boldsymbol{\mu}_- \cdot n(\xi)\frac{\exp\{ik\,(r-\boldsymbol{\mu}_-\cdot\boldsymbol{\xi})\}}{4\pi r} \qquad (2.7.16)$$

These approximations become more accurate, of course, as $|\boldsymbol{\xi}|/|\mathbf{x}|$ gets smaller. If we now use (2.7.11) and (2.7.16) in (2.7.14) we obtain

$$u_s \approx \frac{-ik\,\exp(ikr)}{2\pi r}\,I, \qquad (2.7.17)$$

where

$$I = \int_L \mathbf{n}(\xi)\cdot\boldsymbol{\mu}_\pm \exp\{ik\boldsymbol{\xi}\cdot[\boldsymbol{\mu}_+ - \boldsymbol{\mu}_-]\}\,dA. \qquad (2.7.18)$$

The integral I is essentially our final result. We shall in Chapter 8 study I for "large" k. Because both k and $\boldsymbol{\xi}$ are dimensional quantities, in order to effectively study I in the desired limit it will prove advantageous to introduce dimensionless variables. We let ρ_0 be the average of $|\boldsymbol{\xi}|$ on S and set

$$\boldsymbol{\xi} = \rho_0\,\boldsymbol{\eta}, \qquad \lambda = k\,\rho_0. \qquad (2.7.19)$$

Then (2.7.18) becomes

$$I(\lambda) = \rho_0^2 \int_{n\cdot\boldsymbol{\mu}_+ \leq 0} (\mathbf{n}(\eta)\cdot\boldsymbol{\mu}_\pm)\,\exp\{i\lambda\boldsymbol{\eta}\cdot[\boldsymbol{\mu}_+ - \boldsymbol{\mu}_-]\}\,d\tilde{A}. \qquad (2.7.20)$$

Here $d\tilde{A}$ is the nondimensional differential element of area in the $\boldsymbol{\eta}$ coordinates and the condition $\mathbf{n}\cdot\boldsymbol{\mu}_+ \leq 0$ is equivalent to requiring that $\boldsymbol{\eta}$ lie on the lit side of S.

As we have indicated, the asymptotic behavior of $I(\lambda)$, as $\lambda \to \infty$, will be considered in Chapter 8. We should point out that because (2.7.17) is already an approximation with an inherent error, it would make no sense in the acoustic problem to approximate I to any greater degree of accuracy than that afforded by (2.7.17).

2.8. Fourier Series

The use of Fourier series to approximate solutions of problems in mathematical physics is of fundamental importance. For a function $f(x)$, Riemann integrable on $[-\pi,\pi]$, such a series is defined by

$$f(x) = \frac{a_0}{2} + \sum_{k=1}^{\infty} (a_k \cos kx + b_k \sin kx). \qquad (2.8.1)$$

Here

$$a_k = \frac{1}{\pi}\int_{-\pi}^{\pi} f(\xi)\cos k\xi\,d\xi, \qquad k = 0, 1, 2, \ldots, \qquad (2.8.2)$$

$$b_k = \frac{1}{\pi}\int_{-\pi}^{\pi} f(\xi)\sin k\xi\,d\xi, \qquad k = 1, 2, \ldots. \qquad (2.8.3)$$

Let us introduce the partial sum of (2.8.1)

$$s_n(x) = \frac{a_0}{2} + \sum_{k=1}^{n} (a_k \cos kx + b_k \sin kx). \tag{2.8.4}$$

Upon inserting (2.8.2) and (2.8.3) into (2.8.4) we obtain

$$s_n(x) = \frac{1}{\pi} \int_{-\pi}^{\pi} f(\xi) \left[\frac{1}{2} + \sum_{k=1}^{n} \cos k(x - \xi) \right] d\xi. \tag{2.8.5}$$

If we use the formula for the sum of a finite geometric series, then we readily find that

$$\begin{aligned} D_n(t) &= \frac{1}{2} + \sum_{k=1}^{n} \cos kt = \frac{1}{2} + \mathrm{Re}\left(\sum_{k=1}^{n} e^{ikt} \right) \\ &= \frac{1}{2} - \mathrm{Re}\left(\frac{e^{it/2} - e^{i(n+\frac{1}{2})t}}{2i \sin(t/2)} \right) \\ &= \frac{\sin(n + \frac{1}{2})t}{2 \sin(t/2)}. \end{aligned} \tag{2.8.6}$$

Finally, upon assuming f periodic on $(-\infty, \infty)$ with period 2π, $s_n(x)$ becomes

$$s_n(x) = \frac{1}{\pi} \int_{-\pi}^{\pi} f(x + t)\, D_n(t)\, dt. \tag{2.8.7}$$

We might mention that $D_n(t)$ is usually referred to as the *Dirichlet kernel.*

For any fixed x we want to study the behavior of $s_n(x)$ as $n \to \infty$. Indeed, we want to ascertain under what conditions

$$\lim_{n \to \infty} s_n(x) = f(x). \tag{2.8.8}$$

Sufficient conditions for (2.8.8) to hold can be found in almost any book which discusses Fourier series and will not be discussed here. Another question of interest is, granted that (2.8.8) holds, what is the rate of convergence of $s_n(x)$ to $f(x)$. To answer this question we seek an asymptotic expansion of the integral in (2.8.7) as $n \to \infty$.

As is well known, there are continuous functions f for which (2.8.8) is not true. In such cases, by summing the Fourier series in a special way, a useful approximation can nevertheless be obtained. Indeed, suppose we consider the sequence

$$M_n = \frac{1}{n} \sum_{j=0}^{n-1} s_j, \qquad n = 1, 2, \ldots . \tag{2.8.9}$$

Thus, M_n is the arithmetic mean of the first n partial sums of (2.8.1).

To express M_n as an integral, we first note that

$$F_n(t) = \frac{1}{n} \sum_{j=0}^{n-1} D_j(t) = \frac{1}{n} \sum_{j=0}^{n-1} \frac{\sin(j+\frac{1}{2})t}{2\sin(t/2)}. \tag{2.8.10}$$

If we express the right side of (2.8.10) in terms of exponentials and sum the resulting geometric series, we then obtain

$$F_n(t) = \frac{1}{2n} \left(\frac{\sin(nt/2)}{\sin(t/2)} \right)^2. \tag{2.8.11}$$

From (2.8.7), (2.8.8), and (2.8.10) we have

$$M_n(x) = \frac{1}{\pi} \int_{-\pi}^{\pi} f(x+t)\, F_n(t)\, dt. \tag{2.8.12}$$

Here the function $F_n(t)$ is called the *Féjer kernel.*

The limit of $M_n(x)$ as $n \to \infty$ is called the *Cesàro sum* of the Fourier series (2.8.1). It can be shown that the Cesaro sum of the Fourier series expansion of a continuous function, periodic with period 2π exists and is equal to the function for every x. Again, we must study the asymptotic behavior of the integral in (2.8.12) to determine the rate at which $M_n(x)$ approaches $f(x)$.

2.9. Exercises

2.1. Let

$$w(z) = \exp[-z^2]\operatorname{erfc}(-iz) = \frac{i}{\pi} \int_{-\infty}^{\infty} \frac{\exp[-t^2]\, dt}{z-t}, \qquad \operatorname{Im} z > 0. \tag{2.9.1}$$

(a) Determine the change of variable of integration and the definition of λ under which this integral becomes

$$w(z) = \frac{i}{\pi} \int_{-\infty}^{\infty} g(s;\theta) \exp[-\lambda\, f(s)]\, ds. \tag{2.9.2}$$

Here, $0 < \theta < \pi$, $f(s) = s^2$, $g(s;\theta) = (e^{i\theta} - s)^{-1}$.

(b) Discuss the analytic continuation of $w(z)$ to other values of θ.

2.2. Let the function $\Gamma(\nu, z)$ be defined by the integral (2.2.8) for $|\theta| < \pi$; $\theta = \arg z$, $\lambda = |z|$.

(a) For $\lambda \neq 0$ and any ν, show that

$$\Gamma(\nu, z) = \lambda^{\nu} \int_{e^{i\theta}}^{\infty} s^{-1} \exp[\lambda\, f(s;\alpha)]\, ds, \tag{2.9.3}$$

f given by (2.2.11) and λ^{ν} appropriately defined.

(b) Discuss the analytic continuation of the defining integral (2.2.8) and of (2.9.3) to other values of θ.

2.3. For z not on the real axis between -1 and $+\infty$, the associated Legendre function of the second kind has the following integral representation:

$$Q_\nu^\mu(z) = \frac{e^{i\mu\pi}\sqrt{\pi}\,2^{-\mu}\,\Gamma(\nu+\mu+1)}{\Gamma(\mu+\frac{1}{2})\,\Gamma(\nu-\mu+1)}(z^2-1)^{\mu/2}\,I(z;\mu,\nu); \tag{2.9.4}$$

$$I(z;\mu,\nu) = \int_0^\infty [z+(z^2-1)^{1/2}\cosh t]^{-\nu-\mu-1}(\sinh t)^{2\mu}\,dt, \quad \mathrm{Re}(\nu\pm\mu+1)>0.$$

Recast $I(z;\mu,\nu)$ in a form similar to (2.9.2) with ν playing the role of λ.

2.4. If x and y are independent random variables with probability densities $p_x(s)$ and $p_y(s)$, respectively, and $z = x/y$, then show that

$$p_z(s) = \int_{-\infty}^\infty |\xi|\,p_x(s\xi)\,p_y(\xi)\,d\xi. \tag{2.9.5}$$

2.5. Suppose that $\mathbf{x} = (x,y) = (r\cos\theta, r\sin\theta)$ is a vector in the plane with r and θ independent random variables having probability densities $p_r(s)$ and $p_\theta(s)$, respectively. Let $p_x(s)$ and $p_y(s)$ be the probability densities for x and y, respectively.

(a) Show that

$$p_x(s) = \int_0^\infty p_r(\lambda\rho)\,q(\rho,\mu)\,d\rho, \qquad \lambda = |s|, \qquad \mu = \text{sign } s. \tag{2.9.6}$$

Here

$$q(\rho,\mu) = \begin{cases} \dfrac{p_\theta(\cos^{-1}\mu/\rho) + p_\theta(-\cos^{-1}\mu/\rho)}{\sqrt{\rho^2-1}}, & \rho \ge 1, \\[2ex] 0, & 0 \le \rho < 1. \end{cases} \tag{2.9.7}$$

(b) Find $p_y(s)$.

2.6. (a) In (2.4.1) set $f(t) = (1+\alpha t)^{-3/2}$ and reduce the analysis of the solution to the analysis of an integral of the form (2.4.8).

(b) Repeat for

$$f(t) = \begin{cases} 0, & t \le \alpha^{-1}, \\ (\alpha t - 1)^{-1/2}, & t > \alpha^{-1}. \end{cases}$$

2.7. (a) Apply the generalized Laplace transform method to solve (2.5.15). Obtain the solutions

$$W^{(j)}(z) = a_j \int_{C_j} s^{-(r+1)} \exp\left\{-\frac{s^2}{2} + sz\right\} ds, \qquad j = 1, 2, 3. \tag{2.9.8}$$

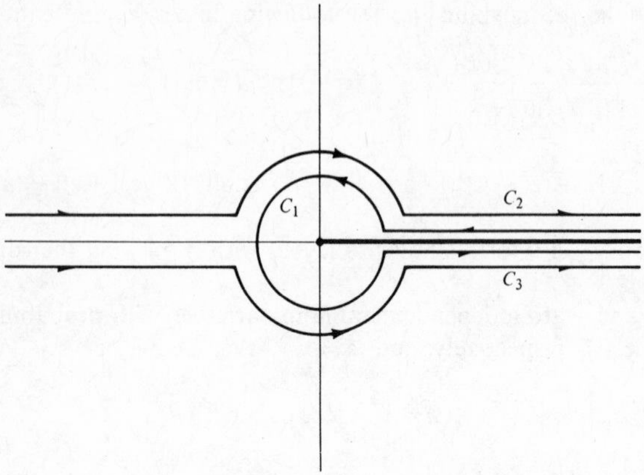

Figure 2.9.1. The Contours of Integration for Exercise 2.7

Here the a_j's are constants and the contours C_j, $j = 1, 2, 3$ are shown in Figure 2.9.1.

(b) Introduce appropriate scalings of s and z to recast (2.9.8) in the form

$$W^{(j)} = a_j\,\lambda^{-r/2} \int_{C_j} g(s)\exp\{\lambda\, f(s;\theta)\}\,ds, \qquad j = 1, 2, 3. \tag{2.9.9}$$

Determine f and g.

(c) Use Cauchy's theorem to show that the three solutions are linearly dependent.

2.8. Consider Bessel's differential equation

$$z\frac{d}{dz}\left(z\frac{dw}{dz}\right) + (z^2 - \nu^2)w = 0. \tag{2.9.10}$$

(a) Introduce $u = z^\alpha w$ and find α such that the equation for u is

$$z\frac{d^2u}{dz^2} + (2\nu + 1)\frac{du}{dz} + zu = 0. \tag{2.9.11}$$

(b) Solve (2.9.11) via the generalized Laplace transform method and obtain the integral representations

$$u^{(j)} = a_j \int_{C_j} (1 + s^2)^{\nu - \frac{1}{2}}\, e^{-sz}\,ds, \qquad j = 1, 2, 3 \tag{2.9.12}$$

with the a_j's being constants and the contours C_j shown in Figure 2.9.2.

(c) Show that the three solutions are linearly dependent by Cauchy's theorem.

2.9. Let $F(\xi)$, $G(\xi)$ be functions whose support is $|\xi| \le 1$; that is, F and G

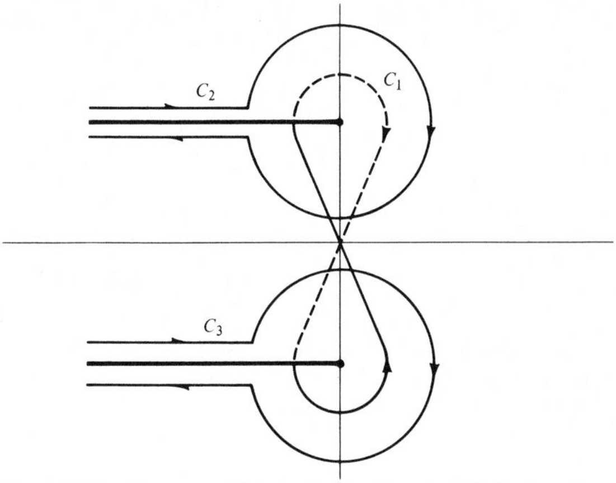

Figure 2.9.2. Contours of Integration for Exercise 2.8, the Bessel Functions (Dashed Curves Are on Lower Riemann Sheet)

are identically zero for $|\xi| > 1$. Let u satisfy (2.6.1) subject to the initial conditions

$$u(\mathbf{x},0) = F\left(\frac{\mathbf{x}}{r}\right); \qquad u_t(\mathbf{x},0) = G\left(\frac{\mathbf{x}}{r}\right), \qquad u \equiv 0, \qquad t < 0. \tag{2.9.13}$$

Then show that the solution u has the following integral representation:

$$U(\lambda;\boldsymbol{\theta}) = u(\mathbf{x},t) = \sum_{\pm}\int B_{\pm}(\boldsymbol{\kappa}) \exp\{i\lambda\,[\boldsymbol{\kappa}\cdot\boldsymbol{\theta} - \nu(\boldsymbol{\kappa})\tau]\}\, d\boldsymbol{\kappa}. \tag{2.9.14}$$

Here

$$\lambda = \frac{|\mathbf{x}|}{r}, \qquad \theta = \frac{\mathbf{x}}{|\mathbf{x}|}, \qquad \nu(\boldsymbol{\kappa}) = \sqrt{\kappa^2 + \beta^2}, \qquad \beta = \frac{rb}{c},$$
$$\tau = \frac{ct}{|\mathbf{x}|}, \tag{2.9.15}$$

and

$$B_{\pm}(\boldsymbol{\kappa}) = \left(\frac{r}{2\pi}\right)^n \int_{|\xi|<1} \left[F(\boldsymbol{\xi}) \mp \left(\frac{r}{ic\ \nu(\boldsymbol{\kappa})}\right) G(\boldsymbol{\xi})\right] \exp\{-i\boldsymbol{\kappa}\cdot\boldsymbol{\xi}\}\, d\boldsymbol{\xi}. \tag{2.9.16}$$

2.10. Obtain a Fourier integral representation of the solution to the inhomogeneous Klein-Gordon equation

$$c^2 u_{xx} - u_{tt} - b^2 u = \delta(x)\, t^{-1/2}\, e^{-\alpha t} \tag{2.9.17}$$

with "zero initial data." Here, $\delta(x)$ is the Dirac delta function. Discuss the scaling to dimensionless variables on the time scale b^{-1} and on the time scale α^{-1}.

2.11. Suppose that $\mathbf{u}$ is an n-vector solution of an initial-value problem for the system of equations

$$\mathbf{u}_t + \sum_{j=1}^{p} A_j \mathbf{u}_{x_j} + B\mathbf{u} = 0. \tag{2.9.18}$$

Here the A_j's and B are constant $n \times n$ matrices, with the A_j's symmetric and B antisymmetric.

(a) Show that the solution $\mathbf{u}$ has the integral representation

$$\mathbf{u} = \sum_{j=1}^{n} \int \mathbf{V}_j(\mathbf{k}) \exp\{i[\mathbf{k}\cdot\mathbf{x} - \omega_j(\mathbf{k})t]\}\, dk_1 \ldots dk_j. \tag{2.9.19}$$

Here $\omega_j(\mathbf{k})$, $j = 1, 2, \ldots, n$ are the eigenvalues of the matrix $\Sigma_{j=1}^{n} k_j A_j - iB$ and the $\mathbf{V}_j$'s are the corresponding eigenvectors with

$$\sum_{j=1}^{n} \mathbf{V}_j(\mathbf{k}) = \frac{1}{(2\pi)^p} \int \mathbf{u}(\mathbf{x},0) \exp\{-i\mathbf{k}\cdot\mathbf{x}\}\, d\mathbf{x}$$

the Fourier transform of the initial data.

(b) Assuming that $x_1, \ldots, x_p$ are spatial coordinates and t is time, introduce dimensionless variables in (2.9.19). Repeat the scaling discussion of Section 2.6 for (2.9.19).

2.12. For the integral (2.7.20), suppose that the body is an ellipsoid of revolution

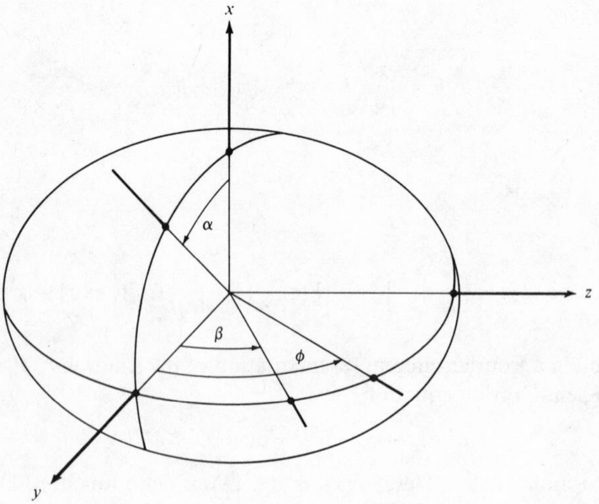

Figure 2.9.3. The Coordinate System for Exercise 2.12

$$x = a \cos \alpha \cos \beta, \qquad y = a \sin \alpha \cos \beta, \qquad z = b \sin \beta. \tag{2.9.20}$$

Here α and β are the polar and azimuthal angles, respectively. Suppose, further, that the field is transmitted and received in the y, z plane at an angle ϕ with the y axis. See Figure 2.9.3. The observed signal is called the back-scattered field. Show that

$$I(\lambda) = \mp \int_{n \cdot \mu_- \geq 0} (\mathbf{n} \cdot \boldsymbol{\mu}_-)(a^2 \cos^2 \beta + b^2 \sin^2 \beta) \sin \alpha \times \exp[-2i\lambda \mathbf{n} \cdot \boldsymbol{\mu}_-]\, d\alpha\, d\beta, \qquad \mathbf{n} \cdot \boldsymbol{\mu}_- \geq 0, \tag{2.9.21}$$

$$\mathbf{n} \cdot \boldsymbol{\mu}_- = \frac{b \sin \alpha \cos \beta \cos \phi + a \sin \beta \sin \phi}{\sqrt{a^2 \sin^2 \beta + b^2 \cos^2 \beta}}.$$

REFERENCES

M. Abramowitz and I. A. Stegun (Eds.). *Handbook of Mathematical Functions* (5th printing). Dover, New York, 1968.
This is a general source book on special functions.

B. B. Baker and E. T. Copson. *The Mathematical Theory of Huygens Principle*. 2nd Ed. Oxford, 1950.
In this book the Kirchhoff method is discussed in detail.

E. A. Coddington and N. Levinson. *The Theory of Ordinary Differential Equations*. McGraw-Hill, New York, 1955.
The generalized Laplace transform is discussed here.

A. Erdélyi (Ed.) *Higher Transcendental Functions, I, II, III*. McGraw-Hill, New York, 1953, 1953, 1955.
These serve as general sources on special functions.

W. Feller. *An Introduction to Probability Theory, II*. Wiley, New York, 1966.
This volume discusses probability theory for continuous random variables.

H. Jeffreys. *Asymptotic Approximations*. Clarendon Press, Oxford, 1962.
In this book an extensive discussion of the integral representations of Bessel functions is presented.

C. Lanczos. *Discourse on Fourier Series*. Oliver and Boyd, Edinburgh, 1966.
See this book for a discussion of the Dirichlet and Fejer kernels.

P. M. Morse and H. Feshbach. *Methods of Theoretical Physics, I*. McGraw-Hill, New York, 1953.
In this volume both the Milne problem and the Klein-Gordon equation are discussed.

E. C. Titchmarsh. *Introduction to the Theory of Fourier Integrals*. Oxford, 1937.
This book gives a general introduction to those integral transforms related to the Fourier transforms.

E. T. Whittaker and G. N. Watson. *A Course in Modern Analysis*. 4th Ed. Cambridge, 1927.
This is another general reference for special functions.

D. V. Widder. *The Laplace Transform*. Princeton, 1941.
This book develops the theory for most of the major integral transforms.

3 Integration by Parts

3.1. General Results

In Section 1.1 we derived an asymptotic expansion, as $x \to \infty$, for the function

$$I(x) = xe^x \int_x^\infty \frac{e^{-t}}{t}\, dt. \tag{3.1.1}$$

There we used the straightforward procedure of repeated integrations by parts. The expansion we obtained in this manner served to motivate the entire theory developed in Chapter 1.

In this and the sections to follow we shall discuss various aspects of the integration by parts technique. In particular, we shall show how this technique can be used to systematically derive asymptotic expansions for integrals in a fairly wide class. Furthermore, the method is fundamental to the development of more sophisticated techniques for the asymptotic expansion of integrals.

Let us begin by considering the integral

$$I(\lambda) = \int_a^b h(t;\lambda) f(t;\lambda)\, dt. \tag{3.1.2}$$

For the present we shall assume that the interval (a,b) is finite. Our objective is to study the behavior of $I(\lambda)$ as λ approaches some prescribed point λ_0.

It will prove convenient in the discussion that follows to introduce, for any function $g(t)$, the symbols $g^{(n)}(t)$ and $g^{(-n)}(t)$, which denote respectively the nth derivative and the nth repeated integral of g. Thus,

$$g^{(0)} \equiv g, \qquad g^{(n)} = \frac{dg^{(n-1)}}{dt}, \qquad g^{(-n+1)} = \frac{dg^{(-n)}}{dt}. \tag{3.1.3}$$

The quantity $g^{(-n)}$ is defined by (3.1.3) only to within n constants of integration. Choosing these constants is equivalent to choosing the n fixed limits of integeration in the n-fold integral $g^{(-n)}$. In many problems the proper choices for these constants are obvious. In general, however, the selections are not so obvious and prove to be pivotal steps in the application of the integration by parts technique. This point will be clarified in the examples below.

Let us now return to (3.1.2) and assume that, for each value of λ in a given interval Λ, $f(t;\lambda)$ has $N+1$ continuous derivatives with respect to t in $[a,b]$ and that $h(t;\lambda)$ is locally integrable on $[a,b]$. Then, upon integrating by parts $m+1$ times in (3.1.2), with $m \le N$, we obtain

$$I(\lambda) = \sum_{n=0}^{m} S_n(\lambda) + R_m(\lambda), \qquad \lambda \varepsilon \Lambda. \tag{3.1.4}$$

Here

$$S_n(\lambda) = (-1)^n \left[f^{(n)}(b;\lambda)\, h^{(-n-1)}(b;\lambda) - f^{(n)}(a;\lambda)\, h^{(-n-1)}(a;\lambda) \right] \tag{3.1.5}$$

and

$$R_m(\lambda) = (-1)^{m+1} \int_a^b f^{(m+1)}(t;\lambda)\, h^{(-m-1)}(t;\lambda)\, dt. \tag{3.1.6}$$

Our goal is to approximate $I(\lambda)$, in an asymptotic sense by the finite sum in (3.1.4) as $\lambda \to \lambda_0$ in Λ. So far we have made no assumptions that would enable us to achieve this goal. Indeed, at this juncture, we have no reason to expect that $R_m(\lambda)$ is "small" compared to $\Sigma_{n=0}^{m} S_n$ for any λ in Λ. Suppose, however, that, as $\lambda \to \lambda_0$, there exists an asymptotic sequence $\{\phi_m(\lambda)\}$ whose terms are such that

$$R_m(\lambda) = o(\phi_m(\lambda)), \qquad m = 0, 1, \ldots, N. \tag{3.1.7}$$

Then by the generalized definition of an asymptotic expansion given in Section 1.5, we can conclude that, as $\lambda \to \lambda_0$,

$$I(\lambda) \sim \sum_{n=0}^{N} S_n(\lambda) \tag{3.1.8}$$

is an asymptotic expansion of $I(\lambda)$ to $N+1$ terms with respect to the auxiliary asymptotic sequence $\{\phi_m(\lambda)\}$.

We have no guarantee that the expansion (3.1.8) is useful in the sense of Section 1.5. That is to say the scale induced by the sequence $\{\phi_m(\lambda)\}$ may not be fine enough to yield accurate estimations. To obtain the desired utility we must require that

$$I(\lambda) \neq o(\phi_0(\lambda)) \tag{3.1.9}$$

and that if $S_{m+1}(\lambda) \not\equiv 0$, then

$$R_m(\lambda) \neq o(\phi_{m+1}(\lambda)). \tag{3.1.10}$$

As an alternative to the use of auxiliary sequences, we might require that, as $\lambda \to \lambda_0$, $\{S_n(\lambda)\}$ itself form an asymptotic sequence. This, in turn, would require that the functions $S_n(\lambda)$ be bounded away from zero in some deleted neighborhood of λ_0. This, however, will not be the case in general.

In any case, to establish the asymptotic nature of (3.1.4), estimates must be made of the remainders $R_m(\lambda)$, $m = 0, 1, 2, \ldots, N$, in the limit $\lambda \to \lambda_0$. Unfortunately, it is difficult to formulate general theorems concerning the behavior of R_m in such a limit and we must be content here with the treatment of special cases. One special case is considered in the following.

THEOREM 3.1. Suppose that $I(\lambda)$ is defined by (3.1.2) with f independent of λ. Suppose further that $f^{(n)}(t)$ is continuous for $n = 0, 1, \ldots, N+1$, while $f^{(N+2)}(t)$ is piecewise continuous in $[a,b]$. Finally suppose that, as $\lambda \to \lambda_0$,

$$|h^{(-n-1)}(t;\lambda)| \leq \alpha_n(t)\, \phi_n(\lambda), \qquad n = 0, 1, \ldots, N+1, \tag{3.1.11}$$

the functions $\alpha_n(t)$ are continuous in $[a,b]$ and the functions $\phi_n(\lambda)$ are elements of an asymptotic sequence. Then, as $\lambda \to \lambda_0$,

$$I(\lambda) \sim \sum_{n=0}^{N} S_n(\lambda), \tag{3.1.12}$$

with $S_n(\lambda)$ given by (3.1.5), represents an asymptotic expansion of $I(\lambda)$ to $N+1$ terms with respect to the auxiliary asymptotic sequence $\{\phi_n(\lambda)\}$.

Simply stated, the theorem asserts that if f is independent of λ and sufficiently smooth, while the iterated integrals of h are bounded by the terms of an asymptotic sequence, then the integration by parts procedure yields an asymptotic expansion with respect to that sequence.

PROOF. In order to establish the theorem, we shall show that in (3.1.4), $R_m(\lambda) = 0(\phi_{m+1}(\lambda)) = o(\phi_m(\lambda))$, $m = 0, 1, \ldots, N$, as $\lambda \to \lambda_0$. For $m \leq N-1$, this is easily accomplished upon integrating by parts once more in (3.1.6). Indeed, we have

$$\begin{aligned} R_m(\lambda) = (-1)^{m+1} &\left[f^{(m+1)}(b)\, h^{(-m-2)}(b;\lambda) - f^{(m+1)}(a)\, h^{(-m-2)}(a;\lambda)\right] \\ &+ (-1)^m \int_a^b f^{(m+2)}(t)\, h^{(-m-2)}(t;\lambda)\, dt, \qquad m \leq N-1. \end{aligned} \tag{3.1.13}$$

From (3.1.11) and (3.1.13) we can immediately conclude that

$$R_m(\lambda) = 0(\phi_{m+1}(\lambda)), \qquad \lambda \to \lambda_0, \qquad m \leq N-1. \tag{3.1.14}$$

For $m = N$ we start with (3.1.6) and decompose $[a,b]$ into subintervals throughout each of which $f^{(N+2)}(t)$ is continuous. We then express R_N as a finite sum of integrals over these subintervals. Finally, upon integrating by parts once more in these integrals and upon using (3.1.11), we obtain the estimate (3.1.14) with $m = N$. This completes the proof.

Remarks. It should be emphasized that the purpose of Theorem 3.1 is to establish the asymptotic nature of (3.1.4) in the case under consideration and not to obtain sharp estimates for the remainders R_m. Also, the theorem yields only a finite expansion since we assumed only a finite number of continuous derivatives for f and estimates (3.1.11) for only a finite number of interated integrals of h. If f is actually infinitely differentiable on $[a,b]$ and if (3.1.11) holds for arbitrarily large N, then we can let N go to infinity in (3.1.12).

Again we have that conditions (3.1.11) do not guarantee that $\{\phi_n\}$ is a useful asymptotic sequence in the sense of Section 1.5. However, if, as $\lambda \to \lambda_0$, either $|h^{(-n-1)}(a,\lambda)| \neq o(\phi_n(\lambda))$ or $|h^{(-n-1)}(b;\lambda)| \neq o(\phi_n(\lambda))$, $n = 0, 1, \ldots, N+1$, then there exists at least one f, satisfying the hypotheses of Theorem 3.1, for which

$$I(\lambda) \neq o(\phi_0(\lambda)),$$

$$I(\lambda) - \sum_{n=0}^{m-1} S_n(\lambda) \neq o(\phi_m(\lambda)), \qquad m = 1, 2, \ldots, N+1. \tag{3.1.15}$$

To illustrate the use of the theorem we offer the following examples.

EXAMPLE 3.1.1. Let

$$I(\lambda) = \int_a^b t^\lambda f(t)\, dt, \qquad 0 \leq a < b, \quad \lambda \to \infty. \tag{3.1.16}$$

Here $h(t;\lambda) = t^\lambda$ and

$$h^{(-n-1)}(t;\lambda) = \frac{t^{\lambda+n+1}}{\prod_{j=0}^{n} (\lambda + j + 1)}, \qquad n = 0, 1, \ldots. \tag{3.1.17}$$

Note that we have selected the fixed limits of integration in the repeated integrals of h to be zero.

The conditions (3.1.11) of Theorem 3.1 are satisfied with $\{\phi_n(\lambda)\}$ defined by

$$\phi_n(\lambda) = \frac{b^{\lambda+n+1}}{\prod_{j=0}^{n} (\lambda + j + 1)} \tag{3.1.18}$$

and $\alpha_n(t) \equiv 1$, $n = 0, 1, 2, \ldots$. Thus, for any f satisfying the hypotheses of the theorem, we have

$$\int_a^b t^\lambda f\, dt \sim \sum_{n=0}^{N} \frac{(-1)^n}{\prod_{j=0}^{n} (j + \lambda + 1)} \{f^{(n)}(b)\, b^{\lambda+n+1} - f^{(n)}(a)\, a^{\lambda+n+1}\} \tag{3.1.19}$$

where, of course, the underlying asymptotic sequence is $\{\phi_n\}$ with ϕ_n given by (3.1.18).

EXAMPLE 3.1.2. We now consider

$$I(\lambda)=\int_a^b e^{-\lambda t}\cos(\lambda\sqrt{2}\,t)\,f(t)\,dt, \qquad a<b, \quad \lambda\to\infty. \tag{3.1.20}$$

We leave it to the exercises to show that

$$h^{(-n-1)}(t;\lambda)=\frac{(-1)^{n+1}}{(\sqrt{3}\,\lambda)^{n+1}}e^{-\lambda t}\cos[\lambda\sqrt{2}\,t+(n+1)\,\theta],$$
$$\theta=\tan^{-1}(\sqrt{2}), \tag{3.1.21}$$

are iterated integrals of $h=e^{-\lambda t}\cos(\lambda\sqrt{2}\,t)$. Here the constant limits of integration are all taken to be $+\infty$.

For any fixed t in $[a,b]$, each of the functions $h^{(-n-1)}(t;\lambda)$ vanishes at a sequence of values of λ having limit $+\infty$. Furthermore, no two of these functions vanish simultaneously. Indeed, the iterated integrals (3.1.21) behave much like the sequence of functions $\{\cos nx/x^n\}$ which illustrated the use of auxiliary sequence in Section 1.5. Consequently, we can expect to obtain an asymptotic expansion here only in the generalized sense. In contrast, the result of Example 3.1.1 could be viewed as the sum of two asymptotic expansions of Poincaré type.

To determine a useful auxiliary asymptotic sequence we use the easily established estimate

$$|h^{(-n-1)}(t;\lambda)|\le\frac{e^{-\mu(t-a)}\,e^{-\lambda a}}{(\sqrt{3}\,\lambda)^{n+1}}, \qquad \lambda>\mu, \quad n=0,1,2,\ldots. \tag{3.1.22}$$

Thus, for any f satisfying the hypotheses of Theorem 3.1 we have that

$$\int_a^b e^{-\lambda t}\cos(\lambda\sqrt{2}\,t)\,f(t)\,dt$$
$$\sim\sum_{n=0}^{N}(\sqrt{3}\,\lambda)^{(-n-1)}\{f^{(n)}(a)\,e^{-\lambda a}\cos[\lambda\sqrt{2}\,a+(n+1)\,\theta]$$
$$-f^{(n)}(b)\,e^{-\lambda b}\cos[\lambda\sqrt{2}\,b+(n+1)\,\theta]\} \tag{3.1.23}$$

represents an asymptotic expansion of (3.1.20), as $\lambda\to\infty$, with respect to the auxiliary asymptotic sequence $\{\phi_n(\lambda)\}$ defined by

$$\phi_n=\frac{e^{-\lambda a}}{(\sqrt{3}\,\lambda)^{n+1}}, \qquad n=0,1,2,\ldots. \tag{3.1.24}$$

EXAMPLE 3.1.3. Let us finally consider

$$I(\lambda)=\int_0^1(1-t^2)^\lambda f(t)\,dt \tag{3.1.25}$$

as $\lambda\to\infty$. Here $h(t;\lambda)=(1-t^2)^\lambda$. The iterated integrals of h now cannot be expressed in terms of simple functions. Indeed, as we shall see, the best we can do is express $h^{(-n-1)}(t;\lambda)$ as an integral between a fixed and variable limit of integration. One problem is to make the proper choice of the fixed limit.

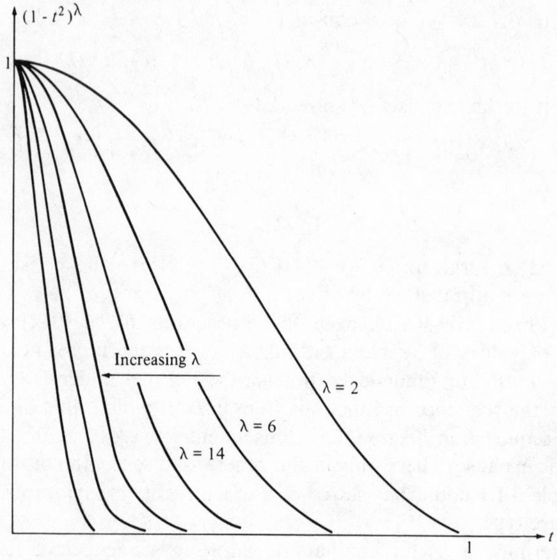

Figure 3.1

Before proceeding, we offer the following heuristic discussion: In Figure 3.1 we depict the graph of $h(t;\lambda)$, $0 \le t \le 1$, for an increasing sequence of λ values. We note that $h(0;\lambda) = 1$ for all λ, but that, as λ increases, the region where $h(t;\lambda) \ge \varepsilon$, for any fixed positive ε, becomes a smaller and smaller half-interval about $t = 0$. Therefore, we might anticipate that the behavior of f only at and near $t = 0$ will be important, that is, $t = 0$ is the only critical point for (3.1.25) as $\lambda \to \infty$.

We now turn to the implementation of the integration by parts technique. The $(n+1)$st repeated integral of h has the general form

$$h^{(-n-1)}(t;\lambda) = (-1)^{n+1} \int_t^{\beta_n} dt_n \int_{t_n}^{\beta_{n-1}} dt_{n-1} \cdots \int_{t_1}^{\beta_0} dt_0 \, (1 - t_0^2)^{\lambda}, \quad n = 0, 1, 2, \ldots . \tag{3.1.26}$$

We note that (3.1.26) is greatly simplified by selecting all of the fixed limits of integration equal to the same constant β. Indeed, in that event, and upon reversing the order of integration in (3.1.26) we obtain

$$h^{(-n-1)}(t;\lambda) = \frac{(-1)^{n+1}}{n!} \int_t^{\beta} (t_0 - t)^n \, (1 - t_0{}^2)^{\lambda} \, dt_0. \tag{3.1.27}$$

The proper selection of β is perhaps not obvious. It should be clear, however, that we should not set $\beta = 0$ in which event $h^{(-n-1)}(0;\lambda) = 0$ for all n. We have anticipated that $t = 0$ is the only critical point for I and if we set $\beta = 0$, then we will get no contribution from $t=0$ at all.

The most natural selection for β is 1. This is so for two reasons. Firstly, with $\beta=1$, $h^{(-n-1)}(1;\lambda)=0$ for all n and hence, as anticipated, there will be no contribution from $t=1$. Secondly, with $\beta=1$, $h^{(-n-1)}(0;\lambda)$ is proportional to a well-studied special function known as the beta function. Indeed,

$$h^{(-n-1)}(0;\lambda)=\frac{(-1)^{n+1}}{2(n)!}\,\frac{\Gamma\left(\frac{n+1}{2}\right)\Gamma(\lambda+1)}{\Gamma\left(\frac{n+3}{2}+\lambda\right)}, \qquad \beta=1, \tag{3.1.28}$$

where we have expressed the beta function in terms of the more familiar gamma functions.

Let us take $\beta=1$ in (3.1.27). The integration by parts procedure then yields the formal expansion

$$I(\lambda)\sim\sum_{n=0}^{\infty}\frac{f^{(n)}(0)}{2(n)!}\,\frac{\Gamma\left(\frac{n+1}{2}\right)\Gamma(\lambda+1)}{\Gamma\left(\frac{n+3}{2}+\lambda\right)}. \tag{3.1.29}$$

Here we have assumed that f is infinitely differentiable on $[0,1]$.

Although the terms in (3.1.29) are rather complicated functions of λ, we can nevertheless obtain a useful and simple underlying asymptotic sequence. In fact, we can show (see Exercise 3.5) that

$$|h^{(-n-1)}(t;\lambda)|\le|h^{(-n-1)}(0;\lambda)|\le C_n\,\lambda^{-(n+1)/2}, \qquad 0\le t\le 1, \tag{3.1.30}$$

$n=0, 1, 2, \ldots$ for some constants C_n. Thus, we can select $\{\phi_n\}=\{\lambda^{-(n+1)/2}\}$ as the auxiliary asymptotic sequence for (3.1.29).

We do not wish to imply that the only possible choice for β is 1. Indeed, if we take $0<\beta<1$, then we have

$$\begin{aligned} h^{(-n-1)}(t;\lambda)=&\frac{(-1)^{n+1}}{n!}\int_t^1 (t_0-t)^n\,(1-t_0{}^2)^\lambda\,dt_0\\ &+\frac{(-1)^n}{n!}\int_\beta^1 (t_0-t)^n\,(1-t_0{}^2)^\lambda\,dt_0. \end{aligned} \tag{3.1.31}$$

But

$$\left|\frac{1}{n!}\int_\beta^1 (t-t_0)^n\,(1-t_0{}^2)^\lambda\,dt_0\right|\le\frac{(1-\beta^2)^\lambda\,(1-\beta)^{n+1}}{n!}=o(\lambda^{-r}) \tag{3.1.32}$$

for all r. Hence, we see that using $0<\beta<1$ yields a result that differs from that obtained with $\beta=1$ by a quantity that is asymptotically zero with respect to the asymptotic sequence $\{\lambda^{-(n+1)/2}\}$.

The above examples illustrate the use of Theorem 3.1. Of course, this theorem yields only sufficient conditions for the asymptotic nature of the integration by parts procedure. Hence, even in cases where Theorem 3.1 is not applicable, integration by parts might still yield an asymptotic expansion.

There are several extensions to Theorem 3.1 that can be made. Indeed, we can allow the interval of integration in (3.1.2) to be semi-infinite or infinite so long as the functions $\alpha_n(t)$ in (3.1.11) are such that $\alpha_n(t)\, f^{(n)}(t)$ are integrable on (a,b). Also, we can allow λ to be complex. Indeed Theorem 3.1 holds as $\lambda \to \lambda_0$ in some sector of the complex λ plane, if the functions $\phi_n(\lambda)$ in (3.1.11) form an asymptotic sequence as $\lambda \to \lambda_0$ in that sector.

Finally, we note that it is unnecessary to assume that f is independent of λ. This assumption was made in Theorem 3.1 so as not to obscure the essential features of the result.

3.2. A Class of Integral Transforms

Everyone who studies linear differential equations beyond an elementary level quickly learns to appreciate the power of the so-called integral transform techniques. Of course, special mention must be made of the Fourier and Laplace transforms and, indeed, we have formulated several problems in Chapter 2 for which integral representations of the solutions were obtained via their use. These representations have the form

$$I(\lambda) = \int_C e^{-\lambda t} f(t)\, dt \tag{3.2.1}$$

where C is a given contour in the complex t plane.

It is important to note that λ appears in (3.2.1) only through the product λt. This suggests that, as a generalization, we might consider the wider class of integrals

$$I(\lambda) = \int_C h(\lambda t)\, f(t)\, dt. \tag{3.2.2}$$

In what follows we shall look upon any function defined by an integral of the form (3.2.2) as an integral transform. Moreover, unless some name is already in use, we shall call (3.2.2) the *h-transform* of f. The function h will be called the *kernel* of the transform.

To completely define any h-transform, the contour of integration C must be specified. Thus, in order to conclude that (3.2.2) is the Fourier transform of f we require that $h = e^{it}$ and C is the real axis.

It should be emphasized that not all h-transforms have proven to be useful in solving problems. In fact, relatively few have been systematically exploited in applications. We shall list some of the more important ones below. Before doing so, however, we wish to point out that integrals of the form (3.2.2) can arise naturally in problems, that is, not through the application of some integral transform. Indeed, we recall the problem discussed in Section 2.3 of determining the cumulative distribution function for the product of two independent random variables.

We now list some of the more useful h-transforms and their inversion formulas. In all cases we denote the function being transformed by $f(t)$ and the transform itself by $I(\lambda)$, in keeping with the present notation. We shall not state conditions on $f(t)$ sufficient for the validity of the inversion formulas stated.

(1) *Complex Fourier transform.*

$$I(\lambda) = \int_{-\infty}^{\infty} e^{\pm i\lambda t} f(t)\, dt; \qquad f(t) = \frac{1}{2\pi} \int_{-\infty}^{\infty} e^{\mp i\lambda t} I(\lambda)\, d\lambda \tag{3.2.3}$$

with the upper signs or lower signs respectively forming a transform pair.

(2) *Fourier sine and cosine transforms.*

$$I(\lambda) = \int_0^{\infty} \begin{Bmatrix} \sin \\ \cos \end{Bmatrix} \lambda t \; f(t)\, dt; \qquad f(t) = \frac{2}{\pi} \int_0^{\infty} \begin{Bmatrix} \sin \\ \cos \end{Bmatrix} \lambda t \; I(\lambda)\, d\lambda. \tag{3.2.4}$$

(3) *Laplace transform.*

$$I(\lambda) = \int_0^{\infty} e^{-\lambda t} f(t)\, dt; \qquad f(t) = \frac{1}{2\pi i} \int_{\sigma - i\infty}^{\sigma + i\infty} e^{\lambda t} I(\lambda)\, d\lambda. \tag{3.2.5}$$

Here, the constant σ must be such that all singularities of $I(\lambda)$ lie to the left of the vertical line $\mathrm{Re}(\lambda) = \sigma$.

(4) *Hankel transform.*

$$I(\lambda) = \int_0^{\infty} J_\nu(\lambda t)\, (\lambda t)^{1/2} f(t)\, dt; \tag{3.2.6}$$

$$f(t) = \int_0^{\infty} J_\nu(\lambda t)\, (\lambda t)^{1/2} I(\lambda)\, d\lambda.$$

Here, the kernel $h = J_\nu(t)$ is the Bessel function of the first kind of order ν.

(5) *Generalized Stieltjes transform.*

$$I(\lambda) = \lambda^\nu \int_0^{\infty} \frac{f(t)}{(1 + \lambda t)^\nu}\, dt. \tag{3.2.7}$$

For this transform we have not exhibited the inversion formula due to its complex nature. For the case $\nu = 1$, the reader is referred to the book by Widder.

There is another important integral transform known as the *Mellin transform.* It is defined by

$$I(\lambda) = \int_0^{\infty} t^{\lambda - 1} f(t)\, dt. \tag{3.2.8}$$

The Mellin transform was not included in the above list because it is *not* an h-transform. Nevertheless, as we shall find in subsequent chapters, the Mellin transform plays a key role in the asymptotic analysis of h-transforms.

It would be difficult to overstate the importance of the specific integral transforms listed above. Motivated by this, we shall now embark on a detailed investigation of the asymptotic behavior of h-transforms in certain limits.

In the remainder of this chapter we shall consider only the application of the integration by parts technique. This will limit the cases treated. In future chapters, more powerful asymptotic methods will be developed to handle cases of greater generality.

As in the previous section, we are unable to formulate an integration by parts theorem of any great generality. However, to indicate the type of result that can be obtained we prove the following.

THEOREM 3.2. Consider

$$I(\lambda) = \int_a^b h(\lambda t)\, f(t)\, dt. \tag{3.2.9}$$

Here the interval (a,b) is a finite segment of the real axis. Suppose that $f(t)$ has $N+1$ continuous derivatives while $f^{(N+2)}(t)$ is piecewise continuous on $[a,b]$. Suppose, further, that the iterated integrals of h satisfy

$$|h^{(-n)}(\lambda t)| \le \alpha_n(t)\, \phi_n(\lambda), \qquad n = 1, 2, \ldots,^{1} \tag{3.2.10}$$

where the functions $\alpha_n(t)$ are continuous on $[a,b]$ and

$$\lambda^{-1}\, \phi_{n+1}(\lambda) = o(\phi_n(\lambda)), \qquad \lambda \to \infty. \tag{3.2.11}$$

[This last assumption implies that $\{\lambda^{-n}\, \phi_n(\lambda)\}$ forms an asymptotic sequence as $\lambda \to \infty$.] Then

$$\begin{aligned} I(\lambda) &\sim \sum_{n=0}^{N} \frac{(-1)^n}{\lambda^{n+1}} [f^{(n)}(b)\, h^{(-n-1)}(\lambda b) - f^{(n)}(a)\, h^{(-n-1)}(\lambda a)] \\ &= \sum_{n=0}^{N} S_n(\lambda) \end{aligned} \tag{3.2.12}$$

represents an asymptotic expansion of I, as $\lambda \to \infty$, to $N+1$ terms, with respect to the auxiliary asymptotic sequence $\{\lambda^{-n}\, \phi_n\}$.

PROOF. Upon integrating by parts m times in (3.2.9) with $m \le N$, we obtain

$$I(\lambda) = \sum_{n=0}^{m-1} S_n(\lambda) + R_{m-1}(\lambda). \tag{3.2.13}$$

Here

$$R_{m-1}(\lambda) = \frac{(-1)^m}{\lambda^m} \int_a^b f^{(m)}(t)\, h^{(-m)}(\lambda t)\, dt, \qquad m = 1, 2, \ldots, N.$$

Another integration by parts yields

$$R_{m-1}(\lambda) = \frac{(-1)^m}{\lambda^{m+1}} [f^{(m)}(b)\, h^{(-m-1)}(\lambda b) - f^{(m)}(a)\, h^{(-m-1)}(\lambda a)]$$

[1] Here $h^{(-n)}(\lambda t) = h^{(-n)}(\tau)\big|_{\tau = \lambda t}$.

$$+\frac{(-1)^{m+1}}{\lambda^{m+1}}\int_a^b f^{(m+1)}(t)\,h^{(-m-1)}(\lambda t)\,dt. \tag{3.2.14}$$

It now follows from (3.2.10) and (3.2.14) that, as $\lambda \to \infty$,

$$R_{m-1}(\lambda) = O(\lambda^{-(m+1)}\,\phi_{m+1}(\lambda)), \qquad m = 1, 2, \ldots, N. \tag{3.2.15}$$

To see that (3.2.15) holds for $m = N + 1$ we must, as in the proof of Theorem 3.1, decompose (a,b) into subintervals throughout each of which $f^{(N+2)}(t)$ is continuous. Then a final integration by parts in R_N yields the desired result. This completes the proof.

Upon comparing Theorems 3.1 and 3.2, we see that they are quite similar and, indeed, their proofs are essentially the same. There is one key difference however. In Theorem 3.1, the functions $\phi_n(\lambda)$ introduced in conditions (3.1.11) are assumed to form an asymptotic sequence as $\lambda \to \lambda_0$. In Theorem 3.2, because of the special way λ appears in the integrand of (3.2.9), namely through the product λt, the weaker assumption that $\{\lambda^{-n}\,\phi_n\}$ is an asymptotic sequence as $\lambda \to \infty$ proves sufficient. We further note that Theorem 3.2 does not guarantee that $\{\lambda^{-n}\,\phi_n\}$ is a useful asymptotic sequence for (3.2.9) in the sense of Section 1.5. To obtain the required utility, at least to leading order, we must further require that $I(\lambda) \neq o(\lambda^{-1}\,\phi_1(\lambda))$, $\lambda \to \infty$.

There are, of course, several generalizations of Theorem 3.2 that can be made. The extensions to complex λ and more general contours of integration are two that come immediately to mind. We shall not discuss these here. There is one extension, however, that is worthy of special consideration.

Let us consider the following.

EXAMPLE 3.2.1.

$$I(\lambda) = \int_{\bar{a}}^{\bar{b}} h(\lambda\,\phi(\tau))\,\bar{f}(\tau)\,d\tau, \tag{3.2.16}$$

where the kernel h satisfies the hypotheses of Theorem 3.2. For ease of discussion, we suppose that ϕ and $\bar{f}$ are both infinitely differentiable on $[\bar{a},\bar{b}]$. Finally, suppose that ϕ is strictly monotonic on $[\bar{a},\bar{b}]$, that is, $\phi'(\tau) \neq 0$, $\tau \in [\bar{a},\bar{b}]$.

This seemingly more general case can, with little effort, be reduced to that treated in Theorem 3.2. Indeed, if we set $t = \phi(\tau)$ in (3.2.16), then we obtain

$$I(\lambda) = \int_a^b h(\lambda t)\,f(t)\,dt. \tag{3.2.17}$$

Here, $a = \phi(\bar{a})$, $b = \phi(\bar{b})$, and

$$f(t) = \left.\frac{\bar{f}(\tau)}{\phi'(\tau)}\right|_{\tau = \phi^{-1}(t)} \tag{3.2.18}$$

Because $\phi'(\tau)$ is nonzero on $[\bar{a},\bar{b}]$, we have that $f(t)$ is infinitely differentiable on $[a,b]$. Thus, an infinite asymptotic expansion of (3.2.17) can be obtained by applying the result of Theorem 3.2 with $N = \infty$. In terms of the original functions ϕ and $\bar{f}$, this expansion is given by

$$I(\lambda) \sim \sum_{n=0}^{\infty} \frac{(-1)^n}{\lambda^{n+1}} \left\{ h^{(-n-1)}(\lambda\phi) \left[\frac{1}{\phi'}\frac{d}{d\tau}\right]^n \left(\frac{\bar{f}}{\phi'}\right) \right\}_{\tau=\bar{a}}^{\tau=\bar{b}} \tag{3.2.19}$$

This last result could have been obtained by directly integrating by parts in (3.2.16). In fact all we need do is multiply and divide the integrand of each remainder integral by ϕ' before integrating by parts. (See Exercise 3.7.)

If $\phi(\tau)$ is strictly monotonic only over subintervals of $[\bar{a},\bar{b}]$, we might be tempted to decompose $[\bar{a},\bar{b}]$ into those subintervals and to then seek an expansion of the form (3.2.19) for each integral that arises. However, because $\phi' = 0$ at the endpoints of the subintervals, $f(t)$ is undefined at these points and Theorem 3.2 cannot be applied. In later chapters we will treat these non-monotonic cases in great detail. At this juncture, we wish merely to point out that the straightforward integration by parts procedure applied to (3.2.16) breaks down whenever ϕ' vanishes at points in $[\bar{a},\bar{b}]$.

In the following examples we illustrate the use of Theorem 3.2.

EXAMPLE 3.2.2. Let us consider

$$I(\lambda) = \int_a^b e^{i\lambda t} f(t)\, dt \tag{3.2.20}$$

which, for obvious reasons, we call an integral of Fourier type. Here $h = e^{it}$ and we choose

$$h^{(-n)}(t) = (i)^{-n} e^{it}, \qquad n = 1, 2, \ldots \tag{3.2.21}$$

as its iterated integrals.[2] Observe that in (3.2.10) we can take $\alpha_n(t) \equiv 1$ and $\phi_n(\lambda) \equiv 1$.

If f satisfies the hypotheses of Theorem 3.2, then we have

$$\int_a^b e^{i\lambda t} f(t)\, dt \sim \sum_{n=0}^{N} \frac{(-1)^n}{(i\lambda)^{n+1}} \left[f^{(n)}(b)\, e^{i\lambda b} - f^{(n)}(a)\, e^{i\lambda a}\right], \qquad \lambda \to \infty. \tag{3.2.22}$$

Here the auxiliary asymptotic sequence is $\{\lambda^{-n}\}$, $n = 1, 2, \ldots$.

We note that if f is piecewise differentiable on $[a,b]$, then we can conclude from a single integration by parts that

$$\lim_{\lambda \to \infty} \int_a^b e^{i\lambda t} f(t)\, dt = 0. \tag{3.2.23}$$

This result is actually a weak version of the *Riemann-Lebesgue lemma* which states that (3.2.23) holds whenever f is absolutely integrable on $[a,b]$.

Let us now suppose that, in addition to the hypotheses of Theorem 3.2,

[2] The fixed limit of integration here is taken to be $i\infty$.

f satisfies the conditions

$$\lim_{t\to a+} f^{(n)}(t) = \lim_{t\to b-} f^{(n)}(t) = 0, \qquad n = 0, 1, \ldots, N-1. \tag{3.2.24}$$

Then it immediately follows from (3.2.22) that, as $\lambda \to \infty$, $I(\lambda) = o(\lambda^{-N})$. We shall exploit this result in the following section.

EXAMPLE 3.2.3. We now consider integrals of Laplace type which have the form

$$I(\lambda) = \int_a^b e^{-\lambda t} f(t)\, dt. \tag{3.2.25}$$

Here the kernel is $h = e^{-t}$ and the obvious choices for its iterated integrals are

$$h^{(-n)} = (-1)^n e^{-t}, \qquad n = 1, 2, \ldots . \tag{3.2.26}$$

In (3.2.10) we can take $\alpha_n = e^{-\mu(t-a)}$ and $\phi_n = e^{-\lambda a}$, with $\lambda > \mu$.

It follows from (3.2.12) and (3.2.26) that if f satisfies the hypotheses of Theorem 3.2, then

$$\int_a^b e^{-\lambda t} f(t)\, dt \sim \sum_{n=0}^{N} \frac{e^{-\lambda a}}{\lambda^{n+1}} f^{(n)}(a), \qquad \lambda \to \infty. \tag{3.2.27}$$

Here the auxiliary asymptotic sequence is $\{e^{-\lambda a}/\lambda^n\}$.

We note that in (3.2.27) the contribution from the upper endpoint $t = b$ is omitted. This is due to the fact that this contribution is asymptotically negligible compared to each term in (3.2.27). Indeed, we have

$$\lim_{\lambda\to\infty} \lambda^n e^{\lambda(a-b)} = 0, \qquad n = 1, 2, \ldots . \tag{3.2.28}$$

We should point out, however, that when doing a physical problem, we often want to retain asymptotically negligible terms because, although small, they may have some meaningful physical interpretation.

If $a = 0$ and $b = \infty$ in (3.2.25), then $I(\lambda)$ is actually the Laplace transform of f. Now, for $I(\lambda)$ to exist for λ sufficiently large, we must require that, as $t \to \infty$,

$$f(t) = O(e^{\alpha t}), \qquad \alpha = \text{const.}$$

The asymptotic expansion of (3.2.25) in this case is obtained by setting $a = 0$ in (3.2.27).

EXAMPLE 3.2.4. In subsequent chapters it will prove important to know an asymptotic expansion of the Γ function for complex argument. In Section 2.1, we found that $\Gamma(z)$ has simple poles at the points $z = -n$, $n = 0, 1, 2, \ldots$, and is otherwise analytic. We therefore seek an asymptotic expansion of $\Gamma(z)$, as $|z| \to \infty$, that remains uniformly valid in the sector $|\arg(z)| < \pi$.

We start from Binet's representation for $\log \Gamma$ given by

$$\log \Gamma(z) = (z - \tfrac{1}{2}) \log z - z + \tfrac{1}{2} \log 2\pi + I(z). \tag{3.2.29}$$

Here $\text{Re}(z) > 0$ and

$$I(z) = \int_0^\infty f(t)\, e^{-zt}\, dt; \qquad f(t) = [(e^t - 1)^{-1} - t^{-1} + \tfrac{1}{2}]\, t^{-1}. \tag{3.2.30}$$

The function $f(t)$ has simple poles at $t = \pm 2n\pi i$, $n = 1, 2, \ldots,$ and is otherwise analytic. Moreover, we have

$$f(t) = \sum_{k=1}^{\infty} \frac{(-1)^{k+1} B_k\, t^{2k-2}}{(2k)!}, \qquad |t| < 2\pi, \tag{3.2.31}$$

where the constants B_k are the Bernoulli numbers. For reference we list the first four Bernoulli numbers:

$$B_1 = \frac{1}{6}, \qquad B_2 = \frac{1}{30}, \qquad B_3 = \frac{1}{42}, \qquad B_4 = \frac{1}{30}.$$

To utilize (3.2.29) for our stated purpose, we must first consider the analytic continuation of the right-hand side into the region $|\arg(z)| < \pi$. Certainly, the continuations of the first three terms are explicit and need not be discussed further. To analytically continue the integral $I(z)$ we set $z = \lambda e^{i\theta}$. As θ increases or decreases from zero, we simultaneously rotate the contour of integration through the angle $-\theta/2$. The justification that this does not alter the value of I is a simple problem in complex function theory and is left to the exercises.

We are thus led to consider

$$I(\lambda e^{i\theta}) = e^{-i\theta/2} \int_0^\infty f(\sigma\, e^{-i\theta/2}) \exp\{-\sigma\lambda\, e^{i\theta/2}\}\, d\sigma, \qquad |\theta| \le \theta_0 < \pi. \tag{3.2.32}$$

For each θ in the indicated range, the integral converges uniformly and absolutely. Furthermore in the complex σ plane, we are justified rotating the contour of integration in (3.2.32) onto the ray $\arg(\sigma) = -\theta/2$. We do so while maintaining $\arg(z)$ fixed at θ. If $|\theta| \ge \pi/2$, this process introduces residues from the poles of $f(\sigma e^{-i\theta/2})$ at the points $\sigma e^{-i\theta/2} = -\text{sgn}\,[\theta]\, 2n\pi i$. However, due to the exponential in the integrand, such residue terms are

$$O(|\exp\{-2n\pi\lambda\, e^{i(\theta - \text{sgn}(\theta)\pi/2)}\}|), \qquad n = 1, 2, \ldots. \tag{3.2.33}$$

Because these terms are all exponentially small, as $\lambda \to \infty$, and because we anticipate that I is algebraic in λ we write

$$I(\lambda e^{i\theta}) = e^{-i\theta} \int_0^\infty f(\tau e^{-i\theta}) \exp\{-\lambda\tau\}\, d\tau, \qquad |\theta| < \pi \tag{3.2.34}$$

to within an exponentially small error. This then is our desired analytic continuation of I.

We now use (3.2.27) with $a = 0$ and $N = \infty$ to obtain

$$\begin{aligned} I(\lambda e^{i\theta}) &\sim e^{-i\theta} \sum_{n=0}^{\infty} \lambda^{-(n+1)} \left(\frac{d}{d\tau}\right)^n f(\tau e^{-i\theta})\Big|_{\tau=0} \\ &= \sum_{n=0}^{\infty} \frac{f^{(n)}(0)}{(\lambda e^{i\theta})^{n+1}}, \qquad \lambda \to \infty, \quad |\theta| < \pi. \end{aligned} \tag{3.2.35}$$

It follows from (3.2.31) that

$$f^{(2k)}(0) = \frac{(-1)^k B_{k+1}}{(2k+2)(2k+1)},$$
$$f^{(2k+1)}(0) = 0, \qquad k = 0, 1, 2, \ldots . \tag{3.2.36}$$

Thus, if we reintroduce $z = \lambda e^{i\theta}$, then we find

$$I(z) \sim \sum_{k=0}^{\infty} \frac{(-1)^k B_{k+1}}{(2k+2)(2k+1)} z^{-2k-1}, \qquad |z| \to \infty, \quad |\arg(z)| < \pi. \tag{3.2.37}$$

Upon combining (3.2.29) and (3.2.37) we obtain an asymptotic expansion of $\log \Gamma(z)$:

$$\log \Gamma(z) \sim \left(z - \frac{1}{2}\right) \log z - z + \frac{1}{2} \log 2\pi + \sum_{k=0}^{\infty} \frac{(-1)^k B_{k+1} z^{-2k-1}}{(2k+2)(2k+1)}. \tag{3.2.38}$$

Finally, if we exponentiate (3.2.38),[3] then we obtain the desired expansion of $\Gamma(z)$ itself. Indeed, we have to leading order,

$$\Gamma(z) = \sqrt{\frac{2\pi}{z}} \left(\frac{z}{e}\right)^z \left[1 + O\left(\frac{1}{z}\right)\right], \qquad |z| \to \infty, \quad |\arg(z)| < \pi. \tag{3.2.39}$$

This last result is often referred to as *Stirling's formula*. We shall rederive this formula in subsequent chapters to illustrate the applications of other asymptotic techniques.

There are two limits that are of particular interest. If we set $z = x + iy$ with x and y real, then these are $x \to \infty$ with $y = 0$ and $|y| \to \infty$ with x finite. The behavior of Γ in the first limit is trivially obtained from (3.2.39).

$$\Gamma(x) = \sqrt{\frac{2\pi}{x}} \left(\frac{x}{e}\right)^x \left[1 + O\left(\frac{1}{x}\right)\right], \qquad x \to \infty. \tag{3.2.40}$$

From this result we, of course, obtain an approximation of $n!$ for large n.

The behavior of Γ in the second limit is somewhat more complicated. After some calculation we find that

$$\Gamma(x+iy) = |y|^{x-\frac{1}{2}} e^{-(\pi/2)|y|} \left\{\sqrt{2\pi}\, e^{(i\pi/2)(x-\frac{1}{2})\operatorname{sign}(y)} \left(\frac{|y|}{e}\right)^{iy}\right\} \left\{1 + O\left(\frac{1}{y}\right)\right\}, \qquad |y| \to \infty. \tag{3.2.41}$$

Thus we see that $\Gamma(z)$ grows quite rapidly along the positive real axis, but decays quite rapidly along vertical lines.

All of the expansions obtained in this section have been for $\lambda \to \infty$. We might argue that it is unreasonable to devote so much effort to this single limit. We recall, however, that the results of any asymptotic analysis are most interesting from a mathematical point of view when one studies a function near its points of nonanalyticity and, in particular, near its points of essential

[3] See Exercise 3.10.

singularity. For h-transforms, $\lambda = 0$ and $\lambda = \infty$ are points at which we would most expect I to have an essential singularity.

Furthermore, $\lambda \to 0+$ and $\lambda \to \infty$ are the limits of greatest practical interest. Indeed, suppose that λ is real and positive. An asymptotic expansion obtained for $\lambda \to \infty$ often yields accurate estimates when λ is of moderate size. Thus, in Section 1.1, we found that the expansion of $I(x) = xe^x\, E_1(x)$, obtained for $x \to \infty$, yields three significant figure accuracy for all $x \geq 10$. Analogous remarks can be made concerning the limit $\lambda \to 0+$.

We could develop an integration by parts formula which, under suitable conditions on h and f, would yield an asymptotic expansion of (3.2.1) as $\lambda \to 0+$. We prefer to delay an investigation of this limit until future chapters where more general cases will be considered.

3.3. Identification and Isolation of Critical Points

When we examine the expansions obtained in Sections 3.1 and 3.2 via the integration by parts procedure, we find that, in each case, they involve iterated integrals and successive derivatives of the integrand functions evaluated at one or both of the endpoints of integration. It will be recalled that, when discussing integrals of the form

$$I(\lambda) = \int_a^b H(x;\lambda)\, dx,$$

we indicated that in some instances the major contributions to $I(\lambda)$, as $\lambda \to \lambda_0$, are determined by the behavior of H in arbitrarily small neighborhoods of certain points in $[a,b]$. These points we called critical points. We might therefore conclude that whenever an infinite asymptotic expansion is generated by the integration by parts procedure, the endpoints of integration are the only critical points.

There is one difficulty, however. It is readily seen that the iterated integrals of a function f, even when evaluated at a particular point, can actually involve the *global* behavior of f. [See, for example, (3.1.27).] For this reason we alter somewhat our definition of critical point. Indeed, we shall say that whenever an asymptotic expansion of an integral involves the local behavior of at least one of the integrand functions at a point,[4] then that point is critical for the expansion.

As we have seen, the integration by parts technique does not, in general, yield an infinite expansion; that is, the process must cease after a finite number of steps. We claim here and shall later show that, in many cases, this is merely a reflection of the simple way in which we have applied the method. In fact, our methods so far enable us to derive contributions to the asymptotic expansion of a given integral only from endpoints of integration. That other points can be critical is seen from the following simple example.

[4] Here we have in mind that $H(x;\lambda)$ is often a product of functions.

EXAMPLE 3.3. Consider

$$I(\lambda) = \int_a^b h(\lambda t) f(t)\, dt \tag{3.3.1}$$

as $\lambda \to \infty$. Here $h(t)$ satisfies the hypotheses of Theorem 3.2 and $f(t)$ is infinitely differentiable on $[a,b]$ except at $t = c$, $a < c < b$, where $f^{(n+1)}$ has a jump discontinuity for $n \geq N$.[5]

It follows from Theorem 3.2 that

$$I(\lambda) \sim \sum_{n=0}^{N-1} \frac{(-1)^n}{\lambda^{n+1}} \left[f^{(n)}(b)\, h^{(-n-1)}(\lambda b) - f^{(n)}(a)\, h^{(-n-1)}(\lambda a) \right] \tag{3.3.2}$$

is an asymptotic expansion of I, as $\lambda \to \infty$, to N terms. Here the underlying asymptotic sequence is $\{\phi_n(\lambda)/\lambda^n\}$ with the functions $\phi_n(\lambda)$ as defined in the hypotheses of the theorem. We note that $t = a$ and $t = b$ are the only critical points for the finite expansion (3.3.2).

Now suppose we write

$$I(\lambda) = \int_a^c h(\lambda t)\, f(t)\, dt + \int_c^b h(\lambda t)\, f(t)\, dt. \tag{3.3.3}$$

By assumption, f is infinitely differentiable throughout each of the subintervals $[a,c]$ and $[c,b]$. Upon applying Theorem 3.2 to both of the integrals in (3.3.3) we obtain the infinite expansion

$$\begin{aligned} I(\lambda) \sim \sum_{n=0}^{\infty} \frac{(-1)^n}{\lambda^{n+1}} &\left[f^{(n)}(b)\, h^{(-n-1)}(\lambda b) - f^{(n)}(a)\, h^{(-n-1)}(\lambda a) \right] \\ + \sum_{n=N+1}^{\infty} \frac{(-1)^n}{\lambda^{n+1}} &\left[h^{(-n-1)}(\lambda c) \left\{ f^{(n)}(c_-) - f^{(n)}(c_+) \right\} \right]. \end{aligned} \tag{3.3.4}$$

Here

$$f^{(n)}(c_\pm) = \lim_{t \to c\pm} f^{(n)}(t). \tag{3.3.5}$$

We see that an infinite asymptotic expansion of (3.3.1) can be obtained only so long as the contribution from the point $t = c$ is included. Because this contribution involves the local behavior of f at $t = c$, we must conclude that $t = c$ is a critical point for the infinite expansion (3.3.4).

The above example suggests that any point in $[a,b]$ at which f ceases to be infinitely differentiable will generate a contribution to the asymptotic expansion of (3.3.1) as $\lambda \to \infty$, and hence should be considered critical. Although this is indeed the case, we cannot determine by the methods discussed so far contributions corresponding to singularities of f more severe than the one treated in the example. In subsequent chapters more general techniques will be developed that will enable us to obtain such contributions.

[5] More precisely, we mean that $\lim_{t \to c\pm} f^{(n)}(t)$ exist for all n, but are not necessarily equal for $n \geq N$.

In the previous section we considered integrals of the form

$$I(\lambda) = \int_a^b h(\lambda\ \phi(t))\ f(t)\, dt \tag{3.3.6}$$

and observed that, if ϕ is strictly monotonic on $[a,b]$, then (3.3.6) is immediately reduced to the form (3.3.1) by introducing $\tau = \phi(t)$ as a new variable of integration. Indeed, in this manner we obtain

$$I(\lambda) = \int_\alpha^\beta h(\lambda\tau)\ F(\tau)\, d\tau, \tag{3.3.7}$$

$$F(\tau) = \left.\frac{f(t)}{\phi'(t)}\right|_{t=\phi^{-1}(\tau)} \tag{3.3.8}$$

Here $\alpha = \phi(a)$ and $\beta = \phi(b)$.

Because points in $[a,b]$ at which ϕ or any of its derivatives is discontinuous correspond to points at which $F(\tau)$ is not infinitely differentiable, it is reasonable to anticipate that such points will be critical for the asymptotic expansion of (3.3.6) as $\lambda \to \infty$. Furthermore, as we shall see in future chapters, points in (a,b) at which ϕ' vanishes are also critical. This is strongly suggested by the fact that when the change of variable $\tau = \phi(t)$ is formally introduced in (3.3.6) points at which ϕ' vanishes correspond to (integrable) singularities of $F(\tau)$.

As Example 3.3 indicates, contributions from interior critical points cannot be obtained via the standard integration by parts procedure. Moreover, this example and the proof of Theorem 3.2 suggest that the best way to proceed when interior critical points are present is to first partition the original interval of integration into subintervals so that, in each, only the endpoints are critical. We would then have a finite sum of integrals to consider, none having interior critical points.

The next problem of course would be to obtain asymptotic expansions for each of the integrals that arise because of the partition. It may or may not be possible to derive such expansions via integration by parts. Let us suppose for the present that this can be done. Then our result for each integral would consist of two series both involving functions evaluated at one or the other endpoint of integration.

The separation of the contributions from the critical points allows us to study them independently. This is very desirable in applied problems where a specific physical phenomenon is associated with each critical point. It therefore seems reasonable to seek a means of isolating the contributions from critical points even in cases where the integration by parts technique is not applicable. We shall see in later chapters that not only is this useful for the purpose of physical interpretation, in addition an isolation of the critical points often facilitates the asymptotic analysis in more complex situations.

As a result of the decomposition described above, we find that we need only concern ourselves with isolating the endpoints of integration. We shall now describe a mathematical device, due to Van der Corput, which results in such an isolation. The essential step in the process is the introduction of certain appropriately constructed functions called *neutralizers.*

The simplest type of neutralizer is a function $q = q(x,\alpha_1,\alpha_2)$ which depends on two parameters α_1 and α_2 with $\alpha_2 > \alpha_1$, which belongs to $C^\infty(-\infty,\infty)$, and which is such that either

$$q(x,\alpha_1,\alpha_2) = \begin{cases} 0, & x \le \alpha_1, \\ 1, & x \ge \alpha_2, \end{cases} \tag{3.3.9}$$

or

$$q(x,\alpha_1,\alpha_2) = \begin{cases} 1, & x \le \alpha_1, \\ 0, & x \ge \alpha_2. \end{cases} \tag{3.3.10}$$

Neutralizer functions can be constructed in a variety of ways. However, because we shall not make use of any of their properties other than those already mentioned, it suffices to exhibit a single example. We first introduce

$$p(x) = \begin{cases} 0, & x \le 0, \\ e^{-1/x}, & x > 0, \end{cases} \tag{3.3.11}$$

which we note belongs to $C^\infty(-\infty,\infty)$. We then set

$$\bar{q}(x,\alpha_1,\alpha_2) = \frac{p(x-\alpha_1)}{p(x-\alpha_1) + p(\alpha_2 - x)}. \tag{3.3.12}$$

It is readily seen that $\bar{q} \,\varepsilon\, C^\infty(-\infty,\infty)$ and that (3.3.9) is satisfied with $q = \bar{q}$. Also, we have that

$$\bar{\bar{q}}(x,\alpha_1,\alpha_2) = 1 - \bar{q}(x,\alpha_1,\alpha_2) \tag{3.3.13}$$

belongs to $C^\infty(-\infty,\infty)$ and (3.3.10) is satisfied with $q = \bar{\bar{q}}$. In Figure 3.3 we depict the graphs of these two functions.

To illustrate the role played by neutralizers in the isolation process, let us consider (3.3.1) as $\lambda \to \infty$ in the case where $t = a$ and $t = b$ are the only critical points. In particular, $f \,\varepsilon\, C^\infty(a,b)$. We first observe that we can

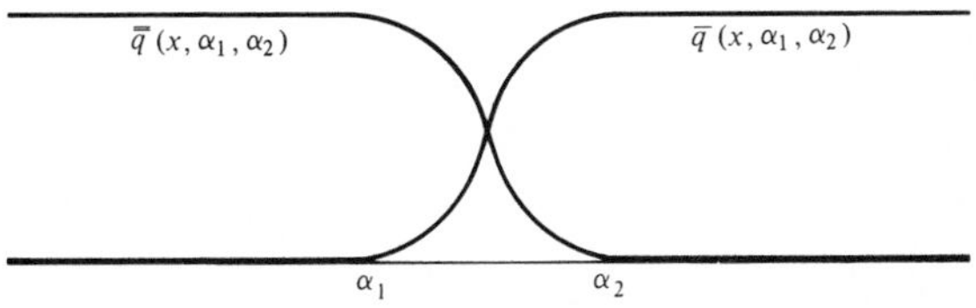

Figure 3.3

rewrite (3.3.1) as

$$I(\lambda) = I_a(\lambda) + I_b(\lambda), \tag{3.3.14}$$

where

$$I_a(\lambda) = \int_a^b \bar{\bar{q}}(t,\alpha_1,\alpha_2)\, f(t)\, h(\lambda t)\, dt \tag{3.3.15}$$

and

$$I_b(\lambda) = \int_a^b \bar{q}(t,\alpha_1,\alpha_2)\, f(t)\, h(\lambda t)\, dt. \tag{3.3.16}$$

Here α_1 and α_2 are such that $a < \alpha_1 < \alpha_2 < b$.

We now write

$$\begin{aligned} f_a(t,\alpha_1,\alpha_2) &= \bar{\bar{q}}(t,\alpha_1,\alpha_2)\, f(t), \\ f_b(t,\alpha_1,\alpha_2) &= \bar{q}(t,\alpha_1,\alpha_2)\, f(t). \end{aligned} \tag{3.3.17}$$

Both f_a and f_b belong to $C^\infty(a,b)$ so that no new critical points have been introduced. Moreover, $f_b \equiv 0$ in $[a,\alpha_1]$ and $f_a \equiv 0$ in $[\alpha_2,b]$.[6] Hence, we can conclude that $t = a$ is the only critical point for I_a while $t = b$ is the only critical point for I_b. In other words, we have succeeded in isolating the two critical points. The asymptotic expansion of I_a yields the contribution to the asymptotic expansion of I from $t = a$, while the asymptotic expansion of I_b yields the contribution to the asymptotic expansion of I from $t = b$.

In the special case where $f \in C^\infty[a,b]$ we have that for $n = 0, 1, 2, \ldots$

$$\begin{aligned} &\lim_{t\to a+} f_a^{(n)}(t) = \lim_{t\to a+} f^{(n)}(t), \qquad &&\lim_{t\to b-} f_a^{(n)}(t) = 0, \\ &\lim_{t\to a+} f_b^{(n)}(t) = 0, &&\lim_{t\to b-} f_b^{(n)}(t) = \lim_{t\to b-} f^{(n)}(t). \end{aligned} \tag{3.3.18}$$

Moreover, if the integrands of I_a and I_b satisfy the hypotheses of Theorem 3.2 perhaps with a different sequence $\{\phi_n(\lambda)\}$ for each of them, then we can apply that theorem to these integrals to obtain

$$\begin{aligned} I_a(\lambda) &\sim \sum_{n=0}^{\infty} \frac{(-1)^{n+1} h^{(-n-1)}(\lambda a)}{\lambda^{n+1}} f^{(n)}(a), \\ I_b(\lambda) &\sim \sum_{n=0}^{\infty} \frac{(-1)^n h^{(-n-1)}(\lambda b)}{\lambda^{n+1}} f^{(n)}(b), \qquad \lambda \to \infty. \end{aligned} \tag{3.3.19}$$

We note that the neutralizers $\bar{q}$ and $\bar{\bar{q}}$ do not enter the asymptotic expansions (3.3.19) in any way. Indeed their roles in the derivations of these expansions are qualitative rather than quantitative in nature. This is the reason we stated

[6] In our future discussions we will have occasion to say that f_a and f_b represent the function f *neutralized* about $t = a$ and $t = b$, respectively.

above that we need only be concerned with the existence of neutralizer functions and not with detailed descriptions of their behavior. This will always be the case even when considering critical points of a more complex nature than those already discussed. Again we emphasize that the major advantage of the neutralization process lies in the fact that it enables us to study the contributions to the asymptotic expansion of an integral from the various critical points separately and independently and, in particular, to apply for each critical point, whatever technique is appropriate for the determination of the corresponding contribution.

At times it is practical to obtain only an asymptotic estimate for one of the integrals, say I_b. In a case where I_b is asymptotically negligible compared to $I_a(\lambda)$, such as for the integral $I(\lambda)$ given by (3.2.25), this would indeed be the case. It is our general experience that neutralization is most useful in dealing with integrals in which $h(t)$ is an *oscillatory* function such as in $I(\lambda)$ defined by (3.2.20).

There is one final point that we wish to discuss. For any given integral, we do not, in general, know a priori that a particular point is critical for the asymptotic expansion of interest. Indeed, we can only ascertain which points are critical after the expansion has been derived. What we do, however, is determine a priori a set of *possible* critical points, that is, the set of points which are likely to yield contributions to the asymptotic expansion. We then proceed to apply the neutralization process based on the assumption that these are, in fact, the only critical points. Finally, asymptotic expansions are sought for each of the neutralized integrals. The remainder of this book will be essentially devoted to the identification of the critical points for certain classes of integrals and the development of techniques for the determination of the corresponding asymptotic contributions.

3.4. An Extension of the Integration by Parts Procedure

When considering the Fourier-type integral

$$I(\lambda) = \int_a^b e^{i\lambda t} f(t)\, dt \tag{3.4.1}$$

as $\lambda \to \infty$, we found that the derivation of the asymptotic expansion of $I(\lambda)$ via integration by parts relied heavily on the differentiability of the function $f(t)$. Suppose, however, that

$$f(t) = (t-a)^{\alpha-1}(b-t)^{\beta-1} F(t), \tag{3.4.2}$$

with $F(t) \in C^\infty[a,b]$ and α, β not integers. Then the best we can achieve by our straightforward integration by parts process is the estimate

$$I(\lambda) = 0(\lambda^{-N}), \tag{3.4.3}$$

where N is the largest integer less than or equal to $\min(\alpha,\beta)$. In particular, if $\alpha < 1$ and $\beta < 1$, then we can only conclude that $I(\lambda) = 0(1)$ as $\lambda \to \infty$, whereas in fact $I(\lambda) = o(1)$ in this limit.

It is possible to extend the standard integration by parts procedure and to obtain thereby an infinite asymptotic expansion of (3.4.1) with $f(t)$ of the form (3.4.2). This extension, however, has somewhat limited applicability and indeed we shall consider here Fourier-type integrals only. In the next chapter we shall present a more extensive technique. We wish to point out that the present method is due to Erdélyi.

We begin by isolating the contributions from the endpoints $t = a$, $t = b$ via the neutralization process introduced in the preceding section. In this manner we obtain

$$I(\lambda) = I_a(\lambda) + I_b(\lambda), \tag{3.4.4}$$

where

$$I_a(\lambda) = \int_a^b (t-a)^{\alpha-1}\, e^{i\lambda t}\, F_a(t)\, dt \tag{3.4.5}$$

and

$$I_b(\lambda) = \int_a^b (b-t)^{\beta-1}\, e^{i\lambda t}\, F_b(t)\, dt. \tag{3.4.6}$$

Here $F_a(t) \varepsilon C^\infty[a,b]$ vanishes infinitely smoothly as $t \to b-$ and identically equals $(b-t)^{\beta-1} F(t)$ in some positive half-neighborhood of $t = a$. Similarly, $F_b(t) \varepsilon C^\infty[a,b]$ vanishes infinitely smoothly as $t \to a+$ and identically equals $(t-a)^{\alpha-1} F(t)$ in some negative half-neighborhood of $t = b$.[7]

Let us first consider $I_a(\lambda)$. We define

$$h(t;\lambda) = (t-a)^{\alpha-1}\, e^{i\lambda t}. \tag{3.4.7}$$

For reasons that will become clear below we select

$$h^{(-n-1)}(t;\lambda) = (-1)^{n+1} \int_t^{t+i\infty} dt_n \int_{t_n}^{t_n+i\infty} dt_{n-1} \cdots \int_{t_1}^{t_1+i\infty} dt_0\, (t_0-a)^{\alpha-1}\, e^{i\lambda t_0},$$
$$n = 0, 1, 2, \ldots, \tag{3.4.8}$$

as the iterated integrals of h to be used in our integration by parts analysis. Here the paths of integration are the rays

$$t_m = t_{m+1} + i\,\sigma_m, \qquad \sigma_m \geq 0, \qquad m = 0, 1, \ldots, n, \qquad t_{n+1} = t. \tag{3.4.9}$$

Upon interchanging the order of integration in (3.4.8) we obtain

$$h^{(-n-1)}(t;\lambda) = \frac{(-i)^{n+1}}{n!} \int_0^\infty (i\sigma + t - a)^{\alpha-1}\, \sigma^n\, e^{-\lambda\sigma + i\lambda t}\, d\sigma, \quad n = 0, 1, 2, \ldots, \tag{3.4.10}$$

which is seen to converge absolutely due to the exponential decay of the integrand as $\sigma \to \infty$. If we set $t = a$ in (3.4.10), then we find (see Exercises 3.18,19)

[7] Note that without loss of generality we can assume $0 < \alpha < 1$ and $0 < \beta < 1$ because the smoothness of F is unaffected by factors of the form $(t-a)^i\,(b-t)^j$ with i and j positive integers.

$$h^{(-n-1)}(a;\lambda)=\frac{(-1)^{n+1}}{n!}\frac{\Gamma(n+\alpha)}{\lambda^{n+\alpha}}e^{(\pi i/2)(n+\alpha)+i\lambda a}. \tag{3.4.11}$$

Now, repeated integration by parts in (3.4.5) yields the formal result

$$I_a(\lambda)\sim\sum_{n=0}^{\infty}\frac{d^n}{da^n}([b-a]^{\beta-1}F(a))\frac{\Gamma(n+\alpha)}{n!\,\lambda^{n+\alpha}}e^{(\pi i/2)(n+\alpha)+i\lambda a}. \tag{3.4.12}$$

Note that no contributions from the upper endpoint $t=b$ arise in (3.4.12) due to the fact that $F_a^{(n)}(b)=0$ for $n=0, 1, 2, \dots$. It is readily seen (see Exercise 3.18) that

$$|h^{(-n-1)}(t;\lambda)|\leq(t-a)^{\alpha-1}\lambda^{-n-1} \tag{3.4.13}$$

and hence the asymptotic nature of the result follows from Theorem 3.1.

The integral $I_b(\lambda)$ can be treated in a completely analogous manner. In this case we have

$$h(t;\lambda)=(b-t)^{\beta-1}e^{i\lambda t} \tag{3.4.14}$$

and[8]

$$h^{(-n-1)}(t;\lambda)=\frac{(-1)^{n+1}}{n!}\int_t^{t+i\infty}(b-\tau)^{\beta-1}(\tau-t)^n e^{i\lambda\tau}\,d\tau. \tag{3.4.15}$$

In particular,

$$h^{(-n-1)}(b;\lambda)=\frac{(-1)^n\Gamma(n+\beta)}{n!\quad\lambda^{n+\beta}}e^{(\pi i/2)(n-\beta)+i\lambda b} \tag{3.4.16}$$

so that

$$I_b(\lambda)\sim\sum_{n=0}^{\infty}\frac{d^n}{db^n}([b-a]^{\alpha-1}F(b))\frac{\Gamma(n+\beta)}{n!\,\lambda^{n+\beta}}e^{(\pi i/2)(n-\beta)+i\lambda b} \tag{3.4.17}$$

Finally, the asymptotic expansion of I is obtained by combining (3.4.12) and (3.4.17).

The above procedure is also applicable to integrals of the form

$$I_a(\lambda)=\int_a^b(t-a)^{\alpha-1}e^{i\lambda(t-a)^{\mu}}F_a(t)\,dt,\qquad \mu>0, \tag{3.4.18}$$

with $F_a(t)$ as above. The crucial difference here is the form of the exponent. We shall find that integrals of the form (3.4.18) are fundamental to the analysis of Fourier integrals with general nonmonotonic exponents. Indeed, we shall reconsider them in Chapter 6 in connection with our discussion of the method of stationary phase.

Suppose now we consider (3.4.18) and define

$$h(t;\lambda)=(t-a)^{\alpha-1}e^{i\lambda(t-a)^{\mu}}. \tag{3.4.19}$$

[8] By setting $\tau-t=i\sigma$, we would obtain exactly the form (3.4.10).

We take as its iterated integrals

$$h^{(-n-1)}(t;\lambda)=\frac{(-1)^{n+1}}{n!}\int_t^{t+\infty e^{i\pi/2\mu}}(\tau-a)^{\alpha-1}(\tau-t)^n\, e^{i\lambda(\tau-a)^\mu}\,d\tau. \tag{3.4.20}$$

To evaluate $h^{(-n-1)}(a;\lambda)$, we introduce a new variable of integration σ defined by

$$(\tau-a)=\lambda^{-1/\mu}\,\sigma^{1/\mu}\,e^{i\pi/2\mu}. \tag{3.4.21}$$

We then find that

$$h^{(-n-1)}(a;\lambda)=\frac{(-1)^{n+1}}{n!}\,\frac{\Gamma\left(\dfrac{n+\alpha}{\mu}\right)}{\mu\,\lambda^{(n+\alpha)/\mu}}\,e^{i\pi(n+\alpha)/2\mu}. \tag{3.4.22}$$

The integration by parts formula (3.1.5) then yields

$$I_a(\lambda)\sim\sum_{n=0}^{\infty}\frac{1}{n!\mu}\frac{d^n}{da^n}([b-a]^{\beta-1}F(a))\,\frac{\Gamma\left(\dfrac{n+\alpha}{\mu}\right)}{\lambda^{(n+\alpha)/\mu}}\,e^{i\pi(n+\alpha)/2\mu}. \tag{3.4.23}$$

Again the asymptotic nature of the result can be established via Theorem 3.1. We leave the details, however, to the exercises.

Also left to the exercises is establishing the validity of the asymptotic expansion

$$I_b(\lambda)=\int_a^b(b-t)^{\beta-1}\,e^{i\lambda(b-t)^\mu}\,F_b(t)\,dt \tag{3.4.24}$$

$$\sim\sum_{n=0}^{\infty}\frac{(-1)^n}{\mu n!}\frac{d^n}{db^n}([b-a]^{\alpha-1}F(b))\,\frac{\Gamma\left(\dfrac{n+\beta}{\mu}\right)}{\lambda^{(n+\beta)/\mu}}\,e^{i\pi(n+\beta)/2\mu}.$$

We note from (3.4.23) and (3.4.24) that as the order of vanishing of the exponent in the corresponding integral increases, the order of vanishing of the integral, as $\lambda\to\infty$, increases.

In summary, an extension of the standard method of integration by parts has been developed at least for integrals of Fourier type. This extension allows for the amplitude function $f(t)$ in (3.4.1) to have singularities at the endpoints of integration and involves the incorporation of these singularities in the definition of the kernel function h.

3.5. Exercises

3.1. (a) For the function $I(x)$ defined by (1.8.3) show that $I(x)=0(1)$, $x\to\infty$ but $I(x)\neq o(1)$ as $x\to\infty$.

(b) For $\mathscr{E}(x,N)$ defined by (1.1.10), show that $\mathscr{E}(x,N)=0(x^{-N})$ but $\mathscr{E}(x,N)\neq o(x^{-N})$, $x\to\infty$.

(c) Identify the functions f, h, $h^{(-n-1)}$ of Theorem 3.1 for this integral and verify that the conditions of the theorem are satisfied.

3.2. Suppose that $h^{(-n-1)}(t;\lambda)$ satisfies the conditions of Theorem 3.1 and, in addition, suppose that $h^{(-n-1)}(a;\lambda) \neq o(\phi_n(\lambda))$. Let

$$f(t) = -e^{-(t-a)}\,\bar{\bar{q}}(t,\alpha_1,\alpha_2). \tag{3.5.1}$$

Here $\bar{\bar{q}}$ is the neutralizer defined by (3.3.13) with $a < \alpha_1 < \alpha_2 < b$. Then show that

$$I(\lambda) \sim \sum_{n=0}^{\infty} h^{(-n-1)}(a;\lambda) \tag{3.5.2}$$

and hence, for this choice of $f(t)$, (3.1.15) is true.

3.3. (a) In Example 3.1.1, suppose that

$$|f^{(n)}(t)| \leq M, \qquad n = 0, \ldots, N. \tag{3.5.3}$$

Then show that

$$\left| I(\lambda) - \sum_{n=0}^{N-1} \frac{(-1)^n}{\prod_{j=0}^{n} (\lambda + j + 1)} f^{(n)}(b)\, b^{\lambda+n+1} \right|$$

$$< M \left\{ \frac{b^{\lambda+N+1}}{\prod_{j=0}^{N} (\lambda + j + 1)} + \frac{a^{\lambda+1}}{\lambda+1} \cdot \frac{1 - \left(\dfrac{a}{\lambda+1}\right)^N}{1 - \dfrac{a}{\lambda+1}} \right\}. \tag{3.5.4}$$

(b) Show that

$$\frac{a^{\lambda+n}}{\lambda^j} = o\left(\frac{b^{\lambda+m}}{\lambda^k}\right), \qquad \lambda \to \infty \quad \text{for} \quad b > a, \tag{3.5.5}$$

and any fixed m, n, j, and k. Thus verify the claim that $\phi_n(\lambda)$ is given by (3.1.18) for this example.

3.4. Let

$$h(t;\lambda) = e^{-\lambda t} \cos \lambda \sqrt{2}\, t = \frac{1}{2} \sum_{\pm} e^{-(1 \pm i\sqrt{2})\lambda t}. \tag{3.5.6}$$

(a) Show that

$$h^{(-n-1)}(t;\lambda) = \frac{1}{2} \sum_{\pm} \left[\frac{-1}{1 \pm i\sqrt{2}}\right]^{n+1} \frac{e^{-(1 \pm i\sqrt{2})\lambda t}}{\lambda^{n+1}}. \tag{3.5.7}$$

(b) Define θ by $\tan\theta = \sqrt{2}$, $0 < \theta < \pi/2$, and rewrite $h^{(-n-1)}$ as

$$h^{(-n-1)}(t;\lambda) = \frac{1}{2}\sum_{\pm}\left[\frac{-1}{\sqrt{3}\lambda}\right]^{n+1} e^{-(1\pm i\sqrt{2})\lambda t \mp i(n+1)\theta} \qquad (3.5.8)$$

and verify (3.1.21).

(c) Verify (3.1.22); that is, show that

$$|h^{(-n-1)}(t;\lambda)| \leq \frac{e^{-\mu(t-a)-\lambda a}}{(\sqrt{3}\,\lambda)^{n+1}}, \qquad \lambda > \mu, \qquad n = 0, 1, 2, \ldots . \qquad (3.5.9)$$

3.5. (a) With $h^{(-n-1)}(t;\lambda)$ given by (3.1.27), show that $|h^{(-n-1)}(t;\lambda)|$ is a monotonically decreasing function of t and hence that

$$|h^{(-n-1)}(t;\lambda)| \leq |h^{(-n-1)}(0;\lambda)|. \qquad (3.5.10)$$

(b) Take $\beta = 1$ and use the estimates

$$e^{-2t^2} \leq 1 - t^2 \leq e^{-t^2}, \qquad 0 \leq t \leq \tfrac{1}{2} \qquad (3.5.11)$$

to show that

$$h^{(-n-1)}(0;\lambda) = O(\lambda^{-(n+1)/2}), \qquad h^{(-n-1)}(0,\lambda) \neq o(\lambda^{-(n+1)/2}). \qquad (3.5.12)$$

Note, here we must verify that

$$h^{(-n-1)}(\tfrac{1}{2};\lambda) = o(\lambda^{-k}) \qquad (3.5.13)$$

for all k.

3.6. Derive an asymptotic expansion of

$$I(\lambda) = \int_0^a |1 - t^2|^{\lambda} f(t)\,dt, \qquad a > 1. \qquad (3.5.14)$$

3.7. Consider

$$I(\lambda) = \int_0^1 \frac{f(t)}{1 + \lambda t}\,dt \qquad (3.5.15)$$

with $f \in C^{\infty}\,[0,1]$. Show that a "useful" asymptotic expansion of I as $\lambda \to \infty$ cannot be obtained via the integration by parts procedure.

3.8. Each of the integrals below is of the form (3.2.16). Obtain three terms of the asymptotic expansion in each case, as $\lambda \to \infty$, by multiplying and dividing by ϕ' and integrating by parts.

(a) $\displaystyle\int_1^2 \frac{e^{i\lambda t^2}}{1+t}\,dt.$

(b) $\int_0^{\pi/4} e^{-\lambda \tan t} \cos t\,dt.$

(c) $\int_1^{10} \mathrm{Ai}(\lambda t^3)\,\mathrm{Ai}(-t)\,dt.$

3.9. Justify the rotation of contour used to deduce (3.2.32). In particular, show that for

$$J = \int_C f(t)\, e^{-zt}\, dt \tag{3.5.16}$$

with C an arc of radius R on which $\arg(t)$ ranges from 0 to $-\theta/2$, $\lim_{R\to\infty} |J| = 0$.

3.10. Justify the exponentiation of the result (3.2.38) to obtain (3.2.39) and show that we can obtain in this manner as many terms as desired in the asymptotic expansion of $\Gamma(z)$.

3.11. The complementary error function is defined by

$$\operatorname{erfc}(z) = \frac{2}{\sqrt{\pi}} \int_z^{\infty} e^{-u^2}\, du. \tag{3.5.17}$$

(a) From this definition, derive the representation

$$\operatorname{erfc}(z) = \sqrt{\frac{\lambda}{\pi}} \int_1^{\infty} \frac{e^{-\lambda t}}{\sqrt{t}}\, dt, \qquad \lambda = z^2. \tag{3.5.18}$$

(b) Use the method of Theorem 3.2 (or Example 3.2.2) to derive the asymptotic expansion

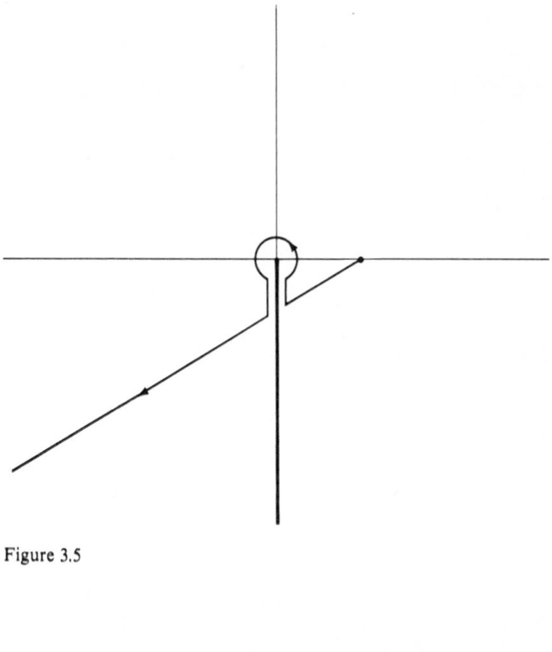

Figure 3.5

$$\operatorname{erfc}(z) \sim \frac{e^{-z^2}}{\sqrt{\pi}\, z} \sum_{n=0}^{\infty} \frac{(-1)^{n+1}}{(2z^2)^n} \prod_{j=0}^{n} (2j-1), \qquad z \to \infty. \tag{3.5.19}$$

(c) Show that for λ complex, we may rotate the contour in (3.5.18) in a manner similar to that used in Example 3.2.4 to obtain in general

$$\operatorname{erfc}(z) = \frac{\sqrt{\lambda}\, e^{-\lambda - i\theta}}{\sqrt{\pi}} \int_0^{\infty} \frac{e^{-|\lambda|\sigma}}{\sqrt{1+\sigma e^{-i\theta}}}\, d\sigma, \qquad \theta = \arg \lambda = 2 \arg z; \qquad |\theta| < \pi. \tag{3.5.20}$$

Thus, conclude that (3.5.19) is valid for $|\arg(z)| < \pi/2$.

(d) Show that to consider erfc(z) in the wider sector $-3\pi/4 < \arg(z) \le -\pi/2$, we need only replace the contour of integration in (3.5.18) by the contour of Figure 3.5. [A similar result holds for the sector $\pi/2 \le \arg(z) < 3\pi/4$.]

(e) Show that the integral over the "keyhole contour" in Figure 3.5 is asymptotically zero with respect to the asymptotic sequence $\{e^{-z^2}/z^{2k+1}\}$, $\pi/2 \le \arg(z) < 3\pi/4$. Hence, conclude that (3.5.19) is valid for $|\arg(z)| < 3\pi/4$.

(f) What happens when $\arg(z) = 3\pi/4$?

3.12. The digamma function $\psi(z) = \Gamma'(z)/\Gamma(z)$ has the integral representation

$$\psi(z) = \log z - \frac{1}{2z} - \int_0^{\infty} [(e^t - 1)^{-1} - t^{-1} + \tfrac{1}{2}] \exp[-tz]\, dt. \tag{3.5.21}$$

(a) Use the method of Example 3.2.3 to obtain an asymptotic expansion of $\psi(z)$ as $|z| \to \infty$, $|\arg z| < \pi$.

(b) Justify integrating this result term-by-term to obtain an asymptotic expansion of $\log \Gamma(z)$, valid in the same open sector, modulo a constant of integration.

(c) Use the identity $\Gamma(z)\,\Gamma(1-z) = \pi \csc \pi z$, for $\arg z = \pi/2$ and $|z| \to \infty$, to find the constant of integration.

3.13. Let

$$I(\lambda) = \int_0^1 t^{s-1} (1-t)^{\lambda-1} f(t)\, dt, \qquad s > 0. \tag{3.5.22}$$

(a) Set

$$h(t;\lambda) = t^{s-1} (1-t)^{\lambda-1} \tag{3.5.23}$$

and introduce

$$h^{(-n-1)}(t;\lambda) = \frac{(-1)^{n+1}}{n!} \int_t^1 \tau^{s-1} (\tau - t)^n (1-\tau)^{\lambda-1}\, d\tau, \qquad n = 0, 1, \ldots. \tag{3.5.24}$$

(b) Show that

$$|h^{(-n-1)}(t;\lambda)| \le |h^{(-n-1)}(0;\lambda)|. \tag{3.5.25}$$

(c) Use the result (see the Appendix)

$$h^{(-n-1)}(0;\lambda) = \frac{(-1)^{n+1}}{n!} \frac{\Gamma(s+n)\,\Gamma(\lambda)}{\Gamma(s+n+\lambda)} \tag{3.5.26}$$

and the asymptotic expansion of the Γ function to show that $\{h^{(-n-1)}(0;\lambda)\}$ forms an asymptotic sequence.

(d) Derive the asymptotic expansion of $I(\lambda)$ for $f(t)$ a C^∞ function.

3.14. Identify the set of possible critical points for each of the following integrals:

(a) $\int_0^1 \log(1+t)\exp\{i\lambda \sin^2 t\}\, dt$.

(b) $\int_0^\infty \mathrm{Ai}(-\lambda\cos^2 t)\,|1-t^2|^{-1/2}\, dt$.

(c) $\displaystyle\int_{-\infty}^{\infty} \cos\left[\lambda\left(\frac{t^3}{3}-t\right)\right]dt$.

3.15. The Bessel function of order ν has the integral representation

$$J_\nu(\lambda) = \left[\frac{\lambda}{2}\right]^\nu \frac{I(\lambda;\nu)}{\sqrt{\pi}\,\Gamma\left(\nu+\frac{1}{2}\right)}, \tag{3.5.27}$$

where

$$I(\lambda;\nu) = \int_{-1}^{1} e^{i\lambda t}(1-t^2)^{\nu-\frac{1}{2}}\, dt, \qquad \mathrm{Re}(\nu) > -\tfrac{1}{2}. \tag{3.5.28}$$

(a) Suppose that the neutralizers $\bar{q}(t)$ and $\bar{\bar{q}}(t)$ introduced in Section 3.3 are chosen so that $\bar{\bar{q}}(-t) = \bar{q}(t)$. Then show that the neutralized integrals

$$I_{-1}(\lambda;\nu) = \int_{-1}^{1} \bar{\bar{q}}\, e^{i\lambda t}(1-t^2)^{\nu-\frac{1}{2}}\, dt, \qquad \mathrm{Re}(\nu) > -\tfrac{1}{2}, \tag{3.5.29}$$

$$I_{1}(\lambda;\nu) = \int_{-1}^{1} \bar{q}\, e^{i\lambda t}(1-t^2)^{\nu-\frac{1}{2}}\, dt, \qquad \mathrm{Re}(\nu) > -\tfrac{1}{2}, \tag{3.5.30}$$

have the property that $I_{-1}(\lambda;\nu) = I_1(-\lambda;\nu)$.

(b) Following the method of Section 3.4 for $I_{-1}(\lambda;\nu)$ introduce the iterated kernels

$$h^{(-n-1)}(t;\lambda) = \frac{(-1)^{n+1}}{n!}\int_t^{i\infty} (\tau+1)^{\nu-\frac{1}{2}}(\tau-t)^n e^{i\lambda(\tau+1)}\, d\tau, \qquad n = 0, 1, 2, \ldots, \tag{3.5.31}$$

and derive the asymptotic expansion

$$I_{-1}(\lambda;\nu) = -\exp\left\{-i\lambda + \frac{i\pi}{4} + \frac{i\nu\pi}{2}\right\} \tag{3.5.32}$$

$$\times\left\{\sum_{n=1}^{N-1} \frac{(i)^n\,\Gamma(\nu+\frac{1}{2})\,\Gamma(\nu+\frac{1}{2}+n)}{2^{n-\nu+\frac{1}{2}}\,\lambda^{\nu+n+\frac{1}{2}}\, n!\,\Gamma(\nu+\frac{1}{2}-n)} + 0(\lambda^{-\nu-N-\frac{1}{2}})\right\}.$$

(c) Derive a similar expansion for $I_1(\lambda;\nu)$ and thus obtain the expansion

$$J_\nu(\lambda) \doteq \sqrt{\frac{2}{\pi\lambda}}\left\{\cos\left\{\lambda-\frac{\pi}{4}-\frac{\nu\pi}{2}\right\}\left[\sum_{n=0}^{N-1}\frac{(-1)^n\,\Gamma(\nu+\frac{1}{2}+2n)}{(2n)!\,(2\lambda)^{2n}\,\Gamma(\nu+\frac{1}{2}-2n)}\right.\right.$$

$$\left.+\,0(\lambda^{-2N})\right]-\sin\left\{\lambda-\frac{\pi}{4}-\frac{\nu\pi}{2}\right\}$$

$$\left.\times\left[\sum_{n=0}^{N-1}\frac{(-1)^n\,\Gamma(\nu+\frac{3}{2}+2n)}{(2n+1)!\,(2\lambda)^{2n+1}\,\Gamma(\nu-\frac{1}{2}-2n)}+0(\lambda^{-2N-1})\right]\right\}. \tag{3.5.33}$$

3.16. The Bessel function of the third kind (Hankel function) and order zero has the integral representation

$$H_0^{(1)}(t)=\left(\frac{2}{t}\right)^{1/2}\frac{\exp\left\{i\left[t-\frac{\pi}{4}\right]\right\}}{\pi}\int_0^\infty e^{-u}\,u^{-1/2}\left(1+\frac{iu}{2t}\right)^{-1/2}du,$$

$$-\frac{\pi}{2}<\arg t<\frac{3\pi}{2}, \tag{3.5.34}$$

with the second square root defined to be positive when $\arg t=\pi/2$. Consider the integral

$$I(\lambda)=\textstyle\int_0^\infty h(\lambda t)\,f(t)\,dt \tag{3.5.35}$$

with $h(t)=H_0^{(1)}(t)$ and define the iterated integrals of $h(t)$ by

$$h^{(-n-1)}(t)=\frac{(-1)^{n+1}}{n!}\int_t^{i\infty}(\tau-t)^n\,H_0^{(1)}(\tau)\,d\tau,\qquad n=0,1,\ldots. \tag{3.5.36}$$

(a) From the representation (3.5.34) show that

$$|H_0^{(1)}(t)|\le\sqrt{\frac{2}{\pi|t|}}\,e^{-\mathrm{Im}(t)},\qquad 0\le\arg t\le\pi, \tag{3.5.37}$$

and thus conclude that the integrals (3.5.36) converge absolutely and that $h^{(-n-1)}(t)\to 0$ as $t\to\infty$.

(b) Suppose that $f(t)$ is a C^∞ function on $[0,\infty)$, identically zero for $t\ge t_0$. Then show that

$$I(\lambda)\sim\sum_{n=0}^{\infty}\frac{(-1)^{n+1}}{\lambda^{n+1}}f^{(n)}(0)\,h^{(-n-1)}(0). \tag{3.5.38}$$

(c) From (3.5.36), show that

$$(-1)^{n+1}\,h^{(-n-1)}(0)=\frac{2i^n}{\pi\,n!}\int_0^\infty s^n\,K_0(s)\,ds \tag{3.5.39}$$

with $K_0(s)$ the modified Bessel function, related to $H_0^{(1)}$ by $K_0(s)=(\pi i/2)$

$H_0^{(1)}(se^{i\pi/2})$. Then use the Appendix to show that

$$(-1)^{n+1}\,h^{(-n-1)}(0)=\frac{(2i)^n}{\pi\, n!}\,\Gamma^2\left(\frac{n+1}{2}\right). \tag{3.5.40}$$

(d) Finally conclude that

$$I(\lambda)\sim\sum_{n=0}^{\infty}\frac{1}{\pi\, n!}\frac{(2i)^n}{\lambda^{n+1}}f^{(n)}(0)\,\Gamma^2\left(\frac{n+1}{2}\right). \tag{3.5.41}$$

(e) If $R_N(\lambda)$ is the remainder after N terms of the sum, show that

$$R_N(\lambda)\le\frac{1}{\lambda^N}\sqrt{\frac{2\pi}{\lambda}}\int_0^{\infty}\frac{|f^{(N)}(t)|}{\sqrt{|t|}}\,dt. \tag{3.5.42}$$

3.17. Repeat Exercise 3.6 for the kernels

(a) $H_0^{(2)}(t)=\bar{H}_0^{(1)}(t)$.

(b) $H_0^{(1)}(t^\alpha)$.

3.18. In (3.4.10), show that

$$|h^{(-n-1)}(t;\lambda)|\le\frac{(t-a)^{\alpha-1}}{n!}\int_0^{\infty}\sigma^n\,e^{-\lambda\sigma}\,d\sigma \tag{3.5.43}$$

and perform the indicated integration to obtain (3.4.13).

3.19. Consider the integral

$$h^{(-n-1)}(t;\lambda)=\frac{(-1)^{n+1}}{n!}\int_t^{t+\infty e^{i\pi/2\mu}}(\tau-a)^{\alpha-1}(\tau-t)^n \times\exp\{i\lambda(\tau-a)^{\mu}\}\,d\tau,\qquad 0<\alpha\le 1, \tag{3.5.44}$$

where the path of integration is taken to be the ray $\tau-t=\sigma\,e^{i\pi/2\mu},\sigma$ real.

(a) Show that

$$\operatorname{Re}\left\{\frac{i}{\mu}\int_{\tau-a}^{\tau-t}s^{\mu-1}\,ds\right\}>0$$

for τ on the path of integration. Then use this result to show that

$$|\exp\{i\lambda(\tau-a)^{\mu}\}|\le|\exp\{i\lambda(\tau-t)^{\mu}\}|,\qquad \lambda>0,\qquad t\ge a,\qquad \mu\ge 1. \tag{3.5.45}$$

(b) Show that

$$|h^{(-n-1)}(t;\lambda)|\le\frac{(t-a)^{\alpha-1}}{n!}\int_0^{\infty}\sigma^n\,e^{-\lambda\sigma^{\mu}}\,d\sigma \tag{3.5.46}$$

and evaluate the integral to obtain

$$|h^{(-n-1)}(t;\lambda)|\le\frac{(t-a)^{\alpha-1}}{\mu\, n!\,\lambda^{(n+1)/\mu}}\,\Gamma\left(\frac{n+1}{\mu}\right). \tag{3.5.47}$$

(c) Use the estimate above to prove that (3.4.23) is the asymptotic expansion of $I_a(\lambda)$ with respect to the auxiliary sequence $\{\lambda^{-(n+1)/\mu}\}$ for $0 < \alpha \leq 1$, $\mu \geq 1$.

(d) Why need we only consider this restricted range on α?

3.20. The purpose of this exercise is to verify (3.4.24).

(a) Let

$$f(t) = (b - t)^{\mu} \tag{3.5.48}$$

with the branch cut of $f(t)$ lying on the real axis from b to $+\infty$. Show that if $\arg(b - t) = \alpha$, then $\arg t \to \pi + \alpha$ as $|t| \to \infty$ and thus verify that

$$h^{(-n-1)}(t;\lambda) = \frac{(-1)^{n+1}}{n!} \times \int_t^{\infty \exp[i(\pi + \pi/2\mu)]} (b - \tau)^{\beta - 1} (\tau - t)^n \exp\{i\lambda(b - \tau)^{\mu}\}\, d\tau \tag{3.5.49}$$

serve as iterated integrals of the kernel

$$h(t;\lambda) = (b - t)^{\beta - 1} \exp\{i\lambda(b - t)^{\mu}\}. \tag{3.5.50}$$

(b) Show that

$$h^{(-n-1)}(b;\lambda) = \frac{1}{\mu n!} \frac{\Gamma\left(\dfrac{n+\beta}{\mu}\right)}{\lambda^{(n+\beta)/\mu}} e^{i\pi(n+\beta)/2\mu}. \tag{3.5.51}$$

(c) Verify (3.4.24).

REFERENCES

E. T. Copson. *Asymptotic Expansions*. University Press, Cambridge, 1965.
The integration by parts technique is extensively discussed and applied to obtain asymptotic expansions of various special functions.

J. G. van der Corput. On the method of critical points. *Nederl. Akad. Wetensch. Proc.* **51,** 650–658, 1948.
This is a fundamental paper in which the use of neutralizer functions in deriving asymptotic expansions is introduced.

A. Erdélyi. *Asymptotic Expansions*. Dover, New York, 1956.
The ordinary integration by parts technique is considered in detail and the extension described in Section 3.4 of the text is developed.

J. B. Rosser. Explicit remainder terms for some asymptotic series. *J. Rat. Mech. Anal.* **4,** 317–324, 1955.
In this paper the author obtains error bounds on the remainder terms for various asymptotic expansions obtained via integration by parts.

E. C. Titchmarsh. *Theory of Fourier Integrals*. 2nd Ed. Clarendon Press, Oxford, 1958.
Integral transforms related to the Fourier transform are considered.

D. V. Widder. *The Laplace Transform*. Princeton, 1941.
In this book, also, transforms related to the Fourier transform are considered in detail.

4 h-Transforms with Kernels of Monotonic Argument

4.1. Laplace Transforms and Watson's Lemma

Throughout Chapter 3, asymptotic expansions were derived exclusively via the integration by parts procedure. We saw there, however, that the procedure is somewhat limited in its application and, in particular, required certain smoothness properties of the integrand functions. In Section 3.2, for example, under the assumption that $f(t)$ is infinitely differentiable at the origin, we obtained an asymptotic expansion of its Laplace transform

$$I(\lambda) = \int_0^\infty e^{-\lambda t} f(t)\, dt \tag{4.1.1}$$

as $\lambda \to \infty$. Indeed by Example 3.2.3, this expansion is given by

$$I(\lambda) \sim \sum_{n=0}^{\infty} \frac{f^{(n)}(0)}{\lambda^{n+1}}.$$

Suppose, however, that $f(t)$ is not infinitely differentiable at the origin. Then the straightforward integration by parts procedure breaks down and (3.2.27) is not valid. There are several ways we might attempt to handle the more general situation. This entire chapter shall be devoted to the development of a systematic asymptotic theory for the specific integral (4.1.1) and further for a wide class of h-transforms in cases where the smoothness conditions required for the integration by parts method are not satisfied. The present section will serve as a partial motivation for this development.

Here we shall only consider integrals of the form (4.1.1). Moreover, we shall restrict our considerations to the behavior of I in the two limits $\lambda \to \infty$ and $\lambda \to 0+$ because, as we have remarked in Section 3.2, these limits are the most interesting from the point of view of asymptotic analysis. For the present, we shall focus our attention on the case $\lambda \to \infty$. We shall assume that $f(t)$ is locally absolutely integrable[1] on $(0, \infty)$ and that, as $t \to \infty$,

$$f(t) = O(e^{at}) \tag{4.1.2}$$

for some real number a.

In order to study $I(\lambda)$ for large λ, let us further suppose that, as $t \to 0+$,

$$f(t) \sim \sum_{m=0}^{\infty} c_m t^{a_m}. \tag{4.1.3}$$

Here $\mathrm{Re}(a_m)$ increases monotonically to $+\infty$ as $m \to \infty$ [a condition we shall henceforth abbreviate by $\mathrm{Re}(a_m) \uparrow \infty$] and, of course, $\mathrm{Re}(a_0) > -1$. If the exponents a_m are not all positive integers, then $f(t)$ is not infinitely differentiable as $t \to 0+$. In that event we cannot expect to obtain an infinite asymptotic expansion by the integration by parts procedure. An infinite expansion can be obtained, however, by a different process whose validity is established in the following.

WATSON'S LEMMA. In (4.1.1) let $f(t)$ be a locally integrable function on $(0, \infty)$ bounded for finite t and let (4.1.2) and (4.1.3) hold. Then, as $\lambda \to \infty$,

$$I(\lambda) \sim \sum_{m=0}^{\infty} c_m \int_0^{\infty} e^{-\lambda t} t^{a_m} \, dt = \sum_{m=0}^{\infty} \frac{c_m \Gamma(a_m + 1)}{\lambda^{a_m + 1}}. \tag{4.1.4}$$

Remarks. Under the stated hypotheses, this lemma says that an infinite asymptotic expansion of I is obtained by replacing $f(t)$ in (4.1.1) by the expansion (4.1.3) and then integrating term-by-term. Thus, we see from (4.1.4) that the asymptotic expansion of $f(t)$, as $t \to 0+$, is directly related to the asymptotic expansion of its Laplace transform, as $\lambda \to \infty$. We shall find in the sections to follow that this type of relationship holds, in some sense, for a class of h-transforms.

PROOF OF WATSON'S LEMMA. Let R be a fixed positive number and set

$$\begin{aligned} I(\lambda) &= \textstyle\int_0^R e^{-\lambda t} f(t) \, dt + \int_R^{\infty} e^{-\lambda t} f(t) \, dt \\ &= I_1(\lambda) + I_2(\lambda). \end{aligned} \tag{4.1.5}$$

It follows from (4.1.2) and the boundedness of f that there exists a positive

[1] By "locally integrable" on (a,b) we mean that the function in question is absolutely integrable on all closed finite subintervals of (a,b).

constant k such that

$$|f| \le ke^{at}$$

for all $t \ge R$. Hence,

$$|I_2(\lambda)| \le \int_R^\infty e^{-\lambda t}\,|f(t)|\,dt \le \frac{k}{\lambda - a} e^{-(\lambda - a)R}$$
$$= o(e^{-\lambda R}), \qquad \lambda \to \infty. \tag{4.1.6}$$

Now for each positive integer N we set

$$f(t) = \sum_{m=0}^{N} c_m\, t^{a_m} + \rho_N(t) \tag{4.1.7}$$

and observe that (4.1.3) implies

$$\rho_N(t) = O(t^{\mathrm{Re}(a_{N+1})}), \qquad t \to 0+. \tag{4.1.8}$$

If we write

$$I_1(\lambda) = \sum_{m=0}^{N} c_m \int_0^R t^{a_m}\, e^{-\lambda t}\,dt + \int_0^R \rho_N\, e^{-\lambda t}\,dt \tag{4.1.9}$$

and note that

$$\int_0^R t^{a_m}\, e^{-\lambda t}\,dt = \int_0^\infty t^{a_m}\, e^{-\lambda t}\,dt - \int_R^\infty t^{a_m}\, e^{-\lambda t}\,dt$$
$$= \frac{\Gamma(1 + a_m)}{\lambda^{1+a_m}} + O(e^{-\lambda R}), \qquad \lambda \to \infty, \tag{4.1.10}$$

then it follows from (4.1.5), (4.1.6), and (4.1.9) that

$$I(\lambda) = \sum_{m=0}^{N} c_m\, \Gamma(a_m + 1)\, \lambda^{-(1+a_m)} + \int_0^R \rho_N(t)\, e^{-\lambda t}\,dt + O(e^{-\lambda R}). \tag{4.1.11}$$

To complete the proof we observe that because (4.1.8) holds and because $\rho_N(t)$ is bounded in $(0, R)$, there exists a constant k_N such that $|\rho_N| \le k_N t^{\mathrm{Re}(a_{N+1})}$ for all t in $[0, R]$. Hence,

$$\left|\int_0^R \rho_N(t)\, e^{-\lambda t}\,dt\right| \le k_N \int_0^\infty t^{\mathrm{Re}(a_{N+1})}\, e^{-\lambda t}\,dt = \frac{k_N\, \Gamma(\mathrm{Re}(a_{N+1}) + 1)}{\lambda^{\mathrm{Re}(a_{N+1})+1}}. \tag{4.1.12}$$

It now follows from (4.1.11) and (4.1.12) that

$$I(\lambda) = \sum_{m=0}^{N} c_m\, \Gamma(a_m + 1)\, \lambda^{-(1+a_m)} + O(\lambda^{-(\mathrm{Re}(a_{N+1})+1)}), \qquad \lambda \to \infty. \tag{4.1.13}$$

Because (4.1.13) holds for any nonnegative integer N, the lemma is proved.

The above result can be generalized in several ways. For example, we can show that it holds for λ complex. Indeed, (4.1.4) is valid as $|\lambda| \to \infty$ in the sector $|\arg \lambda| < \pi/2$. Also, $f(t)$ can be allowed to have an asymptotic expansion, as

$t \to 0+$, with respect to a more general asymptotic sequence than that assumed in (4.1.3). Finally, as we have previously remarked, analogous results can be obtained for other integral transforms. These generalizations will be considered in the remaining sections of this chapter.

Now let us turn to the study of $I(\lambda)$ as $\lambda \to 0+$. We might expect, in analogy with the preceding results, that the asymptotic expansion of I, as $\lambda \to 0+$, is determined by the asymptotic expansion of f, as $t \to +\infty$. That this is only partially true will be shown in subsequent sections. For the present, let us try to obtain an expansion by the formal process of expanding the kernel $e^{-\lambda t}$ in a Taylor series about the origin and then integrating term-by-term. This yields the rather dubious result

$$I(\lambda) \approx \sum_{n=0}^{\infty} \frac{(-\lambda)^n}{n!} \int_0^{\infty} t^n f(t)\, dt, \qquad \lambda \to 0. \tag{4.1.14}$$

Here we have used the symbol $\approx$ rather than $\sim$ to indicate that, in general, (4.1.14) is not a rigorous asymptotic statement. In fact, in most cases (4.1.14) makes no sense at all. This is so because the coefficients involve integer moments of $f(t)$, none of which need exist, because $f(t)$ need not be integrable on $(0,\infty)$. Actually, we can only conclude here that (4.1.14) is a correct asymptotic result if $f(t) = o(t^{-R})$, as $t \to \infty$, for all R.

Despite the fact that the utility of (4.1.14) is rather limited, it does suggest that the integer moments of f will play a role in the asymptotic analysis of $I(\lambda)$ as $\lambda \to 0+$. We recall that the expansion of I for large λ involves the quantities

$$\Gamma(a_m + 1) = \int_0^{\infty} e^{-t}\, t^{a_m}\, dt. \tag{4.1.15}$$

If a_m is a nonnegative integer, then clearly $\Gamma(a_m + 1)$ is a moment of the function e^{-t}. Even if a_m is not a nonnegative integer it is convenient to view $\Gamma(a_m + 1)$ as the a_mth moment of e^{-t}. This leads us to conjecture that when studying the asymptotic behavior of the general h-transform,

$$I(\lambda) = \int_0^{\infty} h(\lambda t) f(t)\, dt \tag{4.1.16}$$

as either $\lambda \to 0+$ or $\lambda \to +\infty$, moments of the integrand functions h and f will come into play.

Any moment of a function can be viewed as the Mellin transform of that function evaluated at a particular point. Indeed, as we have indicated in Section 3.2, the Mellin transform of $f(t)$ is given by

$$M[f;z] = \int_0^{\infty} t^{z-1} f(t)\, dt \tag{4.1.17}$$

which can be directly interpreted as the $(z-1)$st moment of f. Thus, we can restate the above conjecture and anticipate that the Mellin transforms of f and h will play important roles in the asymptotic analyses of the h-transform (4.1.16) in the limits $\lambda \to \infty$ and $\lambda \to 0+$.

4.2. Results on Mellin Transforms

Because the following sections rely heavily on the theory of Mellin transforms, we shall present here a summary of the needed results. As we shall see, the Mellin transform is intimately related to the Laplace transform whose basic theory is assumed familiar to the reader. Because of this, much of the present section will consist of results stated without proof.

It is well known that if $g(\tau)$ is such that

$$\int_0^\tau |g(\xi)|\, d\xi < \infty \tag{4.2.1}$$

for all finite positive τ, and

$$g(\tau) = O(e^{\alpha^* \tau}), \tag{4.2.2}$$

as $\tau \to \infty$ for some real constant α^*, then the one-sided Laplace transform

$$L^+ [g;z] = \int_0^\infty e^{-z\tau}\, g(\tau)\, d\tau \tag{4.2.3}$$

converges absolutely and is holomorphic in the right half-plane $\mathrm{Re}(z) > \alpha^*$. Similarly, if

$$\int_0^\tau |g(-\xi)|\, d\xi < \infty \tag{4.2.4}$$

for all finite positive τ and

$$g(\tau) = O(e^{\beta^* \tau}), \tag{4.2.5}$$

as $\tau \to -\infty$ for some real constant β^*, then the one-sided Laplace transform

$$L^-[g;z] = \int_{-\infty}^0 e^{-z\tau}\, g(\tau)\, d\tau = \int_0^\infty e^{z\tau}\, g(-\tau)\, d\tau \tag{4.2.6}$$

converges absolutely and is holomorphic in the left half-plane $\mathrm{Re}(z) < \beta^*$.

Let us now define

$$\begin{aligned} \alpha &= \inf\{\alpha^* \,|\, g(\tau) = O(e^{\alpha^* \tau}), \text{ as } \tau \to +\infty\}, \\ \beta &= \sup\{\beta^* \,|\, g(\tau) = O(e^{\beta^* \tau}), \text{ as } \tau \to -\infty\}. \end{aligned} \tag{4.2.7}$$

We observe that, if $\beta > \alpha$, then the two-sided (bilateral) Laplace transform

$$\mathscr{L}[g;z] = \int_{-\infty}^\infty e^{-z\tau}\, g(\tau)\, d\tau \tag{4.2.8}$$

converges absolutely and is holomorphic in the vertical strip $\alpha < \mathrm{Re}(z) < \beta$. However, if $\alpha > \beta$, then $\mathscr{L}[g;z]$ does not exist for any z.

We first assume $\alpha < \beta$ and set $\tau = -\log t$, $g(-\log t) = f(t)$ in (4.2.8). This yields

$$\begin{aligned} \mathscr{L}[g;z] &= \int_0^\infty t^{z-1} f(t)\, dt \\ &= M[f;z]. \end{aligned} \tag{4.2.9}$$

Thus we see that, when it exists, the bilateral Laplace transform of $g(t)$ is the Mellin transform of $f(t) = g(-\log t)$. It then follows from (4.2.7) that the Mellin transform of a function $f(t)$ converges absolutely and is holomorphic in the strip $\alpha < \mathrm{Re}(z) < \beta$ where

$$\alpha = \inf\{\alpha^* \mid f = O(t^{-\alpha^*}), \text{ as } t \to 0+\},$$
$$\beta = \sup\{\beta^* \mid f = O(t^{-\beta^*}), \text{ as } t \to +\infty\}. \tag{4.2.10}$$

From (4.2.10) we can conclude that when the Mellin transform of a function f converges absolutely, it does so in a vertical strip whose boundaries are determined by the asymptotic behavior of f in the limits $t \to 0+$ and $t \to +\infty$.

The inversion formula for Mellin transforms follows directly from that for the two-sided Laplace transform. Indeed, if $\mathscr{L}[g;z]$ is defined by (4.2.8), then we have

$$g(\tau) = \frac{1}{2\pi i} \int_{c-i\infty}^{c+i\infty} e^{z\tau} \mathscr{L}[g;z]\, dz \tag{4.2.11}$$

wherever $g(\tau)$ is continuous. Here $\alpha < c < \beta$ with α and β defined by (4.2.7). Upon setting $\tau = -\log t$ in (4.2.11), we find from (4.2.9) that

$$f(t) = g(-\log t) = \frac{1}{2\pi i} \int_{c-i\infty}^{c+i\infty} t^{-z}\, M[f;z]\, dz. \tag{4.2.12}$$

Equation (4.2.12) is the desired inversion formula valid at all points $t \geq 0$ where $f(t)$ is continuous.

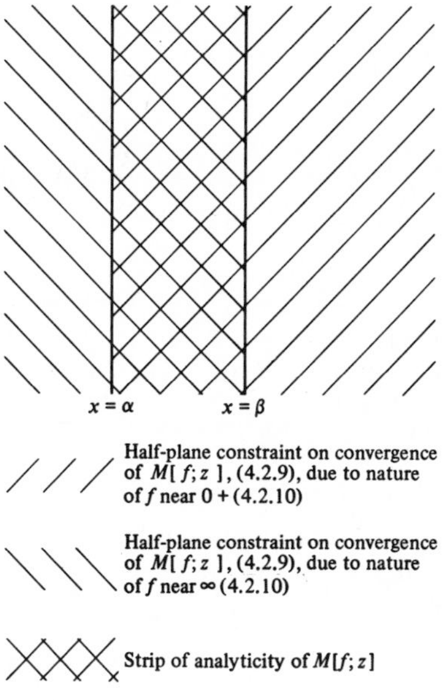

Figure 4.2.

In our future work, we shall need to estimate the behavior of $M[f;z] = M[f;x+iy]$, as $y \to \pm\infty$ with x fixed; that is, along vertical lines. It can be shown that for any x, with $\alpha < x < \beta$,

$$\lim_{|y|\to\infty} |M[f;x+iy]| = 0.^{2} \tag{4.2.13}$$

In other words, $M[f;z]$ goes to zero as z goes to infinity along any vertical line within the strip of absolute convergence. As we shall see in later sections, when additional assumptions are made concerning the behavior of f, more specific statements concerning the rate of decay can be made.

A result whose importance cannot be overemphasized is an identity known as the *Parseval formula for Mellin transforms.* We shall first obtain this identity formally and shall then state conditions sufficient for its validity. Thus, suppose that the functions $f(t)$ and $h(t)$ are such that

$$I = \int_0^\infty f(t)\, h(t)\, dt \tag{4.2.14}$$

exists. Suppose further that $M[f;1-z]$ and $M[h;z]$ are holomorphic each in some vertical strip. We shall assume that these vertical strips overlap which will be the case whenever I is absolutely convergent. If $\operatorname{Re}(z) = c$ lies in the overlapping strip, then we have

$$\begin{aligned} &\frac{1}{2\pi i}\int_{c-i\infty}^{c+i\infty} M[h;z]\, M[f;1-z]\, dz \\ &\quad = \frac{1}{2\pi i}\int_{c-i\infty}^{c+i\infty} M[f;1-z] \int_0^\infty h(t)\, t^{z-1}\, dt\, dz. \end{aligned} \tag{4.2.15}$$

Let us suppose that we can interchange the order of integration in (4.2.15) to obtain

$$\begin{aligned} &\frac{1}{2\pi i}\int_{c-i\infty}^{c+i\infty} M[h;z]\, M[f;1-z]\, dz \\ &\quad = \frac{1}{2\pi i}\int_0^\infty h(t)\, dt \int_{c-i\infty}^{c+i\infty} M[f;1-z]\, t^{z-1}\, dz \end{aligned} \tag{4.2.16}$$

which, upon using the inversion formula (4.2.12), yields the desired Parseval formula

$$\int_0^\infty f(t)\, h(t)\, dt = \frac{1}{2\pi i}\int_{c-i\infty}^{c+i\infty} M[h;z]\, M[f;1-z]\, dz. \tag{4.2.17}$$

In order to validate (4.2.17) we must justify the above interchange in the order of integration. There are several sets of conditions that are sufficient for this purpose, but we will state only two.

(1) If $M[f;1-c-iy]$ is in $L(-\infty < y < \infty)$ while $t^{c-1}\, h(t)$ is in

[2] This actually follows from the Riemann-Lebesgue lemma stated on page 80.

$L(0 \le t < \infty)$, then the interchange is justified by absolute convergence.

(2) Suppose that, as $a \to \infty$,

$$\mathscr{F}(z,a) = \int_{1/a}^{a} f(t)\, t^{z-1}\, dt, \qquad \mathscr{H}(z,a) = \int_{1/a}^{a} h(t)\, t^{z-1}\, dt$$

tend to $M[f;z]$, $M[h;z]$ at $x = 1 - c$, $x = c$, respectively, for all y and in such a way that $e^{-d|z|}\mathscr{F}(z,a)$, $e^{-d|z|}\mathscr{H}(z,a)$ are for some positive d bounded independently of a. Let

$$\frac{\xi^{1-c}}{\xi^{d} + \xi^{-d}} \int_{a}^{b} h(t) f(t)\, dt$$

be bounded for all a, b, ξ and as $a \to 0$, $b \to \infty$ converge to a continuous limit in a neighborhood of $\xi = 1$. Then

$$I = \int_{0}^{\infty} f(t)\, h(t)\, dt = \frac{1}{2\pi i} \lim_{\sigma \to \infty} \int_{c-i\sigma}^{c+i\sigma} M[f;1-z]\, M[h;z]\, dz \tag{4.2.18}$$

whenever the right-hand side exists. We note that I need not be absolutely convergent here.

All of the results in this section concern properties of the Mellin transform $M[f;z]$ only within its strip of analyticity. In the next section we shall investigate the possibility of analytically continuing the transform outside of this strip. Indeed we shall obtain conditions on f that are sufficient to not only guarantee the existence of such a continuation but also to allow for the locating and classifying of any singularities that might arise. The results obtained in the following section are extremely important to the development of our asymptotic technique. To motivate some of these results, we wish to consider here three particular functions and their Mellin transforms. Each of these functions serves as a prototype for a class of functions with which we shall be concerned. The three functions are e^{-t}, e^{it}, and $(1+t)^{-1}$. We note that, as $t \to +\infty$, e^{-t} decays exponentially, e^{it} oscillates, and $(1+t)^{-1}$ decays algebraically and monotonically.

From the Appendix we have that

$$M[e^{-t}; z] = \Gamma(z), \tag{4.2.19}$$

$$M[e^{it}; z] = e^{\pi i z/2}\, \Gamma(z), \tag{4.2.20}$$

$$M[(1+t)^{-1}; z] = \frac{\pi}{\sin \pi z}. \tag{4.2.21}$$

Moreover, we find that $M[e^{-t};z]$ is analytic in the right half-plane $\mathrm{Re}(z) > 0$ while $M[e^{it};z]$ and $M[(1+t)^{-1};z]$ are both analytic in the strip $0 < \mathrm{Re}(z) < 1$. In this strip $M[e^{it};z]$ exists only as a conditionally convergent integral.

The main point of interest here is that both $M[e^{it};z]$ and $M[(1+t)^{-1};z]$ can be analytically continued into the right half-plane $\mathrm{Re}(z) \ge 1$. Indeed, these continuations can be accomplished via formulas (4.2.20) and (4.2.21) directly.

We note that the analytic continuation of $M[e^{it};z]$ is *holomorphic* in $\mathrm{Re}(z) \geq 1$ while that of $M[(1+t)^{-1};z]$ is *meromorphic* in this region. In fact, $M[(1+t)^{-1};z]$ has simple poles at the positive integers.

As we shall show in Section 4.3 the above continuation properties carry over to the Mellin transforms of the functions in the corresponding classes. In particular we shall show that

(1) If f decays exponentially as $t \to +\infty$, then $M[f;z]$ is holomorphic in a right half-plane.
(2) If f is oscillatory as $t \to +\infty$, then $M[f;z]$ can be analytically continued into a right half-plane as a holomorphic function.
(3) If f decays algebraically and monotonically as $t \to +\infty$, then $M[f;z]$ can be analytically continued into a right half-plane as a meromorphic function.

4.3. Analytic Continuation of Mellin Transforms

Let us suppose that $M[f;z]$, the Mellin transform of the function f, is absolutely convergent and analytic in the vertical strip $\alpha < \mathrm{Re}(z) < \beta$. We wish to consider first its analytic continuation into the right half-plane $\mathrm{Re}(z) \geq \beta$. For this purpose we assume that, as $t \to \infty$, $f(t)$ has an asymptotic expansion of the form

$$f(t) \sim \exp(-dt^{\nu}) \sum_{m=0}^{\infty} \sum_{n=0}^{N(m)} c_{mn} (\log t)^n \, t^{-r_m}. \tag{4.3.1}$$

Here $\mathrm{Re}(d) \geq 0$, $\nu > 0$, and $\mathrm{Re}(r_m) \uparrow \infty$. It is also assumed that the nonnegative integers $N(m)$, $m = 0, 1, 2, \ldots$ are all finite. We wish to emphasize that the form (4.3.1) allows f to have almost any behavior, as $t \to \infty$, that might reasonably occur in applications. (Exponential growth of f, however, has been excluded because otherwise $M[f;z]$ would not exist for any z.) We also note that (4.3.1) includes the important special case where $f(t)$ has an asymptotic power series as $t \to \infty$.

Our results will naturally fall into three cases depending on the constant d. We shall present these cases in three lemmas which are proven immediately below.

LEMMA 4.3.1. If $\mathrm{Re}(d) > 0$, then $\beta = +\infty$ so that $M[f;z]$ is absolutely convergent and hence holomorphic for $\mathrm{Re}(z) > \alpha$.

PROOF. We clearly have

$$|f(t)| = O\{\exp[-\mathrm{Re}(d)\, t^{\nu}]\, t^{-\mathrm{Re}(r_0)} (\log t)^{N(0)}\} \tag{4.3.2}$$

as $t \to \infty$ and hence $\beta = +\infty$ in (4.2.10). We might point out that in this case (4.2.13) holds for any $x > \alpha$.

In the proofs of the next two lemmas, it is convenient to make use of the functions $s_k(t)$ and $f_k(t)$ which, for any $k > \mathrm{Re}(r_0)$, are defined by

$$s_k(t) = \begin{cases} 0, & 0 \le t < 1, \\ \exp(-dt^\nu) \displaystyle\sum_{\substack{m \\ \mathrm{Re}(r_m) < k}} \sum_{n=0}^{N(m)} c_{mn}\, t^{-r_m} (\log t)^n, & 1 \le t < \infty, \end{cases} \tag{4.3.3}$$

and

$$f_k(t) = f(t) - s_k(t). \tag{4.3.4}$$

We note that, for $t \ge 1$, $s_k(t)$ involves only a finite number of terms and that, as $t \to \infty$, $f_k(t) = O(e^{-dt^\nu}\, t^{-r_j} (\log t)^{N(j)})$ where j is the least integer such that $\mathrm{Re}(r_j) \ge k$.

LEMMA 4.3.2. If $d = -i\omega$ with ω real and nonzero, in which event f is *oscillatory* at infinity, then $M[f;z]$ can be analytically continued into $\mathrm{Re}(z) \ge \beta$ as a *holomorphic* function. Moreover, for the continued Mellin transform we have

$$M[f;x+iy] = O(|y|^{n(x)}), \qquad |y| \to \infty, \qquad x > \mathrm{Re}(r_0), \tag{4.3.5}$$

where $n(x)$ is the smallest integer greater than or equal to $\mathrm{Re}([x - r_0]/\nu)$.[3]

PROOF. It is readily seen that $M[f;z]$ is holomorphic in the strip $\alpha < \mathrm{Re}(z) < \mathrm{Re}(r_0)$ with (4.2.13) satisfied in this strip. Furthermore, in a similar manner, $f_k(t)$ defined by (4.3.3) and (4.3.4) has a Mellin transform holomorphic in the strip $\alpha < \mathrm{Re}(z) < k$ with

$$\lim_{|y| \to \infty} M[f_k;x+iy] = 0 \tag{4.3.6}$$

for any x in the interval (α,k). Because

$$M[f;z] = M[f_k;z] + M[s_k;z], \tag{4.3.7}$$

in order to continue $M[f;z]$ into the strip $\alpha < \mathrm{Re}(z) < k$, we need only continue $M[s_k;z]$ into this same strip.

The desired continuation is accomplished via an iterative process which we shall now outline in some detail. Let us first set

$$t^{-r_0}\, \sigma_1(t) = \sum_{\substack{m \\ \mathrm{Re}(r_m) < k}} \sum_{n=0}^{N(m)} c_{mn}\, t^{-r_m} (\log t)^n \tag{4.3.8}$$

so that $\sigma_1(t) = O((\log t)^{N(0)})$, $t \to \infty$, and

[3] Actually, we can show that the continuous estimate, $M[f;x+iy] = O(|y|^{\mathrm{Re}(x-r_0)/\nu - 1/2})$, $|y| \to \infty$, $x > \mathrm{Re}(r_0)$ holds. See Exercises 7.24 and 7.25.

$$M[s_k;z] = \int_1^\infty e^{i\omega t^\nu}\, t^{z-r_0-1}\, \sigma_1(t)\, dt$$
$$= \int_1^\infty (i\omega\nu t^{\nu-1}\, e^{i\omega t^\nu})\, \frac{t^{z-r_0-\nu}\, \sigma_1(t)}{i\omega\nu}\, dt. \tag{4.3.9}$$

Upon integrating by parts in (4.3.9) we obtain

$$M[s_k;z] = \left.\frac{t^{z-r_0-\nu}\, e^{i\omega t^\nu}\, \sigma_1(t)}{i\omega\nu}\right|_1^\infty$$
$$- \int_1^\infty t^{z-r_0-\nu-1}\, e^{i\omega t^\nu} \left[(z-r_0-\nu)\,\sigma_1 + t\,\frac{d\sigma_1}{dt}\right] \frac{dt}{i\omega\nu}. \tag{4.3.10}$$

For $\mathrm{Re}(z) < \mathrm{Re}(r_0) + \nu$ we have expressed $M[s_k;z]$ as the sum of a constant term and an integral which we write as

$$M_1[s_k;z] = \int_1^\infty t^{z-r_0-\nu-1}\, e^{i\omega t^\nu}\, \sigma_2(t)\, dt;$$
$$\sigma_2 = \frac{\left[(z-r_0-\nu)\,\sigma_1 + t\,\dfrac{d\sigma_1}{dt}\right]}{i\omega\nu}. \tag{4.3.11}$$

We observe that $M_1[s_k;z]$ is holomorphic in the left half-plane $\mathrm{Re}(z) < \mathrm{Re}(r_0) + \nu$ and moreover is an integral of the same form as (4.3.9). [Note that, as $t \to \infty$, $\sigma_2(t)$ is of the same order as $\sigma_1(t)$.] In this manner we have continued $M[s_k;z]$ into the strip $\mathrm{Re}(r_0) \le \mathrm{Re}(z) < \mathrm{Re}(r_0) + \nu$ as a holomorphic function and hence, through (4.3.7), the same result holds for $M[f;z]$. In addition, because σ_2 is a polynomial of first degree in z, we have that $M[f;z] = O(|y|)$ as $|y| \to \infty$ in this strip.

The same process may now be applied to $M_1[s_k;z]$. Indeed, it is clear from a simple inductive argument that the process can be applied repeatedly. Each integration by parts yields an analytic continuation of $M[s_k;z]$ and hence of $M[f;z]$ a distance ν to the right. Moreover, the continued functions are all holomorphic in the extended strips. Because at each step terms in the continuation are multiplied by factors linear in z, we can conclude that, after p steps, the continued $M[f;z]$ is $O(|y|^p)$ as $|y| \to \infty$ in the strip $\mathrm{Re}(r_0) + (p-1)\,\nu \le \mathrm{Re}(z) < \mathrm{Re}(r_0) + \nu p$. The lemma now follows from the fact that, in order to continue $M[f;z]$ into the strip $\alpha < \mathrm{Re}(z) < k$, we need only apply the procedure j times where j is the smallest integer greater than or equal to $\mathrm{Re}(k - r_0)/\nu$.

LEMMA 4.3.3. If $d = 0$ in (4.3.1), in which event we say that f is *algebraic* at infinity, then $\beta = \mathrm{Re}(r_0)$ and $M[f;z]$ can be analytically continued into the right half-plane $\mathrm{Re}(z) \ge \mathrm{Re}(r_0)$ as a *meromorphic* function, with poles at the points $z = r_m$, $m = 0, 1, \ldots$. Moreover, about $z = r_m$, $M[f;z]$ has a Laurent expansion with singular part

$$\sum_{n=0}^{N(m)} \frac{(-1)^{n+1}\, c_{mn}\, n!}{(z-r_m)^{n+1}}. \tag{4.3.12}$$

Finally, for $x > \alpha$,

$$\lim_{|y| \to \infty} M[f;x+iy] = 0.$$

PROOF. For any $k > \text{Re}(r_0)$ we have that (4.3.7) holds where $M[f_k;z]$ is holomorphic in $\alpha < \text{Re}(z) < k$. In addition, we have that (4.3.6) holds for $\alpha < x < k$. If $\text{Re}(z - r_m) < 0$, then we find directly that

$$\int_1^\infty t^{z - r_m - 1} (\log t)^n \, dt = \frac{n!\,(-1)^{n+1}}{(z - r_m)^{n+1}}. \tag{4.3.13}$$

For all other z, (4.3.13) defines a meromorphic function by analytic continuation. Thus, by computation we have

$$\begin{aligned} M[f;z] &= M[s_k;z] + M[f_k;z] \\ &= \sum_{\substack{m \\ \text{Re}(r_m) < k}} \sum_{n=0}^{N(m)} \frac{(-1)^{n+1} c_{mn}\, n!}{(z - r_m)^{n+1}} + M[f_k;z]. \end{aligned} \tag{4.3.14}$$

Because k is arbitrary, the lemma is proved. (Note that $\lim_{|y| \to \infty} M[s_k;x+iy] = 0$ for all x.)

Under the single assumption (4.3.1) and through Lemmas 4.3.1, 4.3.2, and 4.3.3, we have succeeded in analytically continuing $M[f;z]$ into the right half-plane $\text{Re}(z) > \alpha$ as a meromorphic function at worst. Moreover, we have been able to locate the poles, if any, of the continued function and have determined the singular parts of the corresponding Laurent expansions about these poles solely in terms of the constants that appear in the assumed asymptotic expansion of f as $t \to \infty$.

The analytic continuation of $M[f;z]$ into the left half-plane $\text{Re}(z) \le \alpha$ is accomplished in a completely analogous manner. Indeed, if we assume that, as $t \to 0+$,

$$f(t) \sim \exp(-qt^{-\mu}) \sum_{m=0}^{\infty} \sum_{n=0}^{\bar{N}(m)} p_{mn} (\log t)^n \, t^{a_m} \tag{4.3.15}$$

with $\text{Re}(q) \ge 0$, $\mu > 0$, $\text{Re}(a_m) \uparrow \infty$, and $0 \le \bar{N}(m)$ finite for each m, then this continuation is described in the following three lemmas. We shall omit the proofs of these lemmas because they are essentially the same as those of Lemmas 4.3.1, 4.3.2, and 4.3.3.

LEMMA 4.3.4. If in (4.3.15) $\text{Re}(q) > 0$, in which event f decays exponentially as $t \to 0+$, then $\alpha = -\infty$ in (4.2.10) and $M[f;z]$ is holomorphic in $\text{Re}(z) < \beta$. Moreover, (4.2.13) holds for any $x < \beta$.

LEMMA 4.3.5. If in (4.3.15) $q = -ir$ with r real and nonzero, in which event

f is oscillatory as $t \to 0+$, then $\alpha = -\operatorname{Re}(a_0)$ and $M[f;z]$ can be analytically continued as a holomorphic function into the left half-plane $\operatorname{Re}(z) \leq -\operatorname{Re}(a_0)$. Furthermore, for the continued Mellin transform we have

$$M[f;x+iy] = O(|y|^{n(x)}), \qquad |y| \to \infty, \qquad x < \operatorname{Re} a_0, \qquad (4.3.16)$$

where $n(x)$ is the smallest integer greater than or equal to $-\operatorname{Re}((x+a_0)/\mu)$.[4]

LEMMA 4.3.6. If $q = 0$ in (4.3.15), in which event f is algebraic as $t \to 0+$, then $\alpha = -\operatorname{Re}(a_0)$ and $M[f;z]$ can be analytically continued as a meromorphic function into the left half-plane $\operatorname{Re}(z) \leq \alpha$ with poles at the points $z = -a_m$. The continued $M[f;z]$ has a Laurent expansion about $z = -a_m$ with singular part

$$\sum_{n=0}^{\bar{N}(m)} \frac{p_{mn}(-1)^n n!}{(z+a_m)^{n+1}}. \qquad (4.3.17)$$

Also, for any $x < \beta$ $\lim_{|y| \to \infty} M[f;x+iy] = 0$.

Upon combining the six lemmas of this section we find that under the assumptions $\beta > \alpha$, (4.3.1) and (4.3.15) we have defined, by analytic continuation, the Mellin transform $M[f;z]$ on the entire z plane as a meromorphic function at worst. In addition, we have been able to locate all of the poles that arise in the continuation and have determined the corresponding singular parts of the Laurent expansions about these poles in terms of the constants that appear in the assumed asymptotic forms (4.3.1) and (4.3.15). We wish to emphasize that whenever f has exponential behavior (either exponential decay or oscillation) in both of the limits $t \to +\infty$, $t \to 0+$, there are no finite singularities of the continued Mellin transform. On the other hand, if f is algebraic in either or both of these limits, that is, if either d, q, or both vanish in (4.3.1) and (4.3.15), then poles arise in the continuation of $M[f;z]$. Moreover, the order of any given pole is determined solely by the highest power of $\log t$ that appears in the corresponding terms of the relevant asymptotic expansion of f.

Let us now suppose that $\alpha > \beta$ in (4.2.10) so that the defining integral of $M[f;z]$, namely (4.2.9), does not exist anywhere. We might point out that this can occur even when (4.3.1) and (4.3.15) are both satisfied. Indeed for the function

$$f(t) = (1+t)^{\nu} \qquad (4.3.18)$$

with ν any constant with real part greater than zero, we have $\alpha = 0$ and $\beta = -\operatorname{Re}(\nu)$. Hence (4.3.18) does not have a Mellin transform in the ordinary sense. Because we anticipate that the Mellin transforms of the integrand functions in (4.1.16) are going to play a significant role in the asymptotic analysis

[4] As in Lemma 4.3.2 we can show that the continuous estimate $M[f;x+iy] = O(|y|^{-\operatorname{Re}(x+a_0)/\mu - 1/2})$, $|y| \to \infty, x < -\operatorname{Re}(a_0)$, holds. See Exercises 7.24 and 7.25.

of that integral, it is disturbing that we cannot, at present, include such simple functions as (4.3.18). The following discussion shall show, however, that by using the preceding results on analytic continuation we can introduce a generalization so that the Mellin transforms of functions such as (4.3.18) can be meaningfully defined.

Thus, let us assume that $f(t)$ is locally integrable on $(0,\infty)$ and that (4.3.1) and (4.3.15) are satisfied. If we set

$$f_1(t) = \begin{cases} f(t), & t \in [0,1), \\ 0, & t \in [1,\infty), \end{cases} \tag{4.3.19}$$

and

$$f_2(t) = \begin{cases} 0, & t \in [0,1), \\ f(t), & t \in [1,\infty), \end{cases} \tag{4.3.20}$$

then clearly

$$f(t) = f_1(t) + f_2(t).^5 \tag{4.3.21}$$

Furthermore, we have that $M[f_1;z]$ is holomorphic in $\text{Re}(z) > \alpha$ while $M[f_2;z]$ is holomorphic in $\text{Re}(z) < \beta$ with α and β defined by (4.2.10). Of course, if $\alpha < \beta$, then

$$M[f;z] = M[f_1;z] + M[f_2;z] \tag{4.3.22}$$

is defined and holomorphic in the strip $\alpha < \text{Re}(z) < \beta$. We are concerned now with the case $\alpha > \beta$ so that (4.3.22) does not directly define an analytic function anywhere. We can, however, under the assumption that (4.3.15) holds, analytically continue $M[f_1;z]$ into the left half-plane $\text{Re}(z) \leq \alpha$ and under the assumption that (4.3.1) holds analytically continue $M[f_2;z]$ into the right half-plane $\text{Re}(z) \geq \beta$. Indeed, from Lemmas 4.3.1 through 4.3.6, we can conclude that these continued functions are at worst meromorphic functions in the entire z plane.

Our generalized definition of the Mellin transform is now immediate. For any locally integrable function with asymptotic expansions (4.3.1) and (4.3.15), its Mellin transform is given by (4.3.22) where we mean by the right-hand side the analytic continuation of $M[f_1;z]$ plus that of $M[f_2;z]$. Thus, although the Mellin transform of such an f need not exist for any z as originally defined, with our new definition it now exists in the entire z plane. From now on we shall interpret the symbol $M[f;z]$ as the Mellin transform in this generalized sense unless otherwise stated. We might further point out that, because estimates have been made concerning the behavior of $M[f_1;z]$ and $M[f_2;z]$ as z goes to

[5] The selection of $t = 1$ as the point at which to truncate f is completely arbitrary. Indeed, any positive value of t could be used. We shall find, however, that for many purposes, $t = 1$ is a particularly convenient point.

infinity along vertical lines, corresponding estimates for the behavior of $M[f;z]$ in such limits follow.

We should point out that the analytic continuations necessary to define $M[f;z]$ can often be obtained directly from known formulas for $M[f_1;z]$ and $M[f_2;z]$. This, of course, would result in a simplification and should be exploited whenever possible. In order to clarify this and some of the other points discussed above, let us consider the following.

EXAMPLE 4.3. Suppose that

$$f(t) = |1 - t|^{\nu}, \qquad \mathrm{Re}(\nu) > 0. \tag{4.3.23}$$

Hence $\alpha = 0$ and $\beta = -\mathrm{Re}(\nu)$ in (4.2.10) so that $M[f;z]$ does not exist in the ordinary sense. We have

$$f_1(t) = \begin{cases} (1-t)^{\nu}, & 0 \le t < 1, \\ 0, & 1 \le t < \infty, \end{cases}$$

$$f_2(t) = \begin{cases} 0, & 0 \le t < 1, \\ (t-1)^{\nu}, & 1 \le t < \infty, \end{cases} \tag{4.3.24}$$

from which we can define $M[f;z]$ in the generalized sense. The lemmas of this section predict that $M[f_1;z]$ has simple poles at $z = 0, -1, -2, \ldots$ and $M[f_2;z]$ has simple poles at $z = -\nu + n$, $n = 0, 1, 2, \ldots$. From (4.3.22) we see that the same statements hold for the generalized Mellin transform of f. Indeed we have from the Appendix

$$M[f_1;z] = \int_0^1 t^{z-1}(1-t)^{\nu}\,dt = \frac{\Gamma(\nu+1)\,\Gamma(z)}{\Gamma(\nu+z+1)}, \qquad \mathrm{Re}(z) > 0, \tag{4.3.25}$$

$$M[f_2;z] = \int_1^{\infty} t^{z-1}(t-1)^{\nu}\,dt = \frac{\Gamma(\nu+1)\,\Gamma(-\nu-z)}{\Gamma(1-z)}, \qquad \mathrm{Re}(z) < -\mathrm{Re}(\nu). \tag{4.3.26}$$

The analytic continuations of $M[f_1;z]$ and $M[f_2;z]$ are immediately obtained from the known analytic continuations of the gamma functions that appear in (4.3.25) and (4.3.26). Moreover, because the gamma function $\Gamma(z)$ has simple poles at the nonpositive integers, and does not vanish at any finite point, the results of our lemmas are substantiated. Finally, we have from (4.3.22), (4.3.25), and (4.3.26) that

$$M[|1-t|^{\nu};z] = \Gamma(\nu+1)\left\{\frac{\Gamma(z)}{\Gamma(\nu+z+1)} + \frac{\Gamma(-\nu-z)}{\Gamma(1-z)}\right\}$$

for *all* z.

We conclude this section by noting an important feature of these results from the point of view of practical application. Suppose we have a function f for which the Mellin transform is not known in closed form and for which the value $w_0 = M[f;z_0]$ is desired. If z_0 is in the strip of analyticity of $M[f;z]$, we can calculate w_0 directly on a computer. If z_0 is not in the strip of analyticity of $M[f;z]$ or, alternatively, $M[f;z]$ exists only in a generalized sense, then we first perform the decomposition of f as indicated in the relevant lemma of this section and compute the Mellin transforms of each of the decomposed functions, again by computer, if necessary. We claim that, in this sense, the results of this section are *constructive*.

4.4. Asymptotic Expansions for Real λ

The theory of the Mellin transform outlined in the preceding sections will now be used to systematically develop a method for the asymptotic analysis of

$$H[f;\lambda] = \int_0^\infty h(\lambda t) f(t)\, dt, \tag{4.4.1}$$

the h-transform of $f(t)$. In Section 4.7 we shall allow λ to be complex, but here we assume λ real and seek asymptotic expansions of $H[f;\lambda]$ in the limit $\lambda \to \infty$. The limit $\lambda \to 0+$ will be considered in Section 4.6. Although we shall require below that the integrand functions $f(t)$ and $h(t)$ satisfy certain conditions, for the present we merely assume that these functions are locally integrable on the open interval $(0,\infty)$ and that $H[f;\lambda]$ exists for sufficiently large λ.

Before proceeding with the development of the technique we observe that upon setting $\lambda = \varepsilon^{-1}$ and $\lambda t = \tau$ in (4.4.1) we obtain

$$H[f;\varepsilon^{-1}] = \varepsilon \int_0^\infty f(\varepsilon\tau)\, h(\tau)\, d\tau = \varepsilon\, F[h;\varepsilon] \tag{4.4.2}$$

so that the h-transform of f equals ε times the f-transform of h. From this last identity we can conclude that if, in the development of an asymptotic technique to determine the large λ behavior of $H[f;\lambda]$, the conditions placed on f and h allow us to interchange their roles, then the same technique can also be made to yield the asymptotic expansion of $H[f;\lambda]$ for small λ. Therefore, we shall concentrate on the limit $\lambda \to \infty$ and shall find it a simple matter to obtain analogous results when $\lambda \to 0+$.

Let us first suppose that $h(t)$ and $f(t)$ have Mellin transforms which are initially holomorphic in the vertical strips $\alpha < \mathrm{Re}(z) < \beta$ and $\gamma < \mathrm{Re}(z) < \delta$, respectively. We also suppose that the Parseval formula

$$\begin{aligned} H[f;\lambda] &= \frac{1}{2\pi i}\int_{r-i\infty}^{r+i\infty} \lambda^{-z}\, M[h;z]\, M[f;1-z]\, dz \\ &= \frac{1}{2\pi i}\int_{r-i\infty}^{r+i\infty} \lambda^{-z}\, G(z)\, dz \end{aligned} \tag{4.4.3}$$

is valid. (Sufficient conditions for these suppositions to be true are given in Section 4.2.) The constant r, of course, is such that $\mathrm{Re}(z) = r$ lies in the common strip of analyticity of the Mellin transforms $M[h;z]$ and $M[f;1-z]$. Also, in deriving (4.4.3) we have used the fact that

$$M[h(\lambda t);z] = \int_0^\infty h(\lambda t)\, t^{z-1}\, dt = \lambda^{-z} \int_0^\infty h(\tau)\, \tau^{z-1}\, d\tau$$
$$= \lambda^{-z} M[h;z]. \qquad (4.4.4)$$

Although the actual procedure for obtaining the asymptotic expansion of $H[f;\lambda]$ for $\lambda \to \infty$ is quite straightforward, we shall first describe it in formal terms and shall then proceed to rigorously establish the validity of the formal results. Thus, suppose that $G(z)$ in (4.4.3) is actually defined in the right half-plane $\text{Re}(z) \geq r$ as a meromorphic function at worst. Also, suppose that the behavior of $G(z)$ is such that we can displace the contour of integration to the right, say to the vertical line $\text{Re}(z) = R > r$ and apply Cauchy's integral theorem to conclude

$$H[f;\lambda] = -\sum_{r < \text{Re}(z) < R} \text{res}\{\lambda^{-z} G(z)\} + \frac{1}{2\pi i} \int_{R-i\infty}^{R+i\infty} \lambda^{-z} G(z)\, dz. \qquad (4.4.5)$$

It is assumed here that no poles of $G(z)$ lie on $\text{Re}(z) = R$.

If the above process is legitimate, then it is reasonable to expect that

$$\frac{1}{2\pi i} \int_{R-i\infty}^{R+i\infty} \lambda^{-z} G(z)\, dz = \frac{1}{2\pi} \int_{-\infty}^{\infty} \lambda^{-R-iy} G(R+iy)\, dy = 0(\lambda^{-R}) \qquad (4.4.6)$$

as $\lambda \to \infty$, and hence that the sum of residues in (4.4.5) represents a finite asymptotic expansion of $H[f;\lambda]$ in this limit. Of course, if we can allow R to go to $+\infty$, then an infinite expansion would be obtained.

Our immediate objective is to rigorize the process just described. Let us begin, however, by outlining what has to be shown.

(1) The function $G(z) = M[h;z]\, M[f;1-z]$ is defined as a holomorphic function only in the strip $\max(\alpha, 1-\delta) < \text{Re}(z) < \min(\beta, 1-\gamma)$. In particular, with no further information, it is not known to be a meromorphic function in the region $\text{Re}(z) \geq \min(\beta, 1-\gamma)$. Our first problem then is to continue $G(z)$ into this half-plane as a meromorphic function at worst. We note that this, in turn, requires that $M[h;z]$ be so continued into the *right* half-plane $\text{Re}(z) \geq \beta$, while $M[f;z]$ must be continued into the *left* half-plane $\text{Re}(z) \leq \gamma$.

(2) Assuming that the analytic continuation has been accomplished our next task will be to justify the displacement of the contour of integration in (4.4.3) and, in particular, to establish the validity of (4.4.5).

(3) Finally, to establish the asymptotic nature of (4.4.5) we must show that the error estimate holds and we must examine the behavior of the residue terms as $\lambda \to \infty$.

To accomplish the necessary analytic continuations we shall make use of Lemmas 4.3.1 to 4.3.6. Indeed, we first assume that, as $t \to +\infty$,

$$h(t) \sim \exp(-dt^{\nu}) \sum_{m=0}^{\infty} \sum_{n=0}^{N(m)} c_{mn}\, t^{-r_m} (\log t)^n.^6 \tag{4.4.7}$$

Here $\operatorname{Re}(d) \geq 0$, $\nu > 0$, $\operatorname{Re}(r_m) \uparrow \infty$, and $N(m)$ is finite for each m. Then by Lemmas 4.3.1 to 4.3.3 we can conclude that $M[h;z]$ can be analytically continued into $\operatorname{Re}(z) \geq \beta$ as a meromorphic function at worst. Moreover, poles arise only when $d = 0$, in which event they are located at the points $z = r_m$, $m = 0, 1, \ldots$. The singular part of the Laurent expansion of the continued $M[h;z]$ about $z = r_m$ is given by (4.3.12), namely

$$\sum_{n=0}^{N(m)} \frac{c_{mn}\, n!\, (-1)^{n+1}}{(z - r_m)^{n+1}}. \tag{4.4.8}$$

In order to continue $M[f;z]$ into the region $\operatorname{Re}(z) \leq \gamma$ we assume that, as $t \to 0+$,

$$f \sim \exp(-qt^{-\mu}) \sum_{m=0}^{\infty} \sum_{n=0}^{\bar{N}(m)} p_{mn}\, t^{a_m} (\log t)^n. \tag{4.4.9}$$

Here $\operatorname{Re}(q) \geq 0$, $\mu > 0$, $\operatorname{Re}(a_m) \uparrow \infty$, and $\bar{N}(m)$ is finite for each m. Now, by Lemmas 4.3.4 to 4.3.6, $M[f;z]$ can be analytically continued into $\operatorname{Re}(z) \leq \gamma$ as a meromorphic function at worst. Of course, we are interested in the poles, if any, of $M[f;1-z]$ in the region $\operatorname{Re} z \geq 1 - \gamma$. Such poles arise only when $q = 0$, in which event they are located at the points $z = a_m + 1$, $m = 0, 1, 2, \ldots$. The singular part of the Laurent expansion of $M[f;1-z]$ about $z = a_m + 1$ is, according to (4.3.17), given by

$$-\sum_{n=0}^{\bar{N}(m)} \frac{p_{mn}\, n!}{(z - a_m - 1)^{n+1}}. \tag{4.4.10}$$

Thus we may return to (4.4.3) and can, under the assumptions (4.4.7) and (4.4.9), consider $G(z)$ to be meromorphic in $\operatorname{Re}(z) \geq r$. We would now like to conclude that (4.4.5) follows from a simple application of Cauchy's integral theorem. A sufficient condition for this to be so is readily seen to be

$$\lim_{|y| \to \infty} G(x + iy) = 0 \tag{4.4.11}$$

for all x in the interval $[r,R]$. In Sections 4.3 and 4.4 we have discussed the behavior of Mellin transforms as $|z| \to \infty$ along vertical lines. Here we shall merely assume that f and h are such that (4.4.11) holds for $x \,\varepsilon\, [r,R]$.

In order to establish that (4.4.6) is valid, we shall further assume that

$$\int_{-\infty}^{\infty} |G(R + iy)|\, dy < \infty \tag{4.4.12}$$

[6] More generally, we can allow the asymptotic expansion of h to be a finite linear combination of forms such as (4.4.7). It will be clear how to alter the results below to allow for this additional generality.

since in that event

$$\left|\int_{R-i\infty}^{R+i\infty} \lambda^{-z}\, G(z)\, dz\right| \le \lambda^{-R} \int_{-\infty}^{\infty} |G(R+iy)|\, dy = O(\lambda^{-R}).^{7} \qquad (4.4.13)$$

As we shall see, the above assumptions are sufficient for us to show that (4.4.5) yields an asymptotic expansion of $I(\lambda)$ as $\lambda \to \infty$. Conditions (4.4.11) and (4.4.12), however, are not placed directly on the functions f and h and hence are not easily verified. This is aggravated by the fact that they are placed on the function $G(z)$ in a region into which it has been analytically continued. In a problem where the continuation has been accomplished by inspection, these conditions can often be directly verified. Nevertheless, it would be convenient to have explicit conditions on f and h which, when satisfied, would in turn ensure the satisfaction of (4.4.11) and (4.4.12). Although such conditions can be obtained, we delay deriving them until later sections where they become essential to the development of the theory.

Below we shall describe in detail the terms which appear in the residue series (4.4.5). Now, however, we wish merely to consider their qualitative behavior as $\lambda \to \infty$. Under the assumed asymptotic forms (4.4.7) and (4.4.9), the points in $\mathrm{Re}(z) \ge \min(\beta, 1-\gamma)$ at which poles of $G(z)$ occur are precisely determined. Moreover, the abscissas of those points can be arranged in a sequence $\{\alpha_j\}$ in such a manner that $\alpha_{j+1} > \alpha_j$, $j = 0, 1, 2, \ldots$. We observe that the set $\{\alpha_j\}$ has no finite accumulation point and that for each j there are at most two poles of G along the line $\mathrm{Re}(z) = \alpha_j$.

Let us denote the higher order of the poles of G on $\mathrm{Re}(z) = \alpha_j$ by $n_j + 1$. We now introduce the sequence

$$\Phi_{j,m}(\lambda) = \{\lambda^{-\alpha_j} (\log \lambda)^{n_j - m}\}, \qquad m = 0, 1, \ldots, n_j, \qquad j = 0, 1, 2, \ldots, \qquad (4.4.14)$$

where $\Phi_{i,k}(\lambda)$ lies to the right of $\Phi_{j,m}(\lambda)$ if either $i > j$ or $i = j$ and $k > m$. It is clear that $\{\Phi_{j,m}\}$ is an asymptotic sequence as $\lambda \to \infty$. Finally, we note that the residue of $\lambda^{-z}\, G(z)$ evaluated at the point $z = \alpha_j + iy_j$, where G has a pole of order $n_j + 1$, is a linear combination of the terms $\lambda^{-\alpha_j} (\log \lambda)^{n_j}$, $\lambda^{-\alpha_j} (\log \lambda)^{n_j - 1}, \ldots, \lambda^{-\alpha_j}$.

Upon combining the above results, we find that we have proven the following.

THEOREM 4.4. Let $h(t)$ and $f(t)$ be locally integrable functions on $(0,\infty)$ having asymptotic forms (4.4.7) and (4.4.9), respectively. Let $M[h;z]$ and $M[f;z]$ be holomorphic in the respective strips $\alpha < \mathrm{Re}(z) < \beta$, $\gamma < \mathrm{Re}(z) < \delta$. Suppose that these strips overlap and that the Parseval formula (4.4.3) holds. If conditions (4.4.11) and (4.4.12) are satisfied, then

$$H[f;\lambda] \sim -\sum_{r < \mathrm{Re}(z) < R} \mathrm{res}(\lambda^{-z}\, M[h;z]\, M[f;1-z]) \qquad (4.4.15)$$

[7] Actually, the assumption $G(R+iy) \in L(-\infty < y < \infty)$ is stronger than necessary because G goes to zero as $|y| \to \infty$ and λ^{-iy} is an oscillatory factor. Indeed, if we knew that the Fourier transform of $G(R+iy)$ existed and was bounded at $\pm\infty$, then we could drop (4.4.12).

represents a finite asymptotic expansion of $H[f;\lambda]$ as $\lambda \to \infty$ with respect to the asymptotic sequence (4.4.14) with error $0(\lambda^{-R})$. In (4.4.15) $\max(\alpha, 1-\delta) < r < \min(\beta, 1-\gamma)$. Furthermore, if the above assumptions hold for arbitrarily large R, then (4.4.15) yields an infinite asymptotic expansion of $H[f;\lambda]$ with respect to the sequence (4.4.14).[8]

Before considering examples to illustrate the use of this theorem, let us express the residue series (4.4.15) in more explicit terms. The results are conveniently separated into four distinct cases depending on the constants d and q in the asymptotic forms (4.4.7) and (4.4.9). In each case, we shall assume that the hypotheses of Theorem 4.4 hold for arbitrarily large R.

Case I. If in (4.4.7) $d \neq 0$ while in (4.4.9) $q \neq 0$, then as $\lambda \to \infty$,

$$H[f;\lambda] = o(\lambda^{-R}) \tag{4.4.16}$$

for any real R. This follows from the fact that, in this case, $G(z)$ has no poles in the right half-plane $\text{Re}(z) \geq r$.

Case II. If in (4.4.7) $d \neq 0$ while in (4.4.9) $q = 0$, then any residues in (4.4.15) must arise from poles of $M[f;1-z]$. It then follows from (4.4.10), (4.4.15), and the residue theorem that, as $\lambda \to \infty$,

$$H[f;\lambda] \sim \sum_{m=0}^{\infty} \lambda^{-1-a_m} \sum_{n=0}^{\bar{N}(m)} p_{mn} \sum_{j=0}^{n} \binom{n}{j} (-\log \lambda)^j M^{(n-j)}[h;z]\Big|_{z=1+a_m}, \tag{4.4.17}$$

where $M^{(n)}[h;z] = (d/dz)^n M[h;z]$. In the special case when there are no logarithmic terms in (4.4.9) so that $p_{mn} = 0$ for $n \geq 1$, (4.4.17) simplifies to

$$H[f;\lambda] \sim \sum_{m=0}^{\infty} \lambda^{-1-a_m} p_{m0} M[h;1+a_m]. \tag{4.4.18}$$

Case III. If in (4.4.7) $d = 0$ while in (4.4.9) $q \neq 0$, then any residues in (4.4.15) must arise from poles of $M[h;z]$. Thus we have from (4.4.8) and (4.4.15) that, as $\lambda \to \infty$,

$$H[f;\lambda] \sim \sum_{m=0}^{\infty} \lambda^{-r_m} \sum_{n=0}^{N(m)} c_{mn} \sum_{j=0}^{n} \binom{n}{j} (\log \lambda)^j M^{(n-j)}[f;z]\Big|_{z=1-r_m}, \tag{4.4.19}$$

which, in the special case $c_{mn} = 0$ for $n \geq 1$, reduces to

$$H[f;\lambda] \sim \sum_{m=0}^{\infty} \lambda^{-r_m} c_{m0} M[f;1-r_m]. \tag{4.4.20}$$

Case IV. If in (4.4.7) $d = 0$ while in (4.4.9) $q = 0$, then the residues arise from

[8] We note that we have also shown that (4.4.15) yields a generalized asymptotic expansion of $H[f;\lambda]$ with respect to the auxiliary asymptotic sequence $\{\lambda^{-\alpha_j}\}$.

poles of both $M[h;z]$ and $M[f;1-z]$. Now, however, there are two subcases that must be considered.

(1) $r_m \neq a_n + 1$ for any pair of nonnegative integers m and n. Then the poles of $M[f;1-z]$ are distinct from those of $M[h;z]$. The result in this subcase is simply the sum of the expansions (4.4.17) and (4.4.19). In particular, if no logarithmic terms occur in the asymptotic forms (4.4.7) and (4.4.9), then

$$H[f;\lambda] \sim \sum_{m=0}^{\infty} \lambda^{-r_m} c_{m0} M[f;1-r_m] + \sum_{m=0}^{\infty} \lambda^{-1-a_m} p_{m0} M[h;1+a_m]. \tag{4.4.21}$$

(2) $r_m = a_n + 1$ for one or more pairs of nonnegative integers n, m. Now poles of $M[h;z]$ will coincide with poles of $M[f;1-z]$ so that logarithmic terms will appear in (4.4.15) even when no logarithmic terms appear in either (4.4.7) or (4.4.9). We leave the derivation of a formula for this general case to Exercise 4.16. Suppose, however, as an example, we assume that $c_{mn} = p_{mn} = 0$ for $n \geq 1$, $r_0 = a_0 + 1$ but $r_m \neq a_n + 1$ otherwise. Then we find that, as $\lambda \to \infty$,

$$\begin{aligned} H[f;\lambda] \sim{} & \lambda^{-r_0} \log \lambda \, c_{00} \, p_{00} \\ & + \lambda^{-r_0} \lim_{z \to r_0} \frac{d}{dz} \{(z - r_0)(p_{00} M[h;z] + c_{00} M[f;1-z])\} \\ & + \sum_{m=1}^{\infty} \lambda^{-1-a_m} p_{m0} M[h;1+a_m] + \sum_{m=1}^{\infty} \lambda^{-r_m} c_{m0} M[f;1-r_m]. \end{aligned} \tag{4.4.22}$$

When we consider the expansions (4.4.17) through (4.4.22) it becomes apparent that there is one feature common to all of them. Each sum in these expansions consists of terms that involve either the constants c_{mn} times derivatives of the Mellin transform of f evaluated at some point or the constants p_{mn} times derivatives of the Mellin transform of h evaluated at some point. The constants c_{mn} reflect the *local* behavior of h at $t = \infty$, while the constants p_{mn} reflect the local behavior of f at $t = 0$. The Mellin transforms $M[f;1-z]$ and $M[h;z]$ involve, respectively, the values of f and h for all $t \,\varepsilon\, [0,\infty)$ and hence reflect the *global* behavior of these functions. Thus, each sum in the above expansions reflects local properties of either f or h and global properties of the other. In subcase (2) of Case IV, however, where coalescences of poles of the two Mellin transforms occur, there are terms which involve only the local behavior of the functions f and h.

Let us now illustrate our results by considering several examples.

EXAMPLE 4.4.1. Let $h(t) = e^{-t}$ so that $H[f;\lambda]$ is simply the Laplace transform of f. If in (4.4.9) $q = 0$, then Case II is relevant to this example. Furthermore, we have

$$M[e^{-t};z] = \int_0^\infty t^{z-1} e^{-t} \, dt = \Gamma(z) \tag{4.4.23}$$

and hence by (4.4.17)

$$\int_0^\infty e^{-\lambda t} f(t)\, dt$$
$$\sim \sum_{m=0}^{\infty} \lambda^{-1-a_m} \sum_{n=0}^{\bar{N}(m)} p_{mn} \sum_{j=0}^{n} \binom{n}{j} (-\log \lambda)^j \left(\frac{d}{dz}\right)^{n-j} \Gamma(z) \bigg|_{z=1+a_m} \tag{4.4.24}$$

Here we have assumed that the hypotheses of Theorem 4.4 hold. From (3.2.41) we have that

$$\Gamma(x+iy) = 0(|y|^{x-\frac{1}{2}}\, e^{-(\pi/2)|y|}), \qquad |y| \to \infty.$$

Because Lemmas 4.3.4 to 4.3.6 show that $M[f;1-x-iy]$ can at worst grow algebraically as $|y| \to \infty$, we can conclude that (4.4.11) and (4.4.12) hold for all positive R. Finally, if $p_{mn} = 0$ for $n \geq 1$, then we have

$$f \sim \sum_{m=0}^{\infty} t^{a_m}\, p_{m0} \tag{4.4.25}$$

as $t \to 0+$ and (4.4.24) becomes

$$\int_0^\infty e^{-\lambda t} f(t)\, dt \sim \sum_{m=0}^{\infty} \lambda^{-1-a_m}\, p_{m0}\, \Gamma(1+a_m) \tag{4.4.26}$$

which will be recognized as the result predicted by Watson's lemma.

EXAMPLE 4.4.2. Now suppose that $h(t) = (1+t)^{-1}$ in which event

$$H[f;\lambda] = \int_0^\infty \frac{f(t)}{1+\lambda t}\, dt \tag{4.4.27}$$

is a constant multiple of the Stieltjes transform of f. In addition, we assume that f has the simple asymptotic form (4.4.25). We find from the Appendix

$$M[(1+t)^{-1};z] = \int_0^\infty t^{z-1}(1+t)^{-1}\, dt = \frac{\pi}{\sin \pi z} \tag{4.4.28}$$

which has simple poles at the integers $z = 1, 2, \ldots$ as predicted by Lemma 4.3.3. We note that, as $|y| \to \infty$, $\pi/\sin\pi(x+iy) = 0(e^{-\pi|y|})$ for *all* x. If we assume that $a_m \neq 0, 1, 2, \ldots$, then subcase (1) of Case IV holds and hence from (4.4.21) we have

$$\int_0^\infty \frac{f(t)}{1+\lambda t}\, dt \sim -\sum_{m=0}^{\infty} \lambda^{-1-a_m}\, p_{m0} \frac{\pi}{\sin \pi a_m}$$
$$+ \sum_{m=0}^{\infty} (-1)^m \lambda^{-1-m} M[f;-m]. \tag{4.4.29}$$

Thus we see that, in the case of an algebraically decaying kernel, there are, in general, two sums in the asymptotic expansion of $H[f;\lambda]$. Upon comparing this with the result of Example 4.4.1, where the kernel was exponentially decaying, we see that the first sum in (4.4.29) is analogous to Watson's lemma but that the second sum has no counterpart in the expansion of the Laplace transform.

Suppose that in (4.4.27) we specialize to the case where

$$f(t) = \frac{1}{t} e^{-1/t}. \tag{4.4.30}$$

Then, according to the result for Case III above, the first sum in (4.4.29) vanishes. Of course, from the explicit formula

$$\begin{aligned} M[f;1-z] &= \int_0^\infty t^{-z-1} e^{-1/t}\, dt = \int_0^\infty t^{z-1} e^{-t}\, dt \\ &= \Gamma(z) \end{aligned} \tag{4.4.31}$$

we have that $M[f;1-z]$ is holomorphic for $\mathrm{Re}(z) > 0$. Now it follows from (4.4.29) that

$$\int_0^\infty \frac{e^{-1/t}}{t(1+\lambda t)}\, dt \sim \sum_{m=0}^\infty \frac{(-1)^m \Gamma(1+m)}{\lambda^{1+m}} = \sum_{m=0}^\infty \frac{(-1)^m m!}{\lambda^{1+m}}. \tag{4.4.32}$$

It is of interest to observe that

$$\int_0^\infty \frac{e^{-1/t}}{t(1+\lambda t)}\, dt = \int_0^\infty \frac{e^{-t}}{(\lambda + t)}\, dt = e^\lambda \int_\lambda^\infty \frac{e^{-\tau}}{\tau}\, d\tau \tag{4.4.33}$$

and (4.4.32) agrees with the expansion (1.1.6) obtained via successive integrations by parts in this last integral.

Finally, suppose that in (4.4.27)

$$f(t) \sim f(0), \qquad t \to 0+. \tag{4.4.34}$$

Then $a_0 = 0$ and hence subcase (2) of Case IV is applicable. Indeed, we find from (4.4.22) that now

$$\int_0^\infty \frac{f(t)}{1+\lambda t}\, dt \sim \frac{\log \lambda}{\lambda} f(0). \tag{4.4.35}$$

EXAMPLE 4.4.3. If $h(t) = t^{1/2} J_\nu(t)$, where J_ν is the Bessel function of the first kind of order $\nu > -\frac{1}{2}$, then

$$H[f;\lambda] = \int_0^\infty (\lambda t)^{1/2} J_\nu(\lambda t) f(t)\, dt \tag{4.4.36}$$

is the Hankel transform of f. Let us assume that f has the asymptotic form (4.4.25).

From the table in the Appendix we have

$$M[h;z] = \frac{2^{z-\frac{1}{2}} \Gamma([2\nu + 2z + 1]/4)}{\Gamma([2\nu - 2z + 3]/4)}. \tag{4.4.37}$$

We have from the result of Exercise 3.15 that $J_\nu(t)$ is oscillatory as $t \to \infty$. Thus, it is not surprising that (4.4.37) is holomorphic in the right half-plane

$\operatorname{Re} z > -\nu - \frac{1}{2}$. Furthermore, we have from the estimate (3.2.41) that

$$M[t^{1/2} J_\nu(t); x+iy] = 0(|y|^{x-\frac{1}{2}}), \qquad |y| \to \infty. \tag{4.4.38}$$

Thus, if $M[f; 1-x-iy] = o(|y|^{-n})$, $|y| \to \infty$ for all n, then we obtain from (4.4.18)

$$\int_0^\infty (\lambda t)^{1/2} J_\nu(\lambda t) f(t)\, dt \sim \sum_{m=0}^\infty \left(\frac{2}{\lambda}\right)^{\frac{1}{2}+a_m} p_{m0} \frac{\Gamma([2\nu+2a_m+3]/4)}{\Gamma([2\nu-2a_m+1]/4)}. \tag{4.4.39}$$

4.5. Asymptotic Expansions for Real λ : Continuation

As discussed in Section 4.3 there is a wide class of functions, including all polynomials, whose elements are not Mellin transformable in the usual sense. However, we have in that section defined a generalized Mellin transform which requires only that the function in question be locally integrable on $(0,\infty)$ and have appropriate asymptotic forms at $0+$ and $+\infty$. We shall now extend the results of the last section to include the cases where $M[h;z]$ or $M[f;z]$ exists only in this generalized sense.

It would be convenient if we could merely say that all of the results of the previous section hold when $M[h;z]$ and $M[f;z]$ are interpreted as generalized Mellin transforms. Unfortunately things are not quite so simple because the Parseval formula, which is crucial to our method, is not immediately applicable. We shall now determine what amounts to a generalized Parseval formula and then use it to derive the desired asymptotic expansions.

Thus let us again consider (4.4.1) in the limit $\lambda \to \infty$. As before, we assume that h and f are both locally integrable on $(0,\infty)$ and set

$$\begin{aligned} \alpha &= \inf\{\alpha^*;\ h = O(t^{-\alpha^*}),\ t \to 0+\}, \\ \beta &= \sup\{\beta^*;\ h = O(t^{-\beta^*}),\ t \to +\infty\}, \\ \gamma &= \inf\{\gamma^*;\ f = O(t^{-\gamma^*}),\ t \to 0+\}, \\ \delta &= \sup\{\delta^*;\ f = O(t^{-\delta^*}),\ t \to +\infty\}. \end{aligned} \tag{4.5.1}$$

Of course, if $\alpha < \beta$ and $\gamma < \delta$, then $M[h;z]$ and $M[f;z]$ are holomorphic in $\alpha < \operatorname{Re}(z) < \beta$ and $\gamma < \operatorname{Re}(z) < \delta$, respectively. Here, however, we no longer require that these inequalities hold.

Although it is not necessary, we shall assume that $H[f;1]$ is absolutely convergent which implies

$$\alpha + \gamma < 1, \qquad \beta + \delta > 1. \tag{4.5.2}$$

We observe that (4.5.2) in turn implies

$$(\beta - \alpha) + (\delta - \gamma) > 0 \tag{4.5.3}$$

and therefore at least one of the inequalities $\alpha < \beta$, $\gamma < \delta$ must hold. For most purposes, because we look upon (4.4.1) with the kernel h fixed, as a "black box" into which functions f of a given class are fed, it is convenient to assume $\alpha < \beta$ so that $M[h;z]$ exists in the ordinary sense. This allows for greater generality in the functions f.

From the results of the previous two sections we know that we shall need to analytically continue $M[h;z]$ into a right half-plane and to define $M[f;z]$ in some left half-plane. Moreover, as Lemmas 4.3.1 to 4.3.6 suggest, to accomplish this, it is sufficient to assume that, as $t \to \infty$, $h(t)$ has the asymptotic form (4.4.7) and that, as $t \to 0+$, $f(t)$ has the asymptotic form (4.4.9).

Let us now introduce the truncated functions f_1 and f_2 defined by

$$f_1(t) = \begin{cases} f(t), & t \,\varepsilon\, [0,1), \\ 0, & t \,\varepsilon\, [1,\infty), \end{cases} \qquad f_2(t) = \begin{cases} 0, & t \,\varepsilon\, [0,1), \\ f(t), & t \,\varepsilon\, [1,\infty), \end{cases} \tag{4.5.4}$$

and set

$$H[f_j;\lambda] = I_j(\lambda) = \int_0^\infty h(\lambda t) f_j(t)\, dt, \qquad j = 1, 2. \tag{4.5.5}$$

We then have the exact result

$$H[f;\lambda] = I_1(\lambda) + I_2(\lambda). \tag{4.5.6}$$

Suppose now we define the functions

$$G_j(z) = M[h;z]\, M[f_j;1-z], \qquad j = 1, 2, \tag{4.5.7}$$

and denote by D_j the domain of analyticity of G_j. Then, under the assumptions (4.5.1), we have that D_j, $j = 1, 2$, are the vertical strips defined by

$$\begin{aligned} D_1:& \qquad \alpha < \operatorname{Re}(z) < \min(1-\gamma,\beta), \\ D_2:& \qquad \max(1-\delta,\alpha) < \operatorname{Re}(z) < \beta.^9 \end{aligned} \tag{4.5.8}$$

Furthermore, upon assuming the asymptotic forms (4.4.7) and (4.4.9) we find that each G_j can be analytically continued into a right half-plane as a meromorphic function at worst.

We observe that in the region $\operatorname{Re}(z) > 1 - \delta$, the generalized Mellin transform $M[f;1-z]$ exists and is given by

$$M[f;1-z] = M[f_1;1-z] + M[f_2;1-z]. \tag{4.5.9}$$

Hence, in this region we can define

$$\begin{aligned} G(z) &= M[h;z]\, M[f;1-z] \\ &= G_1(z) + G_2(z). \end{aligned} \tag{4.5.10}$$

Let us now suppose that for $j = 1, 2$ there exists a real number r_j for which

[9] Note that (4.5.2) and the assumption $\alpha < \beta$ guarantee that D_1 and D_2 are not empty.

the ordinary Parseval formula

$$I_j(\lambda) = \frac{1}{2\pi i} \int_{r_j - i\infty}^{r_j + i\infty} \lambda^{-z} \, G_j(z) \, dz \tag{4.5.11}$$

is valid. Here, of course, $z = r_j$ lies in D_j. Then, by (4.5.6),

$$H[f;\lambda] = \frac{1}{2\pi i} \sum_{j=1}^{2} \int_{r_j - i\infty}^{r_j + i\infty} \lambda^{-z} \, G_j(z) \, dz. \tag{4.5.12}$$

Suppose, further, that the following conditions hold:

(1) $\lim_{|y| \to \infty} G_1(x + iy) = 0, \qquad r_1 \le x \le r_2,$

(2) $G_j(r_2 + iy)$ lies in $L(-\infty < y < \infty), \qquad j = 1, 2.$

Then, by displacing the contour of integration in the integral I_1 until it coincides with $\mathrm{Re}(z) = r_2$ we obtain by Cauchy's integral theorem the *generalized Parseval formula*

$$H[f;\lambda] = \sum_{r_1 < \mathrm{Re}(z) < r_2} \mathrm{res}(-\lambda^{-z} \, G_1(z)) + \frac{1}{2\pi i} \int_{r_2 - i\infty}^{r_2 + i\infty} \lambda^{-z} \, G(z) \, dz. \tag{4.5.13}$$

Moreover, by condition (2) we have that the last integral is $0(\lambda^{-r_2})$ as $\lambda \to \infty$.[10]

By deriving (4.5.13) we have essentially reduced our problem to the one already treated in Section 4.4. Although (4.5.13) is itself a finite asymptotic expansion of $H[f;\lambda]$, we wish to investigate the possibility of obtaining further terms and perhaps an infinite expansion. To accomplish this we need only justify moving the contour of integration in (4.5.13) still further to the right. This is done in the following.

THEOREM 4.5. Let f and h satisfy the conditions leading to the generalized Parseval formula (4.5.13). Furthermore, suppose there exists a real number $R > r_2$ such that $G(R + iy)$ lies in $L(-\infty < y < \infty)$ and

$$\lim_{|y| \to \infty} G(x + iy) = 0, \qquad r_2 \le x \le R. \tag{4.5.14}$$

Then

$$H[f;\lambda] \sim \sum_{r_1 < \mathrm{Re}(z) < r_2} \mathrm{res}(-\lambda^{-z} \, G_1(z)) + \sum_{r_2 < \mathrm{Re}(z) < R} \mathrm{res}(-\lambda^{-z} \, G(z)) \tag{4.5.15}$$

represents a finite asymptotic expansion of $H[f;\lambda]$ as $\lambda \to \infty$. Moreover, if the above hypotheses hold for arbitrarily large R, then (4.5.15) yields an infinite asymptotic expansion upon setting $R = \infty$.[11]

[10] Note that if $r_1 > r_2$, then $M[f;1 - z]$ exists in the ordinary sense and the first sum in (4.5.13) is empty.

[11] The underlying asymptotic sequence is of the form (4.4.14), that is, it involves powers of λ^{-1} multiplied by nonnegative integer powers of $\log \lambda$.

PROOF. The proof follows from the Cauchy integral theorem and the estimate

$$\left|\int_{R-i\infty}^{R+i\infty} \lambda^{-z}\, G(z)\, dz\right| = 0(\lambda^{-R}), \qquad \lambda \to \infty .$$

Remark. It appears that it is unnecessary to define $G(z)$ in the region $\operatorname{Re}(z) > 1 - \delta$ because the final result could be expressed as two residue series, one involving residues of $G_1(z)\lambda^{-z}$ and the other involving residues of $G_2(z)\lambda^{-z}$. Although this is quite true, we point out that it often occurs that the function $G(z)$ has "nicer" properties than either $G_1(z)$ or $G_2(z)$ and is indeed, in some cases, a simpler function. As a trivial example of this last point, suppose that $f(t) = 1$, $0 \le t < \infty$. Then we readily find that

$$G_1(z) = M[h;z](1-z)^{-1}, \qquad G_2(z) = -M[h;z](1-z)^{-1},$$

and

$$G(z) = 0 .$$

We wish now to consider various special cases which, as in Section 4.4, are determined by the constants d and q that appear in the asymptotic forms (4.4.7) and (4.4.9). We shall not express the residue terms explicitly here because they have been so expressed in the cases treated after Theorem 4.4. Also we assume that Theorem 4.5 holds for arbitrarily large R.

Case I. If in (4.4.7) $d \neq 0$ while in (4.4.9) $q \neq 0$, then

$$H[f;\lambda] = o(\lambda^{-R}) \tag{4.5.16}$$

for all R. This follows from the fact that the analytic continuations of $M[h;z]$ and $M[f_1;1-z]$ are both holomorphic in the relevant right half-planes.

Case II. If in (4.4.7) $d = 0$ while in (4.4.9) $q \neq 0$, then $M[f_1;1-z]$ is holomorphic in a right half-plane so that the first sum in (4.5.15) is empty and we have

$$H[f;\lambda] \sim \sum_{r_2 < \operatorname{Re}(z) < \infty} \operatorname{res}(-\lambda^{-z}\, G(z)). \tag{4.5.17}$$

Here all residues that appear must arise from poles of $M[h;z]$ and hence the coefficients in (4.5.17) reflect local properties of h near $+\infty$ and global properties of f. Finally, any logarithms that appear are due to logarithms which occur in the expansion of h at $+\infty$.

Case III. If in (4.4.7) $d \neq 0$ while in (4.4.9) $q = 0$, then $M[h;z]$ is holomorphic in a right half-plane. Thus, all residues must arise from poles of $M[f_1;1-z]$

and (4.5.15) can now be written

$$H[f;\lambda] \sim \sum_{r_1 < \mathrm{Re}(z) < \infty} \mathrm{res}(-\lambda^{-z} G_1(z)). \tag{4.5.18}$$

Here the coefficients reflect local properties of f near $0+$ and global properties of h. Also, any logarithms that appear are due to logarithms that occur in the expansion of f at $0+$.

Case IV. If in (4.4.7) $d = 0$ while in (4.4.9) $q = 0$, then the residues that appear in (4.5.15) arise from poles of both $M[f_1;1-z]$ and $M[h;z]$. Hence, the coefficients reflect local and global properties of both h and f. Furthermore, if in (4.4.7) and (4.4.9) $1 + a_m \neq r_n$ for any m, n and if $c_{mn} = p_{mn} = 0$ for $n \geq 1$, then no logarithms appear in (4.5.15). Alternatively, if poles of $M[f_1;1-z]$ and $M[h;z]$ coincide, then by virtue of this coalescence, (4.5.15) will contain logarithmic terms even when no logarithms appear in either (4.4.7) or (4.4.9).

Before doing an example, we wish to point out that a further generalization can be obtained in Case III when $\mathrm{Re}(d) > 0$. Indeed in that event we have

$$h(t) = O[t^{r_0} (\log t)^{N(0)} \exp(-\mathrm{Re}(d)\, t^{\nu})], \qquad t \to \infty.$$

Now we can allow $f(t)$ to grow quite rapidly at ∞ and still have $H[f;\lambda]$ absolutely convergent. In fact all we need require is that

$$f(t) = O(\exp(rt^{\nu})), \qquad t \to \infty \tag{4.5.19}$$

for some finite r.

To illustrate some of the results of this section, let us consider the following.

EXAMPLE 4.5. Suppose

$$H[f;\lambda] = \int_0^{\infty} \frac{|1-t|^{\nu}\, dt}{(1+\lambda t)^{\rho}}, \tag{4.5.20}$$

where $0 < \mathrm{Re}(\nu) < \rho - 1$. Here $h(t) = (1+t)^{-\rho}$ and we have from the Appendix

$$M[h;z] = \frac{\Gamma(z)\,\Gamma(\rho - z)}{\Gamma(\rho)}. \tag{4.5.21}$$

Thus, we see directly that $M[h;z]$ has simple poles at the points $z = \rho + m$, $m = 0, 1, 2, \ldots$, as predicted by our theory. The function $f = |1-t|^{\nu}$ however has no Mellin transform in the ordinary sense. Nevertheless, we do have from the Appendix

$$M[f_1;1-z] = \frac{\Gamma(\nu+1)\,\Gamma(1-z)}{\Gamma(\nu - z + 2)},$$

$$M[f_2;1-z] = \frac{\Gamma(\nu+1)\,\Gamma(z-\nu-1)}{\Gamma(z)}, \tag{4.5.22}$$

which we note are holomorphic in the disjoint half-planes $\mathrm{Re}(z) < 1$ and $\mathrm{Re}(z) > \mathrm{Re}(1+\nu)$ respectively. The analytic continuation of $M[f_1;1-z]$ into $\mathrm{Re}(z) \geq 1$ has simple poles at the positive integers.

If ρ is not a positive integer, then the resulting expansion does not contain any logarithms. Indeed, from (4.5.15) we find after some calculation that, in this case,

$$\int_0^\infty \frac{|1-t|^\nu\, dt}{(1+\lambda t)} \sim \frac{\Gamma(\nu+1)}{\Gamma(\rho)} \left\{ \sum_{n=0}^\infty \frac{(-1)^n \Gamma(\rho-n-1)}{\lambda^{n+1}\, \Gamma(\nu+n+1)} + \sum_{n=0}^\infty \frac{(-1)^n}{\lambda^{n+\rho}} \frac{\Gamma(n+\rho-\nu-1)}{\Gamma(n+1)} \left[1 + \frac{\sin \pi(\nu-\rho)}{\sin \pi\rho}\right] \right\}. \tag{4.5.23}$$

Furthermore, if ν is a positive integer so that f_1 and f_2 are polynomials, then the first sum in (4.5.23) will be finite because $M[f_1;1-z]$ and $M[f_2;1-z]$ have poles only at $z = 1, 2, \ldots, \nu+1$.

If ρ is a positive integer but ν is not, then poles of $M[f_1;1-z]$ coincide with poles of $M[h;z]$ at $z = \rho + m$, $m = 0, 1, \ldots$. In this case the expansion will contain logarithms and, again after some calculation, we find that (4.5.15) yields

$$\int_0^\infty \frac{|1-t|^\nu\, dt}{(1+\lambda t)^\rho} \sim \frac{\Gamma(\nu+1)}{\Gamma(\rho)} \left\{ \sum_{n=0}^{\rho-2} \frac{(-1)^n\, \Gamma(\rho-n-1)}{\lambda^{n+1}\, \Gamma(\nu-n+1)} + \sum_{n=0}^\infty \frac{(-1)^\rho}{\lambda^{n+\rho}} \left[\frac{-\log\lambda - \psi(n+1) + \Gamma(\nu+2-\rho-n) + \pi\csc(\pi\nu)}{\Gamma(\nu+2-\rho-n)\,\Gamma(n+1)} \right] \right\}. \tag{4.5.24}$$

Here $\psi(z) = \Gamma'(z)/\Gamma(z)$ and is usually referred to as the digamma function. It should be noted that all of the conditions necessary for the application of Theorem 4.5 to this example are satisfied.

4.6. Asymptotic Expansions for Small Real λ

As we have previously indicated, the expansion of

$$H[f;\lambda] = \int_0^\infty h(\lambda t) f(t)\, dt \tag{4.6.1}$$

in the limit $\lambda \to 0+$ is, under certain circumstances, recoverable from our results on the expansion of such transforms in the limit $\lambda \to \infty$. Indeed, upon using (4.4.2), we find that all we need do is determine the expansion of

$$F[h;\varepsilon] = \int_0^\infty f(\varepsilon t)\, h(t)\, dt \tag{4.6.2}$$

as $\varepsilon \to \infty$, multiply the result by ε, and then set $\varepsilon = \lambda^{-1}$.

In order for us to use the results of the previous two sections to obtain the asymptotic expansion of $F[h;\varepsilon]$ as $\varepsilon \to \infty$, the conditions placed there on the functions f and h would have to be interchangeable. This is the case for the conditions of Section 4.4 but, unfortunately, not for those of Section 4.5. In Section 4.5, $M[h;z]$ is assumed to exist in the ordinary sense, whereas $M[f;1-z]$ need exist only in the generalized sense. Because we still wish to assume that $M[h;z]$ exists in the ordinary sense, we cannot make direct use of (4.6.2) and hence shall proceed differently.

Thus, let us consider (4.6.1) as $\lambda \to 0+$. We first suppose that $H[f;1]$ is absolutely convergent so that if (4.5.1) holds, then we must have

$$\alpha + \gamma < 1, \qquad \beta + \delta > 1. \tag{4.6.3}$$

We further suppose that $\alpha < \beta$ and hence $M[h;z]$ is holomorphic in the strip $\alpha < \mathrm{Re}(z) < \beta$.

We now introduce the functions f_1, f_2, G_1, and G_2 defined by (4.5.4) and (4.5.7), and assume that the ordinary Parseval formulas (4.5.11) hold which yield the exact result

$$H[f;\lambda] = \frac{1}{2\pi i} \sum_{j=1}^{2} \int_{r_j - i\infty}^{r_j + i\infty} \lambda^{-z} G_j(z)\, dz. \tag{4.6.4}$$

We now expect to derive the desired asymptotic expansion as a residue series by displacing the contours of integration in (4.6.4) to the *left* and applying Cauchy's integral theorem. This, of course, requires that we analytically continue $G_1(z)$ and $G_2(z)$ into left half-planes as meromorphic functions at worst. Indeed, we shall follow this procedure except first we find it convenient to define

$$K_j(z) = M[h;1-z]\, M[f_j;z] = G_j(1-z), \qquad j = 1, 2, \tag{4.6.5}$$

which are seen to be holomorphic in the vertical strips

$$D_1: \qquad \max(\gamma, 1-\beta) < \mathrm{Re}(z) < 1-\alpha,$$

$$D_2: \qquad 1-\beta < \mathrm{Re}(z) < \min(\delta, 1-\alpha),$$

respectively.

If we replace z by $1-z$ in (4.6.4), then we obtain

$$H[f;\lambda] = \frac{1}{2\pi i} \sum_{j=1}^{2} \int_{\rho_j - i\infty}^{\rho_j + i\infty} \lambda^{z-1} K_j(z)\, dz, \tag{4.6.6}$$

where $\rho_j = 1 - r_j$. Here $z = \rho_j$ lies in D_j. Now our asymptotic expansion will, as in previous sections, be obtained by displacing the contours of integration in (4.6.6) to the right and hence $K_j(z)$, $j = 1, 2$ must be continued into right half-planes.

To accomplish the continuations we assume

$$f(t) \sim \exp(-dt^{\nu}) \sum_{m=0}^{\infty} \sum_{n=0}^{N(m)} c_{mn}\, t^{-r_m} (\log t)^n, \qquad t \to \infty, \tag{4.6.7}$$

$$h(t) \sim \exp(-q\, t^{-\mu}) \sum_{m=0}^{\infty} \sum_{n=0}^{\bar{N}(m)} p_{mn}\, t^{a_m} (\log t)^n, \qquad t \to 0+. \tag{4.6.8}$$

Here the constants d, q, v, μ, r_m, and a_m satisfy the conditions stated below (4.4.7) and (4.4.9). Under these assumptions, we have by Lemmas 4.3.1 to 4.3.6 that $M[h;1-z]$ and $M[f_2;z]$ can be analytically continued into right half-planes as meromorphic functions at worst. Indeed, we have that if in (4.6.7) $d \neq 0$, then $M[f_2;z]$ is holomorphic for $\text{Re}(z) \geq \delta$. However, if $d = 0$, then for $m = 0, 1, 2, \ldots$, $M[f_2;z]$ has a pole at the point $z = r_m$ and a Laurent expansion about this point having singular part

$$\sum_{n=0}^{N(m)} \frac{n!\, c_{mn}\, (-1)^{n+1}}{(z - r_m)^{n+1}}. \tag{4.6.9}$$

Similarly, if $q \neq 0$ in (4.6.8), then $M[h;1-z]$ is holomorphic for $\text{Re}(z) \geq \alpha$, while if $q = 0$, then for $m = 0, 1, 2, \ldots$, $M[h;1-z]$ has a pole at $z = a_m + 1$ and a Laurent expansion about this pole having singular part

$$-\sum_{n=0}^{\bar{N}(m)} \frac{n!\, p_{mn}}{(z - a_m - 1)^{n+1}}. \tag{4.6.10}$$

We note that in the region $\text{Re}(z) \geq \gamma$, $M[f;z]$ exists at least in the generalized sense and hence we can set

$$\begin{aligned} K(z) &= M[h;1-z]\, M[f;z] \\ &= K_1(z) + K_2(z). \end{aligned} \tag{4.6.11}$$

We now state and prove the main expansion theorem.

THEOREM 4.6. Let f and h be such that (4.6.6) holds. Suppose that f and h have the asymptotic expansions (4.6.7) and (4.6.8), respectively, and that $M[h;z]$ is holomorphic for $\alpha < \text{Re}(z) < \beta$. Suppose further that the following conditions hold:

(1) $\lim_{|y| \to \infty} K_2(x + iy) = 0, \qquad \rho_2 \leq x < \rho_1,$

(2) $\lim_{|y| \to \infty} K(x + iy) = 0, \qquad \rho_1 \leq x \leq R,$

(3) $K(R + iy)$ lies in $L(-\infty < y < \infty)$.

Then

$$H[f;\lambda] \sim \sum_{\rho_2 < \text{Re}(z) < \rho_1} \text{res}(-\lambda^{z-1} K_2(z)) + \sum_{\rho_1 \leq \text{Re}(z) < R} \text{res}(-\lambda^{z-1} K(z)) \tag{4.6.12}$$

represents a finite asymptotic expansion of $H[f;\lambda]$ as $\lambda \to 0+$ with error $O(\lambda^{R-1})$. Furthermore, if conditions (2) and (3) hold for arbitrarily large R, then (4.6.12) yields an infinite asymptotic expansion of $H[f;\lambda]$ upon setting $R = \infty$.[12]

[12] As we shall show, the underlying asymptotic sequence involves powers of λ multiplied by nonnegative integer powers of $\log \lambda$.

PROOF. The proof follows from Cauchy's integral theorem and the estimate

$$\left| \int_{R-i\infty}^{R+i\infty} \lambda^{z-1} K(z)\, dz \right| = 0(\lambda^{R-1}), \qquad \lambda \to 0+.$$

Remark. If in (4.6.6) ρ_1 can be taken less than ρ_2, then $M[f;z]$ exists in the ordinary sense and the first sum is empty.

We now wish to consider in detail the residue terms that appear in the expansion (4.6.12). Again we find that there are four distinct and exhaustive cases that depend on the constants d and q in the asymptotic expansions (4.6.7) and (4.6.8). In what follows we shall assume that the results of Theorem 4.6 hold for arbitrarily large R.

Case I. If in (4.6.7) $d \neq 0$ while in (4.6.8) $q \neq 0$, then

$$H[f;\lambda] = o(\lambda^R), \qquad \lambda \to 0+ \tag{4.6.13}$$

for all R.

Case II. If in (4.6.7) $d = 0$ while in (4.6.8) $q \neq 0$, then

$$H[f;\lambda] \sim \sum_{m=0}^{\infty} \lambda^{r_m - 1} \sum_{n=0}^{N(m)} c_{mn} \sum_{j=0}^{n} \binom{n}{j} (-\log \lambda)^j M^{(n-j)}[h;z] \Big|_{z=1-r_m} \tag{4.6.14}$$

as $\lambda \to 0+$. In the special case where $c_{mn} = 0$ for $n \geq 1$, (4.6.14) reduces to

$$H[f;\lambda] \sim \sum_{m=0}^{\infty} \lambda^{r_m - 1} c_{m0} M[h;1-r_m], \qquad \lambda \to 0+. \tag{4.6.15}$$

Case III. If in (4.6.7) $d \neq 0$ while in (4.6.8) $q = 0$, then

$$H[f;\lambda] \sim \sum_{m=0}^{\infty} \lambda^{a_m} \sum_{n=0}^{\bar{N}(m)} p_{mn} \sum_{j=0}^{n} \binom{n}{j} (\log \lambda)^j M^{(n-j)}[f;z] \Big|_{z=1+a_m} \qquad \lambda \to 0+. \tag{4.6.16}$$

In the special case where $p_{mn} = 0$ for $n \geq 1$, (4.6.16) reduces to

$$H[f;\lambda] \sim \sum_{m=0}^{\infty} \lambda^{a_m} p_{m0} M[f;1+a_m], \qquad \lambda \to 0+. \tag{4.6.17}$$

Case IV. If in (4.6.7) $d = 0$ while in (4.6.8) $q = 0$, then we must consider two subcases:

(1) $a_m + 1 \neq r_n$ for any n, m. In this event

$$\begin{aligned} H[f;\lambda] \sim {} & \sum_{m=0}^{\infty} \lambda^{r_m - 1} \sum_{n=0}^{N(m)} c_{mn} \sum_{j=0}^{n} \binom{n}{j} (-\log \lambda)^j M^{(n-j)}[h;z] \Big|_{z=1-r_m} \\ & + \sum_{m=0}^{\infty} \lambda^{a_m} \sum_{n=0}^{\bar{N}(m)} p_{mn} \sum_{j=0}^{n} \binom{n}{j} (\log \lambda)^j M^{(n-j)}[f;z] \Big|_{z=1+a_m}, \qquad \lambda \to 0+. \end{aligned} \tag{4.6.18}$$

(2) $a_m + 1 = r_n$ for one or more pairs m, n. Here we obtain logarithmic terms even in cases where no logarithms appear in either (4.6.7) or (4.6.8). Indeed, for example, suppose that $a_0 + 1 = r_0$ and that $p_{0n} = c_{0n} = 0$ for $n \geq 1$. Then

$$H[f;\lambda] \sim -\lambda^{a_0} \log \lambda \, p_{00} \, c_{00}$$

$$+ \lambda^{a_0} \lim_{z \to a_0 + 1} \frac{d}{dz} \{(z - a_0 - 1)(p_{00} \, M[h;1-z] + c_{00} \, M[f;z])\} + o(\lambda^{a_0}). \tag{4.6.19}$$

EXAMPLE 4.6.1. Here we shall consider the Laplace transform

$$H[f;\lambda] = \int_0^\infty e^{-\lambda t} f(t) \, dt. \tag{4.6.20}$$

We have directly

$$M[h;1-z] = \int_0^\infty e^{-t} t^{-z} \, dt = \Gamma(1-z) \tag{4.6.21}$$

which has simple poles at the positive integers.

Let us assume for simplicity that

$$f(t) \sim \sum_{m=0}^{\infty} c_m \, t^{-r_m}, \qquad t \to \infty. \tag{4.6.22}$$

If no r_m is a positive integer, then by (4.6.18) and (4.6.21) we have

$$\int_0^\infty e^{-\lambda t} f(t) \, dt \sim \sum_{m=0}^{\infty} c_m \, \lambda^{r_m - 1} \, \Gamma(1 - r_m) + \sum_{m=0}^{\infty} \lambda^m \frac{(-1)^m}{m!} M[f;m+1]. \tag{4.6.23}$$

This formula shows that the asymptotic expansion of the Laplace transform at the origin in general involves the global behavior of f through its Mellin transform.

It is of interest to compare the rigorous result (4.6.23) with the formal expansion (4.1.14). We observe that the latter is precisely the second sum in the former when we interpret the nth moment of f as $M[f;n+1]$. The first sum in (4.6.23), however, is absent from (4.1.14) and hence this expansion is, in general, false as was anticipated. On the other hand, if $f(t) = o(t^{-n})$ for all n, then $c_m = 0$ for $m = 0, 1, 2, \ldots$, and the first sum in (4.6.23) vanishes. Moreover, in that event, $M[f;n+1]$ is convergent for all n and does indeed represent the nth moment of f. Thus, when all of the integer moments of f exist, the two expansions (4.1.14) and (4.6.23) are identical and yield the standard expansion by moments of the Laplace transform. We can therefore look upon (4.6.23) as a generalization of the expansion by moments which recovers the latter whenever it is valid.

As a last consideration let us suppose that in (4.6.20)

$$f(t) = \begin{cases} 0, & t \,\varepsilon\, [0,1), \\ \lambda^{-\rho} \, t^{-\rho-1}, & t \,\varepsilon\, [1,\infty), \end{cases} \tag{4.6.24}$$

with ρ a nonnegative integer. We see from (2.2.8) that the Laplace transform of f is the incomplete gamma function $\Gamma(-\rho,\lambda)$.

We have that $M[f;z]=M[f_2;z]=-[(z-\rho-1)\lambda^{\rho}]^{-1}$ which has a single pole at $z=1+\rho$. If ρ is a nonnegative integer, then this pole coincides with a pole of $\Gamma(1-z)$. After a simple calculation we find that for $\rho=0, 1, 2, \ldots$,

$$\Gamma(-\rho,\lambda)\sim\frac{[\log\lambda-\psi(\rho+1)](-1)^{\rho+1}}{\rho!}-\lambda^{-\rho}\sum_{\substack{m=0\\ m\neq\rho}}^{\infty}\frac{(m-\rho)^{-1}(-\lambda)^m}{m!}. \tag{4.6.25}$$

EXAMPLE 4.6.2. Now suppose that

$$H[f;\lambda]=\int_0^{\infty}\frac{(1+t^2)^{\nu-\frac{1}{2}}}{(1+\lambda t)^{\rho}}\,dt, \qquad \rho>2\nu. \tag{4.6.26}$$

Hence $H[f;\lambda]$ is proportional to the generalized Stieltjes transform of $f(t)=(1+t^2)^{\nu-\frac{1}{2}}$. We have from (4.5.21)

$$M[(1+t)^{-\rho};1-z]=\frac{\Gamma(1-z)\,\Gamma(z+\rho-1)}{\Gamma(\rho)} \tag{4.6.27}$$

which has simple poles at the positive integers.

For $\nu<\frac{1}{2}$, $M[f;z]$ exists in the ordinary sense in the region $0<\mathrm{Re}(z)<1-2\nu$. For $\nu\geq\frac{1}{2}$, however, $M[f;z]$ exists only in the generalized sense. To determine this generalized Mellin transform, we argue as follows. From the Appendix we have that, when $\mathrm{Re}\,\nu<\frac{1}{2}$,

$$M[f;z]=\frac{\Gamma\left(\dfrac{z}{2}\right)\Gamma([1-z-2\nu]/2)}{2\Gamma(\frac{1}{2}-\nu),}, \tag{4.6.28}$$

which, as noted above, is holomorphic in $0<\mathrm{Re}(z)<\mathrm{Re}(1-2\nu)$. Furthermore, we see that for fixed z, (4.6.28) is meromorphic in ν for complex ν with poles at $\nu=m+(1-z)/2$. Thus (4.6.28) must hold for all ν by analytic continuation except at the singular points. In particular, (4.6.28) holds for $\nu\geq\frac{1}{2}$ and hence must be the generalized Mellin transform of f in the region $\mathrm{Re}(z)>0$.

As is easily seen

$$f(t)=t^{2\nu-1}\sum_{m=0}^{\infty}\binom{\nu-\frac{1}{2}}{m}t^{-2m}, \qquad t\to\infty, \tag{4.6.29}$$

where

$$\binom{\nu-\frac{1}{2}}{m}=\frac{\Gamma(\nu+\frac{1}{2})}{\Gamma(\nu-m+\frac{1}{2})\,m!}.$$

Hence $M[f_2;z]$ has simple poles at $z=2m+1-2\nu$, $m=0, 1, 2, \ldots$. Thus, if

2ν is not an integer, then we have

$$\int_0^\infty \frac{(1+t^2)^{\nu-\frac{1}{2}}}{(1+\lambda t)^\rho}\,dt \sim \sum_{m=0}^{\infty} \binom{\nu-\frac{1}{2}}{m} \frac{\Gamma(2\nu-2m)\,\Gamma(\rho+2m-2\nu)}{\Gamma(\rho)}\lambda^{2m-2\nu}$$
$$+\sum_{m=0}^{\infty} \frac{\Gamma(\rho-1+m)\,\Gamma\left(\frac{m}{2}+\frac{1}{2}\right)\Gamma\left(\frac{-m}{2}-\nu\right)}{\Gamma(\rho)\,\Gamma(\frac{1}{2}-\nu)\,m!}(-\lambda)^m, \qquad \lambda\to 0+, \tag{4.6.30}$$

which is valid whenever $\rho > 2\nu > 1$.

4.7. Asymptotic Expansions for Complex λ

In the preceding sections we developed a technique for obtaining asymptotic expansions of

$$H[f;\lambda] = \int_0^\infty h(\lambda t)\,f(t)\,dt, \tag{4.7.1}$$

as $\lambda\to\infty$ and as $\lambda\to 0+$ through real values. As we know, asymptotic expansions when valid usually hold in some sector of the complex plane. Thus, we shall now investigate the possibility of extending the results of Sections 4.5 and 4.6 to the case of complex λ.

The question naturally arises: Why, since in most applications λ is a real nondimensional physical quantity, is this extension of any interest? One reason is that, even in cases where complex λ is not physically reasonable, it is often an important step in the analysis of a given problem to extend an asymptotic expansion into the complex plane. Furthermore, there are problems in which complex λ has a legitimate physical interpretation. Indeed, when we consider the propagation of "high-frequency" waves in certain dissipative media, the relevant λ is complex. Finally we remark that there are several areas of analysis, for example, number theory, where complex λ occurs quite naturally.

Let us consider (4.7.1) with $f(t)$ locally integrable on $(0,\infty)$ and $h(t)$ analytic in a sector, say $|\arg t| < \theta_0$. We assume the asymptotic forms (4.4.7) on the line and (4.4.9) in the sector, respectively. Furthermore, we again suppose that the constants α, β, γ, and δ defined by (4.5.1) satisfy the inequalities (4.5.2) and that $\alpha < \beta$ so that $M[h;z]$ exists in the ordinary sense. We then have that D_j, the domain of analyticity of $G_j = M[h;z]\,M[f_j;1-z]$, is still defined by (4.5.8).

It is a simple matter to extend the results of Theorem 4.5 to complex λ. Indeed we now prove the following.

THEOREM 4.7.1. Let r and R be given real numbers greater than $1-\delta$ with $r < R$. Let $\mathrm{Re}(z) = r_1$ lie in the domain D_1, the domain of analyticity of G_1. Suppose that

$$M[h;x+iy] = O[\exp(-\theta_0|y|)], \qquad |y|\to\infty, \tag{4.7.2}$$

for some $\theta_0 > 0$ and for all x in the interval (α, R).[13] If no singularity of $G(z) = M[h;z]\, M[f;1-z]$ lies along either $\mathrm{Re}(z) = r$ or $\mathrm{Re}(z) = R$, then

$$H[f;\lambda] \sim \sum_{r_1 < \mathrm{Re}(z) < r} \mathrm{res}(-\lambda^{-z} G_1(z)) + \sum_{r < \mathrm{Re}(z) < R} \mathrm{res}(-\lambda^{-z} G(z)) \quad (4.7.3)$$

represents a finite asymptotic expansion of (4.7.1) that is valid as $\lambda \to \infty$ in the sector defined by $|\arg(\lambda)| < \theta_0$. Moreover, if the above hypotheses hold for arbitrarily large R, then by setting $R = \infty$ in (4.7.2) an infinite expansion is obtained.

PROOF. If we set $\arg(\lambda) = \phi$, then we have

$$|\lambda^{-x-iy}| = |\lambda|^{-x} e^{\phi y}. \quad (4.7.4)$$

Because $M[f_j; x+iy]$, $j = 1, 2$ can grow at worst algebraically as $|y| \to \infty$, we have that whenever $|\phi| < \theta_0$, each of the functions $\lambda^{-x-iy} G_j(x+iy)$, $j = 1, 2$ is $o(|y|^{-n})$, as $|y| \to \infty$ for all n and for all x in (α, R). Moreover, each of these functions is at worst meromorphic in a right half-plane containing the corresponding domain D_j. Thus, the contours of integration in the still valid representation (4.5.12) can be displaced to the right until they both coincide with $\mathrm{Re}(z) = R$. Then upon applying Cauchy's integral theorem and using the fact that $G(z)$ exists at least in the generalized sense for $\mathrm{Re}(z) > 1 - \delta$, we obtain

$$H[f;\lambda] = \sum_{r_1 < \mathrm{Re}(z) < r} \mathrm{res}(-\lambda^{-z} G_1(z)) + \sum_{r < \mathrm{Re}(z) < R} \mathrm{res}(-\lambda^{-z} G(z)) + \frac{1}{2\pi i} \int_{R-i\infty}^{R+i\infty} \lambda^{-z} G(z)\, dz. \quad (4.7.5)$$

The asymptotic nature of (4.7.3) is finally established by the estimate

$$\left| \int_{R-i\infty}^{R+i\infty} \lambda^{-z} G(z)\, dz \right| \le |\lambda|^{-R} \int_{-\infty}^{\infty} e^{\phi|y|} |M[h;R+iy]\, M[f;1-R-iy]|\, dy \le K|\lambda|^{-R}, \quad (4.7.6)$$

which by (4.7.2) is valid for some constant K whenever $|\phi| < \theta_0$.

Remarks. In Theorem 4.7.1, the desired extension was obtained by assuming an exponential decay of $M[h;x+iy]$ as $|y| \to \infty$. Suppose now that $M[f;1-z]$ exists in the ordinary sense. It should be clear that if

$$M[f;1-x-iy] = O[\exp(-\psi_0 |y|)], \qquad |y| \to \infty, \quad (4.7.7)$$

for some $\psi_0 > 0$, then an analogous result could be obtained with no decay of $M[h;z]$. Moreover:

[13] The choices of α, δ, r_1, and R here are made to coincide with their usage in Section 4.5 where the generalized Mellin transform is introduced.

COROLLARY 4.7.1. If both (4.7.2) and (4.7.7) hold, then the sector of validity of the expansion (4.7.3) is increased to $|\arg(\lambda)| < \psi_0 + \theta_0$.

We wish to stress that the utility of Theorem 4.7.1 is limited by the fact that, as yet, we have no a priori way of establishing the assumed decay of $M[h;z]$. We recall that a similar observation was made concerning the results of Theorem 4.4. Thus, we shall devote the remainder of this section to obtaining conditions on $h(t)$ which, when satisfied, guarantee the exponential decay of the functions $G_j(z)$ along vertical lines.

In a given problem, where the relevant Mellin transforms are known explicitly, the requisite behavior can often be established by inspection. Indeed, as an example, consider the case of the Laplace transform. Then the kernel function is $h(t) = \exp(-t)$ and $M[h;z] = \Gamma(z)$. We have

$$\Gamma(x+iy) = O\{\exp[-\left(\frac{\pi}{2} - \varepsilon\right)|y|]\}$$

as $|y| \to \infty$, where ε is any small positive real number. Therefore, we can immediately conclude that Watson's lemma, originally proved for λ real, actually holds as $|\lambda| \to \infty$ in the right half-plane $\mathrm{Re}(\lambda) > 0$. Unfortunately, the Mellin transform of h cannot, in general, be represented in terms of well-studied functions and hence its asymptotic behavior as $|y| \to \infty$ will not be immediately apparent.

Our discussion below will be greatly facilitated by the introduction of the following.

DEFINITION. A function $h(t)$ is said to lie in the class $K(x_0,\theta_0)$ [abbreviated $h(t) \varepsilon K(x_0,\theta_0)$] if, for any $\varepsilon > 0$ and all $x > x_0$,

$$M[h;x+iy] = O[\exp(-[\theta_0 - \varepsilon]|y|)], \qquad |y| \to \infty. \tag{4.7.8}$$

Here, for greater generality, we only require that $M[h;z]$ exist in the generalized sense.

We also introduce, for any $\theta_0 > 0$, the open sector defined by

$$s(\theta_0) = \{t \mid t \neq 0, |\arg(t)| < \theta_0\}. \tag{4.7.9}$$

We wish to determine sufficient conditions for a function to lie in some class $K(x_0,\theta_0)$. Our main result in this regard is given in the following.

THEOREM 4.7.2. Suppose that in the sector $s(\theta_0)$

(1) $h(t)$ is analytic.

(2) $h(t) = O(t^{\alpha}), \qquad |t| \to 0+$.

(3) $h(t) \sim \exp\{-dt^{\nu}\} \sum_{m=0}^{\infty} \sum_{n=0}^{N(m)} c_{mn} (\log t)^n t^{-r_m}, \qquad t \to \infty$.

Here $\text{Re}(d) \geq 0$, $v > 0$, $\text{Re}(r_m) \uparrow \infty$, and $N(m)$ is finite for each m.

(a) If $d = 0$, then $h(t) \in K(-\text{Re}(\alpha), \theta_0)$.

(b) If $\text{Re}(d) > 0$, then $h(t) \in K(-\text{Re}(\alpha), \theta)$ where

$$\theta = \min\left(\theta_0, \frac{\pi - 2\,|\arg(d)|}{2v}\right). \tag{4.7.10}$$

We remark that in (a), $M[h;z]$ need only exist in the generalized sense because $-\text{Re}(\alpha)$ can exceed $\text{Re}(r_0)$.

Before proving this theorem, we shall consider several examples to make the conclusions more meaningful.

EXAMPLE 4.7.1. Suppose

$$h(t) = \sin\psi\,[(t - \cos\psi)^2 + \sin^2\psi]^{-1}, \qquad 0 < \psi \leq \pi. \tag{4.7.11}$$

Then the hypotheses of Theorem 4.7.2 are satisfied with $\theta_0 \leq \psi$ [note that $h(t)$ has poles at $t = e^{\pm i\psi}$], $\alpha = 0$, and $d = 0$. Consequently, conclusion (1) predicts that $h(t) \in K(0,\psi)$. Indeed, for this example, we have the explicit result

$$M[h;z] = -\frac{\pi \sin[(\pi - \psi)(z-1)]}{\sin \pi z} = O(e^{-\psi|y|}), \qquad |y| \to \infty, \tag{4.7.12}$$

which verifies this prediction. We note that $M[h;z]$ satisfies (4.7.8) with $\theta_0 = \psi$, $\varepsilon = 0$, and for all x. This is a consequence of the fact that $h(t)$ belongs to a particular subset of the class of functions which satisfy conditions (1)–(3) of Theorem 4.7.2.

EXAMPLE 4.7.2. Suppose now that

$$h(t) = t^{1/2} K_\mu(t), \tag{4.7.13}$$

where μ is real and K_μ is the modified Bessel function of the second kind. From well-known properties of the Bessel function[14] we have that, in the complex t plane slit along the negative real axis, $t^{1/2} K_\mu$ is analytic and $O(\exp\{-t\})$ as $t \to \infty$. Furthermore, in this domain, if $\mu \neq 0$, then $t^{1/2} K_\mu = O(|t|^{\frac{1}{2}-|\mu|})$ as $t \to 0$, while $t^{1/2} K_0(t) = O(|t|^{1/2} \log|t|)$ in this limit.

By Lemma 4.3.1, $M[t^{1/2} K_\mu(t);z]$ is holomorphic for $x > |\mu| - \frac{1}{2}$. Furthermore, the criteria of Theorem 4.7.2 are satisfied with $\alpha = \frac{1}{2} - |\mu|$, $d = 1$, and $v = 1$. We conclude then that

$$t^{1/2} K_\mu(t) \in K\left(|\mu| - \frac{1}{2}, \frac{\pi}{2}\right). \tag{4.7.14}$$

Indeed, we have from the Appendix that

[14] See Exercise 4.3.

$$M[t^{1/2} K_\mu(t);z] = 2^{z-\frac{3}{2}} \Gamma\left(\frac{z+\frac{1}{2}+\mu}{2}\right) \Gamma\left(\frac{z+\frac{1}{2}-\mu}{2}\right), \tag{4.7.15}$$

and (4.7.14) follows from known properties of the gamma function.

EXAMPLE 4.7.3. Now let us suppose that

$$h(t) = \frac{P(t)}{Q(t)}, \tag{4.7.16}$$

where $P(t)$ and $Q(t)$ are polynomials of degree p and q, respectively. Also suppose that $Q(t)$ has no real zeros except possibly at $t=0$ and that $h(t) = O(t^\alpha)$ as $t \to 0$, $-q \le \alpha \le p$. Then $h(t)$ satisfies the hypotheses of Theorem 4.7.2 with $d=0$ and $r_0 = q-p$. Furthermore, $\theta_0 = \min(|\arg(t_n)|)$ where $t = t_n$, $n = 1, \ldots, n_0 \le q$ are the complex zeros of $Q(t)$. Then Theorem 4.7.2 predicts that

$$h(t) \; \varepsilon \; K(-\alpha, \theta_0). \tag{4.7.17}$$

We leave the direct verification of this result to Exercise 4.32 but now consider an important special case.

EXAMPLE 4.7.4. If

$$h(t) = (a+t)^{-r}, \tag{4.7.18}$$

then (4.7.1) is the generalized Stieltjes transform of $f(t)$. Now the hypotheses of Theorem 4.7.2 are satisfied with $d=0$, $\theta_0 = \pi - |\arg(a)|$, $\alpha = 0$, and $r_0 = r$. Hence, the theorem predicts

$$h(t) \; \varepsilon \; K(0, \pi - |\arg(a)|). \tag{4.7.19}$$

For this particular $h(t)$ we have from the Appendix that

$$M[h;z] = a^{z-r} \frac{\Gamma(z)\,\Gamma(r-z)}{\Gamma(r)} = O[\exp\{-(\pi - |\arg(a)| - \varepsilon)\,|y|\}], \tag{4.7.20}$$

as $|y| \to \infty$, for any $\varepsilon > 0$ and for all $x > 0$. This explicit result verifies (4.7.19).

We shall consider additional examples below; but now we turn to the following.

PROOF OF THEOREM 4.7.2. We consider first case (b). Here

$$M[h;z] = \int_0^\infty h(t)\, t^{z-1}\, dt \tag{4.7.21}$$

is an analytic function in the half-plane $\mathrm{Re}(z) > -\mathrm{Re}(\alpha)$. If we define

$$\tilde{\theta} = \theta - \varepsilon, \qquad \theta = \min\left(\theta_0, \frac{\pi - 2\,|\arg(d)|}{2v}\right), \qquad \varepsilon > 0, \tag{4.7.22}$$

then from conditions (1)–(3) we find that we can rotate the path of integration in (4.7.21) onto either of the two rays

$$t = \sigma\, e^{\pm i\tilde{\theta}}, \qquad 0 \le \sigma < \infty.\text{[15]} \tag{4.7.23}$$

By introducing σ as the new variable of integration on the rotated paths, we find from (4.7.21) that

$$M[h;z] = \exp\{\pm i\tilde{\theta} z\}\, M[g(\sigma);z], \qquad g(\sigma) = h(\sigma\, e^{\pm i\tilde{\theta}}). \tag{4.7.24}$$

We now apply Lemma 4.3.1 to conclude that $M[h(\sigma e^{\pm i\tilde{\theta}});z]$ is analytic for $\operatorname{Re}(z) > -\operatorname{Re}(\alpha)$ and approaches zero, as $|y| \to \infty$, for each fixed $x > -\operatorname{Re}(\alpha)$. If we interpret the $\pm$ sign in (4.7.24) as representing the sign of y, that is, if we rotate up (down) when y is positive (negative), then it follows that

$$M[h(t);z] = O(\exp\{-\tilde{\theta}\,|y|\}), \qquad |y| \to \infty, \qquad x > -\operatorname{Re}(\alpha). \tag{4.7.25}$$

From (4.7.22) and (4.7.25) we have

$$h(t) \,\varepsilon\, K(-\operatorname{Re}(\alpha),\theta).$$

To prove the result for case (a), we first introduce

$$\begin{aligned} M_1[h(t);z] &= \textstyle\int_0^1 h(t)\, t^{z-1}\, dt, \\ M_2[h(t);z] &= \textstyle\int_1^\infty h(t)\, t^{z-1}\, dt, \end{aligned} \tag{4.7.26}$$

and note that these transforms are analytic in the regions $\operatorname{Re}(z) > -\operatorname{Re}(\alpha)$ and $\operatorname{Re}(z) < \operatorname{Re}(r_0)$, respectively. Here we propose to replace each of the integrals in (4.7.26) by an integral along a segment of the ray

$$t = \sigma\, e^{\pm i\theta}; \qquad \theta = \theta_0 - \varepsilon, \qquad \varepsilon > 0, \tag{4.7.27}$$

plus an integral along the arc of the unit circle connecting 1 with $e^{\pm i\theta}$.[16] In this manner we obtain

$$M_1[h;z] = I_1(z) + I_2(z); \qquad M_2[h;z] = I_3(z) + I_4(z), \tag{4.7.28}$$

where

$$\begin{aligned} I_1(z) &= \exp\{\pm i\theta z\}\, M_1[g(\sigma);z], \qquad g(\sigma) = h(\sigma\, e^{\pm i\theta}), \\ I_2(z) &= i\textstyle\int_{\pm\theta}^{0} h(e^{i\psi})\, e^{i\psi z}\, d\psi, \\ I_3(z) &= \exp\{\pm i\theta z\}\, M_2[g(\sigma);z], \\ I_4(z) &= i\textstyle\int_0^{\pm\theta} h(e^{i\psi})\, e^{i\psi z}\, d\psi. \end{aligned} \tag{4.7.29}$$

We first note that $I_4(z)$ is an entire function and that

$$I_2(z) + I_4(z) = 0. \tag{4.7.30}$$

[15] To justify this step, it must be shown that the integral on an arc of large radius becomes vanishingly small as the radius increases. (See Exercise 4.25.)

[16] As in the proof of case (6) we have in mind that the $\pm$ sign in (4.7.27) corresponds to the sign of y. See also footnote 15 and Exercise 4.26.

By Lemma 4.3.3, $M_2[g(\sigma);z]$ can be analytically continued into the right half-plane $\text{Re}(z) > -\text{Re}(\alpha)$ and this continuation goes to zero as $|y| \to \infty$. Hence,

$$I_3(z) = O[\exp\{-(\theta_0 - \varepsilon)|y|\}], \qquad |y| \to \infty \tag{4.7.31}$$

for $\text{Re}(z) > -\text{Re}(\alpha)$.

Finally, we consider $M_1[h;z]$ which is already analytic in the region $\text{Re}(z) > -\text{Re}(\alpha)$. The same is true, therefore, of $M_1[g;z]$ which must go to zero, as $|y| \to \infty$ in this region. Thus, we have that, as $|y| \to \infty$ with $\text{Re}(z) > -\text{Re}(\alpha)$,

$$I_1(z) = O[\exp\{-(\theta_0 - \varepsilon)|y|\}]. \tag{4.7.32}$$

Because the generalized Mellin transform of h is given by

$$M[h;z] = M_1[h;z] + M_2[h;z] \tag{4.7.33}$$

we have, upon combining (4.7.28), and (4.7.30) to (4.7.32), that

$$M[h;z] = O[\exp\{-(\theta_0 - \varepsilon)|y|\}], \tag{4.7.34}$$

as $|y| \to \infty$, for all $x > -\text{Re}(\alpha)$. Hence $M[h;z] \; \varepsilon \; K(-\text{Re}(\alpha),\theta_0)$ which completes the proof.

In the last theorem we established criteria for determining in which class $K(x_0,\theta_0)$, if any, a given function belongs. In the following lemma, whose proof we leave to the exercises, are collected some useful closure properties of $K(x_0,\theta_0)$. If we call functions which belong to some class $K(x_0,\theta_0)$ "good" functions, then these properties will enable us to generate additional good functions from those already known.

LEMMA 4.7.1. If $h(t) \; \varepsilon \; K(x_0,\theta_0)$, then

(a) $r\, h(t) \; \varepsilon \; K(x_0,\theta_0)$ for any complex constant r.
(b) $h(rt) \; \varepsilon \; K(x_0,\theta_0)$ for any real positive constant r.
(c) $h(t^r) \; \varepsilon \; K(rx_0,\theta_0/r)$ for any real positive constant r.
(d) $t^r\, h(t) \; \varepsilon \; K(x_0 + \text{Re}(r),\theta_0)$ for any complex constant r.
(e) If $h_j(t) \; \varepsilon \; K(x_j,\theta_j)$ for $j = 1, 2$, then $h_1(t) + h_2(t) \; \varepsilon \; K(\max[x_1,x_2], \min[\theta_1,\theta_2])$..
(f) If $h_j(t) \; \varepsilon \; K(x_j,\theta_j)$ for $j = 1, 2$ and, in addition, $M[h_j;z]$ converges absolutely in a vertical strip S_j to the right of x_j, then $h_1(t)\, h_2(t) \; \varepsilon \; K(x_1 + x_2, \min[\theta_1,\theta_2])$ provided $h_1\, h_2$ is locally integrable on $(0,\infty)$.

EXAMPLE 4.7.5. To illustrate the application of result (f) of Lemma 4.7.1, we consider

$$h(t) = \prod_{j=1}^{n} (t + a_j)^{\rho_j}. \tag{4.7.35}$$

Indeed this result predicts

$$h(t) \in K(0, \pi - \max_j |\arg(a_j)|) \tag{4.7.36}$$

whenever each a_j is nonzero, and

$$h(t) \in K(-[\rho_1 + \cdots + \rho_m], \pi - \max_{j>m} |\arg(a_j)|) \tag{4.7.37}$$

whenever $a_1 = \cdots = a_m = 0, a_j \neq 0, j > m$.

We, of course, wish to consider continuations of Mellin transforms into left half-planes and, in particular, the analog of Theorem 4.7.2. Such an analysis is greatly facilitated by the easily established relation

$$M[h(t); z] = M\left[h\left(\frac{1}{t}\right); -z\right] \tag{4.7.38}$$

which also holds for functions whose Mellin transforms exist only in the generalized sense. Upon using this relation we immediately obtain the following.

LEMMA 4.7.2. If $h(1/t) \in K(x_0,\theta_0)$, then

$$M[h(t);z] = O[\exp\{-\theta|y|\}], \qquad x < -x_0, \qquad \theta = \theta_0 - \varepsilon, \qquad \varepsilon > 0. \tag{4.7.39}$$

To clarify this last result we consider the following.

EXAMPLE 4.7.6. Suppose

$$h(t) = \sin\psi\,[(t - \cos\psi)^2 + \sin^2\psi]^{-1}, \qquad 0 < \psi \leq \pi.$$

Then $h(1/t)$ satisfies the conditions of Theorem 4.7.2 with $\theta_0 = \psi$, $\alpha = 2$, $r_0 = 0$, and $d = 0$. (Note that here α and r_0 have interchanged the roles they played in Example 4.7.1.) Thus,

$$h\left(\frac{1}{t}\right) \in K(-2,\psi) \tag{4.7.40}$$

and hence by Lemma 4.7.2,

$$M[h(t);z] = O[\exp\{-\theta|y|\}], \qquad x < 2, \qquad \theta = \psi - \varepsilon, \qquad \varepsilon > 0. \tag{4.7.41}$$

We previously noted from the explicit formula (4.7.12) that (4.7.41) holds for *all* x so that

$$\sin\psi\,[(t - \cos\psi)^2 + \sin^2\psi]^{-1} \in K(-\infty,\psi). \tag{4.7.42}$$

This specific result is generalized in the following.

LEMMA 4.7.3. Suppose that $h(t) \in K(x_0,\theta_0)$ and $h(1/t) \in K(x_1;\theta_0)$ If $x_0 < -x_1$, then

$$h(t) \in K(-\infty,\theta_0). \tag{4.7.43}$$

EXAMPLE 4.7.7. Let us again consider $t^{1/2} K_\mu(t)$ with $K_\mu(t)$ the modified Bessel function. From previously stated properties we have that $t^{-1/2} K_\mu(1/t)$ is analytic in the complex plane slit along the negative real axis. Moreover, we have that $t^{-1/2} K_\mu(1/t)$ is algebraic as $t \to \infty$ and $0(|t|^\alpha)$, as $t \to 0$, for all α, with these limits taken in the sector $s(\theta_0 = \pi/2)$. Thus, it follows from Theorem 4.7.2 that

$$t^{-1/2} K_\mu\left(\frac{1}{t}\right) \varepsilon K\left(-\infty, \frac{\pi}{2}\right). \tag{4.7.44}$$

Finally (4.7.14) and Lemma 4.7.3 yield

$$t^{1/2} K_\mu(t) \varepsilon K\left(-\infty, \frac{\pi}{2}\right), \tag{4.7.45}$$

a result which is verified by the explicit formula (4.7.15).

In Example 4.7.1, we observed that (4.7.8) held with $\theta_0 = \psi$ and $\varepsilon = 0$. We now want to consider conditions sufficient for this sharper estimate to hold. To accomplish this we first introduce the following.

DEFINITION. A function $h(t)$ is said to lie in the class $\bar{K}(x_0, \theta_0)$ $[h(t) \varepsilon \bar{K}(x_0, \theta_0)]$ if (4.7.8) holds with $\varepsilon = 0$.

A subset of $\bar{K}(x_0, \theta_0)$ is identified in the following.

LEMMA 4.7.4. Suppose that $h(t)$ satisfies the hypotheses of Theorem 4.7.2 in the sector $s(\theta_1)$ with $\theta_1 > \theta_0$, but for the following possible exceptions on the bounding rays of $s(\theta_0)$:

(a) $h(t)$ has an algebraic branch point of order greater than -1.
(b) $h(t)$ has a logarithmic branch point.
(c) $h(t)$ has a pole.
(d) $h(t)$ has one of the above types of singularities at a finite number of points.

If, in addition, either $d = 0$ or $(\pi - 2|\arg(d)|)/2v > \theta_0$, then $h(t) \varepsilon \bar{K}(-\mathrm{Re}(\alpha), \theta_0)$.

PROOF. The proof follows closely that of Theorem 4.7.2 and is outlined in the exercises.

We shall conclude this section by considering the asymptotic expansions of two explicit integral transforms for complex λ.

EXAMPLE 4.7.8. Let

$$H[f;\lambda] = \int_0^\infty (\lambda t)^{1/2} K_\mu(\lambda t) f(t)\, dt, \tag{4.7.46}$$

which is the K_μ transform of $f(t)$. We wish to study $H[f;\lambda]$ as $|\lambda|\to\infty$. From Example 4.7.2, we have that the resulting asymptotic expansion will be valid for $|\arg(\lambda)|<\pi/2$.

We shall assume that $f(t)$ is locally integrable on $(0,\infty)$, satisfies (4.4.9) with $q=0$ and $p_{mn}=0$ for $n\geq 1$, and has the bound

$$f(t)=O[\exp(\rho t)]$$

as $t\to\infty$ for any real ρ. Then we can apply (4.4.17) and (4.7.15) to conclude

$$\int_0^\infty (\lambda t)^{1/2} K_\mu(\lambda t) f(t)\,dt \sim \sum_{m=0}^\infty \frac{p_{m0}}{\lambda^{1+a_m}} 2^{a_m-\frac{1}{2}}\,\Gamma\left(\frac{a_m+\frac{3}{2}+\mu}{2}\right)\Gamma\left(\frac{a_m+\frac{3}{2}-\mu}{2}\right) \tag{4.7.47}$$

as $\lambda\to\infty$ in the sector $|\arg(\lambda)|<\pi/2$.

EXAMPLE 4.7.9. Let

$$H[f;\lambda]=\int_0^\infty \frac{f(t)}{(1+\lambda t)^r}\,dt, \qquad r>0. \tag{4.7.48}$$

We recall that $\lambda^r H[f;\lambda]$ is the generalized Stieltjes transform of argument λ^{-1}. Thus, the asymptotic expansion of $H[f;\lambda]$ for $\lambda\to\infty$ is proportional to the expansion of the generalized Stieltjes transform as $\lambda\to 0$.

From Theorem 4.7.1 and Lemma 4.7.4 we conclude that the resulting asymptotic expansion will be valid in the sector $|\arg(\lambda)|<\pi$ when $r\geq 1$ and $|\arg(\lambda)|\leq\pi$ when $r<1$. For simplicity, we assume that $f(t)$ is as in Example 4.7.8 except now we require $\mathrm{Re}(a_0)>-1$ in (4.4.9) and that

$$\delta=\sup\{\delta^*\,|\,f=O(t^{-\delta^*}),\quad t\to\infty\}>1. \tag{4.7.49}$$

Then $M[h;1-z]$ is analytic in the strip $1-\delta<\mathrm{Re}(z)<1+\mathrm{Re}(a_0)$.

If we set $a=1$ in (4.7.20), then we have

$$M[(1+t)^{-r};z]=\frac{\Gamma(z)\,\Gamma(r-z)}{\Gamma(r)}. \tag{4.7.50}$$

This function has simple poles at $z=r+n, n=0, 1, 2, \ldots$. Thus, if $r+n\neq 1+a_m$ for any pair of nonnegative integers n, m, then by Theorem 4.7.1,

$$\int_0^\infty \frac{f(t)}{(1+\lambda t)^r}dt \sim \sum_{m=0}^\infty \lambda^{-1-a_m}\, p_{m0}\,\frac{\Gamma(1+a_m)\,\Gamma(r-1-a_m)}{\Gamma(r)} + \sum_{m=0}^\infty (-1)^m \frac{M[f;1-r-m]\,\Gamma(r+m)\,\lambda^{-r-m}}{m!\,\Gamma(r)} \tag{4.7.51}$$

as $\lambda\to\infty$ in $|\arg(\lambda)|<\pi$.

All of the results of this section yield sectors of validity that are of the form

$|\arg(\lambda)| < \theta_0$. It is readily seen that asymmetric sectors can be obtained as well. Indeed, we have the following.

COROLLARY 4.7.2. Suppose that in Theorem 4.7.1, the estimate (4.7.2) is replaced by

$$M[h;x+iy] = O[\exp(-\theta_0^{\pm}|y|)], \qquad \pm y \to \infty, \qquad x > x_0. \tag{4.7.52}$$

Suppose further that

$$M[f;1-x-iy) = O[\exp(-\psi_0^{\pm}|y|)], \qquad \pm y \to \infty, \qquad x > x_0. \tag{4.7.53}$$

Then (4.7.3) is valid for

$$-(\theta_0^- + \psi_0^-) < \arg(\lambda) < \theta_0^+ + \psi_0^+. \tag{4.7.54}$$

The extension afforded by Corollary 4.7.2 is particularly important when the kernel h is oscillatory because in that event θ_0^+ and θ_0^- are most often different. To illustrate this let us consider $h = \exp(it)$. We have

$$M[h;z] = \exp\left(\frac{i\pi z}{2}\right)\Gamma(z) = O\left[|y|^{x-\frac{1}{2}} \exp\left(\frac{-\pi}{2}(y+|y|)\right)\right], \qquad |y| \to \infty. \tag{4.7.55}$$

Thus, in this case

$$\theta_0^+ = \pi, \qquad \theta_0^- = 0. \tag{4.7.56}$$

4.8. Electrostatics

Let us consider the idealized situation consisting of an infinitely thin conducting plane with a hole of unit radius cut out of it. Suppose that a disc, made of the same material as the plane, fills the hole, but is electrically insulated from the rest of the plane. Finally, let the disc be maintained at a fixed potential V while the remainder of the plane is kept at zero potential. Our problem is to find the resulting electrostatic potential field of the system.

We introduce the cylindrical coordinate system r, θ, z as depicted in Figure 4.8. If $\phi = \phi(r, \theta, z)$ is the desired potential at any point in space, then it must satisfy the following boundary-value problem:

$$\Delta\phi = 0, \tag{4.8.1}$$

$$\phi = V, \qquad z = 0, \qquad 0 \le r < 1, \tag{4.8.2}$$

$$\phi = 0, \qquad z = 0, \qquad r > 1, \tag{4.8.3}$$

$$\phi(r,z) \text{ bounded as } \sqrt{r^2+z^2} \to \infty. \tag{4.8.4}$$

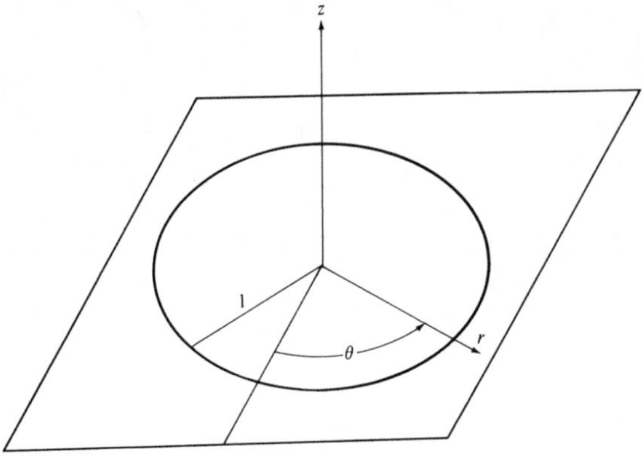

Figure 4.8. Plate with Insulated Disc.

To solve this problem, we first note that the symmetry of the system implies that ϕ does not depend explicitly on θ and that ϕ must be symmetric in z about $z=0$. Therefore, we shall assume that $\phi=\phi(r,z)$ and shall restrict our considerations to the half-space $z \geq 0$. We now introduce

$$\bar{\phi}(p,z)=\int_0^\infty r\, J_0(pr)\, \phi(r,z)\, dr \tag{4.8.5}$$

which is the Hankel transform of ϕ with respect to r. As was pointed out in Section 3.2, the inversion formula for this transform is

$$\phi(r,z)=\int_0^\infty p\, J_0(pr)\, \bar{\phi}(p,z)\, dp. \tag{4.8.6}$$

Thus, if we can determine $\bar{\phi}$, then insertion of the result in (4.8.6) will yield an integral representation of the solution to our problem.

If we multiply (4.8.1) by $r\,J_0(pr)$ and integrate the result with respect to r from 0 to ∞, then after two integrations by parts we have

$$0=\int_0^\infty r\, J_0(rp)\;\; \Delta\phi\, dr = r\frac{\partial\phi}{\partial r} J_0(pr)\bigg|_0^\infty - p\phi r\, J_0{}'(pr)\bigg|_0^\infty + \left(\frac{\partial^2}{\partial z^2}-p^2\right)\int_0^\infty r\, J_0(pr)\, \phi\, dr. \tag{4.8.7}$$

It follows from (4.8.4) that the boundary terms in (4.8.7) vanish and hence

$$\left(\frac{\partial^2}{\partial z^2}-p^2\right)\bar{\phi}(p,z)=0. \tag{4.8.8}$$

The general solution to (4.8.8) is

$$\bar{\phi}(p,z)=A(p)\, e^{-pz}+B(p)\, e^{pz}. \tag{4.8.9}$$

Because z is positive, (4.8.4) and (4.8.6) imply that $B(p)\equiv 0$ and hence we have

$$\phi(r,z) = \int_0^\infty p\, J_0(pr)\, e^{-pz}\, A(p)\, dp. \tag{4.8.10}$$

To complete the solution we need only determine $A(p)$. From (4.8.2) and (4.8.3) we obtain

$$\begin{aligned} V &= \int_0^\infty p\, J_0(pr)\, A(p)\, dp, \qquad 0 \le r < 1, \\ 0 &= \int_0^\infty p\, J_0(pr)\, A(p)\, dp, \qquad 1 < r < \infty, \end{aligned} \tag{4.8.11}$$

which are dual integral equations for the determination of $A(p)$.

From (3.2.6) we see that $p^{1/2} A(p)$ is the Hankel transform of order zero of the function

$$f(r) = \begin{cases} r^{1/2} V, & 0 \le r < 1, \\ 0, & 1 < r < \infty, \end{cases}$$

By the inversion formula we have

$$A(p) = V \int_0^1 r\, J_0(pr)\, dr = \frac{V}{p} J_1(p) \tag{4.8.12}$$

so that

$$\phi(r,z) = V \int_0^\infty J_0(pr)\, J_1(p)\, e^{-pz}\, dp. \tag{4.8.13}$$

At this juncture we have a choice. We can either stop at (4.8.13), being satisfied with having obtained the exact solution in closed form, or we can attempt to approximate the integral in various limits of interest. We, of course, prefer the latter alternative. There are several limits that might be of interest. We shall not, however, try to exhaust them all. Indeed, we shall only consider $\phi(r,z)$ as $r \to \infty$, $r \to 0+$, $z \to 0$, and $z \to \infty$. In all of these limits the second variable is assumed to remain finite.

Case I. $r \to \infty$. Here (4.8.13) is an integral of the form (4.4.1) with $h(p) = J_0(p)$ and $f(p) = V\, J_1(p)\, e^{-pz}$. From the table in the Appendix we find that[17]

$$M[J_0(p);\xi] = \frac{2^{\xi-1}\,\Gamma\!\left(\frac{\xi}{2}\right)}{\Gamma\!\left(1-\frac{\xi}{2}\right)}, \tag{4.8.14}$$

$$M[V\, J_1(p)\, e^{-pz}; 1-\xi] = \frac{V\,\Gamma(2-\xi)}{2} F\!\left(\frac{2-\xi}{2}, \frac{3-\xi}{2}; 2; -\frac{1}{z^2}\right) z^{\xi-2}. \tag{4.8.15}$$

Here $F(a,b;c;x)$ is Gauss' hypergeometric function which is a solution to the

[17] Our use of r and z in their traditional roles as cylindrical coordinates conflicts with our use of z as argument of the Mellin transform and r as a constraint on its real part. In this example we replace the latter z by ξ and make no reference to the latter r.

differential equation

$$x(1-x)\,w'' + (c - [a+b+1]x)w' - abw = 0. \tag{4.8.16}$$

The fundamental properties of $F(a,b;c;x)$ are discussed in most books dealing with special functions.

We see from (4.8.14) that $M[J_0;\xi]$ is analytic in $\mathrm{Re}(\xi) > 0$. For fixed c and x, because $F(a,b;c;x)$ is analytic in a, b, it follows that $M[V\,J_1\,e^{-pz};\,1-\xi]$ has simple poles at the integers $\xi = 2, 3, \ldots$. However, $M[J_0;\xi]$ has zeros at the even integers so that residue contributions will be obtained from the odd integers greater than 1 only. Indeed, we find upon applying (4.4.18) that

$$\phi = \frac{Vz}{2r^3}\,F\left(-\frac{1}{2},0;2;-\frac{1}{z^2}\right) - \frac{3}{4}\frac{Vz^3}{r^5}\,F\left(-\frac{3}{2},-1;2;-\frac{1}{z^2}\right) + O(r^{-7}). \tag{4.8.17}$$

Because $F(a,b;c;x) \equiv 1$ whenever either a or b is zero, we have the simple result

$$\phi = \frac{Vz}{2r^3} + O(r^{-5}), \qquad r \to \infty. \tag{4.8.18}$$

Case II. $r \to 0+$. Now we must consider $M[J_0;1-\xi]$ which has simple poles at the positive integers and $M[J_1\,e^{-pz};\xi]$ which is analytic for $\mathrm{Re}\,\xi > -1$. The desired expansion is obtained by applying (4.6.17) which yields

$$\phi = \frac{V}{2z^2}\,F\left(1,\frac{3}{2};2;-\frac{1}{z^2}\right) - \frac{3\,Vr^2}{4z^4}F\left(2,\frac{5}{2};2;-\frac{1}{z^2}\right) + O(r^4). \tag{4.8.19}$$

This last expression can be simplified by noting that the hypergeometric functions that appear are actually algebraic in z. Indeed we have

$$F\left(1,\frac{3}{2};2;-\frac{1}{z^2}\right) = 2z^2\left(1 - \frac{z}{\sqrt{1+z^2}}\right), \tag{4.8.20}$$

$$F\left(2,\frac{5}{2};2;-\frac{1}{z^2}\right) = \frac{2z^2}{3}\left[1 - \frac{z^3}{(1+z^2)^{3/2}}\right], \tag{4.8.21}$$

so that (4.8.19) becomes

$$\phi = V\left(1 - \frac{z}{\sqrt{1+z^2}}\right) - \frac{Vr^2}{2z^2}\left[1 - \frac{z^3}{(1+z^2)^{3/2}}\right] + O(r^4), \qquad r \to 0+. \tag{4.8.22}$$

Case III. $z \to 0$. Before deriving any result for this case we wish to point out that we might expect difficulty in the region about $r = 1$. This, of course, corresponds to the edge of the disc and the potential is discontinuous there. We should anticipate therefore that our expansion will exhibit a lack of "uniformity" in this region.

By referring to (4.8.13) we see that now the kernel function is $h = e^{-p}$ while the transformed function is $f = V\,J_0(pr)\,J_1(p)$. Again by using the table in the Appendix we find that the relevant Mellin transforms are given by

$$M[e^{-p};1-\xi]=\Gamma(1-\xi), \tag{4.8.23}$$

$$M[V\,J_0(pr)\,J_1(p);\xi]=\begin{cases}\dfrac{V\,\Gamma\left(\frac{1}{2}+\frac{\xi}{2}\right)F\left(\frac{\xi+1}{2},\frac{\xi-1}{2};1;r^2\right)}{2^{1-\xi}\,\Gamma\left(\frac{3}{2}-\frac{\xi}{2}\right)}, & 0\le r<1,\\[2ex] \dfrac{V\,2^{\xi-1}\,\Gamma(1-\xi)\,\Gamma\left(\frac{1}{2}+\frac{\xi}{2}\right)}{\Gamma^2\left(\frac{3}{2}-\frac{\xi}{2}\right)\Gamma\left(\frac{1}{2}-\frac{\xi}{2}\right)}, & r=1,\\[2ex] \dfrac{V\,\Gamma\left(\frac{1}{2}+\frac{\xi}{2}\right)F\left(\frac{\xi+1}{2},\frac{\xi+1}{2};2;\frac{1}{r^2}\right)}{2^{1-\xi}\,r^{\xi+1}\,\Gamma\left(\frac{1}{2}-\frac{\xi}{2}\right)}, & 1<r.\end{cases} \tag{4.8.24}$$

Suppose first that $0\le r<1$. Upon using (4.8.23) and the first of formulas (4.8.24) we find from (4.6.17)[18] that

$$\phi = V\,F(1,0;1;r^2)-V\,z\,F(\tfrac{3}{2},\tfrac{1}{2};1;r^2)+O(z^3). \tag{4.8.25}$$

There is no term $0(z^2)$ because $M[V\,J_0(pr)\,J_1(p);\xi]$ vanishes for $\xi=3$ and $0\le r\le 1$. The first hypergeometric function in (4.8.25) is identically 1 and the second is directly related to $E(r)$, the complete elliptic integral of the second kind. Indeed, we have

$$F(\tfrac{3}{2},\tfrac{1}{2};1;r^2)=\frac{2}{\pi(1-r^2)}\,E(r)=\frac{2}{\pi(1-r^2)}\int_0^{\pi/2}\sqrt{1-r^2\sin^2\theta}\,d\theta \tag{4.8.26}$$

so that

$$\phi = V-\frac{2Vz}{\pi(1-r^2)}E(r)+O(z^3),\qquad z\to 0,\qquad 0\le r<1. \tag{4.8.27}$$

Now suppose that $r=1$ in which event both of the relevant Mellin transforms have poles. $M[e^{-p};1-\xi]$ has a simple pole at each positive integer while $M[V\,J_0(p)\,J_1(p);\xi]$ has a simple pole at each positive even integer. [We point out that this could be predicted from the asymptotic form of $J_0(p)\,J_1(p)$ as $p\to\infty$ which has some terms in its expansion which are not oscillatory as is every term in the expansion of $J_0(pr)\,J_1(p)$ for $r\ne 1$.] The coalescence of poles at $\xi=2$ produces a logarithmic term. In fact we find upon applying (4.6.19)

[18] In Example 4.6.1 we discussed the small parameter Laplace transform, hence we could alternatively apply (4.6.23).

that

$$\phi(1;z)=\frac{V}{2}\left[1+\frac{z\log(z/8)}{\pi}\right]+O(z^3\log z). \tag{4.8.28}$$

Finally we suppose that $r>1$. Here we are content to obtain the leading term. The first contribution arises from the pole of $\Gamma(1-\xi)$ at $\xi=2$. [The pole at $\xi=1$ is cancelled by the zero of $M[V\,J_0(pr)\,J_1(p);\xi]$ at this point.] Thus, we have

$$\phi=\frac{Vz}{2r^3}\,F\left(\frac{3}{2},\frac{3}{2};2;\frac{1}{r^2}\right)+O(z^2). \tag{4.8.29}$$

Here again the hypergeometric function can be expressed in terms of elliptic functions. In fact we have

$$\phi=\frac{2Vz}{\pi r^3}\left[\frac{r^2}{r^2-1}E\left(\frac{1}{r}\right)-K\left(\frac{1}{r}\right)\right]+O(z^2). \tag{4.8.30}$$

as $z\to 0$. Here $K(r)$ is the complete elliptic integral of the first kind and is defined by

$$K(r)=\int_0^{\pi/2}\frac{d\theta}{\sqrt{1-r^2\sin^2\theta}}.$$

It should be noted that both (4.8.27) and (4.8.30) achieve the correct boundary values when $z=0$.

Case IV. $z\to\infty$. The fact that $M[V\,J_0(pr)\,J_1(p);\xi]$ has separate Mellin transforms in the three regions $0\le r<1$, $r=1$, $r>1$ would indicate a similar lack of uniformity in this limit also. This would be disturbing, however, because the potential is perfectly smooth away from the disc and there would be no physical explanation for any nonuniformity. Fortunately there is no problem as we shall show.[19] Restricting our considerations to $0\le r<1$ and using the Mellin transforms $M[e^{-p};\xi]$ and $M[V\,J_0(pr)\,J_1(p);1-\xi]$ given above, we find from (4.4.17)[20] that

$$\phi=\frac{V}{2z^2}\,F(0,-1;1;r^2)-\frac{3}{8}\frac{V}{z^4}F(-1,-2;1;r^2)+O(z^{-6}). \tag{4.8.31}$$

But

$$F(0,-1;1;r^2)=1 \qquad \text{and} \qquad F(-1,-2;1;r^2)=1+2r^2$$

so that

$$\phi=\frac{V}{2z^2}-\frac{3V}{8z^4}(1+2r^2)+O(z^{-6}), \qquad z\to\infty, \qquad 0\le r<1. \tag{4.8.32}$$

[19] See also Exercise 4.29.
[20] Or (4.4.26).

In the region $r > 1$ we find that

$$\phi = \frac{V}{2z^2} - \frac{3Vr^2}{4z^4} F\left(-1, -1; 2; \frac{1}{r^2}\right) + O(z^{-6})$$

$$= \frac{V}{2z^2} - \frac{3}{4}\frac{Vr^2}{z^4}\left(1 + \frac{1}{2r^2}\right) + O(z^{-6}), \qquad (4.8.33)$$

which agrees with (4.8.32). Finally the expansion obtained for $r = 1$ is

$$\phi = \frac{V}{2z^2} - \frac{9}{8}\frac{V}{z^4} + O(z^{-6}) \qquad (4.8.34)$$

the limit obtained as $r \to 1$ in both (4.8.32) and (4.8.33).

4.9. Heat Conduction in a Nonlinearly Radiating Solid

Let us consider the following initial boundary-value problem for the heat equation:

$$u_{xx} - u_t = 0, \qquad 0 < x < \infty, \qquad t > 0, \qquad (4.9.1)$$

$$u(x,0) = 0, \qquad 0 < x < \infty, \qquad (4.9.2)$$

$$\lim_{x \to \infty} u(x,t) = 0, \qquad t > 0, \qquad (4.9.3)$$

$$u_x(0,t) = \alpha[u(0,t)]^n - f(t), \qquad t > 0. \qquad (4.9.4)$$

The solution u to this system represents the temperature in a one-dimensional semi-infinite bar which is initially at zero temperature. The boundary condition (4.9.4) states that, at the end of the bar, there is a heat input of amount $f(t)$ while simultaneously heat is being radiated away at a rate proportional to the nth power of the temperature.

Both α and n are fixed constants. If $n \neq 0$ or 1, then the problem as stated is nonlinear. The particular value $n = 4$ is of special interest because it corresponds to the well-known Stefan-Boltzmann law of radiation.

The standard technique for solving (4.9.1)–(4.9.4) is to first employ the Green's function for the problem to obtain an integral representation for u. Indeed in this manner we obtain

$$u(x,t) = \frac{1}{\sqrt{\pi}} \int_0^t (t-s)^{-1/2} \exp\left\{-\frac{x^2}{4(t-s)}\right\} [f(s) - \alpha\, u^n(0,s)]\, ds, \qquad t \geq 0,$$

$$x \geq 0. \qquad (4.9.5)$$

Thus, we see that to determine $u(x,t)$ anywhere in the region $t \geq 0$, $x \geq 0$, we need only know $u(0,t)$, the temperature of the end of the bar, as a function of time. If in (4.9.5) we set $x = 0$ and $u(0,t) = y(t)$, then that equation becomes

$$y(t) = \frac{1}{\sqrt{\pi}} \int_0^t \frac{[f(s) - \alpha\, y^n(s)]}{\sqrt{t-s}}\, ds, \tag{4.9.6}$$

which, in turn, is a nonlinear integral equation for the determination of y. Our objective here is to obtain an asymptotic expansion of $y(t)$ as $t \to \infty$. For simplicity, throughout our discussion we shall assume that $n \geq 1$ and $f(t) \geq 0$ for all $t \geq 0$. Finally, without loss of generality we shall set $\alpha = 1$.

In (4.9.6) let us set $s = \sigma t$ so that we have

$$y(t) = \sqrt{t} \int_0^\infty g(\sigma)\, h(t\sigma)\, d\sigma. \tag{4.9.7}$$

Here

$$h(t\sigma) = f(t\sigma) - y^n(t\sigma) \tag{4.9.8}$$

and

$$g(\sigma) = \begin{cases} (\pi\,[1-\sigma])^{-1/2}, & 0 \leq \sigma < 1, \\ 0, & 1 < \sigma. \end{cases} \tag{4.9.9}$$

We see from (4.9.7) that to obtain an asymptotic expansion of y we need only determine one for the integral in that equation and then multiply the result by $\sqrt{t}$. We observe, however, that the integral is precisely of the form considered in this chapter and hence we can apply the Mellin transform technique.

We know that to determine the asymptotic expansion of the integral in (4.9.7), it is sufficient to have appropriate asymptotic forms for $h(\sigma)$ and $g(\sigma)$ as $\sigma \to \infty$ and $\sigma \to 0+$, respectively. The behavior of $g(\sigma)$ near $0+$ is easily obtained from (4.9.9). The behavior of h near $+\infty$, however, involves the behavior of y near $+\infty$ which is precisely the result we have set out to obtain. Thus, although we cannot proceed to directly determine the desired expansion of y, the following self-consistent approach to our problem is suggested.

We begin by assuming asymptotic expansions, as $t \to \infty$, for both $y(t)$ and $f(t)$. Because $f(t)$ is a given function, the parameters that appear in its expansion are known. Our problem is to determine the unknown parameters in the expansion of y. Our asymptotic expansion of the right-hand side of (4.9.7) can then be derived by our Mellin transform procedure. It will, of course, involve these unknown parameters. Finally these parameters will be determined by requiring the derived expansion to coincide with the assumed expansion of y.

To pursue our program, let us assume that, as $t \to \infty$,

$$f(t) \sim \gamma_0\, t^{-a_0}, \qquad a_0 \geq 0. \tag{4.9.10}$$

This information will prove sufficient to determine the asymptotic expansion of y to leading order. Although a_0 is fixed, we might anticipate that there will be several cases, depending on the values of a_0 and n, that must be treated separately. Indeed, this turns out to be so. Rather than attempt to consider all of the possibilities, we shall be content here with treating two representative cases in detail.

Case I. $0 \leq a_0 < \frac{1}{2}$, $n > 2$. Let us suppose that, as $t \to \infty$,

$$y(t) \sim c_0\, t^{-r_0} \tag{4.9.11}$$

and seek to determine c_0 and r_0. It follows from (4.9.10) and (4.9.11) that

$$h(t) \sim \gamma_0\, t^{-a_0} - (c_0)^n\, t^{-nr_0}, \qquad t \to \infty, \tag{4.9.12}$$

where we have included both terms because we have not as yet determined the relative magnitudes of a_0 and nr_0. Also, we have explicitly that

$$M[g;1-z] = \frac{\Gamma(1-z)}{\Gamma(\frac{3}{2}-z)}. \tag{4.9.13}$$

Hence, it follows from (4.9.12), (4.9.13), and (4.4.21) that

$$y(t) \sim t^{\frac{1}{2}-a_0}\, \gamma_0 \frac{\Gamma(1-a_0)}{\Gamma(\frac{3}{2}-a_0)} - (c_0)^n\, t^{\frac{1}{2}-nr_0} \frac{\Gamma(1-nr_0)}{\Gamma(\frac{3}{2}-nr_0)}. \tag{4.9.14}$$

By hypothesis, because (4.9.11) must also hold, there are only three possibilities:

(1) $r_0 = a_0 - \frac{1}{2}$,
(2) $a_0 = n\, r_0$,
(3) $r_0 = n\, r_0 - \frac{1}{2}$.

Clearly (1) cannot hold because it implies $r_0 < 0$ which in turn implies that the dominant term on the right side of (4.9.14) is $0(t^{\frac{1}{2}-nr_0})$. This term would then dominate the leading term in the assumed expansion of y which yields a contradiction. If (3) holds, then $r_0 = [2(n-1)]^{-1}$, $\frac{1}{2} - nr_0 < 0$, and the dominant term on the right side of (4.9.14) is $0(t^{\frac{1}{2}-a_0})$ which again dominates the leading term in the assumed asymptotic expansion of y. Thus (2) must hold and hence

$$c_0 = (\gamma_0)^{1/n}.$$

Therefore, in this case, as $t \to \infty$,

$$y(t) \sim (\gamma_0)^{1/n}\, t^{-a_0/n}. \tag{4.9.15}$$

Case II. $a_0 > 2$, $n > 2$. Again let us suppose that (4.9.11) holds. Then it is quite simple to show that to leading order

$$y(t) \sim \{M[f;1] - M[y^n;1]\} \frac{t^{-1/2}}{\sqrt{\pi}} - (c_0)^n \frac{\Gamma(1-nr_0)}{\Gamma(\frac{3}{2}-nr_0)}\, t^{\frac{1}{2}-nr_0}. \tag{4.9.16}$$

In deriving (4.9.16) we have tacitly assumed that $nr_0 \neq 1$ because otherwise the second term on the right would have to be replaced by one $0(t^{-1/2} \log t)$.

However, $nr_0 = 1$ can be eliminated as a possibility because if it held, then $c_0 t^{-r_0}$ would dominate both terms on the right of (4.9.16). Thus, there remains the two possibilities

(1) $r_0 = \frac{1}{2}$,
(2) $r_0 = (2[n-1])^{-1}$.

If (2) holds, then $c_0 = -(c_0)^n \cdot$const., possible only for real c_0 when n is an even integer. For f positive, the maximum principal for diffusion equations would preclude this possibility. Hence (1) holds, so that

$$y(t) \sim \frac{\{M[f;1] - M[y^n;1]\}\, t^{-1/2}}{\sqrt{\pi}} = \frac{\mathscr{E}_0}{\sqrt{\pi}} t^{-1/2}. \tag{4.9.17}$$

We can also write

$$\mathscr{E}_0 = \int_0^\infty [f(s) - y^n(s)]\, ds \tag{4.9.18}$$

which exists under the assumptions made for the result (4.9.17). We have introduced the symbol $\mathscr{E}_0$ because, in our original heat conduction problem, it represents the net energy flux through the end of the bar.

There is an obvious defect in the result just obtained in that the coefficient $\mathscr{E}_0$ depends on the global behavior of the unknown solution y. Furthermore, because $\mathscr{E}_0$ is unknown, we cannot conclude immediately that $\mathscr{E}_0 \neq 0$. By independent arguments, however, we can show that $\mathscr{E}_0 > 0$ and hence, even though we do not know the leading coefficient explicitly, we have that in the present case, $y(t) = 0(t^{-1/2})$ as $t \to \infty$.

We could continue our discussion in several directions. In addition to the remaining cases to be treated we might also seek further terms in the various expansions. We could also attempt to adapt our procedure to the study of y in the limit $t \to 0+$. We shall leave these considerations to the exercises, however.

There is one point we feel worth some elaboration. The analysis given above is formal in that we have not established the asymptotic nature of the results obtained. Although we shall not attempt to rigorize our results here, we wish to briefly indicate what would be sufficient to do so. Suppose that by independent arguments we could establish that $y(t)$ has an asymptotic expansion to leading order of the form (4.9.11), at least in the two cases considered. Then we claim that (4.9.15) and (4.9.17) must be valid. This is so because all of the steps following the initial assumption (4.9.11) are rigorous. Thus, to prove the asymptotic nature of our results, it is sufficient to prove the existence of an asymptotic expansion of y of the form assumed.

4.10. Fractional Integrals and Integral Equations of Abel Type

Let us consider the quantity

$$I^\mu[f;\lambda] = \frac{1}{\Gamma(\mu)} \int_0^\lambda f(\xi)\, (\lambda - \xi)^{\mu-1}\, d\xi. \tag{4.10.1}$$

Here f is a given locally integrable function and $\mathrm{Re}(\mu) > 0$. It is easy to see that when μ is the positive integer n, we have

$$\frac{d^n}{d\lambda^n} I^n[f;\lambda] = f(\lambda), \tag{4.10.2}$$

that is, $I^n[f;\lambda]$ is an nth repeated integral of f.

Because the right side of (4.10.1) is well defined for all μ such that $\mathrm{Re}(\mu) > 0$, it is reasonable to consider $I^\mu[f;\lambda]$ as a fractional integral of f whenever μ is not a positive integer. Indeed, $I^\mu[f;\lambda]$ is usually called the Riemann fractional integral of f of order μ.[21] Of particular importance is the relation

$$I^\mu(I^\nu[f;\xi];\lambda) = I^{\mu+\nu}[f;\lambda]. \tag{4.10.3}$$

This can be established by interchanging the order of integration in

$$I^\mu(I^\nu[f;\xi];\lambda) = \frac{1}{\Gamma(\mu)\,\Gamma(\nu)} \int_0^\lambda \left(\int_0^\xi f(\tau)\,(\xi - \tau)^{\mu-1}\,d\tau \right) (\lambda - \xi)^{\nu-1}\,d\xi \tag{4.10.4}$$

and observing that

$$\int_\tau^\lambda (\xi - \tau)^{\mu-1}\,(\lambda - \xi)^{\nu-1}\,d\xi = (\lambda - \tau)^{\mu+\nu-1}\,\frac{\Gamma(\nu)\,\Gamma(\mu)}{\Gamma(\mu+\nu)}. \tag{4.10.5}$$

Although the use of fractional integrals is rather widespread in analysis, we shall be content here with studying their application to integral equations of Abel type. These integral equations are of the form

$$f(\lambda) = \int_0^\lambda \frac{h(\xi)}{(\lambda - \xi)^\alpha}\,d\xi. \tag{4.10.6}$$

Here $\alpha < 1$, f is a given function, and h is to be determined. The key to solving (4.10.6) is the recognition that f is proportional to the fractional integral of h of order $1 - \alpha$. Indeed we have

$$f(\lambda) = \Gamma(1-\alpha)\, I^{1-\alpha}[h;\lambda]. \tag{4.10.7}$$

If $0 < \alpha < 1$, then upon using (4.10.3) with $\mu = 1 - \alpha$ and $\nu = \alpha$ we find that

$$\frac{1}{\Gamma(1-\alpha)} I^\alpha[f;\lambda] = I^1[h;\lambda] = \int_0^\lambda h(\xi)\,d\xi. \tag{4.10.8}$$

Now upon differentiating (4.10.8) with respect to λ we obtain

$$h(\lambda) = \frac{1}{\Gamma(1-\alpha)} \frac{d}{d\lambda} I^\alpha[f;\lambda] = \frac{1}{\Gamma(1-\alpha)\,\Gamma(\alpha)} \frac{d}{d\lambda} \int_0^\lambda f(\xi)\,(\lambda - \xi)^{\alpha-1}\,d\xi. \tag{4.10.9}$$

If f is continuously differentiable, then we can integrate by parts in (4.10.9)

[21] The quantity $I^\mu[f;\lambda] = (1/\Gamma(\mu)) \int_\lambda^\infty f(\xi)\,(\xi - \lambda)^{\mu-1}\,d\xi$ is called the Weyl fractional integral of f of order μ.

which yields

$$h(\lambda) = \frac{\sin \pi\alpha}{\pi} \frac{d}{d\lambda} \left\{ \frac{f(0)\,\lambda^\alpha}{\alpha} + \frac{1}{\alpha} \int_0^\lambda f'(\xi)(\lambda - \xi)^\alpha \, d\xi \right\}$$

$$= \frac{\sin \pi\alpha}{\pi} \left\{ f(0)\,\lambda^{\alpha-1} + \int_0^\lambda f'(\xi)(\lambda - \xi)^{\alpha-1} \, d\xi \right\}. \tag{4.10.10}$$

Here we have used the relation $\Gamma(\alpha)\,\Gamma(1-\alpha) = \pi/\sin \pi\alpha$.

When $\alpha < 0$, we still have (4.10.7) but $I^\alpha[f;\lambda]$ must be defined through differentiation. Thus, for $n - 1 < -\mathrm{Re}(\alpha) < n$, with n a positive integer, we define

$$I^\alpha[f;\lambda] = \frac{d^n}{d\lambda^n} I^{n+\alpha}[f;\lambda]. \tag{4.10.11}$$

Therefore, in this case, it follows from (4.10.8) that

$$h(\lambda) = \frac{\sin \pi\alpha}{\pi} \frac{d^{n+1}}{d\lambda^{n+1}} \int_0^\lambda f(\xi)\,(\lambda - \xi)^{n-1+\alpha} \, d\xi. \tag{4.10.12}$$

Finally, if α is a negative integer, say $\alpha = -n$, then (4.10.6) reduces to the simple relation

$$f(\lambda) = n!\; I^{n+1}[h;\lambda] \tag{4.10.13}$$

so that

$$h(\lambda) = \frac{1}{n!} f^{(n+1)}(\lambda). \tag{4.10.14}$$

With the above discussion as motivation, let us now turn to the study of the fractional integral (4.10.1) in each of the two limits $\lambda \to 0+$ and $\lambda \to +\infty$. We first note that upon setting $\xi = \lambda t$ in (4.10.1) we obtain

$$I^\mu[f;\lambda] = \frac{\lambda^\mu}{\Gamma(\mu)} \int_0^1 f(\lambda t)\,(1-t)^{\mu-1} \, dt, \tag{4.10.15}$$

which is a transform of the type considered in Section 4.4. Indeed, the fractional integral of f is seen to be the h-transform of the function

$$g_1(t) = \begin{cases} \dfrac{\lambda^\mu}{\Gamma(\mu)} (1-t)^{\mu-1}, & 0 \le t < 1, \\ 0, & 1 < t < \infty. \end{cases} \tag{4.10.16}$$

Here the kernel function h is f itself.

We naturally wish to apply the expansion theorems of Section 4.4. Because we have the explicit result (see the Appendix),

$$M[g_1;z] = \frac{\lambda^\mu}{\Gamma(\mu)} B(\mu, z) = \frac{\lambda^\mu\,\Gamma(z)}{\Gamma(\mu+z)}, \tag{4.10.17}$$

we need only assume appropriate asymptotic forms for f as $t \to 0+$ and as $t \to \infty$ to determine asymptotic expansions of $I^{\mu}[f;\lambda]$ in the desired limits. In what follows we shall assume that the behavior of $M[f;z]$ as $|y| \to \infty$ is such that the Bromwich contours in the relevant Parseval formulas can be displaced arbitrarily far to the right and that the estimates in Theorems 4.4. and 4.6 are valid.

Suppose then we consider (4.10.15) as $\lambda \to 0+$. For simplicity we assume that, as $t \to 0+$,

$$f(t) \sim \sum_{m=0}^{\infty} b_m t^{a_m - 1} \tag{4.10.18}$$

with $\text{Re}(a_m) \uparrow +\infty$, as $m \to \infty$. Then $M[f;1-z]$ has simple poles at the points $z = a_m$ and corresponding singular parts

$$\frac{-b_m}{z - a_m}. \tag{4.10.19}$$

Because $M[g_1;z]$ is analytic in $\text{Re}(z) > 0$, we immediately obtain from the results of Section 4.6 that, in this case,

$$I^{\mu}[f;\lambda] \sim \sum_{m=0}^{\infty} b_m \frac{\Gamma(a_m)}{\Gamma(\mu + a_m)} \lambda^{a_m + \mu - 1}, \qquad \lambda \to 0+. \tag{4.10.20}$$

In order to consider (4.10.15), as $\lambda \to +\infty$, we assume that

$$f(t) \sim e^{-\alpha t} \sum_{m=0}^{\infty} d_m t^{-r_m}, \tag{4.10.21}$$

as $t \to \infty$. Here $\alpha \geq 0$ and $\text{Re}(r_m) \uparrow +\infty$ as $m \to \infty$. We first suppose that $\alpha > 0$ in which event $M[f;z]$ is analytic in $\text{Re}(z) > x_0$ for some x_0. $M[g_1;1-z]$ has simple poles at $z = n$, $n = 1, 2, \ldots$ with corresponding singular parts

$$\frac{(-1)^n \lambda^{\mu}}{\Gamma(\mu + 1 - n)\,(n-1)!\,(z - n)}. \tag{4.10.22}$$

Hence it follows [see (4.4.18)] that, as $\lambda \to \infty$,

$$I^{\mu}[f;\lambda] \sim \sum_{n=1}^{\infty} \frac{M[f;n]\,\lambda^{\mu - n}\,(-1)^{n+1}}{\Gamma(\mu + 1 - n)\,(n-1)!}. \tag{4.10.23}$$

Now suppose that $\alpha = 0$ in (4.10.21). Then $M[f;z]$ has simple poles at the points $z = r_m$ with corresponding singular parts

$$-\frac{d_m}{z - r_m}. \tag{4.10.24}$$

Thus, if r_m is not a positive integer for any m, then

$$I^{\mu}[f;\lambda] \sim \sum_{n=1}^{\infty} \frac{M[f;n]\,(-1)^{n+1}\,\lambda^{-n+\mu}}{\Gamma(\mu + 1 - n)\,(n-1)!} + \sum_{m=0}^{\infty} \frac{d_m\,\Gamma(1 - r_m)\,\lambda^{\mu - r_m}}{\Gamma(\mu + 1 - r_m)}. \tag{4.10.25}$$

Finally, if $\alpha = 0$ and some r_m is a positive integer, then a logarithmic term appears in the resulting expansion. If, for definiteness, we assume that r_0 equals the positive integer m, then we have

$$I^{\mu}[f;\lambda] = \sum_{n=1}^{m-1} \frac{M[f;n](-1)^{n+1}\lambda^{-n+\mu}}{\Gamma(\mu+1-n)(n-1)!} + \frac{(-1)^{m+1} d_0 \lambda^{\mu-m} \log\lambda}{(m-1)!\,\Gamma(\mu+1-m)} + O(\lambda^{\mu-m}). \qquad (4.10.26)$$

4.11. Renewal Processes

Imagine that we have a population of components such as light bulbs, batteries, and so on. Suppose that each component is numbered and that the lifetime of the nth component is represented by the number T_n. In general, there is no way of knowing a priori the value of T_n and therefore we must think of it in probabilistic terms. Thus we shall consider T_n to be a random variable having probability density $f_n(t)$ and cumulative distribution $F_n(t)$. These functions are simply related by

$$F_n(t) = \text{Prob}\,(T_n \leq t) = \int_0^t f_n(\tau)\,d\tau. \qquad (4.11.1)$$

In this model $f_n(t) \equiv 0$ for $t < 0$ and hence we say that T_n is a *positively distributed* random variable. We shall further assume that $f_n(t)$ is continuous for $t > 0$ and that the random variables T_n are independent. Thus no concentrations of probability occur at discrete points and the lifetime of any one component cannot affect the lifetime of any other component.

Typically, we use the population by first selecting at random a single component. This component is used until it fails. Upon failure, it is replaced immediately by a second component again chosen at random. The second component is then used until it fails and the process is repeated continuously. A practical question we might want answered is, how many replacements (renewals) can we expect to make in a given time interval? Our objective here is to at least partially answer this question.

Let us suppose then that the above procedure is carried out and that the components are relabeled so that the nth one selected has the number n and hence lifetime T_n. We now introduce a new random variable

$$S_n = T_1 + T_2 + \cdots + T_n, \qquad (4.11.2)$$

which, assuming the process starts at $t = 0$, represents the failure time of the nth component. It is also convenient to define still another variable N_t which represents the number of renewals in the time interval $[0,t]$. Here N_0 is taken to be zero. We readily see that

$$\text{Prob}(N_t \leq n) = 1 - \text{Prob}(S_n \leq t). \qquad (4.11.3)$$

The expected or mean value of N_t is called the *renewal function*. It is usually denoted by $H(t)$. The stated goal of our analysis is the determination of $H(t)$. Another quantity of interest is the so-called *renewal density* $h(t)$ related to $H(t)$ by

$$h(t) = H'(t). \tag{4.11.4}$$

Note that $h(t)$ is not a probability density function and $H(t)$ is not a cumulative distribution function.

Although it is not necessary, a significant simplification is achieved upon assuming that the random variables T_n are identically distributed. That is to say, $f_n(t) = f(t)$ and $F_n(t) = F(t)$ for all n. In that event the process is usually called an ordinary renewal process. Certainly, if the components of our population were all manufactured under the same conditions, then this last assumption is quite reasonable.

The simplification that is achieved lies in the fact that now $h(t)$ satisfies the linear integral equation

$$h(t) = f(t) + \int_0^t h(t-u) f(u)\, du. \tag{4.11.5}$$

We can heuristically justify (4.11.5) by multiplying through by Δt and observing that $f(t)\Delta t$ is the probability that the first component fails in the time interval $(t, t+\Delta t)$ while $\Delta t \int_0^t h(t-u) f(u)\, du$ represents the probability of renewal in $(t, t+\Delta t)$ given that the immediately preceding renewal occurred in $(0, t]$.

Upon integrating (4.11.5) with respect to t from 0 to t we obtain

$$H(t) = F(t) + \int_0^t H(t-u) f(u)\, du, \tag{4.11.6}$$

which is a linear integral equation for the determination of H. We wish to study the asymptotic behavior of H as $t \to \infty$. Our procedure will be to obtain an integral representation of $H(t)$ via Laplace transforms and then to apply the asymptotic methods of the previous sections to this integral representation.

Upon taking the Laplace transform of (4.11.6) we find that

$$\mathscr{L}[H;s] = \frac{1}{s} \frac{\mathscr{L}[f;s]}{1-\mathscr{L}[f;s]}. \tag{4.11.7}$$

Here we have used the basic Laplace convolution theorem and the relation

$$\begin{aligned}\mathscr{L}[F;s] &= \int_0^\infty e^{-st} F(t)\, dt = \frac{-1}{s} e^{-st} F(t)\Big|_0^\infty + \frac{1}{s}\int_0^\infty e^{-st} f(t)\, dt \\ &= \mathscr{L}\frac{1}{s}[f;s],\end{aligned} \tag{4.11.8}$$

which follows from the fact that $F(0) = 0$.

Because

$$F(\infty) = \int_0^\infty f(t)\, dt = \mathscr{L}[f;0] = 1, \tag{4.11.9}$$

we have that $\mathscr{L}[f;s]$ is analytic for $\text{Re}(s) > 0$. If we could show that

$\{1-\mathscr{L}[f;s]\}^{-1}$ is also analytic for $\text{Re}(s)>0$, then it would immediately follow from the Laplace inversion formula (4.2.11) that

$$H(t)=\frac{1}{2\pi i}\int_{c-i\infty}^{c+i\infty}\frac{e^{st}\mathscr{L}[f;s]}{s(1-\mathscr{L}[f;s])}\,ds, \qquad c>0. \tag{4.11.10}$$

Because $f(t)\geq 0$, we have

$$|\mathscr{L}[f;s]|=\left|\int_0^\infty e^{-st}f(t)\,dt\right|\leq\int_0^\infty e^{-\text{Re}(s)t}f(t)\,dt<\int_0^\infty f(t)\,dt=1 \tag{4.11.11}$$

for $\text{Re}(s)>0$. Thus, in this region, $1-\mathscr{L}[f;s]$ is bounded away from zero and (4.11.10) is valid.

We now suppose that $\mathscr{L}[H;s]$, which has been shown analytic in $\text{Re}(s)>0$, can be analytically continued into $\text{Re}(s)\leq 0$. Moreover, we shall assume that the singularities of the continuation that arise in $\text{Re}(s)<0$ are either poles or branch points and that the only singularity on the imaginary axis occurs at the origin.

If we assume that $\mathscr{L}[f;s]\to 0$ as $|s|\to\infty$ in a neighborhood of the imaginary axis,[22] then it follows from the Cauchy integral theorem that we can replace the contour of integration in (4.11.10) by a sum of contours which consists of loops around each branch cut of $\mathscr{L}[H;s]$ and small circles around each pole. We shall agree to draw all branch cuts from the relevant branch points to infinity in the left half-plane in such a manner that no finite singularity lies on the cut except the branch point itself.

It is readily seen that regardless of the nature of the singularity of $\mathscr{L}[H;s]$ that occurs at, say, $s=s_0$, the corresponding contribution to $H(t)$ is $0(t^\alpha\exp\{t\,\text{Re}(s_0)\})$ as $t\to\infty$. Here α is a finite constant and hence the contribution is exponentially small whenever $\text{Re}(s_0)<0$. Thus, in order to determine the asymptotic expansion of $H(t)$, we need only consider the integral corresponding to the singularity of $\mathscr{L}[H;s]$ at the origin. As we shall see, this singularity usually involves a multivaluedness of $\mathscr{L}[H;s]$ so that, in general, $s=0$ is a branch point. Thus, we have

$$H(t)\sim\frac{1}{2\pi i}\int_c e^{st}\mathscr{L}[H;s]\,ds, \qquad t\to\infty, \tag{4.11.12}$$

where c is a semi-infinite loop around the branch cut through $s=0$. If $\mathscr{L}[H;s]$ has only a pole at $s=0$, then c is equivalent to a small circle about the origin. We can anticipate from our results on Laplace-type integrals and in particular from Watson's lemma that in (4.11.12) we need only be concerned with the integral over a small portion of c near the origin. For this reason there is no loss of generality in assuming that $s=0$ is the only singular point of $\mathscr{L}[H;s]$ on the negative real axis, so that this axis can be taken as the corresponding

[22] That this is true for $\text{Re}(s)\geq 0$ follows from the Riemann-Lebesgue lemma.

branch cut. Thus (4.11.12) becomes

$$H(t) \sim \frac{1}{2\pi i} \int_{-\infty}^{0+} e^{st} \mathscr{L}[H;s]\, ds, \tag{4.11.13}$$

where we have adopted the standard notation for a loop contour around the negative real axis.

As we have indicated, our plan is to apply the results previously obtained for Laplace-type integrals. We cannot apply Watson's lemma directly, however, because our contour is a loop rather than a segment of the real axis. Indeed we must now use the following.

WATSON'S LEMMA FOR LOOP INTEGRALS. Consider the integral

$$I(\lambda) = \int_{-\infty}^{0+} g(s)\, e^{\lambda s}\, ds. \tag{4.11.14}$$

Suppose that $g(s)$ is analytic in some domain containing the loop contour in (4.11.14) and cut along the negative real axis. Suppose further that

$$g(s) = 0(e^{as}), \qquad |s| \to \infty, \qquad \theta \le |\arg(s)| \tag{4.11.15}$$

and

$$g(s) \sim \sum_{m=0}^{\infty} d_m\, s^{r_m}, \qquad |s| \to 0, \qquad |\arg(s)| \le \pi - \delta, \qquad \delta < \frac{\pi}{2}. \tag{4.11.16}$$

Here a is real and $\operatorname{Re}(r_m) \uparrow \infty$. Then $I(\lambda)$ exists for $\lambda > a$ and

$$I(\lambda) \sim \sum_{m=0}^{\infty} d_m \int_{-\infty}^{0+} e^{\lambda s}\, s^{r_m}\, ds = \sum_{m=0}^{\infty} \frac{2\pi i\, d_m\, \lambda^{-r_m-1}}{\Gamma(-r_m)}. \tag{4.11.17}$$

Here we have used Hankel's integral representation of the gamma function

$$\Gamma(z) = \frac{1}{2i \sin \pi z} \int_{-\infty}^{0+} s^{z-1}\, e^s\, ds \tag{4.11.18}$$

and the well-known relation

$$\sin \pi z\, \Gamma(z) = \frac{\pi}{\Gamma(1-z)}. \tag{4.11.19}$$

Note that $\operatorname{Re}(r_0) > 0$ is not assumed so that the origin need not be an integrable singularity of g.

PROOF. The proof is analogous to that of the ordinary Watson's lemma proved in Section 4.1 and is left to the exercises.

Let us now return to (4.11.13) and assume that $\mathscr{L}[H;s]$ satisfies the conditions (4.11.15) and (4.11.16). It then follows from (4.11.17) that

$$H(t) \sim \sum_{m=0}^{\infty} \frac{d_m\, t^{-r_m-1}}{\Gamma(-r_m)}, \qquad t \to \infty. \tag{4.11.20}$$

We, of course, want to relate the constants d_m and r_m back to the given probability density function $f(t)$. For this purpose let us assume that, as $t \to \infty$,

$$f(t) \sim e^{-\alpha t} \sum_{m=0}^{\infty} c_m t^{-a_m}. \tag{4.11.21}$$

Suppose first that $\alpha > 0$. Then we have that $\mathscr{L}[f;s]$ is analytic for $\text{Re}(s) > -\alpha$. Moreover we find from (4.6.23), with all $c_m = 0$ in that result, that

$$\mathscr{L}[f;s] = \sum_{n=0}^{\infty} \frac{(-s)^n}{n!} M[f;n+1] = \sum_{n=0}^{\infty} \frac{(-s)^n}{n!} \mu_n, \qquad s \to 0+, \tag{4.11.22}$$

where we have used μ_n to denote the nth moment of f. Thus $\mu_0 = 1$ and μ_1 is the common mean of the random variables T_n.

Upon inserting (4.11.22) into (4.11.7) we obtain

$$\mathscr{L}[H;s] = \frac{1}{s^2 \mu_1} \sum_{m=0}^{\infty} \beta_m s^m. \tag{4.11.23}$$

Here

$$\beta_0 = 1 \qquad \text{and} \qquad \beta_1 = \frac{\mu_2 - 2\mu_1^2}{2\mu_1}. \tag{4.11.24}$$

This expansion, moreover, is valid for all values of $\arg(s)$. In this case, because $r_m = m - 2$, only two terms are nonzero in (4.11.20). Indeed, we find

$$H(t) \sim \frac{t}{\mu_1} + \frac{\mu_2 - 2\mu_1^2}{2\mu_1}, \qquad t \to \infty. \tag{4.11.25}$$

Actually (4.11.25) is correct to within an exponentially small error.

Of more interest, at least from a mathematical point of view, is the case where $\alpha = 0$ in (4.11.21). Because we have not allowed for the appearance of logarithms in (4.11.16), we shall insist that a_m is not a positive integer for any m. Furthermore, $\text{Re}(a_0)$ must be greater than 1 for $\mathscr{L}[f;s]$ to exist. It now follows from (4.6.23) that

$$\mathscr{L}[f;s] \sim \sum_{n=0}^{\infty} \frac{(-s)^n}{n!} M[f;n+1] + \sum_{m=0}^{\infty} s^{a_m - 1} c_m \Gamma(1 - a_m), \tag{4.11.26}$$

which we assume is valid as $|s| \to 0$ in $|\arg(s)| < \pi$. (Sufficient conditions for this to be so are given in Section 4.7.)

There are three subcases we wish to consider separately.

(1) $1 < a_0 < 2$. Here

$$\mathscr{L}[f;s] \sim 1 + c_0 \Gamma(1 - a_0) s^{a_0 - 1} \tag{4.11.27}$$

so that

$$\mathscr{L}[H;s] = \frac{s^{-a_0}}{-c_0 \Gamma(1 - a_0)} + o(s^{-a_0}), \qquad |s| \to 0. \tag{4.11.28}$$

Hence we have from (4.11.20) that, in this case,

$$H(t) = -\frac{t^{a_0-1}}{c_0\,\Gamma(a_0)\,\Gamma(1-a_0)} + o(t^{a_0-1})$$

$$= -\frac{t^{a_0-1}\sin(\pi a_0)}{c_0\,\pi} + o(t^{a_0-1}), \qquad t\to\infty. \tag{4.11.29}$$

[Note that $-\sin\pi a_0 > 0$ for $a_0 \in (1,2)$.]
(2) $2 < a_0 < 3$. Now

$$\mathcal{L}[f;s] = 1 - \mu_1 s + c_0\,\Gamma(1-a_0)\,s^{a_0-1} + o(s^{a_0-1}) \tag{4.11.30}$$

and

$$\mathcal{L}[H;s] = \frac{1}{s^2\mu_1}\left[1 + \frac{c_0}{\mu_1}\Gamma(1-a_0)\,s^{a_0-2} + o(s^{a_0-2})\right]. \tag{4.11.31}$$

Therefore, (4.11.20) yields

$$H(t) = \frac{t}{\mu_1} + \frac{c_0\,\Gamma(1-a_0)}{\mu_1^2\,\Gamma(4-a_0)}\,t^{3-a_0} + o(t^{3-a_0}). \tag{4.11.32}$$

(3) $a_0 > 3$. In this case

$$\mathcal{L}[f:s] = 1 - \mu_1 s + \frac{\mu_2 s^2}{2} + o(s^2), \tag{4.11.33}$$

$$\mathcal{L}[H;s] = \frac{1}{\mu_1 s^2}\left[1 + \frac{(\mu_2 - 2\mu_1^2)\,s}{2\mu_1^2} + o(s)\right] \tag{4.11.34}$$

and hence

$$H(t) = \frac{t}{\mu_1} + \frac{\mu_2 - 2\mu_1^2}{2\mu_1^2} + o(1), \qquad t\to\infty. \tag{4.11.35}$$

Upon considering the above results we can make the following statements: If $f(t)$ has the asymptotic expansion (4.11.21) and if the mean of f exists, then $H(t) \sim t/\mu_1$ as $t\to\infty$. if, in addition, the second moment of f exists, then $H \sim t/\mu_1 + (\mu_2 - 2\mu_1^2)/2\mu_1^2$. In this last approximation the error is exponentially small if f is exponentially small at infinity and algebraically small if f is algebraically small at ∞. If μ_1 does not exist in the ordinary sense, then H is $0(t^{a_0-1})$ where $1 < a_0 < 2$. Finally, when f is exponentially small at ∞, only its first two moments are involved in the asymptotic expansion of H, whereas when f is algebraic at ∞, the expansion of H may involve higher moments of f.

4.12. Exercises

4.1. Use Watson's lemma to calculate the asymptotic expansion, as $\lambda \to \infty$, of

$$I(\lambda) = \int_0^\infty e^{-\lambda t} f(t)\, dt$$

when

(a) $f(t) = t^{-1/2} \sin t^{3/2}$.
(b) $f(t) = J_{1/2}(t)$.
(c) $f(t) = e^{-1/t}(1 + t^2)^{1/2}$.
(d) $f(t) = \log(1 + t)$.

4.2. Suppose that in Watson's lemma λ is complex with $|\arg(\lambda)| \le \pi/2 - \delta$ for any fixed $\delta > 0$.

(a) Show that, in this case, the estimate (4.1.6) can be replaced by

$$|I_2(\lambda)| = o\{\exp[-|\lambda| R \sin \delta]\}, \qquad |\lambda| \to \infty. \tag{4.12.1}$$

(b) Show that (4.1.10) now becomes

$$\int_0^R t^{a_m} e^{-\lambda t}\, dt = \frac{\Gamma(1 + a_m)}{\lambda^{1+a_m}} + o(\exp[-|\lambda| R \sin \delta]). \tag{4.12.2}$$

(c) Show that the estimate (4.1.12) is now replaced by

$$\left| \int_0^R \rho_N(t)\, e^{-\lambda t}\, dt \right| \le \frac{K_N \Gamma(\mathrm{Re}(a_{N+1}) + 1)}{(|\lambda| \sin \delta)^{\mathrm{Re}((a_{N+1}) + 1)}}. \tag{4.12.3}$$

(d) Use parts (a), (b), and (c) to prove that Watson's lemma, stated for real λ in Section 4.1, remains valid for complex λ. In particular, show that (4.1.13) holds as $|\lambda| \to \infty$ with $|\arg(\lambda)| \le \pi/2 - \delta$, $\delta > 0$.

4.3. The modified Bessel function of the second kind has the integral representation

$$K_\nu(\lambda) = \frac{\sqrt{\pi}}{\Gamma(\nu + \frac{1}{2})} \left(\frac{\lambda}{2}\right)^\nu \int_1^\infty e^{-\lambda t}(t^2 - 1)^{\nu - \frac{1}{2}}\, dt. \tag{4.12.4}$$

(a) Use Watson's lemma to show that

$$K_\nu(\lambda) \sim e^{-\lambda} \sqrt{\frac{\pi}{2\lambda}} \sum_{n=0}^\infty \frac{\Gamma(\nu + \frac{1}{2} + n)}{\Gamma(\nu + \frac{1}{2} - n) n!} (2\lambda)^{-n}, \qquad \lambda \to \infty. \tag{4.12.5}$$

[*Hint*: Replace t by $1 + \tau$ in (4.12.4).]

(b) Use the result of Exercise 4.2 to conclude that (4.12.5) is valid as $\lambda \to \infty$ in $|\arg(\lambda)| < \pi/2$.

(c) Rotate the contour of integration in (4.12.4) through angle $-\arg(\lambda)/2$ and thereby conclude that (4.12.5) is valid as $\lambda \to \infty$ in $|\arg(\lambda)| < \pi$. [*Hint*: See

Example 3.2.4 where this technique is used.]

(d) For $\pi \le |\arg(\lambda)| < 3\pi/2$, rotate the contour of integration in (4.12.4) no further than the negative real axis and show that (4.12.5) remains valid

(e) Use the relation

$$\mathrm{Ai}(\lambda) = \frac{1}{\pi}\left(\frac{\lambda}{3}\right)^{1/2} K_{1/3}\left(\frac{2}{3}\lambda^{3/2}\right) \tag{4.12.6}$$

along with (4.12.5) to obtain the following asymptotic expansion for the Airy function:

$$\mathrm{Ai}(\lambda) \sim \frac{e^{-2\lambda^{3/2}/3}}{\sqrt{\pi}2\lambda^{1/4}} \sum_{n=0}^{\infty} \frac{\Gamma(\frac{5}{6}+n)}{\Gamma(\frac{5}{6}-n)n!}\left(\frac{3}{4\lambda^{3/2}}\right)^n, \quad |\lambda| \to \infty, \quad |\arg(\lambda)| < \pi. \tag{4.12.7}$$

4.4. (a) By following the line of proof of Watson's lemma in Section 4.1, show that if

$$I(\lambda) = \int_0^\infty \mathrm{Ai}(\lambda t) f(t)\, dt \tag{4.12.8}$$

with $f(t)$ locally integrable on $(0, \infty)$, $f(t) = 0(e^{at^{3/2}})$, $t \to \infty$, a real, and

$$f(t) \sim \sum_{n=0}^{\infty} c_m t^{a_m}, \qquad t \to 0+, \qquad \mathrm{Re}(a_m) \uparrow +\infty, \qquad \mathrm{Re}(a_0) > -1,$$

then

$$I(\lambda) \sim \sum_{m=0}^{\infty} \frac{c_m}{2\pi} \frac{3^{(2a_m/3)-\frac{1}{2}}}{\lambda^{a_m+1}} \Gamma\left(\frac{a_m+1}{3}\right)\Gamma\left(\frac{a_m+2}{3}\right), \qquad \lambda \to \infty. \tag{4.12.9}$$

(Use the Appendix to calculate the integrals $\int_0^\infty \mathrm{Ai}(t)\, t^{a_m}\, dt$.)

(b) Use the result (4.12.7) and the type of argument used in Exercise 4.2 to show that (4.12.9) is valid for

$$|\arg(\lambda)| \le \frac{\pi}{3} - \delta, \qquad \delta > 0. \tag{4.12.10}$$

4.5. Calculate the asymptotic expansion of

$$I(\lambda) = \int_0^\infty \mathrm{Ai}(\lambda t) f(t)\, dt$$

as $|\lambda| \to \infty$ with $|\arg(\lambda)| < \pi/3$, when

(a) $f(t) = t^{-1/2} \sin t^{3/2}$.

(b) $f(t) = J_{1/2}(t)$.

(c) $f(t) = e^{-1/t}(1+t^2)^{1/2}$.

(d) $f(t) = \log(1+t)$.

4.6. The Weber function $D_\nu(z)$ has the following integral representation for ν

and z in the indicated ranges:

$$D_\nu(z) = \frac{2^{\nu/2}\, e^{-z^2/4}}{\Gamma(-\nu/2)} \int_0^\infty e^{-tz^2/2}\, t^{-1-\nu/2}\,(1+t)^{(\nu-1)/2}\, dt,$$

$$\mathrm{Re}(\nu) < 0, \qquad |\arg(z)| \le \frac{\pi}{4}. \tag{4.12.11}$$

(a) Use Watson's lemma and the result of Exercise 4.2 to show that

$$D_\nu(z) \sim \frac{e^{-z^2/4}}{\Gamma(-\nu/2)} \sum_{n=0}^{\infty} \frac{2^n \Gamma\left(\frac{\nu+1}{2}\right)}{\Gamma\left(\frac{\nu+1}{2}-n\right)} \cdot \frac{\Gamma\left(n-\frac{\nu}{2}\right)}{n!\, z^{2n-\nu}}, \qquad |z| \to \infty,$$

$$|\arg(z)| < \frac{\pi}{4}. \tag{4.12.12}$$

(b) View (4.12.11) as a contour integral. Justify the rotation of the contour of integration through the angle $-\theta$ and of $\arg(z)$ through the angle $\theta/2$ with $|\theta| < \pi$ to obtain

$$D_\nu(z) = \frac{2^{\nu/2}\, e^{-z^2/4 + i\nu\theta/2}}{\Gamma(-\nu/2)} \int_0^\infty e^{-t\zeta^2/2}\, t^{-1-\nu/2}(1 + e^{-i\theta} t)^{(\nu-1)/2}\, dt. \tag{4.12.13}$$

Here $\zeta = ze^{-i\theta/2}$, $\mathrm{Re}(\nu) < 0$.

(c) Show that (4.12.12) is valid for $|\arg(\zeta)| < \pi/4$; that is, for $|\arg(z) + \theta| < \pi/4$ or $|\arg(z) - \theta| < 3\pi/4$.

4.7. (a) Show that $e^{-t} \varepsilon K(-\infty, \pi/2)$, the class of functions $K(x_0, \theta_0)$ being as defined in Section 4.7.

(b) Show that $t^{-1-\nu/2}(1+t)^{(\nu-1)/2} \varepsilon K(-\infty, \pi)$.

(c) Use Corollary 4.7.1 to conclude that $|\arg(z)| < 3\pi/4$ is the sector of validity for (4.12.12).

(d) State and prove a Watson-type lemma for the kernel $D_\nu(\lambda t)$ with λ complex.

4.8. (a) Prove Watson's lemma for loop integrals as stated in Section 4.11 for the integral (4.11.14). [*Hint*: It is useful to replace the loop contour by the contour of Figure 4.12.1 and to then follow the line of proof of Watson's lemma in Section 4.1.]

(b) Suppose that

$$2\pi i\, I(\lambda) = \int_{0+}^{\infty} e^{-\lambda t} f(t)\, dt. \tag{4.12.14}$$

Here $f(t)$ is analytic in the t plane cut along the positive real axis. In addition, suppose that

(i) $f(t) = O(e^{at})$, $\quad |t| \to \infty, \quad |\arg(t)| < \theta.$

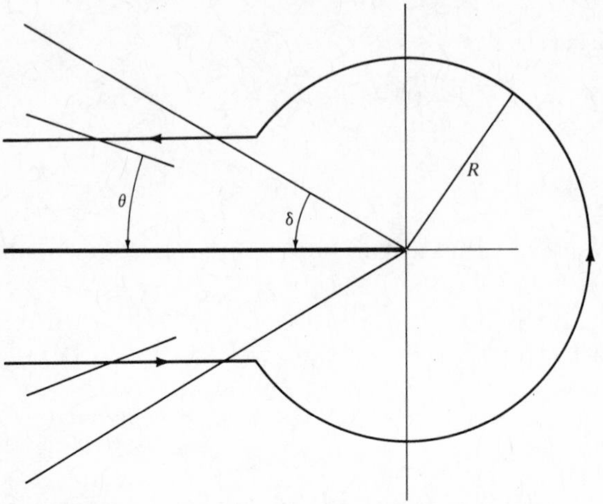

Figure 4.12.1. The Contour of Integration for Exercise 4.8(a).

(ii) $f(t) \sim \sum_{m=0}^{\infty} c_m t^{r_m}, \qquad 2\pi - \delta \geq \arg(t) \geq \delta, \qquad |t| \to 0, \qquad 0 < \delta < \frac{\pi}{2}.$

Then show that

$$I(\lambda) \sim \sum_{m=0}^{\infty} \frac{-c_m e^{i\pi r_m} \lambda^{-r_m-1}}{\Gamma(-r_m)}, \qquad \lambda \to \infty. \tag{4.12.15}$$

[*Hint*: Consider part (a) (which deals with loop integrals around the negative real axis) and find an appropriate change of variable of integration in (4.12.15) to obtain the desired result for loop integrals around the positive real axis.]

(c) Show that the results of parts (a) and (b) are valid for $|\arg(\lambda)| < \pi/2$.

(d) Show that, if we replace the asymptotic expansion in (ii) above by

$$f(t) \sim \sum_{m=0}^{\infty} \sum_{n=0}^{N(m)} c_{mn} t^{r_m} (\log t)^n$$

with $N(m)$ finite for each m, then (4.12.15) is replaced by

$$I(\lambda) \sim \sum_{m=0}^{\infty} \sum_{n=0}^{N(m)} c_{mn} \left(\frac{d}{dz}\right)^n \left\{ \frac{(\lambda e^{-i\pi})^{-z-1}}{\Gamma(-z)} \right\} \Bigg|_{z=r_m} \tag{4.12.16}$$

(e) Prove that the result in (d) is valid for $|\arg(\lambda)| < \pi/2$.

4.9. Define

$$\widetilde{M}[f; z] = \int_0^{\infty} f\left(\frac{1}{t}\right) t^{z-1}\, dt. \tag{4.12.17}$$

(a) Show that

$$M[f;z] = \widetilde{M}[f;-z] \tag{4.12.18}$$

and hence analytically continuing $M[f;z]$ to the *left* is equivalent to analytically continuing $\widetilde{M}[f;z]$ to the *right*.

(b) Prove Lemmas 4.3.4 to 4.3.6 by applying Lemmas 4.3.1 to 4.3.3 to $\widetilde{M}[f;z]$.

4.10. Let $f(t) = (e^t + 1)^{-1}$ and set

$$F(z) = M[f;z] = \int_0^\infty (e^t + 1)^{-1} t^{z-1}\, dt. \tag{4.12.19}$$

(a) Show that for $\mathrm{Re}(z) > 0$, the integral (4.12.19) converges absolutely; while for any finite A, with $A \geq \mathrm{Re}(z) \geq \delta > 0$, it converges absolutely and uniformly.

(b) Use integration by parts to obtain a representation valid for $\mathrm{Re}(z) > -1$ and $\mathrm{Re}(z) > -2$. Identify the singularities of the analytic continuation into these larger domains.

(c) Show that for z not an integer

$$F(z) = \frac{e^{-i\pi z}}{2i \sin \pi z} \int_{0+}^{\infty} (e^t + 1)^{-1} t^{z-1}\, dt. \tag{4.12.20}$$

(d) Show that the representation (4.12.20) has removable singularities at the positive integers and that, as z approaches the positive integer n, (4.12.20) has the limit $F(n)$ as defined by (4.12.19).

(e) Show that $F(z)$ has a simple pole at each nonpositive integer with

$$\mathrm{res}\{F(z); z = -n\} = \frac{1}{n!}\left(\frac{d}{dt}\right)^n (1 + e^t)^{-1}\bigg|_{t=0}, \qquad n = 0, 1, 2, \ldots .$$

(f) Show that $f(t)$ satisfies the conditions of Lemma 4.3.1 with $d = 1$, $\nu = 1$, and the conditions of Lemma 4.3.6 with $q = 0$, $a_m = m$, and $\overline{N}(m) = 0$ for all m. Verify the results of parts (a) and (e) by using these lemmas.

4.11. Apply the Mellin transform technique to rederive asymptotic expansions for the following integrals, as $\lambda \to \infty$. (These integrals were previously considered in Exercise 4.5.)

(a) $I(\lambda) = \int_0^\infty \mathrm{Ai}(\lambda t) t^{-1/2} \sin t^{3/2}\, dt$.
(b) $I(\lambda) = \int_0^\infty \mathrm{Ai}(\lambda t)\, J_{1/2}(t)\, dt$.
(c) $I(\lambda) = \int_0^\infty \mathrm{Ai}(\lambda t) \log(1 + t)\, dt$.

4.12. Show that the expansion obtained in Exercise 4.11 (c) is valid for $|\arg(\lambda)| < 4\pi/3$.

4.13. Calculate the asymptotic expansion of

$$I(\lambda) = \int_0^\infty K_\nu(\lambda t) f(t)\, dt,$$

as $\lambda \to \infty$, when

(a) $f(t) = t^{-1/2} \sin t^{3/2}$.
(b) $f(t) = J_{1/2}(t)$.
(c) $f(t) = \log(1+t)$.

4.14. (a) Show that the result obtained in 4.13 (a) is valid for $|\arg(\lambda)| < \pi$.
(b) Show that the result obtained in 4.13 (b) is valid for $|\arg(\lambda)| < \pi$.
(c) Show that the result obtained in 4.13 (c) is valid for $|\arg(\lambda)| < 2\pi$.

4.15. Calculate asymptotic expansion for the following integrals, as $\lambda \to \infty$:

(a) $\int_0^1 f(t) \log(1+\lambda t)\, dt, \qquad f \varepsilon C^\infty[0,1]$.
(b) $\int_0^\infty e^{-\lambda t} I_\nu(\lambda t) \sin t$ dt. Here $I_\nu(t)$ is the modified Bessel function of the first kind.
(c) $\int_0^\infty e^{-\lambda t} |1-t|^\nu\, dt, \qquad \mathrm{Re}(\nu) > -1, \qquad \nu$ not an integer.
(d) $\displaystyle\int_0^\infty \frac{(\lambda t)^{1/2}}{1+(\lambda t)^2} \sin(\log t)\, dt$.
(e) $\displaystyle\int_0^\infty \frac{\sin \lambda t}{\lambda t} \log|t-1|\, dt$.

4.16. (a) In Case IV, subcase (2) of Section 4.4, suppose that $r_m = a_n + 1$ for some m and n. Show that the contribution to the asymptotic expansion of $H[f;\lambda]$, as $\lambda \to \infty$ arising from the pole of $G(z) = M[h;z]\ M[f;1-z]$ at $z = r_m$ is given by

$$-\operatorname{res}\{\lambda^{-z}\, G(z); z = r_m\}$$
$$= -\lambda^{-r_m} \sum_{j=0}^{L} \frac{(-\log \lambda)^j}{j!\,(L-j)!} \left(\frac{d}{dz}\right)^{L-j} \{(z-r_m)^{L+1} M[h;z]\ M[f;1-z]\}\bigg|_{z=r_m} \tag{4.12.21}$$

Here $L = N(m) + \bar{N}(m) + 1$, with $N(m)$ and $\bar{N}(m)$ as defined in (4.4.7) and (4.4.9), respectively.
(b) Verify that the terms $O(\lambda^{-r_0})$ and $O(\lambda^{-r_0} \log \lambda)$ in (4.4.22) can be obtained from (4.12.21) by setting $r_m = r_0$ and $L = 1$.

4.17. Consider the generalized Stieltjes transform

$$I(\lambda) = \int_0^\infty \frac{t^{-\rho}}{(1+\lambda t)^\rho} e^{-t}\, dt, \qquad \rho < 1.$$

(a) Derive an asymptotic expansion of $I(\lambda)$ as $\lambda \to \infty$ via the Mellin transform technique. Consider all ρ in the indicated region and, in particular, the cases where $\rho = (1-2n)/2$, $n = 0, 1, 2, \dots$.

(b) Derive an asymptotic expansion of $I(\lambda)$ as $\lambda \to 0+$.

(c) Compare the results obtained in parts (a) and (b) with those obtained from the exact relation

$$I(\lambda) = \frac{1}{\sqrt{\pi\lambda}}\,\Gamma(1-\rho)\,e^{1/2\lambda}\,K_{\rho-1/2}\left(\frac{1}{2\lambda}\right).$$

4.18. The hypergeometric function has the integral representation

$$F(a,b;c;1-\lambda) = \frac{\Gamma(c)}{\Gamma(b)\,\Gamma(c-b)}\int_0^\infty \frac{s^{b-1}}{(1+s)^{c-a}}\,\frac{ds}{(1+\lambda s)^a},$$

$$\operatorname{Re}(c) > \operatorname{Re}(b) > 0, \qquad |\arg(\lambda)| < \pi.$$

Assume that $\operatorname{Re}(c) > \operatorname{Re}(a) > 0$ and show

(a)
$$F(a,b;c;1-\lambda) \sim \frac{\Gamma(c)}{\Gamma(a)\,\Gamma(b)\,\Gamma(c-b)\,\Gamma(c-a)}\cdot\sum_{n=0}^{\infty}\frac{(-\lambda)^{-n}}{n!}.$$
$$\times\{\lambda^{-b}\,\Gamma(c-a+n)\,\Gamma(b+n)\,\Gamma(a-b-n)$$
$$+\lambda^{-a}\,\Gamma(c-b+n)\,\Gamma(a+n)\,\Gamma(b-a-n)\},$$
$$\lambda\to\infty, \qquad b-a \text{ not an integer.}$$

(b)
$$F(a,b;c;1-\lambda) \sim \frac{\Gamma(c)}{\Gamma(a)\,\Gamma(b)\,\Gamma(c-b)\,\Gamma(c-a)}\cdot\left\{\sum_{n=0}^{k-1}(-1)^n\lambda^{-a-n}\right.$$
$$\times\,\Gamma(k-n)\,\Gamma(c-a+n-k)\,\Gamma(a+n)/n!$$
$$+(-1)^k\sum_{n=0}^{\infty}\lambda^{-a-n-k}\,\Gamma(c-a+n)\,\Gamma(a+k+n)/n!(n+k)!$$
$$\times\left.[\log\lambda+\psi(c-a+n)+\psi(a+k+n)-\psi(n) - \psi(n+k)]\right\},$$
$$\lambda\to\infty, \qquad b-a=k, \qquad \text{with } k \text{ a positive integer.}$$

4.19. Verify that the following expansions are correct:

(a)
$$\int_0^\infty \frac{\operatorname{Ai}(-\lambda t)}{1+t^2}\,dt \sim \sum_{n=0}^{\infty}(-1)^n\frac{3^{4n/3-1/2}}{\pi\lambda^{2n+1}}\,\Gamma\left(\frac{2n+1}{3}\right)\Gamma\left(\frac{2n+2}{3}\right)\cos\left(\frac{2n\pi}{3}\right),$$
$$\lambda\to\infty.$$

(b)
$$\int_0^\infty \frac{\operatorname{Ai}(-\lambda t)}{1+t} \sim \sum_{n=1}^{\infty}\left\{\frac{\lambda^{3n-3}3^{-2n+4/3}}{\Gamma(n)\Gamma(n-1/3)}\right.$$
$$\times\left[\log\lambda - \frac{2}{3}\log 3 - \frac{\pi}{3}\frac{2}{\sqrt{3}} - \frac{\psi(n)}{3} - \frac{\psi(n-1/3)}{3}\right]$$

$$+\frac{4\pi\lambda^{3n-1}3^{-2n-3/2}}{\Gamma(n+1/3)\Gamma(n+2/3)}-\frac{\lambda^{3n-2}3^{-2n+2/3}}{\Gamma(n+1/3)\Gamma(n)}$$

$$\cdot\left[\log\lambda-\frac{2}{3}\log 3+\frac{\pi}{3}\frac{2}{\sqrt{3}}-\frac{\psi(n+1/3)}{3}-\frac{\psi(n)}{3}\right]\Bigg\},$$

$\lambda\to 0+.$

(c) $$\int_0^\infty \frac{D_\nu(\lambda t)}{1+t}\,dt \sim \sum_{n=1}^{\infty}(-1)^{n+1}\lambda^{-n}\sqrt{\pi}2^{(1+\nu)/2}\Big/\Gamma\left(\frac{n+1-\nu}{2}\right)$$

$$\times\ \Gamma(n)\,F\left(\frac{n+1}{2},\frac{1-\nu}{2};\frac{n+1-\nu}{2};-1\right),\qquad \lambda\to\infty.$$

4.20. (a) Show that the expansion in 4.19 (a) is valid for $|\arg(\lambda)|<\pi/2$. [*Hint*: Use (4.12.7) and (2.5.9) to (2.5.11).]

(b) The integral in 4.19 (a) does not converge for λ complex. How can this be reconciled with part (a)?

(c) Show that the result in 4.19 (b) is valid for $|\arg(\lambda)|<\pi$. [*Hint*: Use (4.12.7) and (2.5.9) to (2.5.11).]

(d) The integral in 4.19 (b) does not converge for λ complex. How can this be reconciled with part (c)?

(e) Show that the result in 4.19 (c) is valid for $|\arg(\lambda)|<3\pi/4$. Assume here that (4.12.12) is valid for any ν.

4.21. The complete elliptic integrals of the first and second kind are given respectively by

$$K(r)=\int_0^1\frac{dt}{(1-t^2)^{1/2}(1-r^2t^2)^{1/2}}\tag{4.12.22}$$

and

$$E(r)=\int_0^1\frac{(1-r^2t^2)^{1/2}}{(1-t^2)^{1/2}}\,dt.\tag{4.12.23}$$

Determine asymptotic expansions for these functions, as $r\to 0$, via the Mellin transform technique.

4.22. Calculate asymptotic expansions for the following integrals in the stated limits:

(a) $$\int_0^\infty\frac{\mathrm{Ai}(-\lambda t)}{1+t^2}\,dt,\qquad \lambda\to 0+.$$

(b) $$\int_0^\infty\frac{\mathrm{Ai}(-\lambda t)}{1+t}\,dt,\qquad \lambda\to\infty.$$

(c) $$\int_0^\infty\frac{D_\nu(\lambda t)}{1+t}\,dt,\qquad \lambda\to 0+.$$

4.23. Derive an expansion, as $\lambda \to 0+$, for the exponential integral

$$E_1(\lambda) = \int_\lambda^\infty \frac{e^{-t}}{t}\, dt. \tag{4.12.24}$$

Compare with the stated result (1.1.3). Here, set $t = \lambda + \tau$ in (4.12.24) to obtain

$$\lambda e^\lambda E_1(\lambda) = \int_0^\infty \frac{e^{-t}}{1 + \lambda^{-1} t}\, dt. \tag{4.12.25}$$

Now apply the Mellin transform method to (4.12.25). Note that $\psi(1) = \Gamma'(z)/\Gamma(z)|_{z=1} = -\gamma$ where γ is the Euler-Mascheroni constant.

4.24. Let

$$I(\lambda) = \frac{1}{\Gamma(\nu)} \int_0^\infty \frac{|1 - t|^{\nu - 1}}{1 + \lambda^2 t^2}\, dt, \qquad 1 < \nu < 2.$$

(a) Show that

$$I(\lambda) \sim \sum_{n=1}^\infty \frac{(-1)^{n+1}\lambda^{-2n+1}\pi}{2\Gamma(\nu - 2n + 2)\Gamma(2n - 1)}$$
$$+ \frac{(-1)^n \lambda^{-2n}\pi}{2\Gamma(\nu - 2n + 1)\Gamma(2n)}\left[\log \lambda - \psi(\nu - 2n + 1) + \psi(2n) + \frac{\pi}{\sin \pi \nu}\right], \quad \lambda \to \infty.$$

(b) Show that the sector of validity in part (a) is defined by $|\arg(\lambda)| < \pi/2$.

(c) Show that

$$I(\lambda) \sim \sum_{n=0}^\infty \left\{(-1)^n \lambda^{2n} \frac{\Gamma(2n + 1)}{\Gamma(2n + \nu + 1)} + \frac{(-1)^{n+1}\lambda^{n-\nu}\pi}{2 \sin \pi\left(\dfrac{n - \nu}{2}\right)} \frac{1}{\Gamma(\nu - n)n!}\right\}, \qquad \lambda \to 0+.$$

4.25. Let

$$I(R) = \int_C t^{z+r-1} (\log t)^N \exp\{-dt^\nu\}\, dt, \qquad \nu > 0.$$

Here C is a positively directed arc of the circle of radius R centered at the origin. On C, $0 \le \arg(t) \le \pi/2\nu - \phi/\nu - \varepsilon = \tilde{\theta}$, ν is real, $\pi/2 > \phi = |\arg(d)|$, and $\varepsilon > 0$.

(a) Obtain the estimate

$$|I(R)| \le K R^{x+\mathrm{Re}(r)} (\log R)^N \int_0^{\tilde{\theta}} \exp\{-|d|\, R^\nu \cos[\nu\theta + \phi]\}\, d\theta.$$

(b) Establish the inequality

$$\cos(\nu\theta + \phi) \ge 1 - \frac{2}{\pi}(\nu\theta + \phi), \qquad 0 \le \theta \le \tilde{\theta}.$$

(c) Conclude that

$$|I(R)| \leq KR^{x+\mathrm{Re}(r)} (\log R)^N \int_0^{\theta} \exp\{-|d|\, R^{\nu}[1 - \frac{2}{\pi}(\nu\theta + \phi)]\}\, d\theta,$$

K a constant.

(d) Carry out the integration in (c) and obtain the estimate

$$|I(R)| \leq K' R^{x+\mathrm{Re}(r)-\nu} (\log R)^N \exp\{-\frac{2\nu}{\pi}\varepsilon\, |d|\, R^{\nu}\}, \qquad K' \text{ a constant.}$$

(e) Show how the above result can be used to justify the rotation of the contour of integration in the proof of part (2) of Theorem 4.7.2.

4.26. Consider the integral

$$J(R) = \int_C (\log t)^N\, t^{r+z-1}\, dt, \qquad \mathrm{Re}(z) + \mathrm{Re}(r) < 0.$$

Here C is a positively oriented arc of the circle of radius R centered at the origin. On C, $0 \leq \arg(t) \leq \theta$.

(a) Show that for sufficiently large R

$$|J(R)| \leq K\theta\, (\log R)^N\, R^{\mathrm{Re}(z+r)}.$$

(b) Show that $\lim_{R \to \infty} J(R) = 0$.

(c) How is the result in part (b) used in the proof of Theorem 4.7.2?

4.27. Prove parts (a) to (f) of Lemma 4.7.1.

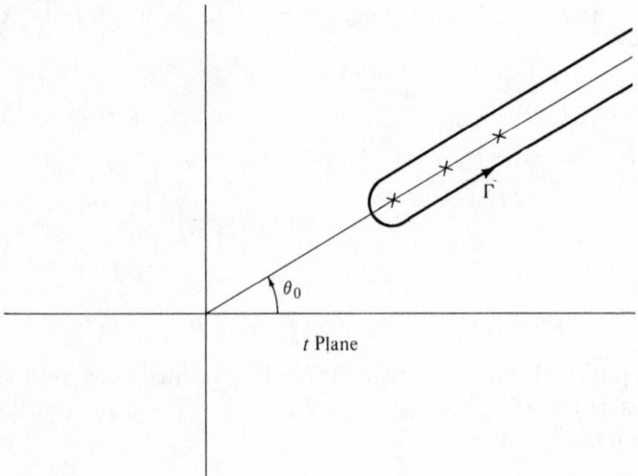

Figure 4.12.2. The Contour Γ for Exercise 4.28 in the Case y Positive. Singularities of h Denoted by x.

4.28. This problem yields an outline of the proof of Lemma 4.7.4.

(a) Show that

$$M[h;z] = e^{i\theta_2 z} \int_0^\infty h(\sigma e^{i\theta_2})\, \sigma^{z-1}\, d\sigma + \int_\Gamma h(t)\, t^{z-1}\, dt.$$

Here,

$$\theta_2 = \pm\theta_3, \qquad \theta_0 < \theta_3 < \text{Min}\left(\theta_1, \frac{\pi - 2|\arg(d)|}{2v}\right).$$

The contour Γ is a loop around the ray defined by $\arg(t) = \text{sign}(y)\,\theta_0$. (See Figure 4.12.2.)

(b) Verify that for the singularities (a) to (d) considered in Lemma 4.7.4, the integral along the contour Γ may be expressed as a residue sum, each having a factor $e^{i\,\text{sign}(y)\,\theta_0 z}$ and integrals in each of which $|\arg(t)| \geq \theta_0$ on the path of integration.

(c) Conclude that $M[h;z] = 0(e^{-\theta_0|y|})$; that is, $h \in \bar{K}(-\text{Re}(\alpha), \theta_0)$ with $\bar{K}$ defined directly above Lemma 4.7.4.

4.29. Rewrite the integral (4.8.13) as

$$\phi(r,z) = V\int_0^\infty J_0(pR\sin\theta)\, e^{-pR\cos\theta}\, J_1(p)\, dp.$$

Here $R^2 = r^2 + z^2$ and θ is the polar angle. Show that

(a) $$\phi(r,z) \sim V\sum_{n=1}^{\infty} (-1)^{n+1}\, 2(2R)^{-2n}\, \Gamma(2n)\, P_{2n-1}(\cos\theta)/n!\,(n-1)!,$$

$$R \to \infty, \qquad 0 < \theta < \frac{\pi}{2}.$$

(b) $$\phi(r,z) \sim V + V\sum_{n=1}^{\infty} (-1)^n\, (2R)^{2n-1}\, P_{2n-1}(\cos\theta)\frac{\Gamma(n+\frac{1}{2})\,\Gamma(n-\frac{1}{2})}{\pi\,\Gamma(2n)},$$

$$R \to 0, \qquad 0 < \theta < \frac{\pi}{2}.$$

In (a) and (b), $P_n(x)$ is the nth Legendre polynomial defined by

$$P_n(x) = \frac{1}{2^n n!}\frac{d^n}{dx^n}(x^2 - 1)^n.$$

4.30. Consider the integral equation of Section 4.9. Verify the following expansions for $t \to \infty$:

(a) $$y(t) \sim \frac{\mathscr{E}_0 t^{-1/2}}{\sqrt{\pi}} + \frac{\mathscr{E}_1 t^{-3/2}}{2\sqrt{\pi}}, \qquad a_0 > 2, \qquad n = 3 \text{ or } n > 4.$$

Here $\mathscr{E}_0$ is defined by (4.9.18) and

$$\mathscr{E}_1 = \int_0^\infty s[f(s) - y^n(s)]\, ds.$$

(b) $y(t) \sim \dfrac{\mathscr{E}_0}{\sqrt{\pi}} t^{-1/2} - \dfrac{(\mathscr{E}_0)^4}{2\pi^{5/2}} t^{-3/2} \log t, \qquad a_0 > 2, \qquad n = 4.$

(c) $y(t) \sim \dfrac{M[f;1]\, t^{-3/2}}{2\sqrt{\pi}}, \qquad a_0 \geq 2, \qquad n = 1.$

(d) $y(t) \sim \dfrac{\gamma_0}{\sqrt{\pi}} t^{-1/2} \log t, \qquad a_0 = 1, \qquad n > 2, \qquad \gamma_0$ defined by (4.9.10).

(e) $y(t) \sim (\gamma_0)^{1/n}\, t^{-1/n}, \qquad a_0 = 1, \qquad 1 \leq n \leq 2.$

4.31. In Section 4.9, suppose that

$$f(t) \sim dt^{\mu - 1}, \qquad t \to 0+.$$

Then verify the following expansions for $t \to 0+$:

(a) $y \sim \dfrac{d\Gamma(\mu)}{\Gamma(\mu + \frac{1}{2})} t^{\mu - 1/2}, \qquad \text{any } n, \qquad \mu \geq \frac{1}{2}.$

(b) $y \sim \dfrac{d\Gamma(\mu)}{\Gamma(\mu + \frac{1}{2})} t^{\mu - 1/2}, \qquad 0 < \mu < \frac{1}{2}, \qquad 1 \leq n < \dfrac{1 - \mu}{\frac{1}{2} - \mu}.$

(c) $y \sim bt^{\mu - 1/2}, \qquad 0 < \mu < \frac{1}{2}, \qquad n = \dfrac{1 - \mu}{\frac{1}{2} - \mu}.$

Here b is a root of $b = \dfrac{\Gamma(\mu)}{\Gamma(\frac{1}{2} + \mu)} [d - b^n].$

4.32. Let

$$h(t) = \frac{P(t)}{Q(t)}$$

with P and Q polynomials of degree p and q, respectively. Furthermore, assume that Q has no real zeros except possibly at $t = 0$.

(a) Show that $h(t)$ can be rewritten as

$$h(t) = R(t) + \sum_{n=1}^{m} \sum_{\ell=1}^{d_n} \frac{a_{\ell n}}{(t - t_n)^\ell}.$$

Here R is a polynomial of degree $p - q$ if $p - q \geq 0$ or $R = 0$ otherwise. The t_n's are the complex zeros of $Q(t)$ and d_n is the multiplicity of t_n.

(b) Show that, except on the vertical lines,

$$x = 0, 1, \ldots, p - q, \qquad M[R; z] = 0.$$

Here M denotes the generalized Mellin transform.

(c) Show that

$$M[(t-t_n)^{-\ell};z] = -(t_n)^{-\ell}\,\pi\,(t_n\,e^{-i\pi})^z \csc \pi z \,\frac{\Gamma(z)}{\Gamma(\ell)\,\Gamma(z-\ell+1)}.$$

(d) If $\theta_n = \arg t_n$, show that

$$M[(t-t_n)^{-\ell};z] = O[\exp\{y[\pi-\theta_n] - |y|(\pi-\varepsilon)\}], \qquad |y| \to \infty,$$

any $\varepsilon > 0$.

(e) If

$$h(t) = O(t^{\alpha}), \qquad t \to 0, \qquad -q \le \alpha < p,$$

verify that

$$M[h(t);z] \in K(-\alpha, \theta_0)$$

with

$$\theta_0 = \min_{1 \le n \le q} (\theta_n).$$

REFERENCES

M. ABRAMOWITZ. On practical evaluation of integrals. *SIAM J. Appl. Math.* **2,** 20–35, 1954.

The author uses Barnes integral representations to obtain asymptotic expansions.

W. A. BEYER and L. HELLER. Analytic continuation of Laplace transforms by means of asymptotic series. *J. Math. Phys.* **8,** 1004–1018, 1967.

In this paper Laplace transforms are analytically continued along lines used in Section 4.3 for Mellin transforms.

C. J. BOUWKAMP. Note on an asymptotic expansion. *Indiana University Math. J.* **21,** 547–549, 1971.

In this paper the author extends Watson's lemma to an example in which the asymptotic expansion of the integrand near the origin involves inverse powers of the log function.

B. L. J. BRAAKSMA. Asymptotic expansions and analytic continuations for a class of Barnes integrals. *Compositio Math.* **15,** 239–341, 1964.

In this paper, the author discusses the general problem of analytic continuation and asymptotic expansions of the types of integrals which arise in this chapter as a consequence of the Mellin-Parseval theorem.

H. E. DANIELS. The minimum of a Stochastic Markov process superimposed on a *U*-shaped trench. *J. Appl. Prob.* **6,** 399–408, 1969.

In this paper, the author uses the Mellin transform method to calculate the asymptotic expansion of a probability density function.

G. DOETSCH. *Handbuch der Laplace Transformation*. Vol. I. Birkhauser Verlag, Basel, 1950.

Chapter 13 of this book presents results similar to Watson's lemma. Paragraph 4 of this chapter contains an elementary result on the analytic continuation of Mellin transforms.

W. FELLER. *An Introduction to Probability Theory*. Vol. II. Wiley, New York, 1966.

Renewal theory is considered in Chapter 14 of this book.

C. FOX. Applications of Mellin's transformation to integral equations. *Proc. Lond. Math. Soc.* **39,** 495–502, 1933.

The author uses Barnes integral representations to obtain asymptotic expansions.

R. A. HANDELSMAN and J. S. LEW. Asymptotic expansions of a class of integral transforms via Mellin transforms. *Arch. Rat. Mech. Anal.* **35,** 382–396, 1969.

Section 2 of this paper discusses the nature of the Mellin transform in its strip of analyticity and the nature of its analytic continuation outside of this strip. Section 3 obtains results, similar to those obtained in Sections 4.4 and 4.5 of the text, for exponentially dominated kernels.

R. A. HANDELSMAN and J. S. LEW. Asymptotic expansions of Laplace transforms near the origin. *SIAM J. Math. Anal.* **1,** 118–130, 1970.

R. A. HANDELSMAN and J. S. LEW. Asymptotic expansions of integral transforms having algebraically dominated kernels. *J.M.A.A.* **35,** 405–433, 1971.

These two papers continue the development of the Mellin transform technique.

R. A. HANDELSMAN and W. E. OLMSTEAD. Asymptotic solution to a class of nonlinear Volterra integral equations. *SIAM J. Appl. Math.* **22,** 373–384, 1972.

Section 4.9 of the text is based on the results of this paper.

D. S. JONES. Generalized transforms and their asymptotic behavior. *Proc. Roy. Soc. London* **265,** 1–43, 1969.

The author considers several problems that are treated in the text. The functions being transformed are allowed to be generalized functions in the sense of Lighthill.

R. F. MILLAR. On the asymptotic behavior of two classes of integrals. *SIAM Review* **8,** 188–195, 1966.

The author derives small parameter expansions of Fourier integrals and Cauchy integrals. The method employed is closely related to the Mellin transform method presented in the text.

F. W. J. OLVER. Error bounds for the Laplace approximation for definite integrals. *J. Approx. Theory* **1,** 293–313, 1968.

In this paper numerical rather than O estimates are obtained for the remainder in Watson's lemma.

F. STENGER. The asymptotic approximation of certain integrals. *SIAM J. Math. Anal.* **1,** 392–404, 1970.

The author presents an extension of the integration by parts procedure which differs greatly from anything appearing in the text.

F. STENGER. Transform methods for obtaining asymptotic expansions of definite integrals. *SIAM J. Math. Anal.* **3,** 20–30, 1972.

An alternative approach is presented for the asymptotic expansion of integral transforms. In this method the Fourier transform of the kernel function plays a crucial role. Results are obtained in terms of a general asymptotic sequence.

E. C. TITCHMARSH. *Theory of Fourier Integrals*. 2nd Ed. Clarendon Press, Oxford, 1948.

Much of the general theory of Mellin transforms is presented in this book including the proofs of validity of the two Mellin-Parseval theorems.

G. N. WATSON. Harmonic functions associated with the parabolic cylinder. *Proc. London Math. Soc.* Series 2, **17,** 116–148, 1918.

The lemma now known as Watson's lemma was first stated in this paper.

D. V. WIDDER. *The Laplace Transform*. Princeton, 1941.

The Mellin transform is considered in some detail.

W. S. WOOLCOCK. Asymptotic behavior of Stieltjes transforms, II. *J. Math. Phys.* **9,** 1350–1356, 1968.

S. ZIMMERING. Some asymptotic behavior of Stieltjes transforms. *J. Math. Phys.* 10, 181–183, 1969.

These last two papers are among the few that consider *h*-transforms with an algebraic kernel.

5 h-Transforms with Kernels of Nonmonotonic Argument

5.1. Laplace's Method

The previous two chapters were devoted to the asymptotic analysis of the h-transform

$$I(\lambda) = \int_C h[\lambda\phi(t)]\, f(t)\, dt \tag{5.1.1}$$

in the case where the contour of integration C is the real interval $[a,b]$, and where the "argument function" $\phi(t)$ is strictly monotonic on this interval. However, when we discussed critical points in Section 3.3, we anticipated that, in certain instances, points in $[a,b]$ at which $\phi'(t)$ vanishes would be critical in the sense that small neighborhoods of them would produce significant contributions to the integral $I(\lambda)$ as $\lambda \to \infty$. Our immediate objective is not only to establish that this is indeed the case, but also to develop techniques for the explicit determination of the contributions to the asymptotic expansion of (5.1.1) corresponding to such points.

There are basically three distinct cases that most often arise in applications:

(1) The contour C in (5.1.1) is all or part of the real line and $h(t)$ either decays exponentially or has purely algebraic (nonoscillatory) behavior as $t \to \pm\infty$.

(2) The contour C is all or part of the real line and $h(t)$ is oscillatory as $t \to \pm\infty$.

(3) The contour C is a curve in the complex t plane while h, f, and ϕ are analytic functions of their arguments in appropriate domains.

In this chapter we shall consider case (1) exclusively and shall delay until the following two chapters our treatment of cases (2) and (3). We do this for two reasons. Firstly, case (1) is significantly simpler to treat than the other two. Secondly, many of the ideas and methods to be introduced in the present chapter will prove useful in the analyses of the more difficult cases.

Perhaps the most important integrals that fall into category (1) are those of Laplace type. They are of the form

$$I(\lambda) = \int_a^b \exp\{-\lambda\phi(t)\} f(t)\, dt. \tag{5.1.2}$$

It will prove fruitful to discuss the behavior of (5.1.2) as $\lambda \to \infty$. If $\phi(t)$ increases monotonically on $[a,b]$ and if both ϕ and f have the appropriate asymptotic behavior as $t \to a+$, then the analysis can be reduced simply to an application of Watson's lemma. Indeed, Watson's lemma implies that under the assumed monotonicity of ϕ, the only critical point for (5.1.2) is the endpoint $t = a$. We might point out that it is not coincidental that, in this case, the minimum of ϕ in $[a,b]$ occurs at $t = a$.

Suppose now that $\phi(t)$ is not monotonic on $[a,b]$. Suppose further that the absolute minimum of ϕ in $[a,b]$ occurs at the point $t = t_0$ where $a < t_0 < b$, $\phi'(t_0) = 0$, and $\phi''(t_0) > 0$. (These assumptions are sometimes expressed verbally by saying that ϕ has a smooth absolute minimum at the interior point $t = t_0$.) Finally, let us assume for simplicity that $\phi'(t) \neq 0$ in $[a,b]$ except at $t = t_0$. This last assumption of course implies the differentiability of ϕ throughout $[a,b]$. In this regard we shall further suppose that both f and ϕ are sufficiently smooth for the operations below.

It is instructive to study the qualitative behavior of the integrand in (5.1.2). First, let us consider $\psi = \exp[-\lambda(\phi(t) - \phi(t_0))]$. This function is plotted in Figure 5.1 for a typical ϕ satisfying the stated conditions and for a sequence of increasing values of λ.

Upon considering Figure 5.1, the critical nature of the point $t = t_0$ becomes apparent. Indeed, as λ increases, we see that the region where ψ is significantly different from zero becomes a smaller and smaller neighborhood of $t = t_0$. Because $f(t)$ does not depend on λ, this statement must also hold for $f\psi$. Thus, it is reasonable to conclude that, in the determination of the asymptotic behavior of $\exp[\lambda\phi(t_0)]\, I(\lambda)$, as $\lambda \to \infty$, we need only be concerned with the behavior of ϕ and f in an arbitrarily small neighborhood of $t = t_0$.

Let us suppose that the conclusion just stated is correct. If we set

$$\exp\{\lambda\phi(t_0)\}\, I_0(\lambda) = \int_{t_0-\varepsilon}^{t_0+\varepsilon} f(t) \exp[-\lambda(\phi(t) - \phi(t_0))]\, dt, \tag{5.1.3}$$

where ε is a fixed, small, positive constant, then this supposition implies that

$$\lim_{\lambda \to \infty} \frac{I_0(\lambda)}{I(\lambda)} = 1. \tag{5.1.4}$$

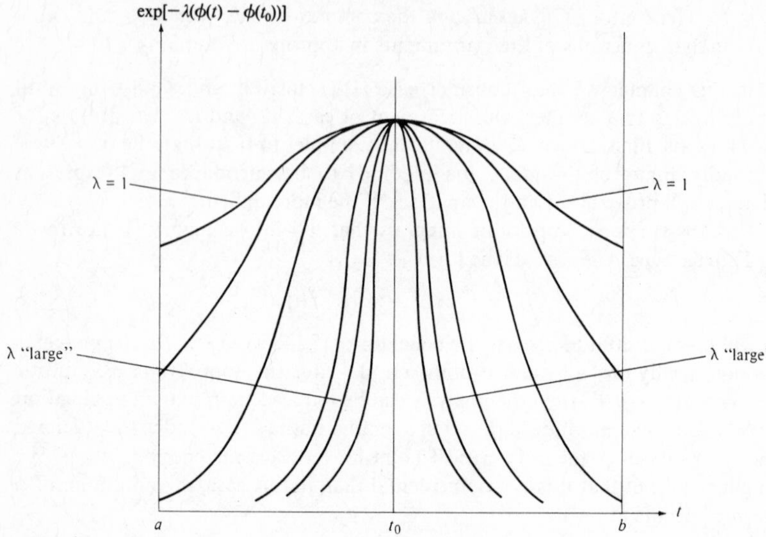

Figure 5.1. $\exp[-\lambda(\phi(t)-\phi(t_0))]$ for Various Values of λ.

Of course, we would expect that the smaller ε is the larger λ would have to be for I_0 to closely approximate I.

Because ε is assumed small, it is reasonable to approximate both $\phi(t)-\phi(t_0)$ and $f(t)$ throughout $[t_0-\varepsilon, t_0+\varepsilon]$ by the first nonvanishing terms of their respective Taylor series expansions about $t=t_0$. [For simplicity we shall assume that $f(t_0) \neq 0$.] Then we have

$$\exp\{\lambda\phi(t_0)\}\, I_0(\lambda) \approx \int_{t_0-\varepsilon}^{t_0+\varepsilon} f(t_0)\exp\left\{-\frac{\lambda}{2}\phi''(t_0)(t-t_0)^2\right\} dt.^1 \qquad (5.1.5)$$

Here, the smaller ε is, the better does the right-hand side of (5.1.5) approximate the left.

It now follows from (5.1.4) that, for λ sufficiently large,

$$I(\lambda) \approx \exp\{-\lambda\phi(t_0)\} f(t_0)\int_{t_0-\varepsilon}^{t_0+\varepsilon} \exp\left\{-\frac{\lambda}{2}\phi''(t_0)(t-t_0)^2\right\} dt.^1 \qquad (5.1.6)$$

In order to complete the estimate, we set

$$\tau = \sqrt{\frac{\lambda}{2}\phi''(t_0)}\,(t-t_0) \qquad (5.1.7)$$

in (5.1.6) which yields

[1] Here the symbol $\approx$ is used to indicate that the right side approximates the left side. Of course, the validity of this statement relies on the validity of (5.1.4).

$$I(\lambda) \approx \exp\{-\lambda\phi(t_0)\} f(t_0) \sqrt{\frac{2}{\lambda\, \phi''(t_0)}} \int_{-\sqrt{\frac{\lambda}{2}\phi''(t_0)}\,\varepsilon}^{\sqrt{\frac{\lambda}{2}\phi''(t_0)}\,\varepsilon} \exp\{-\tau^2\}\, d\tau. \tag{5.1.8}$$

We observe that no matter how small ε is, λ can be taken so large that the value of the integral in (5.1.8) is changed only slightly when the limits of integration are set equal to $\pm\infty$. Thus we finally have

$$\begin{aligned} I(\lambda) &\approx \exp\{-\lambda\phi(t_0)\} f(t_0) \sqrt{\frac{2}{\lambda\, \phi''(t_0)}} \int_{-\infty}^{\infty} \exp(-\tau^2)\, d\tau \\ &= \exp\{-\lambda\phi(t_0)\} f(t_0) \sqrt{\frac{2\pi}{\lambda\phi''(t_0)}}. \end{aligned} \tag{5.1.9}$$

The approximation afforded by (5.1.9) is well known and is often referred to as *Laplace's formula* and the application of this result is called *Laplace's method.* We point out, however, that care was taken not to use the asymptotic symbol $\sim$ in writing this formula. Indeed, because no error estimate has been obtained, the best we can say at present is that it is plausible that the right-hand side of (5.1.9) represents the leading term of an asymptotic expansion of (5.1.2) as $\lambda \to \infty$. There are several ways to rigorously establish the asymptotic nature of (5.1.9). We shall accomplish this in Section 5.2 for a much wider class of integrals. For now, we are content to limit our considerations to Laplace-type integrals and shall attempt to rigorously derive (5.1.9) via Watson's lemma.

Let us reconsider (5.1.2) under the assumptions on ϕ and f leading to (5.1.9) and set

$$\begin{aligned} I(\lambda) &= \int_a^{t_0} \exp\{-\lambda\phi\} f(t)\, dt + \int_{t_0}^{b} \exp\{-\lambda\phi\} f(t)\, dt \\ &= I_a(\lambda) + I_b(\lambda). \end{aligned} \tag{5.1.10}$$

We observe that throughout each of the half-open intervals $[a, t_0), (t_0, b]$, $\phi(t)$ is a monotonic function. Furthermore, upon assuming $\phi \,\varepsilon\, C^4[a,b]$ and $f \,\varepsilon\, C^2[a,b]$ we may write

$$\phi(t) = \phi(t_0) + \tfrac{1}{2}\, \phi''(t_0)\, (t - t_0)^2 + \tfrac{1}{6}\, (t - t_0)^3\, \phi'''(t_0) + O((t - t_0)^4) \tag{5.1.11}$$

and

$$f(t) = f(t_0) + f'(t_0)\, (t - t_0) + O((t - t_0)^2), \qquad t \to t_0 + . \tag{5.1.12}$$

Suppose that we consider $I_b(\lambda)$ first and introduce the change of variable defined by

$$\phi(t) - \phi(t_0) = \tau, \tag{5.1.13}$$

which we observe is one-to-one except at the origin. Then we have

$$I_b(\lambda) = \exp\{-\lambda\phi(t_0)\} \int_0^{\phi(b)-\phi(t_0)} G(\tau) \exp\{-\lambda\tau\}\, d\tau, \tag{5.1.14}$$

where

$$G(\tau) = \left.\frac{f(t)}{\phi'(t)}\right|_{t=\phi^{-1}(\phi(t_0)+\tau)} \tag{5.1.15}$$

In order to apply Watson's lemma, we need only determine an asymptotic power series expansion for $G(\tau)$, as $\tau \to 0+$. This is easily accomplished except for one difficulty which involves an ambiguity in sign introduced by the nonsingle valuedness of the transformation (5.1.13) at the origin. This ambiguity is resolved, however, by observing that, as t increases from t_0 to b, τ increases from 0 to $\phi(b) - \phi(t_0)$. We then find that

$$G(\tau) = \frac{f(t_0)}{\sqrt{2\,\phi''(t_0)}}\tau^{-1/2} + \left\{\frac{f'(t_0)}{\phi''(t_0)} - \frac{f(t_0)\phi'''(t_0)}{3(\phi''(t_0))^2}\right\} + O(\tau^{1/2}), \tag{5.1.16}$$

as $\tau \to 0+$. It then immediately follows from (4.1.3) that

$$I_b(\lambda) = \sqrt{\frac{\pi}{2\lambda\,\phi''(t_0)}} f(t_0) \exp\{-\lambda\phi(t_0)\} + \left\{\frac{f'(t_0)}{\phi''(t_0)} - \frac{f(t_0)\,\phi'''(t_0)}{3(\phi''(t_0))^2}\right\} \frac{\exp\{-\lambda\phi(t_0)\}}{\lambda} + O\left\{\frac{\exp\{-\lambda\phi(t_0)\}}{\lambda^{3/2}}\right\} \tag{5.1.17}$$

as $\lambda \to \infty$.

We can analyze $I_a(\lambda)$ in a completely analogous manner. Indeed, if we again make the change of variable defined by (5.1.13), then we obtain

$$I_a(\lambda) = -\exp\{-\lambda\phi(t_0)\} \int_0^{\phi(a)-\phi(t_0)} G(\tau) \exp(-\lambda\tau)\, d\tau \tag{5.1.18}$$

with $G(\tau)$ still given by (5.1.15). Again an ambiguity in sign arises in the determination of the asymptotic power series expansion of $G(\tau)$ as $\tau \to 0+$. In this case, it is resolved by observing that, as t increases from a to t_0, τ decreases from $\phi(a) - \phi(t_0)$ to 0. The result then is

$$-G(\tau) = \frac{f(t_0)\,\tau^{-1/2}}{\sqrt{2\phi''(t_0)}} - \left\{\frac{f'(t_0)}{\phi''(t_0)} - \frac{f(t_0)\,\phi'''(t_0)}{3(\phi''(t_0))^2}\right\} + O(\tau^{1/2}) \tag{5.1.19}$$

as $\tau \to 0+$, so that by Watson's lemma,

$$I_a(\lambda) = \sqrt{\frac{\pi}{2\lambda\,\phi''(t_0)}} f(t_0) \exp\{-\lambda\phi(t_0)\} - \frac{1}{\lambda}\left\{\frac{f'(t_0)}{\phi''(t_0)} - \frac{f(t_0)\,\phi'''(t_0)}{3(\phi''(t_0))^2}\right\} \exp\{-\lambda\phi(t_0)\} + O\left(\frac{\exp\{-\lambda\phi(t_0)\}}{\lambda^{3/2}}\right). \tag{5.1.20}$$

The desired expansion of $I(\lambda)$ is, of course, given by the sum of (5.1.17) and

(5.1.20) which yields

$$I(\lambda) = \sqrt{\frac{2\pi}{\lambda\phi''(t_0)}} f(t_0) \exp\{-\lambda\phi(t_0)\} + O\left(\frac{\exp\{-\lambda\phi(t_0)\}}{\lambda^{3/2}}\right). \quad (5.1.21)$$

Equation (5.1.21) having been rigorously derived establishes the asymptotic nature of Laplace's formula derived heuristically in (5.1.9). We remark that further terms in the expansion of $I(\lambda)$ can be obtained by determining further terms in the asymptotic expansion of the function $G(\tau)$. This would require, however, additional assumptions about the behavior of f and ϕ near $t = t_0$.

Throughout our discussion we have assumed that $t = t_0$, the point at which ϕ achieves its absolute minimum in $[a,b]$, is such that $a < t_0 < b$. Now suppose that t_0 coincides with one of the endpoints of integration. We wish to determine in what ways the asymptotic expansion of $I(\lambda)$ will be altered. Clearly, we need only compare (5.1.21) with either (5.1.17) or (5.1.20) to obtain the desired information. We then find that there are two ways in which the expansion corresponding to an interior minimum differs from that corresponding to an endpoint minimum. Under the smoothness assumptions made, the leading term for an interior minimum is twice that obtained when the minimum occurs at an endpoint. Also, the second term for an interior minimum is $O(\lambda^{-3/2} \exp\{-\lambda\phi(t_0)\})$ whereas it is $O(\lambda^{-1} \exp\{-\lambda\phi(t_0)\})$ when $t = t_0$ is an endpoint. We might expect that similar disparities occur in higher-order terms as well. That this is indeed the case is to be shown in Exercise 5.5.

In most problems we will not be so fortunate as to have the integral we wish to consider in precisely the form (5.1.2). Nevertheless, it often occurs that with a little ingenuity, the integral can be transformed into the desired type. Indeed, the following two examples will illustrate this point.

EXAMPLE 5.1.1. As we know, the gamma function is defined by the integral

$$\Gamma(\lambda + 1) = \int_0^\infty e^{-t} t^\lambda \, dt. \quad (5.1.22)$$

For the present, we shall restrict our considerations to the case where λ is large and positive. Although (5.1.22) is not of the form (5.1.2) we note that it can be written

$$\Gamma(\lambda + 1) = \int_0^\infty e^{-t} e^{\lambda \log t} \, dt. \quad (5.1.23)$$

At first glance it appears that the results of the section can be applied with $f(t) = \exp(-t)$ and $\phi(t) = -\log t$. Unfortunately, this latter function does not satisfy all of the above stated conditions. Indeed, we note that $\phi'(t) = -t^{-1}$ vanishes only at $t = +\infty$. However, $\phi^{(n)}(t)\big|_{t=\infty} = 0$ for all $n \geq 1$ and in particular for $n = 2$, so that (5.1.21) cannot be applied.

To avoid this difficulty we set[2] $t = \lambda\tau$ in (5.1.22) which then becomes

$$\Gamma(\lambda + 1) = \lambda^{\lambda+1} \int_0^\infty \exp\{\lambda(\log\tau - \tau)\} \, d\tau. \quad (5.1.24)$$

[2] The reader will recall that this is the transformation proposed in Section 2.2.

Now we have an integral of the form (5.1.2) with $f = \lambda^{\lambda+1}$ and $\phi(\tau) = \tau - \log \tau$. Moreover,

$$\phi'(\tau) = 1 - \frac{1}{\tau} \tag{5.1.25}$$

vanishes at $\tau = 1$ and $\phi''(\tau)|_{\tau=1} = +1$. Thus (5.1.21) can be directly applied to obtain

$$\Gamma(\lambda+1) = \sqrt{2\pi\lambda}\left(\frac{\lambda}{e}\right)^{\lambda} + O\left[\left(\frac{\lambda}{e}\right)^{\lambda}\lambda^{-1/2}\right] \tag{5.1.26}$$

as $\lambda \to \infty$.

The result stated in (5.1.26) is, of course, Stirling's formula which we derived in Section 3.2 by a different technique. We might remark that since the sector of validity of any asymptotic expansion derived via Watson's lemma was shown, in general, to be the right half-plane, we can immediately conclude that (5.1.26) is valid as $|\lambda| \to \infty$ with $|\arg \lambda| < \pi/2$.[3]

EXAMPLE 5.1.2. In real analysis, the quantity

$$\| g \|_p = \left(\int_a^b |g(t)|^p \, dt\right)^{1/p} \tag{5.1.27}$$

is known as the L_p norm of the function g, it being assumed that $g(t)$ is defined on the interval $[a,b]$ and such that (5.1.27) exists in the Lebesgue sense. Suppose that we wish to study the behavior of $\| g \|_p$ as $p \to \infty$, in the case where $|g| \varepsilon\, C^4[a,b]$ and where g has a unique absolute maximum at the interior point $t = t_0$.

Let us define

$$I(p) = \int_a^b |g(t)|^p \, dt \tag{5.1.28}$$

so that

$$\| g \|_p = I(p)^{1/p}. \tag{5.1.29}$$

We now write

$$I(p) = \int_a^b e^{p \log |g(t)|} \, dt. \tag{5.1.30}$$

Before we can apply Laplace's formula, we must clear up one technical difficulty. In deriving that formula, we assumed that ϕ, which in this case is $-\log|g(t)|$, is continuously differentiable in $[a,b]$. However, if $g(t)$ vanishes anywhere in $[a,b]$, then so does $|g(t)|$ in which event $\log |g(t)|$ becomes negatively infinite. It should be clear that a small neighborhood of any point in $[a,b]$ at which $g(t)$ vanishes will yield a negligible contribution to $I(p)$ for p large. We therefore claim that any discontinuity introduced by the vanishing of g can be neglected.

[3] It will be recalled that in Section 3.2 we showed that (5.1.26) is valid as $\lambda \to \infty$ in the sector defined by $|\arg(\lambda)| < \pi$.

Upon applying (5.1.21) with $\phi = -\log|g(t)|$ and $f(t) = 1$, we obtain

$$I(p) = \sqrt{\frac{2\pi|g(t_0)|}{p\,|g''(t_0)|}}\,|g(t_0)|^p\,\{1 + O(p^{-1})\} \tag{5.1.31}$$

as $p \to \infty$. Thus, it follows from (5.1.29)[4]

$$\|g\|_p = |g(t_0)|\left\{1 - \frac{\log p}{2p} + O\left(\frac{1}{p}\right)\right\}$$

$$= \max_{t\epsilon[a,b]} |g(t)|\left\{1 - \frac{\log p}{2p} + O\left(\frac{1}{p}\right)\right\}, \qquad p \to \infty. \tag{5.1.32}$$

Equation (5.1.32) expresses the result, well known in real analysis, that for the class of functions under consideration the L_p norm converges to the maximum norm, as $p \to \infty$. Moreover, the rate of convergence is explicit and, we might add, rather slow. As we shall see, analogous results can be obtained under much weaker assumptions on $g(t)$.

Throughout this section we have been overly restrictive in our assumptions. Indeed, there are several ways in which the above results can be extended. The most significant extension would be to h-transforms with kernel functions other than pure exponentials. Because Watson's lemma has played such an important role in the present section and because the extensions to Watson's lemma were accomplished via the techniques of Chapter 4, it is reasonable to expect that our major generalizations will be derived by using these techniques. This shall be the objective of the following sections.

5.2. Kernels of Exponential Type

Here and in the following section we shall consider the asymptotic behavior of functions defined by definite integrals of the form

$$I(\lambda) = \int_a^b h(\lambda\phi(t))\,f(t)\,dt \tag{5.2.1}$$

in the case where h decays exponentially as $t \to \infty$. Section 5.4 is devoted to the case where h is algebraic in this limit, while Chapter 6 is concerned with oscillatory kernels.

Our goal is to obtain an asymptotic expansion of (5.2.1), as $\lambda \to \infty$. We hope to accomplish this by reducing (5.2.1) to an integral of a form already treated in Chapter 4 so that the Mellin transform method can be directly applied. Before we proceed, however, several conditions will be placed on the functions h, ϕ, and f some of which will be relaxed below. We first assume that $f(t)$ and $h(\lambda\phi(t))$ are locally integrable on (a,b) and that (5.2.1) exists for sufficiently large λ.

[4] See Exercise 5.8.

For the present, we assume that the argument function $\phi(t)$ satisfies the following conditions:

(1) $\phi(a) \geq 0$,
(2) $\phi'(t) > 0, \qquad t \in (a,b)$,
(3) $\phi(t) - \phi(a) = \alpha_0(t-a)^{\nu_0} + o((t-a)^{\nu_0}), \qquad t \to a+$.

In (3), α_0 and ν_0 are positive constants. These conditions guarantee that $\phi(t)$ is positive and differentiable on (a,b) and that its absolute minimum on $[a,b]$ is unique and occurs at $t = a$.

It is clear from the assumptions made above that the endpoint of integration $t = a$ will enjoy a special status in our analysis, that is, it will be a critical point in the sense described in Section 3.3. Indeed, this follows from the exponential decay of $h(t)$ as $t \to \infty$ and the fact that $\phi(t)$ has a nonnegative absolute minimum at $t = a$.

In order to proceed with the asymptotic analysis we first set

$$\phi(t) = u \tag{5.2.2}$$

in (5.2.1) which then becomes

$$I(\lambda) = \int_{\phi(a)}^{\phi(b)} h(\lambda u)\, F(u)\, du. \tag{5.2.3}$$

Here

$$F(u) = \left.\frac{f(t)}{\phi'(t)}\right|_{t=\phi^{-1}(u)} \tag{5.2.4}$$

We note that, by hypothesis, $F(u)$ is locally integrable on $(\phi(a), \phi(b))$.

In preparation for applying the results of Chapter 4, we now introduce the function $\bar{F}(u)$ defined by

$$\bar{F}(u) = \begin{cases} 0, & 0 \leq u < \phi(a), \\ F(u), & \phi(a) \leq u \leq \phi(b), \\ 0, & \phi(b) < u < \infty. \end{cases} \tag{5.2.5}$$

In terms of $\bar{F}(u)$ (5.2.3) becomes

$$I(\lambda) = \int_0^\infty h(\lambda u)\, \bar{F}(u)\, du \tag{5.2.6}$$

which is an integral precisely of the form considered in Chapter 4. The advantage gained by introducing $\bar{F}(u)$ lies in the fact that it is always defined as $u \to 0+$, whereas the same need not be true of $F(u)$. It will be recalled that this local behavior plays a vital role in the asymptotic analysis of (5.2.6) as $\lambda \to \infty$.

We found in Section 4.4 that, in order to apply the Mellin transform method, we must require that, as $u \to 0+$, $\bar{F}(u)$ has an asymptotic representation of the form

$$\bar{F}(u) \sim \exp(-qu^{-\mu}) \sum_{m=0}^{\infty} \sum_{n=0}^{\bar{N}(m)} (\log u)^n u^{a_m} p_{mn}. \tag{5.2.7}$$

Here $\operatorname{Re}(q) \geq 0$, $\mu > 0$, $\operatorname{Re}(a_m) \uparrow \infty$, and $\bar{N}(m)$ is finite for each m. [We recall that if only a finite asymptotic expansion is required, then (5.2.7) can be truncated at the appropriate term.]

If $\phi(a)$ is positive, then (5.2.7) is trivially satisfied because all of the coefficients p_{mn} are zero. If, however, $\phi(a) = 0$, then additional restrictions must be placed on ϕ and f to ensure that $\bar{F}(u)$ has the asymptotic form (5.2.7). Conditions sufficient for this to be so are that $f(t)$ itself have such an asymptotic form, as $t \to a+$, and that

$$\phi(t) \sim \sum_{n=0}^{\infty} \alpha_n (t-a)^{\nu_n}, \qquad t \to a+, \qquad \operatorname{Re}(\nu_n) \uparrow \infty. \tag{5.2.8}$$

As in Section 5.1, an ambiguity may arise in the selection of the proper branch of the inverse relation $t = \phi^{-1}(u)$. By requiring that u increase as t increases, this ambiguity is resolved and the expansion of $\bar{F}(u)$, as $u \to 0+$, can be uniquely determined.

We recall the discussion following Case III in Section 4.5. There we indicated that $\bar{F}(u)$ can grow quite rapidly as $u \to \infty$. Indeed, if, as we shall assume, $h(t) = O(\exp\{-dt^{\nu}\})$, as $t \to \infty$, for positive d, ν, then we need only have

$$\bar{F}(u) = O(\exp\{\gamma u^{\nu}\}), \qquad u \to \infty,$$

for some finite γ. Moreover, in that event

$$I(\lambda) = \int_0^{\infty} h(\lambda u)\, \bar{F}_1(u)\, du + o(\lambda^{-R}), \qquad \lambda \to \infty \tag{5.2.9}$$

for all R. Here

$$\bar{F}_1(u) = \begin{cases} \bar{F}(u), & u \in [0,1), \\ 0, & u \in [1,\infty). \end{cases} \tag{5.2.10}$$

We shall assume $G_1(z) = M[h;z]\, M[\bar{F}_1; 1-z]$ is such that the ordinary Parseval formula (4.2.17) holds and that the contour of integration in this formula can be displaced arbitrarily far to the right when the appropriate residues are taken into account.[5]

Because all of the results are implicitly contained in Sections 4.4 and 4.5, we shall be content here with a semiquantitative description of the two cases that can arise.

Case I. Let us first suppose that $\phi(a)$ is positive. Then $\bar{F}(u)$ vanishes identically in a positive half-neighborhood of $u = 0$ and $M[\bar{F}_1; 1-z]$ is an entire function. Because $M[h;z]$ is holomorphic in a right half-plane, we can immediately

[5] If the displacement of the contour can be justified only for $\operatorname{Re}(z) < x_0 < \infty$, then, of course, only a finite asymptotic expansion will be obtained by this process.

conclude

$$I(\lambda) = o(\lambda^{-R}), \qquad \lambda \to \infty \tag{5.2.11}$$

for all R. [This also follows directly from (4.4.17) upon setting all of the coefficients p_{mn} equal to zero.]

The result given by (5.2.11) is rather unsatisfactory especially in the light of the expansion (5.1.17) obtained in the case where $h = e^{-t}$. From this expansion, we see that if $\phi(a) > 0$ and $h = e^{-t}$, then $I(\lambda)$ is actually $O(\exp\{-\lambda\phi(a)\})$ as $\lambda \to \infty$. The reason for the relatively poor estimate afforded by (5.2.11) is that, except under special circumstances, the Mellin transform method does not directly pick up exponentially small terms. When we return to Case I below, we shall show how, with little effort, such terms can be recovered.

Case II. We now suppose that $\phi(a) = 0$. To obtain the asymptotic expansion of $I(\lambda)$ in this case, we need only be concerned with the poles, if any, of $M[\bar{F};1-z]$. If in (5.2.7) $q \neq 0$, then no poles arise and $I(\lambda) = o(\lambda^{-R})$, as $\lambda \to \infty$, for all R. If $q = 0$, then poles occur at the points $z = a_m + 1$, $m = 0, 1, \ldots$ and the expansion is given by (4.4.17) which, for ease of reference, we repeat here.

$$I(\lambda) \sim \sum_{m=0}^{\infty} \lambda^{-1-a_m} \sum_{n=0}^{\bar{N}(m)} p_{mn} \sum_{j=0}^{n} \binom{n}{j} (-\log\lambda)^j M^{(n-j)}[h;z]\Big|_{z=1+a_m}, \qquad \lambda \to \infty. \tag{5.2.12}$$

As (5.2.12) shows, the asymptotic expansion involves the global behavior of the kernel h and the local behavior of $F(u)$ as $u \to 0+$. In this case, as $u \to 0+$, the original variable of integration t goes to $a+$. Hence we find that $t = a$ is a critical point as anticipated.

The utility of (5.2.12) would be greatly increased if we could express the constants that appear there in terms of the constants that appear in the asymptotic expansions of f and ϕ as $t \to a+$. In principle, this can always be accomplished but, in practice, it is extremely tedious to obtain any more than the leading term except in certain simple cases. Indeed, suppose that, as $t \to a+$,

$$f(t) \sim \sum_{n=0}^{\infty} \gamma_n (t-a)^{\mu_n} \tag{5.2.13}$$

with $\mathrm{Re}(\mu_n) \uparrow \infty$ and that $\phi(t)$ has the asymptotic expansion (5.2.8). Then we find that, in (5.2.12),

$$a_0 = \frac{\mu_0 + 1}{\nu_0} - 1, \qquad p_{mn} = 0, \qquad n \geq 1, \qquad m = 0, 1, 2, \ldots,$$
$$p_{00} = (\alpha_0)^{-(\mu_0+1)/\nu_0} \frac{\gamma_0}{\nu_0}. \tag{5.2.14}$$

Hence,

$$I(\lambda) = (\alpha_0\,\lambda)^{-(\mu_0+1)/\nu_0}\,\frac{\gamma_0}{\nu_0}\,M\Big[h;\frac{\mu_0+1}{\nu_0}\Big] + o(\lambda^{-(\mu_0+1)/\nu_0}), \qquad \lambda\to\infty, \qquad (5.2.15)$$

but additional terms are rather awkward to express explicitly.

It is of interest to consider (5.2.15) in some detail. The order of this leading term is $\lambda^{-(\mu_0+1)/\nu_0}$ which increases with increasing ν_0 and decreases with increasing μ_0. Clearly, as μ_0 increases, f vanishes more rapidly as $t\to a+$. Thus we have the perhaps expected result that, as the order of vanishing of f at the critical point $t=a$ increases, the contribution to the asymptotic expansion from this critical point decreases. Because ν_0 is the order of vanishing of ϕ as $t\to a+$, we have that the more rapidly ϕ vanishes as $t\to a+$, the greater is the contribution from $t=a$. This can be understood from the following heuristic argument: As ν_0 increases, the positive half-neighborhood of $t=a$ throughout which ϕ remains "small" increases. In this interval the "largeness" of λ in (5.2.1) is counteracted by the "smallness" of ϕ and hence $h(\lambda\phi)$ is not exponentially small there. This has the effect of increasing the value of $I(\lambda)$ for λ large.

Let us now suppose that in (5.2.8) $\nu_n = n+2$, while in (5.2.13) $\mu_n = n$. Then

$$\alpha_n = \frac{\phi^{(n+2)}(a)}{(n+2)!}, \qquad \gamma_n = \frac{f^{(n)}(a)}{n!}, \qquad (5.2.16)$$

where these quantities are obviously right derivatives. In this simple situation, it is reasonable to calculate further terms in the expansion. We find that

$$a_0 = -\tfrac{1}{2}, \qquad a_1 = 0, \qquad a_2 = \tfrac{1}{2}, \qquad a_3 = 1,$$

$$p_{mn} = 0, \qquad n \ge 1, \qquad m = 0, 1, 2, \ldots,$$

$$p_{00} = \frac{f(a)}{\sqrt{2\phi^{(2)}(a)}}, \qquad p_{10} = \frac{f^{(1)}(a)}{\phi^{(2)}(a)} - \frac{f(a)\,\phi^{(3)}(a)}{3\,[\phi^{(2)}(a)]^2}, \qquad (5.2.17)$$

$$p_{20} = \frac{1}{\sqrt{2}\,[\phi^{(2)}(a)]^{3/2}} \times \left\{ f^{(2)}(a) - \frac{f^{(1)}(a)\,\phi^{(3)}(a)}{\phi^{(2)}(a)} + \frac{f(a)}{4\phi^{(2)}(a)}\left(\frac{5}{3}\,\frac{[\phi^{(3)}(a)]^2}{\phi^{(2)}(a)} - \phi^{(4)}(a)\right)\right\}.$$

It therefore follows from (5.2.12) that the leading three terms in the asymptotic expansion of I are now given by

$$I(\lambda) = (2\phi^{(2)}(a)\lambda)^{-1/2} f(a)\, M[h;\tfrac{1}{2}] + (\lambda\phi^{(2)}(a))^{-1}\left[f^{(1)}(a) - \frac{f(a)\,\phi^{(3)}(a)}{3\phi^{(2)}(a)}\right] M[h;1]$$

$$+\frac{(\lambda\phi^{(2)}(a))^{-3/2}}{\sqrt{2}}\left\{f^{(2)}(a)-\frac{f^{(1)}(a)\,\phi^{(3)}(a)}{\phi^{(2)}(a)}\right.$$
$$\left.+\frac{f(a)}{4\phi^{(2)}(a)}\left(\frac{5[\phi^{(3)}(a)]^2}{3\phi^{(2)}(a)}-\phi^{(4)}(a)\right)\right\}M[h;\tfrac{3}{2}]$$
$$+O(\lambda^{-2}),\qquad \lambda\to\infty. \tag{5.2.18}$$

Before we do any examples, let us return to Case I and attempt to recover the exponentially small terms in the expansion of $I(\lambda)$. We first suppose that $h(t)=e^{-t}$ and observe

$$I(\lambda)=\int_a^b e^{-\lambda\phi}\,f\,dt=e^{-\lambda\phi(a)}\int_a^b e^{-\lambda\psi}\,f\,dt=e^{-\lambda\phi(a)}\,\hat{I}(\lambda). \tag{5.2.19}$$

Here $\psi(t)=\phi(t)-\phi(a)$ so that if ϕ satisfies conditions (1), (2), and (3) on page 188 with $\phi(a)>0$, then $\psi(t)$ satisfies these conditions with $\psi(a)=0$. Therefore, $\hat{I}(\lambda)$ is an integral precisely of the form considered in Case II above and its expansion is given by (5.2.12) under the assumptions that f and ϕ have the proper asymptotic expansions as $t\to a+$. We might further point out that, in this particular case, we need not assume $\phi(a)$ is nonnegative because the sign of $\phi(a)$ does not affect $\hat{I}(\lambda)$. Thus when $h(t)=e^{-t}$ we can recover the exponential terms in the asymptotic expansion quite easily.

In the general situation the problem is somewhat more difficult. Indeed, the device employed for $h(t)=e^{-t}$ does not work because the decomposition

$$h(\lambda\phi)=h(\lambda\phi(a))\,h(\lambda[\phi(t)-\phi(a)]) \tag{5.2.20}$$

already implies that $h(t)$ is precisely a linear combination of exponential functions. If, however, we assume that, as $t\to\infty$, $h(t)$ has the asymptotic expansion (4.4.7), then, because $\phi(a)>0$, we have

$$h(\lambda\phi)\sim e^{-d(\lambda\phi)^{v}}\sum_{m=0}^{\infty}\sum_{n=0}^{N(m)}c_{mn}\,(\lambda\phi)^{-r_m}\,(\log\lambda+\log\phi)^n,\qquad \lambda\to\infty \tag{5.2.21}$$

for *all* t in $[a,b]$. We now claim that an asymptotic expansion of $I(\lambda)$ is obtained by formally replacing $h(\lambda\phi)$ by (5.2.21) in (5.2.1) and integrating term-by-term. This yields

$$I(\lambda)\sim\sum_{m=0}^{\infty}\sum_{n=0}^{N(m)}c_{mn}\,\lambda^{-r_m}\sum_{j=0}^{n}\binom{n}{j}(\log\lambda)^j$$
$$\times\int_a^b e^{-d\lambda^{v}\phi^{v}}\,\phi^{-r_m}\,(\log\phi)^{(n-j)}\,f\,dt. \tag{5.2.22}$$

Upon setting $\lambda^{v}=\Lambda$ we find that each of the integrals in (5.2.22) is reduced to the case where $h(t)=e^{-t}$ which we have already considered in detail. Thus, to complete the expansion, we need only expand these integrals according to the above description and then collect the terms obtained in the proper asymptotic order. This procedure will be illustrated in Example 5.2.3 below.

We shall leave the justification of (5.2.22) to the exercises. We also mention that the reader should be careful not to allow $\phi(a)$ to be negative in (5.2.22) because, in that event, (5.2.21) need not hold throughout the interval of integration.

EXAMPLE 5.2.1. Suppose $h(t) = e^{-t}$ so that (5.2.1) is an integral of Laplace type. Let ϕ satisfy the conditions on page 188 except now allow $\phi(a)$ to be negative. Finally, assume that, as $t \to a+$, $\phi(t) - \phi(a)$ has the asymptotic expansion (5.2.8) and $f(t)$ has the asymptotic expansion (5.2.13). In this case, because $M[h; z] = \Gamma(z)$, we have

$$I(\lambda) = \int_a^b e^{-\lambda\phi} f \, dt$$

$$= e^{-\lambda\phi(a)}(\alpha_0\lambda)^{-(\mu_0+1)/\nu_0} \frac{\gamma_0}{\nu_0} \Gamma\left(\frac{\mu_0+1}{\nu_0}\right) + o(e^{-\lambda\phi(a)} \lambda^{-(\mu_0+1)/\nu_0}). \quad (5.2.23)$$

In the special case where $\nu_n = n + 2$ and $\mu_n = n$, the first three terms in the expansion can be obtained by setting $M[h; z] = \Gamma(z)$ in (5.2.18) and multiplying by $e^{-\lambda\phi(a)}$. In this manner we find that now

$$\int_a^b e^{-\lambda\phi} f \, dt = \exp(-\lambda\phi(a))\left\{\sqrt{\frac{\pi}{2\lambda\phi^{(2)}(a)}} f(a) + (\lambda\phi^{(2)}(a))^{-1}\right.$$

$$\times \left[f^{(1)}(a) - \frac{f(a)\,\phi^{(3)}(a)}{3\phi^{(2)}(a)}\right] + \frac{(\lambda\phi^{(2)}(a))^{-3/2}}{2}$$

$$\times \sqrt{\frac{\pi}{2}} \left[f^{(2)}(a) - \frac{f^{(1)}(a)\,\phi^{(3)}(a)}{\phi^{(2)}(a)}\right.$$

$$\left.\left. + \frac{f(a)}{4\phi^{(2)}(a)}\left(\frac{5[\phi^{(3)}(a)]^2}{3\phi^{(2)}(a)} - \phi^{(4)}(a)\right)\right] + O(\lambda^{-2})\right\}. \quad (5.2.24)$$

This last formula is seen to be in agreement with (5.1.17) to second order.

EXAMPLE 5.2.2. Consider

$$K_\nu(\lambda) = \int_0^\infty e^{-\lambda \cosh t} \cosh(\nu t) \, dt, \quad (5.2.25)$$

which is an integral representation for the modified Bessel function of the second kind. If we set $\phi(t) = \cosh t$ and $f(t) = \cosh \nu t$, then we find that $\phi'(t) > 0$, $t > 0$ and that the expansions (5.2.8) and (5.2.13) hold with ν_n and μ_n positive integers. Hence, we can immediately apply (5.2.24) to obtain

$$K_\nu(\lambda) = \sqrt{\frac{\pi}{2\lambda}}\, e^{-\lambda} \left\{1 + \frac{4\nu^2 - 1}{8\lambda} + O(\lambda^{-3/2})\right\}, \quad (5.2.26)$$

as $\lambda \to \infty$, with ν fixed.

EXAMPLE 5.2.3. Let us now suppose that $h(t) = K_\nu(t)$ with $K_\nu(t)$ defined by (5.2.25). It follows from (5.2.26) that $K_\nu(t)$ is a kernel of the type considered in this section. Moreover, we have explicitly (see the Appendix)

$$M[K_\nu(t);z] = 2^{z-2}\,\Gamma(\tfrac{1}{2}[z+\nu])\,\Gamma(\tfrac{1}{2}[z-\nu]). \tag{5.2.27}$$

Thus, if $\phi(a) = 0$ in (5.2.1), while (5.2.8) and (5.2.13) hold, then we have

$$\int_a^b K_\nu(\lambda\phi)\, f\, dt = \left(\frac{2}{\alpha_0\lambda}\right)^{(\mu_0+1)/\nu_0} \frac{\gamma_0}{4\nu_0}\,\Gamma\left(\frac{1}{2}\left[\frac{\mu_0+1}{\nu_0}+\nu\right]\right)\Gamma\left(\frac{1}{2}\left[\frac{\mu_0+1}{\nu_0}-\nu\right]\right) + o(\lambda^{-(\mu_0+1)/\nu_0}) \tag{5.2.28}$$

as $\lambda \to \infty$.[6]

If $\phi(a) > 0$, then we can make use of (5.2.22) and (5.2.26) and write

$$\int_a^b K_\nu(\lambda\phi)\, f\, dt = \sqrt{\frac{\pi}{2\lambda}}\int_a^b e^{-\lambda\phi}\, f\, \phi^{-1/2}\, dt + \sqrt{\frac{\pi}{2}}\,\frac{(4\nu^2-1)}{8\lambda^{3/2}}\int_a^b e^{-\lambda\phi}\, f\, \phi^{-3/2}\, dt + o\left(\lambda^{-3/2}\textstyle\int_a^b e^{-\lambda\phi}\, f\, \phi^{-3/2}\, dt\right). \tag{5.2.29}$$

Each of the integrals on the right side of (5.2.29) is of Laplace type and can be asymptotically evaluated by using the result of Example 5.2.1.

If we assume $u_n = n$, $n = 0, 1, 2, \ldots$ in (5.2.8), and $\mu_n = n$ in (5.2.13), then we find that, to determine the asymptotic expansion to $O(\exp\{-\lambda\phi(a)\}\,\lambda^{-5/2})$, we must retain three terms in the expansion of the first integral on the right side of (5.2.29) and one term in the expansion of the second integral. In this manner we obtain after some computation

$$\begin{aligned}\int_a^b K_\nu(\lambda\phi) f\, dt = \frac{\exp\{-\lambda\phi(a)\}}{[\phi(a)]^{1/2}}\Bigg\{&\frac{\pi}{2\lambda\,(\phi^{(2)}(a))^{1/2}}\, f(a) \\ &+ \sqrt{\frac{\pi}{2}}\,\frac{\lambda^{-3/2}}{\phi^{(2)}(a)}\left[f^{(1)}(a) - \frac{f(a)\,\phi^{(3)}(a)}{3\phi^{(2)}(a)}\right] \\ &+ \frac{\pi}{2\lambda^2}\Bigg[\frac{f^{(2)}(a)}{2(\phi^{(2)}(a))^{3/2}} - \frac{f(a)}{4\phi(a)\,(\phi^{(2)}(a))^{1/2}} \\ &- \frac{f^{(1)}(a)\,(\phi^{(2)}(a))^{-5/2}}{2}\,\phi^{(3)}(a) \\ &+ \frac{f(a)}{8(\phi^{(2)}(a))^{5/2}}\left\{\frac{5}{3}\,\frac{(\phi^{(3)}(a))^2}{\phi^{(2)}(a)} - \phi^{(4)}(a)\right\} \\ &+ \frac{4\nu^2-1}{8\phi(a)}\,\frac{f(a)}{(\phi^{(2)}(a))^{1/2}}\Bigg] + O(\lambda^{-5/2})\Bigg\}.\end{aligned} \tag{5.2.30}$$

[6] We must require that $(\mu_0 + 1)/\nu_0 - |\nu| > 0$ for the existence of (5.2.28).

The complexity of this last formula is an indication of the difficulties we can encounter in obtaining higher-order terms in the asymptotic expansion of (5.2.1).

5.3. Kernels of Exponential Type: Continuation

We now turn to simple extensions and generalizations of the results obtained in Section 5.2. In that section, we treated (5.2.1) in the case where the absolute minimum of ϕ in $[a,b]$ occurs only at the lower endpoint of integration $t = a$. Thus, as our first extension, we wish to consider (5.2.1), as $\lambda \to \infty$, when $h(t)$ is exponentially decreasing at infinity and when $\phi(t)$ has a unique absolute minimum in $[a,b]$ at $t = b$.

To begin, we assume that

(1) $\phi(b) \geq 0$,
(2) $\phi'(t) < 0 \quad$ in $\quad (a,b)$,
(3) $\phi(t) - \phi(b) = \beta_0 (b-t)^{\rho_0} + o((b-t)^{\rho_0}), \qquad \beta_0, \rho_0 \quad > 0$.

If we set $\phi(t) = u$ in (5.2.1), then we have

$$I(\lambda) = \int_{\phi(b)}^{\phi(a)} h(\lambda u)\, F(u)\, du. \tag{5.3.1}$$

Here

$$F(u) = \left(\frac{-f}{\phi'}\right)_{t=\phi^{-1}(u)}$$

and u increases from $\phi(b)$ as t decreases from b.

To complete the asymptotic analysis of (5.3.1) we need only assume that $F(u)$, defined by (5.2.5) with a and b interchanged, has the asymptotic expansion (5.2.7) as $u \to 0+$. In that event, the results of Section 5.2 are directly applicable. For simplicity, we suppose that, as $t \to b-$,

$$f(t) \sim \sum_{n=0}^{\infty} \delta_n (b-t)^n, \tag{5.3.2}$$

$$\phi(t) \sim \sum_{n=0}^{\infty} \beta_n (b-t)^{n+2}. \tag{5.3.3}$$

Now

$$\beta_n = (-1)^n \frac{\phi^{(n+2)}(b)}{(n+2)!}, \qquad \delta_n = (-1)^n \frac{f^{(n)}(b)}{n!}, \tag{5.3.4}$$

where these quantities are left derivatives. Note that (5.3.3) implies $\phi(b) = 0$.

The leading three terms in the asymptotic expansion of $I(\lambda)$ are readily determined. Indeed, calculations analogous to those leading to (5.2.18) yield

$$I(\lambda) = (2\phi^{(2)}(b)\lambda)^{-1/2} f(b)\, M[h;\tfrac{1}{2}] - (\lambda\phi^{(2)}(b))^{-1} \left[f^{(1)}(b) - \frac{f(b)\,\phi^{(3)}(b)}{3\phi^{(2)}(b)} \right]$$

$$\times M[h;1] + \frac{(\lambda\phi^{(2)}(b))^{-3/2}}{\sqrt{2}}$$

$$\times \left[f^{(2)}(b) - \frac{f^{(1)}(b)\,\phi^{(3)}(b)}{\phi^{(2)}(b)} + \frac{f(b)}{4\phi^{(2)}(b)} \left\{ \frac{5[\phi^{(3)}(b)]^2}{3\phi^{(2)}(b)} - \phi^{(4)}(b) \right\} \right] \tag{5.3.5}$$

$$\times M[h;\tfrac{3}{2}] + O(\lambda^{-2}).$$

Now let us suppose that ϕ has its absolute minimum in $[a,b]$ at an interior point. To obtain the asymptotic expansion in this case, we need only split the interval of integration at that point and treat separately the resulting two integrals. To illustrate the procedure, we again consider the Laplace-type integral

$$I(\lambda) = \int_\alpha^\beta e^{-\lambda\phi} f\, dt. \tag{5.3.6}$$

We assume that the absolute minimum of ϕ in $[\alpha,\beta]$ occurs at $t = c$, where $\alpha < c < \beta$, and that ϕ' is nonzero in both (α,c) and (c,β). We further assume ϕ is four times continuously differentiable at $t = c$ and

$$\begin{aligned} f &= \sum_{n=0}^{2} \gamma_n (t-c)^n + O((t-c)^3), \qquad t \to c+, \\ f &= \sum_{n=0}^{2} \delta_n (c-t)^n + O((c-t)^3), \qquad t \to c-. \end{aligned} \tag{5.3.7}$$

We now write

$$I(\lambda) = \int_\alpha^c e^{-\lambda\phi} f\, dt + \int_c^\beta e^{-\lambda\phi} f\, dt. \tag{5.3.8}$$

To three terms the asymptotic expansion of the first integral in (5.3.8) is given by (5.3.5) multiplied by $\exp\{-\lambda\phi(c)\}$ when b is replaced by $c-$ and $M[h;z]$ is replaced by $\Gamma(z)$. The corresponding expansion of the second integral is given by (5.2.24) when a is replaced by $c+$. Upon adding the two expansions we obtain

$$\int_\alpha^\beta e^{-\lambda\phi} f\, dt = \exp\{-\lambda\phi(c)\} \left[\sqrt{\frac{\pi}{2\lambda\phi^{(2)}(c)}} \{f(c+) + f(c-)\} \right.$$

$$+ \frac{1}{\lambda\phi^{(2)}(c)} \{ [f']_{t=c} - \frac{\phi^{(3)}(c)}{3\,\phi^{(2)}(c)} [f]_{t=c} \}$$

$$+ \frac{(\lambda\phi^{(2)}(c))^{-3/2}}{2} \sqrt{\frac{\pi}{2}} \left\{ f^{(2)}(c+) + f^{(2)}(c-) \right)$$

$$-\frac{\phi^{(3)}(c)}{\phi^{(2)}(c)}(f'(c+)+f'(c-))+\left(\frac{5[\phi^{(3)}(c)]^2}{3\phi^{(2)}(c)}-\phi^{(4)}(c)\right)$$
$$\times\left[\frac{f(c+)+f(c-)}{4\phi^{(2)}(c)}\right]\Bigg\}+O(\lambda^{-2})\Bigg]. \tag{5.3.9}$$

Here we have used the symbol $[g]_{t=c}$ to denote the "jump" of any function g at $t=c$; that is,

$$[g]_{t=c}=g(c+)-g(c-). \tag{5.3.10}$$

If f is continuously differentiable at $t=c$, then the jumps $[f]_{t=c}$ and $[f']_{t=c}$ are both zero and we find that

$$\int_\alpha^\beta e^{-\lambda\phi}\, f\, dt=e^{-\lambda\phi(c)}\sqrt{\frac{2\pi}{\lambda\phi^{(2)}(c)}}\,f(c)+O(\exp(-\lambda\phi(c))\,\lambda^{-3/2}) \tag{5.3.11}$$

which agrees with (5.1.21). On the other hand, if either f or f' (or both) has a jump discontinuity at $t=c$, then the error in (5.3.11) is, in general, $O(e^{-\lambda\phi(c)}\,\lambda^{-1})$.

It is of value to have a formula at hand for the case where h is a general kernel of exponential type and where ϕ has a unique absolute minimum in the interior of $[a,b]$. Thus, suppose that at $t=c$, $a<c<b$, $\phi^{(m)}(c)=0$, $m=0, 1, 2, \ldots, n-1$, while $\phi^{(n)}(c)>0$, where n is an even integer ≥ 2. Suppose further that ϕ' is nonzero in (a,c) and (c,b) and that $f(c)$ is continuous at $t=c$. Then, as is readily shown

$$\int_a^b h(\lambda\phi)\,f\,dt=\frac{2f(c)}{n}\left(\frac{\phi^{(n)}(c)\,\lambda}{n!}\right)^{-1/n} M\left[h;\frac{1}{n}\right]+o(\lambda^{-1/n}). \tag{5.3.12}$$

In general $\phi'(t)$ will vanish at several points in (a,b). In most instances however (a,b) can be subdivided so that ϕ is monotonic throughout each subinterval. Each integral that arises can be analyzed by the method of this and the preceding section. The total asymptotic expansion of I will, of course, be obtained by summing the individual expansions. It should be clear that the major contributions to the total expansion will still come from small neighborhoods of those points in $[a,b]$ at which ϕ achieves its absolute minimum. For this reason such points are called *dominant* critical points while all other critical points are called *subdominant.* Because the contributions from subdominant critical points are exponentially smaller than those from dominant critical points, these contributions need only be retained when they have some special significance.

If only the dominant contributions are to be determined, then we need only require that the functions ϕ and f satisfy appropriate conditions in small neighborhoods of the dominant critical points. Thus, for example, we need not assume that ϕ' is continuous throughout the entire interval (a,b).

EXAMPLE 5.3.1. Consider the integral

$$I(\lambda) = \int_0^{3\pi/2} e^{-\lambda \sin t} f(t)\, dt, \tag{5.3.13}$$

where f is continuously differentiable on $[0, 3\pi/2]$. We first write

$$I(\lambda) = \int_0^{\pi/2} e^{-\lambda \sin t} f(t)\, dt + e^{\lambda} \int_{\pi/2}^{3\pi/2} e^{-\lambda(\sin t + 1)} f(t)\, dt$$

$$= I_1(\lambda) + I_2(\lambda)\, e^{\lambda}.$$

Finite asymptotic expansions of I_1 and I_2 can be obtained from (5.2.15) and (5.3.5), respectively. Indeed, we have

$$I_1(\lambda) = \frac{f(0)}{\lambda} + O(\lambda^{-2}), \tag{5.3.14}$$

$$e^{\lambda} I_2(\lambda) = f\left(\frac{3\pi}{2}\right) e^{\lambda} \sqrt{\frac{\pi}{2\lambda}} + O(e^{\lambda} \lambda^{-1}). \tag{5.3.15}$$

Thus we see that the contribution from the subdominant critical point $t = 0$ is asymptotically negligible compared to that from the dominant critical point $t = 3\pi/2$. Hence,

$$I(\lambda) \sim e^{\lambda} \sqrt{\frac{\pi}{2\lambda}}\, f\left(\frac{3\pi}{2}\right) + O e^{\lambda} \lambda^{-1}). \tag{5.3.16}$$

EXAMPLE 5.3.2. Let

$$I(\lambda) = \int_0^{2\pi} t^{\mu_0} (2\pi - t)^{\mu_1}\, e^{-\lambda \sin^2 t}\, dt, \qquad \mu_0, \mu_1 > -1. \tag{5.3.17}$$

Here $\phi = \sin^2 t$ takes on its minimum value in $[0, 2\pi]$ at the three points $t = 0$, $t = \pi$, and $t = 2\pi$. Thus, we write

$$I(\lambda) = \int_0^{\pi/2} + \int_{\pi/2}^{3\pi/2} + \int_{3\pi/2}^{2\pi} t^{\mu_0} (2\pi - t)^{\mu_1}\, e^{-\lambda \sin^2 t}\, dt$$

$$= I_1(\lambda) + I_2(\lambda) + I_3(\lambda). \tag{5.3.18}$$

In I_1, the absolute minimum of ϕ occurs at the lower endpoint $t = 0$. In I_2 it occurs at the interior point $t = \pi$, while in I_3 it occurs at the upper endpoint $t = 2\pi$. Upon applying the results derived above we readily find

$$I_1(\lambda) = \frac{(2\pi)^{\mu_1}}{2}\, \Gamma\left(\frac{\mu_0 + 1}{2}\right)\, \lambda^{-(\mu_0+1)/2} + o(\lambda^{-(\mu_0+1)/2}),$$

$$I_2(\lambda) = \pi^{\mu_0+\mu_1} \sqrt{\frac{2\pi}{\lambda}} + o(\lambda^{-1/2}), \tag{5.3.19}$$

$$I_3(\lambda) = \frac{(2\pi)^{\mu_0}}{2}\, \Gamma\left(\frac{\mu_1 + 1}{2}\right)\, \lambda^{-(\mu_1+1)/2} + o(\lambda^{-(\mu_1+1)/2}).$$

In this example, each of the points $t=0$, $t=\pi$, and $t=2\pi$ is a dominant critical point. We observe that the contributions to the asymptotic expansion of (5.3.17) from each of these points is algebraic in nature. To find the leading term in the expansion we must know the values of μ_0 and μ_1. Thus, for example, if $-1<\mu_0<0$ while $\mu_1 \geq 0$, then

$$I(\lambda)=\frac{(2\pi)^{\mu_1}}{2}\Gamma\left(\frac{\mu_0+1}{2}\right)\lambda^{-(\mu_0+1)/2}+o(\lambda^{-(\mu_0+1)/2}). \qquad (5.3.20)$$

On the other hand, if both μ_0 and μ_1 are zero, then all three terms in (5.3.19) are comparable and we have

$$I(\lambda)=2\sqrt{\frac{\pi}{\lambda}}+o(\lambda^{-1/2}).$$

Except in the special case where $h(t)=e^{-t}$, we have assumed here and in Section 5.2 that $\phi(t)$ is nonnegative throughout the interval of integration. The reason for this is that we have made no assumptions concerning the behavior of $h(t)$ as $t\to-\infty$. If, however, appropriate assumptions concerning this asymptotic behavior are made, then we could allow ϕ to become negative with few, if any, additional complications. This point is discussed further in Section 5.4.

The extension to complex λ is also readily accomplished. Indeed, because we have reduced (5.2.1) to a sum of integrals of the form treated in Section 4.7, the theory developed in that section can be directly applied. One word of caution is in order, however. We must bear in mind that the procedure of replacing $h(\lambda\phi)$ in (5.2.1) by its asymptotic expansion, when ϕ is bounded away from zero in (a,b), already places a restriction on the sector of validity. In other words, it is possible that the sector of validity of the asymptotic expansion of $I(\lambda)$ is determined by the corresponding sector for the expansion of h, as $t\to\infty$, and not by what we would obtain by applying the results of Section 4.7 to each of the integrals in (5.2.22).

5.4. Kernels of Algebraic Type

We now consider the behavior of the integral

$$I(\lambda)=\int_a^b h(\lambda\phi)f\,dt, \qquad (5.4.1)$$

as $\lambda\to\infty$, in the case where $h(t)$ is algebraic in the limit $t\to\infty$. Indeed, we shall assume that the asymptotic expansion (4.4.7) holds with $d=0$.

The analysis will proceed essentially as in Section 5.2. Nevertheless, we have chosen to treat algebraic kernels separately because, as we shall see, some of the results are markedly different from those for kernels of exponential type. Let us first suppose that $I(\lambda)$ exists for λ sufficiently large and that ϕ satisfies conditions (1), (2), and (3) on page 188. We are tempted to apply arguments similar to those given in Section 5.2, to predict that $t = a$ will be a critical point for the asymptotic expansion of $I(\lambda)$. We shall find, however, that this need not be the case at all. In fact, one objective of this section is to determine which points in $[a,b]$ are critical.

If in (5.4.1) we set $\phi(t) = u$, then we obtain

$$I(\lambda) = \int_0^\infty h(\lambda u)\, \bar{F}(u)\, du. \tag{5.4.2}$$

Here $\bar{F}(u)$ is defined by (5.2.4) and (5.2.5). We shall again require that $\bar{F}(u)$ have an asymptotic expansion, as $u \to 0+$, of the form (5.2.7). Because $h(t)$ is algebraic, as $t \to \infty$, it need not have a Mellin transform in the ordinary sense. However, throughout this section we shall assume, as in Section 4.5, that $M[h;z]$ exists in the ordinary sense but shall allow $M[\bar{F}(u);1-z]$ to exist only in the generalized sense.

It again proves convenient to consider the two cases $\phi(a) > 0$ and $\phi(a) = 0$ separately.

Case I. $\phi(a) > 0$. Here $\bar{F}(u) \equiv 0$ in a positive half-neighborhood of $u = 0$ and hence $M[\bar{F}(u);1-z]$ exists and is holomorphic in a right half-plane.[7] If we assume that h and $\bar{F}$ satisfy the hypotheses of Theorem 4.4, then we find that (4.4.15) holds where the only contributions arise from the poles in the analytic continuation of $M[h;z]$ into the right half-plane. We find that in this case the expansion is given by (4.4.19)

$$I(\lambda) \sim \sum_{m=0}^{\infty} \lambda^{-r_m} \sum_{n=0}^{N(m)} c_{mn} \sum_{j=0}^{n} \binom{n}{j} (\log \lambda)^j\, M^{(n-j)}[\bar{F};z]\Big|_{z=1-r_m} \tag{5.4.3}$$

The coefficients in (5.4.3) involve the quantities

$$M^{(n-j)}[\bar{F};z] = \int_0^\infty (\log u)^{n-j}\, \bar{F}(u)\, u^{z-1}\, du \tag{5.4.4}$$

evaluated at the points $z = 1 - r_m$. Upon setting $\phi(t) = u$ in (5.4.4), that is, upon returning to the original variable of integration in (5.4.1), we find that

$$M^{(n-j)}[\bar{F};z]\Big|_{z=1-r_m} = \int_{\phi(a)}^{\phi(b)} \left(\frac{f}{\phi'}\right)_{t=\phi^{-1}(u)} (\log u)^{n-j}\, u^{-r_m}\, du \tag{5.4.5}$$

$$= \int_a^b (\log \phi)^{n-j}\, \phi^{-r_m}\, f\, dt.$$

Thus, we see that the asymptotic expansion in this case involves the global behavior of f and ϕ on $[a,b]$ and the local behavior of $h(t)$ at $t = \infty$. This shows that $t = a$ is *not* a special critical point for the expansion.

We note that the asymptotic expansion could have been obtained directly from (5.4.1) by replacing $h(\lambda\phi)$ by its asymptotic expansion at ∞ and then integrating term-by-term. This of course succeeds because, in this case, $\phi(t)$

[7] If $\phi(b) = \infty$, then $M[\bar{F};1-z]$ exists only when $\bar{F}(u)$ has an algebraic bound as $u \to \infty$. This, however, is guaranteed by the assumed existence of (5.4.1) and the algebraic behavior of $h(t)$ as $t \to \infty$.

is bounded away from zero throughout $[a,b]$. We finally note that here conditions (1), (2), and (3) on page 188 can be replaced by the single assumption $\phi(t) \geq \delta > 0$ for t in $[a,b]$.

Case II. $\phi(a) = 0$. Now additional conditions must be placed on ϕ and f to ensure that (5.2.7) holds. Sufficient conditions will be given below. If $\mathrm{Re}(q) > 0$ in (5.2.7), then $M[\bar{F}(u);1-z]$ is holomorphic in a right half-plane and the asymptotic expansion of $I(\lambda)$ is still given by (5.4.3) and (5.4.5) If q is purely imaginary then we know that $M[\bar{F};1-z]$ can be continued into a right half-plane as a holomorphic function. Hence (5.4.3) again holds except now the coefficients that appear perhaps involve the generalized Mellin transform of $\bar{F}(u)$ and we cannot use formula (5.4.5). In these last two subcases, we still find that $t = a$ is not a critical point for the expansion.

The most interesting situation arises when in (5.2.7) $q = 0$. Now the analytic continuation of $M[\bar{F}_1;1-z]$ has poles in the right half-plane $\mathrm{Re}\, z \geq a_0 + 1$. In the case where $a_m + 1 \neq r_n$ for any m, n so that poles of $M[h;z]$ and $M[\bar{F}_1;1-z]$ do not coincide, then we find that

$$I(\lambda) \sim \sum_{m=0}^{\infty} \lambda^{-r_m} \sum_{n=0}^{N(m)} c_{mn} \sum_{j=0}^{n} \binom{n}{j} (\log \lambda)^j M^{(n-j)}[\bar{F};z]\Big|_{z=1-r_m} \tag{5.4.6}$$

$$+ \sum_{m=0}^{\infty} \lambda^{-a_m-1} \sum_{n=0}^{\bar{N}(m)} p_{mn} \sum_{j=0}^{n} \binom{n}{j} (-\log \lambda)^j M^{(n-j)}[h;z]\Big|_{z=a_m+1}$$

Here the quantities $M^{(n-j)}[\bar{F};z]\Big|_{z=1-r_m}$ are to be interpreted in the generalized sense.

The important point to observe is that now the asymptotic expansion involves the local behavior of the functions f and ϕ as $t \to a+$ in addition to the global behavior of these functions on $[a,b]$. In other words, $t = a$ is a critical point in this case.

To obtain explicit expressions for the leading terms in (5.4.6) we assume

$$\phi(t) \sim \alpha_0(t-a)^{\nu_0}, \qquad f(t) \sim \gamma_0(t-a)^{\mu_0}, \qquad t \to a+. \tag{5.4.7}$$

Then, we have [see (5.2.14)]

$$F(u) \sim \frac{\gamma_0}{\nu_0}(\alpha_0)^{-(\mu_0+1)/\nu_0}\, u^{(\mu_0+1)/\nu_0 - 1}, \qquad u \to 0+. \tag{5.4.8}$$

If we further assume that in (4.4.7), $r_0 < (\mu_0 + 1)/\nu_0 < r_1$ and $c_{0n} = 0$ for $n \geq 1$, then to second order we obtain

$$I(\lambda) \sim \lambda^{-r_0} c_{00} \int_a^b \phi^{-r_0} f\, dt + M\left[h;\frac{\mu_0+1}{\nu_0}\right] \frac{\gamma_0}{\nu_0}(\alpha_0\, \lambda)^{-(\mu_0+1)/\nu_0}. \tag{5.4.9}$$

Note that under the assumptions made the integral in (5.4.9) is convergent.

Of course, there are many other special cases for which we can obtain explicit expressions for the leading terms in the asymptotic expansion. We cannot possibly hope to exhaust all of the possibilities here. There is one additional case however we would like to consider. Indeed, suppose that (5.4.7) holds with $(\mu_0 + 1)/\nu_0 = r_0$. Then a logarithmic term appears to leading order even when $N(0) = 0$ in (4.4.7). As is readily seen, in this case

$$I(\lambda) \sim \frac{c_{00}\,\gamma_0}{\nu_0} \frac{\log \lambda}{(\alpha_0\,\lambda)^{(\mu_0+1)/\nu_0}}, \qquad \lambda \to \infty. \tag{5.4.10}$$

We have seen in Case II that, whenever ϕ and f are such that $q = 0$ in (5.2.7), $t = a$ is a critical point for the asymptotic expansion of I as $\lambda \to \infty$. We might ask, is the critical nature of $t = a$ in these instances due to its being the point at which ϕ achieves its absolute minimum on $[a,b]$ or due to the vanishing of $\phi(a)$. In Case I above we found that when $\phi(a) > 0$, $t = a$ is not a critical point. We would now like to investigate whether or not the vanishing of $\phi(a)$ is alone sufficient to cause $t = a$ to be a critical point.

To accomplish this, we clearly must allow ϕ to become negative in the interval of integration in which event we must be concerned with the behavior of $h(t)$ as $t \to -\infty$. For definiteness we shall assume that h is algebraic in this limit also. Let us consider the following.

EXAMPLE 5.4.1. Suppose

$$I(\lambda) = \int_{-1}^{1} \frac{f(t)\,dt}{i + \lambda t}. \tag{5.4.11}$$

Here we have introduced the complex number i to avoid problems with the convergence of $I(\lambda)$. We now have $h(t) = (i + t)^{-1}$ and $\phi(t) = t$. The latter vanishes at $t = 0$ but does not have a minimum there.

After simple manipulations, we find that

$$I(\lambda) = \int_0^1 \left\{ \frac{-i[f(t) + f(-t)] + \lambda t[f(t) - f(-t)]}{1 + \lambda^2 t^2} \right\} dt. \tag{5.4.12}$$

If, as $t \to 0\pm$, $f(t)$ has asymptotic expansions of the form (4.4.9) with $q = 0$, then we can immediately conclude that $t = 0$ is a critical point for the asymptotic expansion of (5.4.11). Indeed, if $f(t)$ is continuous and nonzero at $t = 0$, then we have

$$I(\lambda) \sim -\frac{2i\,f(0)}{\lambda} M[(1 + t^2)^{-1}; 1] = -\frac{\pi i\,f(0)}{\lambda}. \tag{5.4.13}$$

In this example, the critical nature of $t = 0$ is due solely to the vanishing of $\phi(0)$. Thus, when considering (5.4.1) when the kernel is algebraic in the two limits $t \to \pm\infty$, we must anticipate that any point at which ϕ vanishes will be critical. Moreover, a point in $[a,b]$ at which ϕ achieves either its absolute minimum or its absolute maximum value will not be critical unless, of course, ϕ vanishes there.

Now that we have established that, in general, points at which ϕ vanishes are critical, we would like to understand just why this is so. Clearly, in order to find the explanation, we must first determine what of any significance happens to $h(\lambda\phi)$ when ϕ vanishes. Consider then any subinterval of $[a,b]$ throughout which ϕ is bounded away from zero. If ϕ is positive in this subinterval, then the behavior of $h(\lambda\phi)$ is determined by the asymptotic expansion (4.4.7) with $d=0$ and t replaced by $\lambda\phi$. If ϕ is negative throughout the subinterval, then presumably a similar, but not necessarily identical, expansion will hold for $h(\lambda\phi)$ as $\lambda\phi \to -\infty$. The important point is that, whatever asymptotic expansion holds, it is valid for all t in the subinterval.

Now let us suppose that $\phi(t)$ vanishes at $t=c$ with $a \le c \le b$. In general, there will be no single asymptotic expansion of $h(\lambda\phi)$, $\lambda \to \infty$, that remains valid throughout any interval having $t=c$ either in its interior or on its boundary. Any contribution to the asymptotic expansion of $I(\lambda)$ determined by the local behavior of ϕ and f near $t=c$ can be construed as a reflection of the change in the asymptotic behavior of $h(\lambda\phi)$ for $\lambda \to \infty$ as t approaches this point. Indeed, assuming ϕ continuous in $[a,b]$, there is a small interval about any zero of ϕ throughout which $\lambda\phi = 0(1)$ as $\lambda \to \infty$. Hence, in this region, which we might term a "boundary layer," we cannot replace $h(\lambda\phi)$ by its asymptotic expansion for large argument. Outside of this boundary layer but away from any other zero of ϕ, $|\lambda\phi| = 0(\lambda)$ as $\lambda \to \infty$, and the asymptotic expansion of $h(\lambda\phi)$ for large argument is valid. Of course, the asymptotic behavior of $h(\lambda\phi)$ undergoes a smooth transition. Nevertheless, this somewhat crude discussion indicates why the behavior of ϕ and f in small boundary layers about the zeros of ϕ plays a significant role in the asymptotic expansion of $I(\lambda)$.

The above discussion also holds for the exponential kernels of Sections 5.2 and 5.3. There we found that when ϕ is restricted to be positive in $[a,b]$, the points at which ϕ achieves its absolute minimum are the dominant critical points. Had we allowed ϕ to be negative and further assumed $h(t)$ to be exponentially decreasing as $t \to -\infty$, we would have found, as in this section, that the dominant critical points are the zeros of ϕ. On the other hand, for kernels such as e^{-t} which increase exponentially, as $t \to -\infty$, the absolute minima of ϕ are the dominant critical points.

We could, as in Section 5.3, consider generalizations of the results obtained so far in this section. For example, we might develop a general formula for the expansion of (5.4.1) in the case where ϕ vanishes at one or more points in the interior of $[a,b]$. We choose, however, to illustrate such generalizations in the context of specific examples.

EXAMPLE 5.4.2. Let

$$I(\lambda)=\int_{-\pi}^{\pi}\frac{f(t)\,dt}{(1+\lambda\sin^2 t)^{\alpha}},\qquad \alpha>0. \tag{5.4.14}$$

For simplicity, we assume that $f(t)$ is infinitely differentiable on $[-\pi,\pi]$. Our first objective is to write (5.4.14) as a sum of integrals so that all possible critical points become endpoints of integration. As we have seen, only the zeros of $\phi(t)=\sin^2 t$ on $[-\pi,\pi]$ can be critical points. Hence, we write

$$\begin{aligned}I(\lambda)&=\int_{-\pi}^{0}\frac{f(t)\,dt}{(1+\lambda\sin^2 t)^{\alpha}}+\int_{0}^{\pi}\frac{f(t)\,dt}{(1+\lambda\sin^2 t)^{\alpha}}\\&=\int_0^{\pi}\frac{[f(t)+f(-t)]}{(1+\lambda\sin^2 t)^{\alpha}}\,dt \qquad (5.4.15)\\&=\int_0^{\pi}\frac{\mathscr{F}(t)\,dt}{(1+\lambda\sin^2 t)^{\alpha}}\end{aligned}$$

We cannot directly apply our procedure to (5.4.15) because $\phi=\sin^2 t$ is not monotonic in $(0,\pi)$. To avoid this difficulty we write

$$\begin{aligned}I(\lambda)&=\int_0^{\pi/2}\frac{\mathscr{F}(t)\,dt}{(1+\lambda\sin^2 t)^{\alpha}}+\int_{\pi/2}^{\pi}\frac{\mathscr{F}(t)\,dt}{(1+\lambda\sin^2 t)^{\alpha}}\\&=\int_0^{\pi/2}\frac{[\mathscr{F}(t)+\mathscr{F}(\pi-t)]\,dt}{(1+\lambda\sin^2 t)^{\alpha}}\\&=\int_0^{\pi/2}\frac{\hat{\mathscr{F}}(t)\,dt}{(1+\lambda\sin^2 t)^{\alpha}}.\end{aligned} \tag{5.4.16}$$

In terms of the original integrand function $f(t)$, we have

$$\hat{\mathscr{F}}(t)=f(t)+f(-t)+f(\pi-t)+f(t-\pi). \tag{5.4.17}$$

We are now in a position to apply the results of this section. The two pieces of information we need are the Mellin transform of the kernel $h=(1+t)^{-\alpha}$ and the asymptotic expansion, as $u\to 0+$, of

$$F(u)=\left.\frac{\hat{\mathscr{F}}(t)}{\sin 2t}\right|_{t=\sin^{-1}\sqrt{u}} \tag{5.4.18}$$

We have from the Appendix

$$M[(1+t)^{-\alpha};z]=\frac{\Gamma(z)\,\Gamma(\alpha-z)}{\Gamma(\alpha)} \tag{5.4.19}$$

and after some calculation we find

$$F(u) = \frac{\{2f(0) + f(\pi) + f(-\pi)\}\, u^{-1/2}}{2} + \frac{1}{2}\{f'(-\pi) - f'(\pi)\} + O(u^{1/2}), \quad u \to 0+. \tag{5.4.20}$$

If $0 < \alpha < \frac{1}{2}$, then we can apply (5.4.9) to obtain

$$I(\lambda) = \lambda^{-\alpha} \int_{-\pi}^{\pi} (\sin^2 t)^{-\alpha} f(t)\, dt + \frac{\sqrt{\pi}\,\Gamma(\alpha - \frac{1}{2})}{2\,\lambda^{1/2}\,\Gamma(\alpha)} \{2\, f(0) + f(\pi) + f(-\pi)\}$$
$$+ \frac{\Gamma(\alpha - 1)}{2\lambda\,\Gamma(\alpha)} \{f'(\pi) - f'(-\pi)\} + O(\lambda^{-1-\alpha}). \tag{5.4.21}$$

If $\alpha = \frac{1}{2}$, then we have a coalescence of poles and it follows from (5.4.10) that

$$I(\lambda) = \frac{1}{2}\frac{\log \lambda}{\lambda^{1/2}} \{2\, f(0) + f(\pi) + f(-\pi)\} + O(\lambda^{-1/2}). \tag{5.4.22}$$

If $\frac{1}{2} < \alpha < 1$, then we find

$$I(\lambda) = \frac{\sqrt{\pi}\,\Gamma(\alpha - \frac{1}{2})}{2\,\lambda^{1/2}\,\Gamma(\alpha)} \{2\, f(0) + f(\pi) + f(-\pi)\} + \lambda^{-\alpha} M[F;1-\alpha]$$
$$+ \frac{\Gamma(\alpha - 1)}{2\lambda\,\Gamma(\alpha)} \{f'(\pi) - f'(-\pi)\} + O(\lambda^{-3/2}) \tag{5.4.23}$$

which involves the analytic continuation of $M[F;1-z]$. Finally, if $\alpha = 1$, then the expansion becomes

$$I(\lambda) = \frac{\pi}{2\,\lambda^{1/2}} \{2\, f(0) + f(\pi) + f(-\pi)\} \tag{5.4.24}$$
$$- \frac{\log \lambda}{2\,\lambda} \{f'(\pi) - f'(-\pi)\} + O(\lambda^{-1}).$$

We note that in all cases the points $t = -\pi$, $t = 0$, and $t = \pi$ are critical.

EXAMPLE 5.4.3. Now suppose that

$$I(\lambda) = \int_{-\pi/2}^{\pi/2} \sinh^{-1}(\lambda \sin t) f(t)\, dt. \tag{5.4.25}$$

Here $\phi = \sin t$ vanishes in $[-\pi/2, \pi/2]$ only at $t = 0$. Upon using the fact that $\sinh^{-1} t$ is odd about the origin, we have

$$I(\lambda) = \int_0^{\pi/2} \sinh^{-1}(\lambda \sin t) \{f(t) - f(-t)\}\, dt. \tag{5.4.26}$$

The kernel $h = \sinh^{-1} t$ can also be written

$$h(t) = \log(t + \sqrt{1 + t^2}), \qquad t \geq 0, \tag{5.4.27}$$

which clearly has the appropriate asymptotic behavior as $t \to \infty$. Indeed

$$h(t) = \log t + \log 2 + \frac{t^{-2}}{4} + O(t^{-4}), \qquad t \to \infty. \tag{5.4.28}$$

Thus, we find that $M[\sinh^{-1} t; z]$ has a double pole at $z = 0$ and simple poles at the even integers.

If $f(t)$ is continuously differentiable at the origin, then

$$F(u) = \left.\frac{f(t) - f(-t)}{\cos t}\right|_{t = \sin^{-1} u} = 2u\, f'(0) + o(u), \qquad u \to 0+, \tag{5.4.29}$$

so that the first pole of $M[F; 1 - z]$ occurs at $z = 2$ and is simple. Now we can write down the first three terms in the asymptotic expansion of (5.4.25). Indeed we find that

$$I(\lambda) \sim \log \lambda\, M[F;1] + \left\{\log 2\, M[F;1] - \frac{d}{dz} M[F;1-z]\bigg|_{z=0}\right\}$$
$$+ \frac{\log \lambda}{2\lambda^2} f'(0). \tag{5.4.30}$$

Here the Mellin transforms are given explicitly by

$$M[F;1] = \int_{-\pi/2}^{\pi/2} \operatorname{sgn} t\, f(t)\, dt,$$
$$\frac{d}{dz} M[F;1-z]\bigg|_{z=0} = -\int_0^{\pi/2} \log(\sin t)\, [f(t) - f(-t)]\, dt. \tag{5.4.31}$$

5.5. Expansions for Small λ

In this section we briefly consider

$$I(\lambda) = \int_a^b h(\lambda\phi(t))\, f(t)\, dt \tag{5.5.1}$$

in the limit $\lambda \to 0+$. For simplicity, we shall assume here that $I(1)$ is absolutely convergent. From the results of Section 4.6 and a heuristic boundary layer-type argument analogous to the one introduced in the previous section, we anticipate that we must be concerned, in particular, with the behavior of $h(t)$ as $t \to 0+$ and of the functions ϕ and f as t approaches any *infinity* of ϕ.

We shall not attempt to be as complete in our discussion here as we were in the preceding sections. Our major objective is to obtain results sufficient to establish the critical nature of the infinities of ϕ.

Let us first suppose that ϕ is nonnegative and finite throughout $[a,b]$ and that, as $t \to 0+$, $h(t)$ has the asymptotic expansion

$$h(t) \sim \sum_{m=0}^{\infty} \sum_{n=0}^{\bar{N}(m)} p_{mn} (\log t)^n\, t^{a_m} \tag{5.5.2}$$

with $\text{Re}(a_m) \uparrow +\infty$ and $\bar{N}(m)$ finite for each m. Here we have assumed that $h(t)$ is algebraic as $t \to 0+$, because this is true of most kernels that arise in applications.

Because ϕ is finite in $[a,b]$, we can replace $h(\lambda\phi)$ in (5.5.1) by its asymptotic expansion for small $\lambda\phi$ and then integrate term-by-term. In this manner we obtain the formal result

$$I(\lambda) \sim \sum_{m=0}^{\infty} \lambda^{a_m} \sum_{n=0}^{\bar{N}(m)} p_{mn} \sum_{j=0}^{n} \binom{n}{j} (\log \lambda)^j \int_a^b (\log \phi)^{n-j} \phi^{a_m} f\, dt. \tag{5.5.3}$$

Note that, under the assumption $I(1)$ is absolutely convergent, all of the integrals in (5.5.3) are finite. We leave to the exercises a proof of the validity of (5.5.3).

Of more interest is the case where ϕ is nonnegative and monotonically increasing in $[a,b]$ with $\phi(b) = \infty$. To analyze I in this case we set $u = \phi(t)$ in (5.5.1) to obtain

$$I(\lambda) = \int_0^\infty h(\lambda u)\, \bar{F}(u)\, du. \tag{5.5.4}$$

Here

$$\bar{F}(u) = \begin{cases} 0, & 0 \le u < \phi(a), \\ \left.\dfrac{f}{\phi'}\right|_{t=\phi^{-1}(u)}, & \phi(a) \le u < \infty. \end{cases} \tag{5.5.5}$$

The desired expansion will be obtained by applying the results of Section 4.6.

Let us suppose that, as $u \to \infty$,

$$\left.\frac{f}{\phi'}\right|_{t=\phi^{-1}(u)} \sim \sum_{m=0}^{\infty} \sum_{n=0}^{N(m)} c_{mn}\, u^{-r_m} (\log u)^n \tag{5.5.6}$$

with $\text{Re}(r_m) \uparrow \infty$ and $N(m)$ finite for each m. We are assuming $F(u)$ [and hence $\bar{F}(u)$] algebraic at infinity to reduce the number of cases to be considered. If we now suppose that the hypotheses of Theorem 4.6 are satisfied, then the results of Case IV in Section 4.6 yield the desired expansion of I.

In particular, if $a_m + 1 \neq r_n$ for any m, n, then (4.6.18) holds. We repeat that result here.

$$\begin{aligned} I(\lambda) \sim & \sum_{m=0}^{\infty} \lambda^{a_m} \sum_{n=0}^{\bar{N}(m)} p_{mn} \sum_{j=0}^{n} \binom{n}{j} (\log \lambda)^j M^{(n-j)}[\bar{F};z]\Big|_{z=1+a_m} \\ & + \sum_{m=0}^{\infty} \lambda^{r_m - 1} \sum_{n=0}^{N(m)} c_{mn} \sum_{j=0}^{n} \binom{n}{j} (-\log \lambda)^j M^{(n-j)}[h;z]\Big|_{z=1-r_m}. \end{aligned} \tag{5.5.7}$$

To illustrate what happens when a coalescence occurs, suppose that $a_0 + 1 = r_0$, while $c_{0n} = p_{0n} = 0$ for $n \ge 1$. Then we find

$$I(\lambda) \sim -\lambda^{a_0} \log \lambda\, p_{00}\, c_{00}. \tag{5.5.8}$$

We note that when $\phi(a) = 0$, the Mellin transforms of $\bar{F}$ in (5.5.7) need exist only in the generalized sense.

To obtain more explicit expressions for the leading terms in (5.5.7) and (5.5.8) we now suppose that, as $t \to b-$,

$$f \sim f(b),$$

$$\phi \sim \alpha_0(b-t)^{-\gamma_0}, \qquad \alpha_0, \gamma_0 > 0. \tag{5.5.9}$$

We then find

$$\bar{F}(u) \sim \frac{f(b)}{\gamma_0} (\alpha_0)^{1/\gamma_0} u^{-(\gamma_0+1)/\gamma_0}, \qquad u \to +\infty. \tag{5.5.10}$$

If $a_0 > 1/\gamma_0$, then it follows from (5.5.7) that

$$I(\lambda) \sim (\alpha_0 \lambda)^{1/\gamma_0} \frac{f(b)}{\gamma_0} M\left[h; -\frac{1}{\gamma_0}\right]. \tag{5.5.11}$$

If $a_0 < 1/\gamma_0$ and $p_{0n} = 0$ for $n \geq 1$, then (5.5.7) yields

$$I(\lambda) \sim \lambda^{a_0} p_{00} M[\bar{F}; 1 + a_0] = \lambda^{a_0} p_{00} \textstyle\int_a^b \phi^{a_0} f\, dt. \tag{5.5.12}$$

Finally, if $a_0 = 1/\gamma_0$ and $p_{0n} = 0$ for $n \geq 1$, then

$$I(\lambda) \sim -(\alpha_0 \lambda)^{a_0} \log \lambda\, p_{00} f(b)/\gamma_0 \tag{5.5.13}$$

We see from (5.5.11)–(5.5.13) that $t = b$ is indeed a critical point for the expansion of I when $\phi(b) = \infty$.

EXAMPLE 5.5. Let us consider

$$I(\lambda) = \int_0^{\pi/2} e^{-\lambda \sec t} (t + \tan t)^\alpha\, dt \tag{5.5.14}$$

in the limit $\lambda \to 0+$. Here $h(t) = e^{-t}$ and $\phi(t) = \sec t$. Because $\phi(\pi/2) = \infty$, the results (5.5.7) and (5.5.8) apply. We find that as $u \to \infty$,

$$F(u) = \frac{\left(\sec^{-1} u + \sqrt{u^2 - 1}\right)^\alpha}{u\sqrt{u^2 - 1}} \sim u^{\alpha - 2}. \tag{5.5.15}$$

Also, we have

$$M[e^{-t}; 1 - z] = \Gamma(1 - z). \tag{5.5.16}$$

Thus, if $1 - \alpha > 0$, then

$$\int_0^{\pi/2} e^{-\lambda \sec t} (t + \tan t)^\alpha\, dt \sim M[\bar{F}; 1] = \int_0^{\pi/2} (t + \tan t)^\alpha\, dt. \tag{5.5.17}$$

If $\alpha = 1$, then we find from (5.5.8) that

$$\int_0^{\pi/2} e^{-\lambda \sec t} (t + \tan t)\, dt \sim -\log \lambda. \tag{5.5.18}$$

Finally, if $\alpha > 1$, then (5.5.7) yields

$$\int_0^{\pi/2} e^{-\lambda \sec t} (t + \tan t)^\alpha\, dt \sim \lambda^{1-\alpha}\, \Gamma(\alpha - 1). \tag{5.5.19}$$

5.6. Exercises

5.1. Determine the leading two terms in the asymptotic expansion of each of the following integrals as $\lambda \to \infty$:

(a) $\int_0^\pi e^{-\lambda t \sin t} \sqrt{t^2+1}\, dt$.

(b) $\int_0^\pi e^{-\lambda(e^t - 2t)} \sin^2 t\, dt$.

(c) $\int_{-1}^1 e^{-\lambda t(\log|t| - 1)}(1+t)^3\, dt$.

(d) $\int_{-1}^1 e^{+\lambda \cos\sqrt{t^2+1}} \cosh t\, dt$.

5.2. Consider the Sievert integral

$$I(\lambda;\theta) = \int_0^\theta e^{-\lambda \sec t}\, dt. \tag{5.6.1}$$

(a) For $0 < \theta \le \pi/2$, verify that

$$I(\lambda;\theta) \sim \sqrt{\frac{\pi}{2\lambda}}\, e^{-\lambda} \{1 + O(\lambda^{-1})\}, \qquad \lambda \to \infty. \tag{5.6.2}$$

(b) Show that (5.6.2) is valid for $|\arg(\lambda)| < \pi/2$.

5.3. Consider the class of integrals

$$I_m(x) = \int_0^\infty t^m e^{-t^2 - x/t}\, dt. \tag{5.6.3}$$

(a) Determine the change of variable which recasts (5.6.3) in the form

$$I_m(x) = \lambda^{(m+1)/2} \int_0^\infty \tau^m e^{-\lambda[\tau^2 + 1/\tau]}\, d\tau, \qquad \lambda = x^{2/3}. \tag{5.6.4}$$

(b) Show that

$$I_m(\lambda^{3/2}) \sim \lambda^{m/2} \sqrt{\frac{\pi}{3}}\, \frac{e^{-3(2^{-2/3})\lambda}}{2^{m/3}} [1 + O(\lambda^{-1})]. \tag{5.6.5}$$

(c) Show that (5.6.5) is valid for $|\arg(x)| < \pi$.

5.4. Consider

$$I(\lambda) = \int_a^b \exp\{-\lambda\phi(t)\} f(t)\, dt, \tag{5.6.6}$$

where both f and ϕ belong to $C^\infty[a,b]$.

(a) Suppose that ϕ has a unique absolute minimum at $t = t_0$, $a < t_0 < b$. Suppose further that

$$\phi^{(j)}(t_0) = 0, \qquad j = 1, 2, \ldots, 2n-1, \qquad \phi^{(2n)}(t_0) > 0. \tag{5.6.7}$$

Show that the relation

$$\phi(t) - \phi(t_0) = \sigma^{2n} \tag{5.6.8}$$

defines $\sigma = \sigma(t)$ in a two-sided neighborhood of $t = t_0$ as a C^∞ function with $d\sigma/dt > 0$ when the ambiguity in the sign of the $2n$th root is resolved by taking $\operatorname{sign}(\sigma) = \operatorname{sign}(t - t_0)$.

(b) Let

$$I_0(\lambda) = \int_{t_0-\varepsilon}^{t_0+\varepsilon} f(t) \exp\{-\lambda\phi(t)\}\, dt \tag{5.6.9}$$

with t_0 as in (a) and the interval $[t_0 - \varepsilon, t_0 + \varepsilon]$ contained in the domain of regularity of $\sigma(t)$. Introduce σ as a new variable of integration in (5.6.9) to obtain

$$I_0(\lambda) = \exp\{-\lambda\phi(t_0)\} \int_{-\varepsilon_1}^{\varepsilon_2} G(\sigma) \exp\{-\lambda\sigma^{2n}\}\, d\sigma. \tag{5.6.10}$$

Determine $G(\sigma)$, ε_1, ε_2 in terms of ϕ, f, and σ. Verify that $G(\sigma)\ \varepsilon\ C^\infty(-\varepsilon_1, \varepsilon_2)$.

(c) By using the method of Section 5.1 show that

$$I_0(\lambda) \sim \exp\{-\lambda\phi(t_0)\} \sum_{j=0}^{\infty} \frac{G^{(2j)}(0)}{n(2j)!} \frac{\Gamma\left(\dfrac{2j+1}{2n}\right)}{\lambda^{(2j+1)/2n}}, \qquad \lambda \to \infty. \tag{5.6.11}$$

(d) Verify that for $n = 1$, this result agrees with (5.1.21).

(e) Show that for any n, the result analogous to (5.1.21) is

$$I_0(\lambda) = \exp\{-\lambda\phi(t_0)\} \left(\frac{(2n)!}{\lambda\phi^{(2n)}(t_0)}\right)^{1/2n} \frac{f(t_0)\,\Gamma\left(\dfrac{1}{2n}\right)}{n} [1 + O(\lambda^{-1/n})]. \tag{5.6.12}$$

(f) Show that $I(\lambda) = I_0(\lambda)$ to within an exponentially small error.

5.5. (a) Show that, if $t_0 = a$ in Exercise 5.4, then

$$I(\lambda) \sim \frac{\exp\{-\lambda\phi(a)\}}{2n} \sum_{j=0}^{\infty} \frac{G^{(j)}(0)}{j!} \frac{\Gamma\left(\dfrac{j+1}{2n}\right)}{\lambda^{(j+1)/2n}}, \qquad \lambda \to \infty. \tag{5.6.13}$$

(b) In particular show that

$$\begin{aligned}
I(\lambda) \sim \frac{\exp\{-\lambda\phi(a)\}}{2n} &\left\{ f(a)\,\Gamma\left(\frac{1}{2n}\right) \left[\frac{(2n)!}{\lambda\phi^{(2n)}(a)}\right]^{1/2n} \right. \\
&+ \Gamma\left(\frac{1}{n}\right) \left[\frac{(2n)!}{\lambda\phi^{(2n)}(a)}\right]^{1/n} \left[f^{(1)}(a) - \frac{f(a)\,\phi^{(2n+1)}(a)}{n(2n+1)\,\phi^{(2n)}(a)} \right] \\
&+ \Gamma\frac{3}{2n} \left[\frac{(2n)!}{\lambda\phi^{(2n)}(a)}\right]^{3/2n} \left[\frac{f^{(2)}(a)}{2} - \frac{3f^{(1)}(a)\,\phi^{(2n+1)}(a)}{2n(2n+1)\,\phi^{(2n)}(a)} \right. \\
&+ \frac{3f(a)}{2n(2n+1)\,\phi(2n)(a)} \left(\left(1 + \frac{3}{2n}\right) \frac{[\phi^{(2n+1)}(a)]^2}{2(2n+1)\,\phi^{(2n)}(a)} \right.
\end{aligned}$$

$$-\frac{\phi^{(2n+2)}(a)}{2(n+1)}\Bigg)\Bigg]+O(\lambda^{-2/n})\Bigg\}. \tag{5.6.14}$$

(c) Verify that one can obtain (5.6.14) formally in the following manner: Set

$$\phi(t)=\phi(a)+c_0\, s^{2n}+c_1\, s^{2n+1}+c_2\, s^{2n+2}+\cdots,$$
$$f(t)=a_0+a_1 s+a_2 s^2+\cdots, \qquad s=t-a. \tag{5.6.15}$$

Substitute (5.6.15) into (5.6.6) and expand $\exp\{-\lambda[c_1 s^{2n+1}+c_2 s^{2n+2}+\cdots]\}$ in a power series. Finally multiply this series by $\exp\{-\lambda[\phi(a)+c_0 s^{2n}]\}$ and the series for f and integrate the resulting series term-by-term from zero to infinity.

5.6. Suppose that in (5.6.6) both ϕ and f belong to $C^\infty(a,b)$. Furthermore, suppose that $\phi'>0$ on (a,b) and that, as $t\to a+$,

$$\phi(t)=\phi(a)+\alpha(t-a)^\beta+O((t-a)^{\beta'}), \qquad \beta'>\beta>0, \qquad \alpha>0,$$
$$f(t)=f_0(t-a)^\gamma+O((t-a)^{\gamma'}), \qquad -1<\gamma<\gamma'. \tag{5.6.16}$$

(a) Show that the change of variable defined by

$$\sigma^\beta=\phi(t)-\phi(a)$$

with σ positive for $t>a$ is one-to-one and differentiable for t in some one-sided neighborhood of a.

(b) Show that

$$I(\lambda)=\exp\{-\lambda\phi(a)\}\frac{f_0}{\beta}\frac{\Gamma((\gamma+1)/\beta)}{(\alpha\lambda)^{(\gamma+1)/\beta}}[1+O(\lambda^{-\delta/\beta})], \tag{5.6.17}$$
$$\delta=\min[\beta'-\beta,\gamma'-\gamma].$$

5.7. Let

$$I(\lambda)=\int_a^b[\phi(t)]^\lambda f(t)\,dt. \tag{5.6.18}$$

Assume that ϕ is nonnegative on $[a,b]$, $\phi\in C^{(4)}[a,b]$, and $f\in C^{(2)}[a,b]$.

(a) Suppose that the absolute maximum of ϕ on $[a,b]$ occurs at $t=a$ with $\phi^{(1)}(a)=0$, $\phi^{(2)}(a)<0$. Show that, as $\lambda\to\infty$,

$$I(\lambda)=\frac{[\phi(a)]^{\lambda+1/2}}{\lambda^{1/2}}f(a)\sqrt{\frac{\pi}{2|\phi^{(2)}(a)|}}$$
$$-\frac{[\phi(a)]^{\lambda+1}}{\lambda}\left\{\frac{f^{(1)}(a)}{\phi^{(2)}(a)}-\frac{f(a)\,\phi^{(3)}(a)}{3(\phi^{(2)}(a))^2}\right\}+O\left(\frac{[\phi(a)]^\lambda}{\lambda^{3/2}}\right). \tag{5.6.19}$$

(b) Suppose that the absolute maximum of ϕ on $[a,b]$ occurs at $t=b$ with $\phi^{(1)}(b)=0$, $\phi^{(2)}(b)<0$. Show that

$$I(\lambda) = \frac{[\phi(b)]^{\lambda+\frac{1}{2}}}{\lambda^{1/2}} f(b) \sqrt{\frac{\pi}{2|\phi^{(2)}(b)|}}$$
$$+ \frac{[\phi(b)]^{\lambda+1}}{\lambda} \left\{ \frac{f^{(1)}(b)}{\phi^{(2)}(b)} - \frac{f(b)\,\phi^{(3)}(b)}{3(\phi^{(2)}(b))^2} \right\} + O\left(\frac{[\phi(b)]^{\lambda}}{\lambda^{3/2}}\right). \tag{5.6.20}$$

(c) Show that if the absolute maximum of ϕ on $[a,b]$ occurs at the interior point $t = c$ with $\phi^{(2)}(c) < 0$, then

$$I(\lambda) = [\phi(c)]^{\lambda+\frac{1}{2}} \sqrt{\frac{2\pi}{\lambda|\phi^{(2)}(c)|}} f(c) + O\left(\frac{[\phi(c)]^{\lambda}}{\lambda^{3/2}}\right). \tag{5.6.21}$$

(d) Finally show that if the absolute maximum of ϕ on $[a,b]$ occurs at $t = a$ and if, as $t \to a+$,

$$\phi(t) = \phi(a) - \alpha(t-a)^{\beta} + O((t-a)^{\beta'}), \qquad \beta' > \beta > 0, \qquad \alpha > 0,$$
$$f(t) = f_0(t-a)^{\gamma} + O((t-a)^{\gamma'}), \qquad \gamma' > \gamma > -1, \tag{5.6.22}$$

then

$$I(\lambda) = [\phi(a)]^{\lambda+(\gamma+1)/\beta} \frac{f_0}{\beta} \frac{\Gamma((\gamma+1)/\beta)}{(\alpha\lambda)^{(\gamma+1)/\beta}} [1 + O(\lambda^{-\delta/\beta})],$$
$$\delta = \min[\beta' - \beta, \gamma' - \gamma, \beta]. \tag{5.6.23}$$

5.8. Let $f(p) = a^{1/p} p^{-1/2p} \{f_1(p)\}^{1/p}$.

(a) Set $a^{1/p} = \exp\{(\log a)/p\}$ and show that

$$a^{1/p} = 1 + \frac{\log a}{p} + O(p^{-2}), \qquad p \to \infty.$$

(b) Set $p^{-1/2p} = \exp\{(-\log p)/2p\}$ and show that

$$p^{-1/2p} = 1 - \frac{\log p}{2p} + O\left(\left(\frac{\log p}{p}\right)^2\right), \qquad p \to \infty.$$

(c) Show that if $f_1(p) = 1 + b/p + O(p^{-2})$, $p \to \infty$, then

$$\{f_1(p)\}^{1/p} = 1 + O(p^{-2}).$$

(d) Verify (5.1.32).

5.9. (a) Suppose that $g(t)$ is a nonnegative function on $[0,1]$ whose absolute maximum occurs at $t = 0$. Furthermore, let $g(t) = g(0) - \alpha t^{\nu} + o(t^{\nu})$, $t \to 0+$, $\alpha > 0$, $\nu > 0$. Let $\|g\|_p$ as defined by (5.1.27) exist for all $p \geq p_0$. Show that

$$\|g\|_p = g(0)\left(1 - \frac{\log p}{\nu p} + o\left(\frac{\log p}{p}\right)\right), \qquad p \to \infty. \tag{5.6.24}$$

(b) If the absolute maximum of g is at $t = 1$, with

$$g(t) = g(1) - \beta(1-t)^\mu + o((1-t)^\mu), \qquad t \to 1-,$$

then show that

$$\|g\|_p = g(1)\left(1 - \frac{\log p}{\mu p} + o\left(\frac{\log p}{p}\right)\right), \qquad p \to \infty. \tag{5.6.25}$$

(c) Suppose that $g \in C^{(2n+1)}[a,b]$ and that the absolute maximum of g occurs at the interior point $t = c$. Suppose further that $g^{(j)}(c) = 0$, $j = 1, 2, \ldots, (2n-1)$, $g^{(2n)}(c) < 0$. Show that

$$\|g\|_p = g(c)\left[1 - \frac{\log p}{2\,np} + o\left(\frac{\log p}{p}\right)\right], \qquad p \to \infty. \tag{5.6.26}$$

5.10. Consider the following integral representation for the modified Bessel function of the second kind:

$$K_\nu(a) = \frac{1}{2}\int_{-\infty}^{\infty} \exp\{-a\cosh t + \nu t\}\, dt. \tag{5.6.27}$$

(a) Show that $\phi = -\nu t + a\cosh t$ attains its minimum when $t = \sinh^{-1} \nu/a$.
(b) Show that under the change of variable defined by

$$\tau = t - \sinh^{-1}\frac{\nu}{a},$$

(5.6.27) becomes

$$K_\nu(a) = \frac{1}{2}\left\{\frac{\sqrt{\nu^2+a^2}+\nu}{a}\right\}^\nu \int_{-\infty}^{\infty} \exp\{-\nu\,[e^\tau - \tau]\}\exp\left\{-\frac{a^2\cosh\tau}{\nu+\sqrt{\nu^2+a^2}}\right\} d\tau. \tag{5.6.28}$$

(c) Formally apply Laplace's method to obtain the leading term of the asymptotic expansion of K_ν as $\nu \to \infty$ with a fixed. Here Exercise 5.8 must be used as well.

(d) With parts (a)–(c) as motivation, develop sufficient conditions for Laplace's method to remain valid when treating integrals of the form

$$I(\lambda) = \int_a^b \exp\{-\lambda\phi(t)\}\, f(t;\lambda)\, dt. \tag{5.6.29}$$

5.11. Consider the following integral representation for the associated Legendre function of the second kind:

$$Q^m_{n-\frac{1}{2}}(\cosh\eta) = \frac{(-1)^m\,\Gamma(n+\frac{1}{2})}{\Gamma(n-m+\frac{1}{2})}\int_0^\infty \frac{\cosh mt\, dt}{(\cosh\eta + \cosh t\sinh\eta)^{n+\frac{1}{2}}}. \tag{5.6.30}$$

Use the method developed in Exercise 5.7 to obtain the expansion

$$Q^m_{n-\frac{1}{2}}(\cosh\eta) \sim (-1)^m (n)^{m-\frac{1}{2}} e^{-n\eta} \sqrt{\frac{\pi}{2\sinh\eta}}, \qquad \eta > 0, \qquad n \to \infty. \quad (5.6.31)$$

5.12. Show that the function $F(u)$ defined by (5.2.4) is locally integrable on $(\phi(a), \phi(b))$.

5.13. Consider (5.2.1) with h a kernel of exponential type. Suppose that, as $t \to a+$, (5.2.8) holds and that

$$f(t) \sim \gamma(t-a)^\mu [\log(t-a)]^n + O((t-a)^\mu [\log(t-a)]^{n-1}). \quad (5.6.32)$$

Show that, to leading order (5.2.1) has the asymptotic expansion

$$I(\lambda) \sim (-1)^n (\alpha_0 \lambda)^{-(\mu+1)/\nu_0} (\log\lambda)^n \frac{\gamma}{\nu_0^{n+1}} M\left[h; \frac{\mu+1}{\nu_0}\right] [1 + O((\log\lambda)^{-1})]. \quad (5.6.33)$$

5.14. Justify the claim made following Equation (5.2.21). In particular, show that when the sum on m is made finite with upper limit M, the error is

$$O\{(\log\lambda)^{N(M+1)} \lambda^{-r_{M+1}} \int_a^b e^{-\alpha\lambda^\nu\phi^\nu} (\log\phi)^{N(M+1)} \phi^{-r_{M+1}} f\,dt\}. \quad (5.6.34)$$

5.15. Verify that the following asymptotic approximations are correct:

(a) $$I(\lambda; a) = \int_0^{\pi/2} \frac{\mathrm{Ai}(\lambda\sqrt{1+a^2-2a\cos\theta})\,d\theta}{(1+a^2-2a\cos\theta)^{1/4}}$$

$$\sim \frac{\exp\{-\frac{2}{3}\lambda^{3/2}(1-a)^{3/2}\}}{2\sqrt{2a(1-a)}\,\lambda}, \qquad 0 < a < 1, \qquad \lambda \to \infty. \quad (5.6.35)$$

(See Exercise 4.3 for the asymptotic expansion of the Airy function Ai.)

(b) $$I(\lambda; 1) \sim \frac{3^{-5/6}}{2\sqrt{\pi\lambda}} \Gamma\left(\frac{1}{6}\right), \qquad \lambda \to \infty. \quad (5.6.36)$$

Here $I(\lambda; a)$ is as in part (a).

(c) $$\int_0^{\pi} D_\nu(\lambda \sin^2\theta)\cos 2\theta\, d\theta \sim \frac{\pi\, 2^{(\nu+1)/2}}{\lambda^{1/2}\,\Gamma\left(\frac{3}{4}-\frac{\nu}{2}\right)} F\left(\frac{3}{4},\frac{1-\nu}{2};\frac{3}{4}-\frac{\nu}{2};-1\right)$$

$$-\pi\, 2^{(\nu+1)/2}$$

$$\times \sum_{n=0}^{\infty} \frac{(-1)^n\,\Gamma(n+\frac{5}{2})\,F\left(\frac{n}{2}+\frac{5}{4},\frac{1-\nu}{2};\frac{n-\nu}{2}+\frac{5}{4};-1\right)}{\lambda^{n+3/2}\,\Gamma(\frac{1}{2}-n)\,(n+1)!\,\Gamma\left(\frac{n-\nu}{2}+\frac{5}{4}\right)}, \qquad \lambda\to\infty. \tag{5.6.37}$$

Here D_ν is the Weber function of order ν and $F(a,b;c;z)$ is the hypergeometric function.

(d) $I(\lambda)=\int_0^{\pi/2} F(a,b;c;-\lambda\sin\theta)\sqrt{\sin 2\theta}\, d\theta$

$$\sim \frac{1}{\sqrt{2}}\frac{\Gamma(c)\,\Gamma(\frac{3}{4})}{\Gamma(a)\,\Gamma(b)}\sum_{n=0}^{\infty}\frac{(-\lambda)^{-n}}{n!}\left\{\frac{\lambda^{-a}\,\Gamma(a+n)\,\Gamma(b-a-n)\,\Gamma\left(\frac{3}{4}-\frac{a+n}{2}\right)}{\Gamma(c-a-n)\,\Gamma\left(\frac{3}{2}-\frac{a+n}{2}\right)}\right.$$

$$+\frac{\lambda^{-b}\,\Gamma(b+n)\,\Gamma(a-b-n)\,\Gamma\left(\frac{3}{4}-\frac{b+n}{2}\right)}{\Gamma(c-b-n)\,\Gamma\left(\frac{3}{2}-\frac{b+n}{2}\right)}$$

$$\left.-\frac{\lambda^{-n-\frac{3}{2}}\,\Gamma(2n+\frac{3}{2})\,\Gamma(a-2n-\frac{3}{2})\,\Gamma(b-2n-\frac{3}{2})}{\Gamma(c-2n-\frac{3}{2})\,\Gamma(\frac{3}{4}-n)}\right\}, \qquad \lambda\to\infty. \tag{5.6.38}$$

Here $a\neq b$ and neither is a half-integer; $b-a$ is not an integer.

(e) Let $a=b$ in part (d) with neither a half-integer. Then

$$I(\lambda)\sim\frac{1}{\sqrt{2}}\frac{\Gamma(c)\,\Gamma(\frac{3}{4})}{\Gamma^2(a)}\sum_{n=0}^{\infty}\frac{(-\lambda)^{-n}}{(n!)^2}\left\{\frac{\lambda^{-a}\,\Gamma(a+n)\,\Gamma\left(\frac{3}{4}-\frac{a+n}{2}\right)}{\Gamma(c-a-n)\,\Gamma\left(\frac{3}{2}-\frac{a+n}{2}\right)}\right.$$

$$\times\left[\log\lambda-\psi(a+n)+2\psi(n+1)-\psi(c-a-n)\right.$$

$$\left.-\frac{1}{2}\psi\left(\frac{3}{2}-\frac{a+n}{2}\right)+\frac{1}{2}\psi\left(\frac{3}{4}-\frac{a+n}{2}\right)\right]$$

$$\left.+\frac{2(-1)^n\lambda^{-n-\frac{3}{2}}\,\Gamma(2n+\frac{3}{2})\,\Gamma^2(a-2n-\frac{3}{2})\,n!}{\Gamma(c-2n-\frac{3}{2})\,\Gamma(\frac{3}{4}-n)}\right\}, \qquad \lambda\to\infty. \tag{5.6.39}$$

(f) Obtain an expansion of $I(\lambda)$ in part (d) for $a=\frac{1}{2}$ and b not a half-integer.

(g) Obtain an expansion of $I(\lambda)$ in part (d) for $a=b=\frac{1}{2}$.

5.16. Let

$$I(\lambda) = \int_0^a |1 - \lambda\phi(t)|^{1/2} f(t)\,dt. \tag{5.6.40}$$

Assume that $\phi(t)$ is nonnegative on $[0, a]$ with its only zero at $t = 0$. Furthermore assume that, as $t \to 0+$,

$$\begin{aligned} f(t) &= \gamma t^{\mu} + o(t^{\mu}), \qquad \mu > 0, \\ \phi(t) &= \alpha t^{\nu} + o(t^{\nu}), \qquad \nu > 0. \end{aligned} \tag{5.6.41}$$

(a) If $(\mu + 1)/\nu < \frac{1}{2}$, then show that

$$I(\lambda) = \lambda^{1/2} M\left[G_1; \frac{3}{2}\right] + (\alpha\lambda)^{-(\mu+1)/\nu} \left(\frac{\gamma}{\nu}\right) \frac{\sqrt{\pi}}{2} \frac{\Gamma\left(\frac{\mu+1}{\nu}\right)}{\Gamma\left(\frac{3}{2} + \frac{\mu+1}{\nu}\right)} \left[1 - \tan\frac{\pi(\mu+1)}{\nu}\right]$$

$$+ o(\lambda^{-(\mu+1)/\nu}), \qquad \lambda \to \infty. \tag{5.6.42}$$

Here

$$M[G_1; z] = \int_0^a [\phi(t)]^{z-1} f(t)\,dt. \tag{5.6.43}$$

(b) If $(\mu + 1)/\nu = \frac{1}{2}$, then show that

$$I(\lambda) = \lambda^{1/2} M\left[G_1; \frac{3}{2}\right] - \frac{\gamma}{2\nu} \frac{\log\lambda}{(\alpha\lambda)^{1/2}} + O(\lambda^{-1/2}), \qquad \lambda \to \infty. \tag{5.6.44}$$

(c) If $(\mu + 1)/\nu > \frac{1}{2}$, then show that

$$I(\lambda) = M\left[G_1; \frac{3}{4}\right] - \frac{1}{2}\lambda^{-1/2} M\left[G_1; \frac{1}{2}\right] + O(\lambda^{-\beta});$$

$$\beta = \min\left(\frac{\mu+1}{\nu}, \frac{3}{2}\right). \tag{5.6.45}$$

5.17. Prove that (5.5.3) is correct under the conditions stated in Section 5.5.

5.18. Verify that the following asymptotic approximations are correct:

(a) $$\int_0^{\pi/2} \frac{K_\nu(\lambda \tan t)\,dt}{\sqrt{1+t^2}} \sim \frac{\lambda^{-\nu}\,\Gamma(\nu)}{2^{1-\nu}} \int_0^{\pi/2} \frac{(\tan t)^{-\nu}\,dt}{\sqrt{1+t^2}}, \qquad \lambda \to 0, \qquad 0 < \nu < 1. \tag{5.6.46}$$

(b) $$\int_0^{\pi/2} \frac{K_0(\lambda \tan t)\,dt}{\sqrt{1+t^2}} \sim -(\log\lambda) \log\left[\frac{\pi}{2} + \sqrt{\frac{\pi^2}{4} + 1}\right] + O(1), \qquad \lambda \to 0. \tag{5.6.47}$$

(c) $$\int_0^{\pi/2} e^{-\lambda \sec t}\, dt \sim \frac{\pi}{2} + \sum_{k=1}^{\infty} \left\{ \frac{\lambda^{2k-1} \pi^{3/2} \Gamma(k - \frac{1}{2})}{(2k-1)!\,(k-1)!} \times [\log \lambda - \psi(2k) - \tfrac{1}{2}\psi(k) + \tfrac{1}{2}\psi(\tfrac{3}{2} - k)] \right\}, + \frac{\pi}{2}, \quad \lambda \to 0. \tag{5.6.48}$$

REFERENCES

E. T. COPSON. *Asymptotic Expansions*. University Press, Cambridge, 1965.
An extensive discussion of Laplace's method is presented in this book.

A. ERDÉLYI. *Asymptotic Expansions*. Dover, New York, 1956.
Laplace's method is discussed and an historical development is presented.

W. FULKS. A generalization of Laplace's method. *Proc. Amer. Math. Soc.* **2,** 613–622, 1951.
In this paper the author considers a Laplace-type integral with two large parameters.

R. A. HANDELSMAN and J. S. LEW. Asymptotic expansions of a class of integral transforms via Mellin transforms. *Arch. Rat. Mech. Anal.* **35,** 382–396, 1969.
The authors extend the Mellin transform method to exponentially dominated kernels of nonmonotonic argument and thus generalize Laplace's method.

R. A. HANDELSMAN and J. S. LEW. On the convergence of the L^p norm to the L^∞ norm. *Amer. Math. Monthly* **79,** 618–622, 1972.
The authors apply Laplace's method to study how the L^p norm of a function converges, as $p \to \infty$, to the L^∞ norm of that function.

D. S. JONES. Asymptotic behavior of integrals. *SIAM Review* **14,** 286–317, 1972.
In this paper, explicit error bounds are obtained in the application of Laplace's method.

P. S. LAPLACE. *Théorie Analytique des Probabilités*. Vol. 1. Paris, 1820.
This is the original work in which the technique now called Laplace's method is developed.

F. W. J. OLVER. Error bounds for the Laplace approximation for definite integrals. *J. Approx. Theory*. **1,** 293–313, 1968.
The author develops explicit rather than order estimates for the error in Laplace's approximation.

D. V. WIDDER. *The Laplace Transform*. University Press, Princeton, 1946.
Laplace's method is developed under rather general conditions.

6 h-Transforms with Oscillatory Kernels

6.1. Fourier Integrals and the Method of Stationary Phase

In this section we shall consider the asymptotic expansion as $\lambda \to \infty$, of integrals of the form

$$I(\lambda) = \int_a^b \exp\{i\lambda\, \phi(t)\} f(t)\, dt \tag{6.1.1}$$

with ϕ real.

First, a heuristic development of an asymptotic formula or leading term will be given, while later in the section a rigorous derivation will be offered. The consideration here of the Fourier-type integrals (6.1.1) will serve to motivate the treatment in the sections to follow of *h*-transforms with general oscillatory kernels.

We know from previous work that, whenever f and ϕ are infinitely differentiable with ϕ monotonic on $[a,b]$, an infinite asymptotic expansion of (6.1.1) can be obtained via the integration-by-parts procedure. Furthermore, we recall that it was anticipated in Section 3.3 that points at which ϕ' vanishes would be critical. Our objective is to derive the leading term of the contribution to the asymptotic expansion from such a critical point. Before doing so, however, let us try to better understand the critical nature of the zeros of $\phi'(t)$.

Suppose that $t = c$ is a point in (a,b) at which ϕ' does *not* vanish. Then there exists a small neighborhood N_c of this point such that, as t varies throughout N_c, $\phi(t)$ is changing. Furthermore, if λ is large, then the change in $\lambda\phi$ is rapid so that the oscillations of the real and imaginary parts of $\exp\{i\lambda\phi\}$ about zero are rapid. Now consider

$$I_c(\lambda) = \int_{N_c} \exp\{i\lambda\phi\} f\, dt. \tag{6.1.2}$$

Because N_c is a small interval about $t = c$, we may closely approximate $f(t)$ in (6.1.2) by $f(c)$. Then, upon assuming λ large, we find that the rapid oscillation of $\exp\{i\lambda\phi\}$ produces cancellations which, in turn, tend to decrease the value of $I_c(\lambda)$.

Now suppose that ϕ' does vanish at $t = c$. Then no matter how large λ is, there exists a neighborhood $N_c(\lambda)$ of $t = c$ throughout which $\lambda\phi$ does not change rapidly. Of course $N_c(\lambda)$ shrinks to the single point $t = c$ as $\lambda \to \infty$. Nevertheless, as t varies in $N_c(\lambda)$, $\exp\{i\lambda\phi\}$ does not oscillate rapidly and cancellation does not occur. We are therefore led to anticipate that the value of $I(\lambda)$, for λ large, depends primarily on the behavior of f and ϕ near points at which ϕ' is zero. In calculus, such points are called *stationary points* of ϕ. Because the length of $N_c(\lambda)$ goes to zero as λ goes to ∞, we might further anticipate that, so long as f is continuous, the contribution to $I(\lambda)$ corresponding to any stationary point of ϕ goes to zero in this limit.

A heuristic derivation of the leading term of the asymptotic expansion of $I(\lambda)$ follows closely that of Laplace's formula given in Section 5.1. Indeed, let us consider (6.1.1) and suppose that $f \in C[a,b]$ while $\phi \in C^2[a,b]$. Suppose further that ϕ' vanishes in $[a,b]$ only at the point $t = c$. Finally, assume that $\phi''(c) \neq 0$. [Note that no assumption has been made concerning the sign of $\phi''(c)$.]

If we believe the heuristic argument given above, then we can apply the formal calculations that led to (5.1.8) to conclude that, for any small positive ε

$$I(\lambda) \approx \exp\{i\lambda\phi(c)\} f(c) \sqrt{\frac{2}{\lambda\,|\phi''(c)|}} \int_{-\varepsilon\sqrt{\lambda|\phi''(c)|/2}}^{\varepsilon\sqrt{\lambda|\phi''(c)|/2}} \exp\{i\mu\tau^2\}\, d\tau. \tag{6.1.3}$$

Here $\mu = \operatorname{sgn} \phi''(c)$. The approximation (6.1.3) presumably improves as $\varepsilon \to 0$ with $\sqrt{\lambda}\,\varepsilon \to \infty$. In that event we can write

$$I(\lambda) \approx \exp\{i\lambda\phi(c)\} f(c) \sqrt{\frac{2}{\lambda\,|\phi''(c)|}} \int_{-\infty}^{\infty} \exp\{i\mu\tau^2\}\, d\tau. \tag{6.1.4}$$

The integral in (6.1.4) can be evaluated explicitly (it exists as an improper Riemann integral), to yield

$$I(\lambda) \approx \exp\{i\lambda\phi(c)\} f(c) \sqrt{\frac{2\pi}{\lambda\,|\phi''(c)|}} \exp\left\{\frac{\pi i\mu}{4}\right\}. \tag{6.1.5}$$

This last formula is the desired result. We conjecture that it represents the leading term of the contribution to $I(\lambda)$ corresponding to the stationary point $t = c$. In many physical problems, especially those which involve the propagation of waves, because ϕ has the interpretation of a phase, (6.1.5) has come to be known as the *stationary phase* formula and the analytical procedure which led to its derivation as the *method of stationary phase.*

We note that the stationary phase formula heuristically derived above predicts that the corresponding contribution is $O(\lambda^{-1/2})$ as $\lambda \to \infty$. If we compare this to the endpoint contributions which are obtained via the process of neutralization and integration by parts and which, in this manner and under the assumed smoothness of f and ϕ are found to be $O(\lambda^{-1})$, then our original prediction that the zeros of ϕ' are the dominant critical points is partially confirmed.

The approximation (6.1.5) can be rigorously derived via the extension of the integration-by-parts procedure described in Section 3.4. Indeed, as we shall find, a significant generalization can be obtained in this manner. Thus, let us again consider (6.1.1) and still suppose that $t = c$ is the only stationary point of ϕ in $[a,b]$. Because we wish to study the contribution corresponding to this critical point, we must first isolate it by neutralization. For our present purposes, however, we need only assume that $f(t)$ vanishes infinitely smoothly at the two endpoints $t = a$ and $t = b$. Although it is not necessary, we shall further assume that both ϕ and f are infinitely differentiable on the half-open intervals $[a,c)$ and $(c,b]$.

We now write

$$\begin{aligned} I(\lambda) &= \int_a^c \exp\{i\lambda\phi\} f\,dt + \int_c^b \exp\{i\lambda\phi\} f\,dt \\ &= I_-(\lambda) + I_+(\lambda). \end{aligned} \tag{6.1.6}$$

Let us first consider $I_+(\lambda)$ and assume that, as $t \to c_+$,

$$\begin{aligned} \phi(t) - \phi(c) &= \alpha(t-c)^\nu + o((t-c)^\nu), \qquad \nu > 0, \\ f(t) &= \gamma_+(t-c)^\delta + o((t-c)^\delta), \qquad \delta > -1. \end{aligned} \tag{6.1.7}$$

In deriving (6.1.5) we assumed $\nu = 2$. In the more general case now to be considered ν can be any positive real number. (Of course, if we insist that ϕ be stationary at $t = c$, then ν must be taken greater than 1.) If ν is an integer ≥ 2, then $t = c$ is said to be a stationary point of ϕ of *order* $\nu - 1$. If $\nu = 2$ (and $\alpha \neq 0$), then $t = c$ is usually referred to as a *simple stationary point*. Even when ν is not an integer, it is convenient to consider $\nu - 1$ as the order of the stationary point at $t = c$.

To analyze $I_+(\lambda)$ we set $\mu = \operatorname{sgn} \alpha$ and

$$\mu u = \phi(t) - \phi(c). \tag{6.1.8}$$

Then we have

$$I_+(\lambda) = \exp\{i\lambda\phi(c)\} \int_0^{|\phi(b)-\phi(c)|} F(u) \exp\{i\lambda\mu u\}\,du. \tag{6.1.9}$$

Here

$$F(u) = \left.\frac{\mu f(t)}{\phi'(t)}\right|_{t=\phi^{-1}(u)} = \frac{\gamma_+ u^{(\delta+1)/\nu-1}}{\nu|\alpha|^{(\delta+1)/\nu}} + o(u^{(\delta+1)/\nu-1}), \qquad u \to 0+. \quad (6.1.10).$$

Furthermore, $F(u)$ vanishes infinitely smoothly as $u \to |\phi(b) - \phi(c)| -$ and any ambiguity that arises in the determination of $t = \phi^{-1}(u)$ is resolved by requiring that u increase as t increases.

In Section 3.4 we considered integrals of the form (6.1.9) in the case where $F(u) = u^{\sigma-1}\,\hat{F}(u)$ with $\hat{F}(u)$ infinitely differentiable at the origin. That method can be extended to the case where $F(u)$ has an asymptotic expansion, as $u \to 0+$, of the form

$$F(u) \sim \sum_{m=0}^{\infty} \sum_{n=0}^{N(m)} c_{mn}\, u^{a_m}\, (\log u)^n. \qquad (6.1.11)$$

In any event, we find from (6.1.10) and (3.4.23) that, as $\lambda \to \infty$,

$$I_+(\lambda) = \exp\{i\lambda\phi(c)\} \left[\frac{\gamma_+ \Gamma\left(\frac{\delta+1}{\nu}\right)}{\nu(\lambda|\alpha|)^{(\delta+1)/\nu}} \exp\{i\mu\pi(\delta+1)/2\nu\} \right] [1 + o(\lambda^{-(\delta+1)/\nu})]. \quad (6.1.12)$$

The leading term in the expansion of $I_-(\lambda)$ is obtained in a completely analogous manner. Indeed, if we assume that, as $t \to c-$,

$$\phi(t) - \phi(c) = \beta(c-t)^\rho + o((c-t)^\rho), \qquad \rho > 0,$$
$$f(t) = \gamma_-(c-t)^\sigma + o((c-t)^\sigma), \qquad \sigma > -1, \qquad (6.1.13)$$

then we find that, as $\lambda \to \infty$,

$$I_-(\lambda) = \exp\{i\lambda\phi(c)\} \left[\frac{\gamma_- \Gamma\left(\frac{\sigma+1}{\rho}\right)}{\rho(\lambda|\beta|)^{(\sigma+1)/\rho}} \exp\{i\eta\pi(\sigma+1)/2\rho\} \right] [1 + o(\lambda^{-(\sigma+1)/\rho})]. \qquad (6.1.14)$$

Here $\eta = \operatorname{sgn} \beta$.

As a special case let us suppose that $t = c$ is a simple interior stationary point of ϕ and that ϕ and f are infinitely differentiable at $t = c$. Then we find

$$\gamma_+ = \gamma_- = f(c), \qquad \delta = \sigma = 0, \qquad \rho = \nu = 2, \qquad \alpha = \beta = \phi''(c)/2. \quad (6.1.15)$$

We note that the sum of (6.1.12) and (6.1.14) in this case agrees exactly with (6.1.5) thereby establishing the validity of that formula. We might point out that whenever the stationary point $t = c$ coincides with an endpoint of integration then, to leading order, either (6.1.12) or (6.1.14) alone yields the asymptotic expansion of I.

From the results just obtained, we find that the algebraic order of $I(\lambda)$, as $\lambda \to \infty$, *increases* as the order of the stationary point increases and *decreases* as the order of vanishing of f at $t = c$ increases. The latter relation is easily understood. To explain the former, we need only appeal to our earlier discussion where we argued that any stationarity of ϕ tends to increase the value of $I(\lambda)$ as $\lambda \to \infty$. This tendency will naturally be enhanced as the order of stationarity increases.

We now consider several examples to illustrate some of the results derived above.

EXAMPLE 6.1.1. If n is an integer, then $J_n(\lambda)$, the Bessel function of the first kind, has the integral representation

$$J_n(\lambda) = \frac{1}{\pi} \int_0^\pi \cos(nt - \lambda \sin t)\, dt \tag{6.1.16}$$

$$= \frac{1}{2\pi} \sum_{\pm} \int_0^\pi \exp\{\pm int\} \exp(\mp i\lambda \sin t)\, dt.$$

Each integral in the last sum is of the form (6.1.1). Indeed we may set $f_\pm(t) = \exp\{\pm int\}$ and $\phi_\pm = \mp \sin t$ so that $t = \pi/2$ is the only stationary point of $\phi_\pm$ in $[0, \pi]$. Moreover, because $\phi_\pm''(\pi/2) = \pm 1$, we have that $t = \pi/2$ is a simple stationary point of $\phi_\pm$. Upon applying (6.1.5), we then obtain

$$J_n(\lambda) \sim \frac{1}{\sqrt{2\pi\lambda}} \sum_{\pm} \exp\left\{\pm i\left(\frac{\pi}{4} + \frac{n\pi}{2} - \lambda\right)\right\}$$

$$= \sqrt{\frac{2}{\pi\lambda}} \cos\left(\lambda - \frac{n\pi}{2} - \frac{\pi}{4}\right), \qquad \lambda \to \infty, \qquad n \text{ an integer.} \tag{6.1.17}$$

Actually (6.1.17) holds even when n is not an integer. Indeed, we have that for real n and positive λ

$$J_n(\lambda) = \frac{1}{\pi} \int_0^\pi \cos(nt - \lambda \sin t)\, dt - \frac{\sin n\pi}{\pi} \int_0^\infty \exp(-nt - \lambda \sinh t)\, dt. \tag{6.1.18}$$

To leading order, the asymptotic expansion of the first integral on the right of (6.1.18) is given by (6.1.17) and, as is readily seen, the second integral is $0(\lambda^{-1})$.

EXAMPLE 6.1.2. Let us now consider the Bessel function as both its order and argument get large. We have from (6.1.16) that

$$J_\lambda(\lambda r) = \frac{1}{2\pi} \sum_{\pm} \int_0^\pi \exp\{\pm i\lambda\, [t - r \sin t]\}\, dt + O(\lambda^{-1}). \tag{6.1.19}$$

We shall assume here that $r > 1$, in which event

$$\phi_\pm = \pm(t - r \sin t) \tag{6.1.20}$$

has a simple stationary point in $[0, \pi]$ at $t = \cos^{-1}(1/r)$. Also

$$\phi_{\pm}''\left(\cos^{-1}\left(\frac{1}{r}\right)\right) = \pm\, r \sin\left(\cos^{-1}\left(\frac{1}{r}\right)\right) = \pm\sqrt{r^2 - 1}. \tag{6.1.21}$$

Upon applying (6.1.5) we now find that

$$J_\lambda(\lambda r) \sim \sqrt{\frac{2}{\pi\lambda}}\,(r^2 - 1)^{-1/4} \cos\left\{\lambda\left(\sqrt{r^2-1} - \cos^{-1}\left(\frac{1}{r}\right)\right) - \frac{\pi}{4}\right\},$$
$$\lambda \to \infty, \qquad r > 1. \tag{6.1.22}$$

EXAMPLE 6.1.3. We again consider $J_\lambda(\lambda r)$ for large λ and note that the expansion (6.1.22) breaks down as $r \to 1+$ due to the presence of the factor $(r^2 - 1)^{-1/4}$. The reason for the difficulty is that (6.1.22) was derived under the assumption that $t = \cos^{-1}(1/r)$ is a simple stationary point of (6.1.20). When $r = 1$, however, this is not true. Indeed $t = \cos^{-1}(1) = 0$ is a stationary point of order 2 of $\phi_{\pm} = \pm[t - \sin t]$.

To obtain an expansion of $J_\lambda(\lambda)$ we note that as $t \to 0+$

$$\phi_{\pm} \sim \pm\frac{t^3}{3!}. \tag{6.1.23}$$

Because $f \equiv 1/2\pi$, it follows from (6.1.12) that

$$J_\lambda(\lambda) \sim \frac{\Gamma(\frac{1}{3})\,(3!)^{1/3}}{6\pi\,\lambda^{1/3}}\left\{\exp\left(\frac{\pi i}{6}\right) + \exp\left(-\frac{\pi i}{6}\right)\right\}$$
$$= \frac{\Gamma(\frac{1}{3})\,2^{1/3}}{2\pi\,3^{1/6}\,\lambda^{1/3}}. \tag{6.1.24}$$

Upon comparing (6.1.22) and (6.1.24) we find that $J_\lambda(\lambda r) = 0(\lambda^{-1/2})$, $\lambda \to \infty$, where $r \geq r_0 > 1$, while $J_\lambda(\lambda) = 0(\lambda^{-1/3})$ in this limit. In actuality, there is a smooth transition as $r \to 1+$. A single uniform expansion which correctly describes this transition will be obtained in Chapter 9.

6.2. Further Results on Mellin Transforms

Throughout the remainder of this chapter we shall consider integrals of the form

$$I(\lambda) = \int_a^b h(\lambda\phi)\, f\, dt \tag{6.2.1}$$

in the case where $h(t)$ is oscillatory as $t \to \infty$. Indeed, we shall assume that $h(t)$ is locally integrable on $(0, \infty)$ and has, as $t \to \infty$, an asymptotic expansion of the form

$$h(t) \sim \exp\{i\omega t^{\nu}\} \sum_{m=0}^{\infty} \sum_{n=0}^{N(m)} c_{mn}\, t^{-r_m} (\log t)^n \tag{6.2.2}$$

or a finite linear combination of such forms. In (6.2.2) ν is positive, ω is real and nonzero, and $\text{Re}(r_m) \uparrow \infty$.

At first glance, we might predict that the analysis of (6.2.1) as $\lambda \to \infty$ with h oscillatory will proceed essentially as in the case where h decreases exponentially because, under the assumption (6.2.2), $M[h;z]$ has no singularities in some right half-plane. We shall see below, however, that there are marked differences between the two cases.

One of these differences can be discussed immediately. We know that the Mellin transform of an exponentially decreasing kernel decays to zero as $z \to \infty$ along vertical lines. Furthermore, it follows from the results of Section 4.7 that if the kernel is analytic in some sector containing the positive real axis and is algebraically bounded as $t \to 0$ in that sector, then the decay is itself exponential. Thus when considering (6.2.1) in this latter situation, an infinite asymptotic expansion can be obtained under mild restrictions on ϕ and f by applying the results of Sections 5.2 and 5.3.

We have seen in Section 4.3 that the Mellin transform of an oscillatory kernel grows algebraically along vertical lines and that the rate of growth worsens as these lines are shifted to the right. As a result, an infinite expansion of (6.2.1) with h oscillatory cannot be obtained by the Mellin transform method unless certain restrictions are placed on f and ϕ. These restrictions will be discussed in detail in the following sections.

In any event, we must anticipate that precise estimates for the behavior of Mellin transforms along vertical lines will be needed for our subsequent asymptotic analysis. Thus, we shall devote the remainder of this section to obtaining such estimates. As we shall see, our problem essentially involves the study of the Fourier transform

$$\hat{g}(y) = \int_{-\infty}^{\infty} e^{iyt}\, g(t)\, dt \tag{6.2.3}$$

as $y \to \pm\infty$. A fundamental result (which we have appealed to before) we state without proof as the following.

LEMMA 6.2.1. (Riemann-Lebesgue). If in (6.2.3) $g(t)$ belongs to $L(-\infty < t < \infty)$, then

$$\hat{g}(y) = o(1), \qquad y \to \pm\infty. \tag{6.2.4}$$

The above lemma states that the Fourier transform of any absolutely integrable function goes to zero as its argument goes to $\pm\infty$. With no further restrictions on g we cannot say how rapid is this decay. To illustrate how more precise estimates can be obtained we prove the following.

LEMMA 6.2.2. Let $g(t)$ be of bounded total variation on $(-\infty, \infty)$ and let $\lim_{|t|\to\infty} g(t) = 0$. Then the Fourier transform (6.2.3), which exists at least for $y \neq 0$, is such that

$$\hat{g}(y) = O(|y|^{-1}), \qquad y \to \pm\infty. \tag{6.2.5}$$

PROOF. For $y \neq 0$ we have

$$\hat{g}(y) = \lim_{a\to\infty} \int_{-a}^{a} e^{iyt}\, g(t)\, dt$$
$$= \frac{1}{iy} \lim \left[e^{iya}\, g(a) - e^{-iya}\, g(-a) - \int_{-a}^{a} e^{iyt}\, dg(t)\right], \tag{6.2.6}$$

through Riemann-Stieltjes integration by parts. Thus,

$$|\hat{g}(y)| \leq \frac{1}{|y|} \lim_{a\to\infty} \int_{-a}^{a} |dg| = \frac{V(g)}{|y|}. \tag{6.2.7}$$

Here $V(g)$ denotes the total variation of g. Because $V(g) < \infty$ we have

$$|\hat{g}| = O(|y|^{-1}), \qquad y \to \pm\infty. \tag{6.2.8}$$

This completes the proof.

We now turn to the Mellin transform itself and first prove the following.

THEOREM 6.2.1. Let $f(t)$ be n times continuously differentiable on $(0, \infty)$. Suppose there exists a real number x_0 such that for all $x > x_0$ and for $p = 0, 1, \ldots, n$

$$f_p(t;x) = \left(t\frac{d}{dt}\right)^p (t^x f) \tag{6.2.9}$$

vanishes as $t \to \infty$ and $t^{-1} f_p$ is absolutely integrable at the origin. Then

$$M[f;z] = o(|y|^{-n}), \qquad y \to \pm\infty \tag{6.2.10}$$

for all $\mathrm{Re}(z) > x_0$.

PROOF. By hypotheses

$$M[f;z] = \int_0^\infty t^{iy-1}(t^x f)\, dt = \int_0^\infty t^{iy-1} f_0(t;x)\, dt \tag{6.2.11}$$

is absolutely convergent for all $\mathrm{Re}(z) > x_0$. Upon integrating by parts n times in (6.2.11) we obtain

$$M[f;z] = (-iy)^{-n} \int_0^\infty t^{iy-1} f_n(t;x)\, dt. \tag{6.2.12}$$

(Note that the assumptions made imply that all boundary terms vanish.) If in (6.2.12) we set $t = e^u$, then we have

$$M[f;z] = (-iy)^{-n} \int_{-\infty}^{\infty} e^{iyu} f_n(e^u;x)\, du. \tag{6.2.13}$$

But for $x > x_0$, $f_n(e^u;x)$ belongs to $L(-\infty < u < \infty)$ and hence it follows by the Riemann-Lebesgue lemma that

$$M[f;z] = o(|y|^{-n}), \qquad y \to \pm\infty, \qquad x > x_0. \tag{6.2.14}$$

COROLLARY 6.2.1. If the hypotheses of Theorem 6.2.1 are satisfied for arbitrarily large n, then

$$M[f;z] = o(|y|^{-R}), \qquad y \to \pm\infty, \qquad x > x_0, \tag{6.2.15}$$

for all R.

The following two theorems yield improvements on the estimate (6.2.10).

THEOREM 6.2.2. Let f satisfy the hypotheses of Theorem 6.2.1. In addition, assume that for $x > x_0$, $f_n(t,x)$ is of bounded total variation on $[0,\infty)$. Then

$$M[f;z] = O(|y|^{-n-1}), \quad y \to \pm\infty, \qquad x > x_0. \tag{6.2.16}$$

PROOF. We have that (6.2.13) still holds. Also, for $x > x_0, f_n(e^u;x)$ is of bounded total variation on $(-\infty < u < \infty)$ and vanishes as $|u| \to \infty$. Thus, by Lemma 6.2.2, the integral in (6.2.13) is $O(|y|^{-1})$ as $y \to \pm\infty$ which completes the proof.

THEOREM 6.2.3. Let f satisfy the hypotheses of Theorem 6.2.1. Suppose that, for $x > x_0, f_{n+1}(t;x)$ is of bounded total variation on $[0,\infty)$ and vanishes as $t \to \infty$ and as $t \to 0+$. Then

$$M[f;z] = O(|y|^{-n-2}), \qquad y \to \pm\infty, \qquad x > x_0. \tag{6.2.17}$$

PROOF. If in (6.2.12) we integrate by parts once more and set $t = e^u$ in the result, then we obtain

$$M[f;z] = (-iy)^{-n-1} \int_{-\infty}^{\infty} e^{iyu} f_{n+1}(e^u;x)\, du. \tag{6.2.18}$$

By Lemma 6.2.2 the integral is $O(|y|^{-1})$ for $x > x_0$ and the theorem is proved.

EXAMPLE 6.2. As a simple example which illustrates the sharpness of Theorem 6.2.3 let us consider

$$f(t) = \begin{cases} (1-t), & 0 \le t \le 1, \\ 0, & t > 1. \end{cases}$$

Because f is continuous on $[0,\infty)$ and has a piecewise continuous first derivative which is of bounded total variation on $[0,\infty)$, Theorem 6.2.3 predicts $M[f;z] = O(|y|^{-2})$, $y \to \pm\infty$. Moreover, $(t\,(d/dt))\,(t^x f)$ vanishes as $t \to 0+$ for $x > 0$ so that this estimate should hold in the right half-plane $\mathrm{Re}(z) > 0$.

Indeed, we have directly

$$M[f;z] = \int_0^1 t^{z-1}(1-t)\,dt = \frac{1}{z(z+1)} = O(|y|^{-2}), \qquad y \to \pm\infty. \quad (6.2.19)$$

We note that not only does the estimate (6.2.19) hold for $x > 0$, it actually holds for all x. That is, in this case, the estimate predicted by Theorem 6.2.3 to hold in the original domain of analyticity of $M[f;z]$, also holds for the analytic continuation of $M[f;z]$ into the entire z plane. In the following theorem we identify a class of functions f for which estimates of $M[f;z]$ as $y \to \pm\infty$ can be obtained for all x.

THEOREM 6.2.4. Let $f(t)$ be n times continuously differentiable on $(0,\infty)$. Suppose that, as $t \to 0+$,

$$f(t) \sim \sum_{m=0}^{\infty} p_m t^{a_m}, \qquad \mathrm{Re}(a_m) \uparrow \infty. \quad (6.2.20)$$

Suppose further that for $j = 1, 2, \ldots, n \ldots$, the asymptotic expansion of $f^{(j)}(t)$ as $t \to 0+$ is obtained by successively differentiating (6.2.20) term-by-term. Finally, let $f_p(t;x) = (t(d/dt))^p (t^x f)$ vanish as $t \to \infty$ for $p = 0, 1, \ldots, n$ and $x > -\mathrm{Re}(a_0)$. Then

$$M[f;z] = O(|y|^{-n}), \qquad y \to \pm\infty \quad (6.2.21)$$

for *all* x. Here by $M[f;z]$ we mean the analytic continuation of this Mellin transform into the entire z plane.

PROOF. If $x > -\mathrm{Re}(a_0)$, then the result follows by Theorem 6.2.1 when we note that the conditions imposed on $f_p(t;x)$ at the origin in that theorem are implied by the assumed differentiability properties of the asymptotic expansion (6.2.20).

Now suppose that ρ is any real number greater than $\mathrm{Re}(a_0)$. Also let $\mu(\rho)$ be a positive integer such that

$$\mathrm{Re}(a_{\mu-1}) < \rho \le \mathrm{Re}(a_\mu) \quad (6.2.22)$$

and let $\delta(\rho)$ be any integer such that

$$\mathrm{Re}(a_0) + \delta > \mathrm{Re}(a_\mu). \quad (6.2.23)$$

We introduce the functions

$$\sigma_\rho(t) = \left(\sum_{m=0}^{\mu-1} p_m t^{a_m}\right) e^{-t^\delta}, \quad (6.2.24)$$

$$\hat{f}(t) = f(t) - \sigma_\rho(t), \quad (6.2.25)$$

and note that $\delta(\rho)$ has been chosen so that, as $t \to 0+$,

$$\hat{f} = O(t^{\mathrm{Re}(a_\mu)}). \quad (6.2.26)$$

We also note that $\hat{f}$ has all of the properties attributed to f in Theorem 6.2.1 with $x_0 = -\operatorname{Re}(a_\mu)$. Hence, by that theorem

$$M[\hat{f};z] = O(|y|^{-n}), \qquad y \to \pm\infty, \qquad x > -\operatorname{Re}(a_\mu). \tag{6.2.27}$$

By direct calculation we have that

$$M[\sigma_\rho;z] = \sum_{m=0}^{\mu-1} \frac{p_m}{\delta} \Gamma\left(\frac{z+a_m}{\delta}\right) \tag{6.2.28}$$

in the region $\operatorname{Re}(z) > -\operatorname{Re}(a_0)$ and by analytic continuation in the entire z plane. Furthermore, we know that each gamma function in (6.2.28) decays exponentially as $y \to \pm\infty$ for all x. Thus, because

$$M[f;z] = M[\hat{f};z] + M[\sigma_\rho;z], \tag{6.2.29}$$

we have that (6.2.21) holds for $\operatorname{Re}(z) > -\operatorname{Re}(a_\mu)$. However, ρ is arbitrary and $\lim_{\rho\to\infty} \operatorname{Re}(a_{\mu(\rho)}) = \infty$, so that the theorem is proved upon letting $\rho \to \infty$.

Theorem 6.2.4 can be extended to include functions $f(t)$ whose asymptotic expansion, as $t \to 0+$, involves integer powers of $\log t$. Indeed, we have the following.

COROLLARY 6.2.2. If in Theorem 6.2.4 we assume that

$$f \sim \sum_{m=0}^{\infty} \sum_{n=0}^{\bar{N}(m)} p_{mn}\, t^{a_m} (\log t)^n, \qquad t \to 0+ \tag{6.2.30}$$

with $\bar{N}(m)$ finite for each m and $\operatorname{Re}(a_m) \uparrow \infty$, then the estimate (6.2.21) remains valid for all x.

PROOF. The proof follows that of Theorem 6.2.4 except now to establish the exponential decay of $M[\sigma_\rho;z]$ as $y \to \pm\infty$ involves showing that derivatives of the gamma function have such decay. Although the last assertion is true, we shall not prove it here.

We finally have the following.

COROLLARY 6.2.3. If $f(t)$ satisfies the hypotheses of Theorem 6.2.4 (or of Corollary 6.2.1) with $n = \infty$, then

$$M[f;z] = o(|y|^{-R}), \qquad y \to \pm\infty \tag{6.2.31}$$

for all R and all x.

6.3. Kernels of Oscillatory Type

We shall now consider the asymptotic expansion of (6.2.1) as $\lambda \to \infty$ in the case where the kernel $h(t)$ is oscillatory as $t \to \infty$. In particular, we shall assume that, as $t \to \infty$, h has an asymptotic expansion given by a finite linear combination of asymptotic forms (6.2.2).

From our previous discussions, we anticipate that the set of possible critical points for the proposed expansion consists of

(1) the endpoints of integration;
(2) points in (a,b) at which either ϕ or f is not infinitely differentiable;
(3) points in (a,b) at which ϕ vanishes;
(4) points in (a,b) at which ϕ' vanishes (stationary points of ϕ).

We wish to study the contributions from each critical point separately. To accomplish this we must first isolate the critical points via the neturalization process described in Section 3.3.

For the present, we shall assume that both ϕ and f belong to $C^\infty(a,b)$ and that ϕ is a positive, strictly monotonic function in (a,b). (Below, ϕ will be allowed to change sign, which will necessitate our making assumptions concerning the asymptotic behavior of h as $t \to -\infty$.) Under these assumptions, we expect that the endpoints $t = a$ and $t = b$ are the only critical points and hence total isolation is readily achieved.

We find from the discussion of Section 3.3 that we can write

$$I(\lambda) = I_a(\lambda) + I_b(\lambda), \tag{6.3.1}$$

where

$$I_a(\lambda) = \int_a^b h(\lambda\phi) f_a(t)\, dt, \tag{6.3.2}$$

$$I_b(\lambda) = \int_a^b h(\lambda\phi) f_b(t)\, dt. \tag{6.3.3}$$

In (6.3.2), $f_a \equiv f$ in some positive half-neighborhood of $t = a$, belongs to $C^\infty(a,b]$ and vanishes for $\gamma \le t$ with $a < \gamma < b$. Also, in (6.3.3), $f_b \equiv f$ in some negative half-neighborhood of $t = b$, belongs to $C^\infty[a,b)$ and vanishes for $t \le \delta$ with $a < \delta < b$.

In Subsections I and II below, we shall consider the asymptotic expansion of the integrals I_a and I_b, respectively. For each of these integrals, there are two distinct cases to be treated depending on the value of ϕ at the relevant endpoint. In Subsection III we shall treat the case where ϕ is negative in (a,b).

I. ANALYSIS OF $I_a(\lambda)$

We first consider the following.

Case (1). $\phi(a) = 0$. Because ϕ is positive and strictly monotonic in (a,b), we must now have $\phi'(t) > 0$. If in (6.3.2), we set

$$u = \phi(t), \tag{6.3.4}$$

then

$$I_a(\lambda) = \int_0^\infty h(\lambda u)\, F_a(u)\, du. \tag{6.3.5}$$

Here

$$F_a(u) = \begin{cases} \left.\dfrac{f_a}{\phi'}\right|_{t=\phi^{-1}(u)}, & 0 \le u \le \phi(\gamma), \\ 0, & u > \phi(\gamma), \end{cases} \tag{6.3.6}$$

which we note lies in $C^\infty(0, \infty)$.

Our plan is to derive the asymptotic expansion of $I_a(\lambda)$ by applying the Mellin transform method. To accomplish this we first assume that, as $u \to 0+$,

$$F_a(u) \sim \sum_{m=0}^{\infty} \sum_{n=0}^{\bar{N}(m)} p_{mn}\, u^{a_m} (\log u)^n. \tag{6.3.7}$$

Here $\text{Re}(a_m) \uparrow \infty$ and $\bar{N}(m)$ is finite for each m. Because we anticipate having to apply Thcorem 6.2.2 and its corollaries, we further assume that, for all n, the asymptotic expansion of $F_a^{(n)}(u)$, as $u \to 0+$, is obtained by successive term-by-term differentiations of (6.3.7).

Although it is not necessary, we shall suppose that $I_a(\lambda)$ is absolutely convergent and that $M[h;z]$ exists in the ordinary sense in some vertical strip. Then the Parseval formula (4.2.17) yields

$$I_a(\lambda) = \frac{1}{2\pi i} \int_{c-i\infty}^{c+i\infty} \lambda^{-z}\, M[h;z]\, M[F_a;1-z]\, dz. \tag{6.3.8}$$

Under the assumption that h has the asymptotic form (6.2.2), because the analytic continuation of $M[h;z]$ into the right half-plane $\text{Re}(z) \ge c$ can grow at worst algebraically as $|y| \to \infty$; and, by Corollary 6.2.3, because $M[F_a;1-z]$ decreases faster than any power of $|y|^{-1}$ in this limit, we can obtain an infinite expansion of I_a as a residue series in the standard way. Moreover, we have essentially reduced our problem to that treated in Case II of Section 5.2 and hence we can freely use results obtained in that section.

The expansion itself is given by (5.2.12) which we repeat here:

$$I_a(\lambda) \sim \sum_{m=0}^{\infty} \lambda^{-1-a_m} \sum_{n=0}^{\bar{N}(m)} p_{mn} \sum_{j=0}^{n} \binom{n}{j} (-\log \lambda)^j\, M^{(n-j)}[h;z]\Big|_{z=1+a_m}; \qquad \lambda \to \infty. \tag{6.3.9}$$

If we assume that, as $t \to a+$,

$$f(t) \sim \gamma_0 (t-a)^{\mu_0}, \qquad \phi(t) \sim \alpha_0 (t-a)^{\nu_0}, \qquad \alpha_0, \nu_0 > 0, \tag{6.3.10}$$

then it follows from (5.2.15) that, to leading order

$$I_a(\lambda) \sim (\alpha_0 \lambda)^{-(\mu_0+1)/\nu_0} \frac{\gamma_0}{\nu_0} M\left[h; \frac{\mu_0+1}{\nu_0}\right]. \tag{6.3.11}$$

We note that, as we found in Section 6.1 for the case where $h = e^{it}$, the algebraic order of $I_0(\lambda)$, as $\lambda \to \infty$, increases as the order of vanishing of ϕ as $t \to a+$ increases, and decreases as the order of vanishing of f as $t \to a+$ increases. The explanation offered in Section 6.1 in terms of cancellation effects remains valid here.

To illustrate what happens when logarithmic terms appear in (6.3.7), we suppose that ϕ is as in (6.3.10) while

$$f(t) \sim \gamma_{01}(t-a)^{\mu_0}\log(t-a) + \gamma_{00}(t-a)^{\mu_0}. \tag{6.3.12}$$

After some calculation we find that in (6.3.7) $\bar{N}(0) = 1$ and

$$a_0 = \frac{\mu_0+1}{\nu_0} - 1, \qquad p_{01} = \frac{\gamma_{01}}{\nu_0{}^2}(\alpha_0)^{-(\mu_0+1)/\nu_0},$$
$$p_{00} = \frac{(\alpha_0)^{-(\mu_0+1)/\nu_0}}{\nu_0}\left[\gamma_{00} - \frac{\gamma_{01}\log\alpha_0}{\nu_0}\right]. \tag{6.3.13}$$

Thus, it follows from (6.3.9) that in this case

$$I_a(\lambda) \sim -\frac{(\lambda\alpha_0)^{-(\mu_0+1)/\nu_0}}{\nu_0}\left\{\frac{\gamma_{01}}{\nu_0}\log\lambda\, M\left[h;\frac{\mu_0+1}{\nu_0}\right]\right. \tag{6.3.14}$$
$$\left. + \left(\frac{\gamma_{01}}{\nu_0}\log\alpha_0 - \gamma_{00}\right) M\left[h;\frac{\mu_0+1}{\nu_0}\right] - \frac{\gamma_{01}}{\nu_0}\frac{d}{dz}M[h;z]\bigg|_{z=(\mu_0+1)/\nu_0}\right\}.$$

EXAMPLE 6.3.1. Let $h(t)$ be one of the complex Fourier kernels $e^{\pm it}$. Then we have explicitly

$$M[e^{\pm it};z] = \exp\left(\frac{\pm\pi iz}{2}\right)\Gamma(z). \tag{6.3.15}$$

If (6.3.10) holds, then we have from (6.3.11)

$$\int_a^b \exp\{\pm i\lambda\phi\} f_a(t)\,dt \sim (\lambda\alpha_0)^{-(\mu_0+1)/\nu_0}\frac{\gamma_0}{\nu_0}\Gamma\left(\frac{\mu_0+1}{\nu_0}\right)\exp\left\{\frac{\pm\pi i}{2}\left(\frac{\mu_0+1}{\nu_0}\right)\right\}, \tag{6.3.16}$$

which is in agreement with the result given by (6.1.12).

Of special interest is the case where $\mu_0 = 0$ and $\nu_0 = 2$ in (6.3.10). Then (6.3.16) reduces to

$$\int_a^b \exp\{\pm i\lambda\phi\} f_a\,dt \sim \frac{1}{2}\{2\pi/\phi''(a)\lambda\}^{1/2} f(a+)\exp\left(\frac{\pm\pi i}{4}\right). \tag{6.3.17}$$

If (6.3.12) holds, then it follows from (6.3.14) and (6.3.15) that

$$\int_a^b \exp(\pm i\lambda\phi) f_a\,dt \sim -\frac{(\lambda\alpha_0)^{-(\mu_0+1)/\nu_0}}{\nu_0}\Gamma\left(\frac{\mu_0+1}{\nu_0}\right)\exp\left\{\frac{\pm\pi i}{2}\left(\frac{\mu_0+1}{\nu_0}\right)\right\}$$

$$\times\left\{\frac{\gamma_{01}}{\nu_0}\log\lambda+\frac{\gamma_{01}}{\nu_0}\log\alpha_0-\gamma_{00}\right.$$
$$\left.-\frac{\gamma_{01}}{\nu_0}\left(\psi\left(\frac{\mu_0+1}{\nu_0}\right)\pm\frac{\pi i}{2}\right)\right\} \tag{6.3.18}$$

with ψ the digamma function.

EXAMPLE 6.3.2. Let us now suppose that $h(t)=J_\nu(t)$ so that $I_a(\lambda)$ is the Hankel transform of $f_a(t)$. From the Appendix we have

$$M[J_\nu\,;z]=\frac{2^{z-1}\,\Gamma\left(\dfrac{z+\nu}{2}\right)}{\Gamma\left(\dfrac{\nu-z}{2}+1\right)}. \tag{6.3.19}$$

In this example we shall assume that (6.3.10) holds so that by (6.3.11) and (6.3.19) we have

$$\int_a^b J_\nu(\lambda\phi)\,f_a(t)\,dt\sim(\lambda\alpha_0)^{-(\mu_0+1)/\nu_0}\,\frac{\gamma_0}{\nu_0}\cdot 2^{((\mu_0+1)/\nu_0-1)}\,\frac{\Gamma\left(\dfrac{\mu_0+1}{2\nu_0}+\dfrac{\nu}{2}\right)}{\Gamma\left(\dfrac{\nu}{2}+1-\dfrac{\mu_0+1}{2\nu_0}\right)}. \tag{6.3.20}$$

In the special case where $\mu_0=0$ and $\nu_0=2$ in (6.3.10), (6.3.20) reduces to

$$\int_a^b J_\nu(\lambda\phi)\,f_a\,dt\sim\frac{1}{2}\,\{\phi''(a)\,\lambda\}^{-1/2}\,\frac{\Gamma\left(\dfrac{\nu}{2}+\dfrac{1}{4}\right)}{\Gamma\left(\dfrac{\nu}{2}+\dfrac{3}{4}\right)}\,f(a+). \tag{6.3.21}$$

We now turn to the consideration of the following.

Case (2). $\phi(a)>0$. In this case ϕ is strictly positive throughout the effective range of integration in (6.3.2) and hence we can replace $h(\lambda\phi)$ by its asymptotic expansion for large argument and then integrate term-by-term. (See Exercise 6.11.) Indeed, upon using (6.2.2) we obtain the formal result

$$I_a(\lambda)\sim\sum_{m=0}^{\infty}\sum_{n=0}^{N(m)}c_{mn}\,\lambda^{-r_m}\sum_{j=0}^{n}\binom{n}{j}(\log\lambda)^j$$
$$\times\int_a^b e^{i\omega\lambda^\nu\phi^\nu}\;\phi^{-r_m}\,(\log\phi)^{n-j}\,f_a\,dt. \tag{6.3.22}$$

To analyze the integrals in (6.3.22) we introduce the quantities

$$\psi(t)=\xi\omega(\phi^\nu(t)-\phi^\nu(a)),\qquad \xi=\operatorname{sgn}(\omega\phi'),\qquad t\in(a,b),$$
$$f_a^{mnj}=f_a\,\phi^{-r_m}\,(\log\phi)^{n-j},\qquad \Lambda=\lambda^\nu. \tag{6.3.23}$$

In terms of these quantities, a typical integral in (6.3.22) becomes

$$\int_a^b e^{i\omega\lambda^\nu\phi^\nu} f_a \, \phi^{-r_m} (\log \phi)^{n-j} \, dt = \exp\{i\omega\lambda^\nu\phi^\nu(a)\} \int_a^b \exp\{i\xi\Lambda\psi\} f_a^{mnj} \, dt. \quad (6.3.24)$$

In (6.3.24), ψ satisfies all of the conditions imposed on ϕ in Case (1) above. In particular, $\psi(a) = 0$ while $\psi(t)$ and $\psi'(t)$ are both positive in (a,b). Thus, we have reduced the current case to the study of a sequence of integrals each of the type considered in Example 6.3.1 above.

To illustrate the procedure, let us suppose that $c_{mn} = 0$ for $n \geq 1$ so that no logarithms appear in (6.2.2). Then (6.3.22) reduces to a single sum whose leading term is given by

$$I_a(\lambda) \sim \lambda^{-r_0} \, c_{00} \exp\{i\omega\lambda^\nu\phi^\nu(a)\} \int_a^b \exp\{i\Lambda\xi\psi\} f_a \, \phi^{-r_0} \, dt. \quad (6.3.25)$$

If we now assume that, as $t \to a+$,

$$\phi(t) - \phi(a) \sim \alpha_0 \, (t-a)^{\nu_0}, \qquad \nu_0 > 0, \quad (6.3.26)$$

$$f(t) \sim \gamma_0 \, (t-a)^{\mu_0}, \qquad \mu_0 > -1, \quad (6.3.27)$$

then it follows from (6.3.16) with the obvious changes in notation that

$$I_a(\lambda) \sim \frac{c_{00} \, \gamma_0}{\nu_0 \, [\phi(a)]^{r_0}} \{|\alpha_0\omega| \, \nu\phi^{\nu-1}(a)\}^{-(\mu_0+1)/\nu_0} \, \lambda^{-(\nu(\mu_0+1)/\nu_0 + r_0)} \, \Gamma\left(\frac{\mu_0+1}{\nu_0}\right)$$

$$\times \exp\left\{i\omega\lambda^\nu\phi^\nu(a) + \frac{\pi i}{2}\left(\frac{\mu_0+1}{\nu_0}\right) \operatorname{sgn}(\alpha_0\omega)\right\}, \qquad \phi(a) > 0. \quad (6.3.28)$$

II. ANALYSIS OF $I_b(\lambda)$

Because the study of I_b follows closely that of I_a above, we shall be brief in our discussion. Again we must consider two distinct cases.

Case (1). $\phi(b) = 0$. Because $\phi > 0$ in (a,b) by assumption, we must now have $\phi'(t) < 0$. Upon setting $u = \phi(t)$ in (6.3.3) we obtain

$$I_b(\lambda) = \int_0^\infty h(\lambda u) \, F_b(u) \, du. \quad (6.3.29)$$

Here

$$F_b(u) = \begin{cases} \left(\dfrac{f}{-\phi'}\right)_{t=\phi^{-1}(u)}, & 0 \leq u \leq \phi(\delta), \\ 0, & u > \phi(\delta). \end{cases} \quad (6.3.30)$$

Upon assuming that, as $u \to 0+$,

$$F_b(u) \sim \sum_{m=0}^{\infty} \sum_{n=0}^{\tilde{N}(m)} p_{mn} \, u^{b_m} (\log u)^n \quad (6.3.31)$$

and that the asymptotic expansion of $F_b^{(j)}(u)$, $j = 1, 2, \ldots$, as $u \to 0+$, can be obtained by differentiating (6.3.31) term-by-term, we find that Corollary 6.2.3

holds and an infinite asymptotic expansion of $I_b(\lambda)$ can be obtained. Indeed we have that, as $\lambda \to \infty$,

$$I_b(\lambda) \sim \sum_{m=0}^{\infty} \lambda^{-(1+b_m)} \sum_{n=0}^{\tilde{N}(m)} p_{mn} \sum_{j=0}^{n} \binom{n}{j} (-\log \lambda)^j M^{(n-j)}[h;z]\Big|_{z=1+b_m}. \quad (6.3.32)$$

In the special case where, as $t \to b-$,

$$\begin{aligned} \phi(t) &\sim \beta_0 (b-t)^{\delta_0}, \qquad \beta_0, \delta_0 > 0, \\ f(t) &\sim \xi_0 (b-t)^{\eta_0}, \end{aligned} \quad (6.3.33)$$

we find that, to leading order,

$$I_b(\lambda) \sim \frac{\xi_0}{\delta_0} (\beta_0 \lambda)^{-(\eta_0+1)/\delta_0} M\left[h; \frac{\eta_0+1}{\delta_0}\right]. \quad (6.3.34)$$

Case (2). $\phi(b) > 0$. Calculations similar to those made above for $I_b(\lambda)$ with $\phi(a) > 0$ yield

$$\begin{aligned} I_b(\lambda) \sim \sum_{m=0}^{\infty} \sum_{n=0}^{N(m)} c_{mn} \lambda^{-r_m} \sum_{j=0}^{n} \binom{n}{j} (\log \lambda)^j \exp\{i\omega\lambda^{\nu}\phi^{\nu}(b)\} \\ \times \int_a^b \phi^{-r_m} (\log \phi)^{n-j} f_b \exp\{-i\xi\lambda^{\nu}\psi\}\, dt. \end{aligned} \quad (6.3.35)$$

Here

$$\psi = \xi\omega[\phi^{\nu}(b) - \phi^{\nu}(t)] \quad (6.3.36)$$

while ξ is defined by (6.3.23).

Each of the integrals in (6.3.35) can be asymptotically evaluated by using the results of Example 6.3.1. Indeed, if we assume that in (6.2.2) $c_{mn} = 0$ for $n > 0$ and that

$$\begin{aligned} \phi(t) - \phi(b) &\sim \beta_0 (b-t)^{\delta_0}, \qquad \delta_0 > 0, \\ f(t) &\sim \xi_0 (b-t)^{\eta_0} \end{aligned} \quad (6.3.37)$$

then

$$\begin{aligned} I_b(\lambda) \sim \frac{c_{00}\,\xi_0}{\delta_0 [\phi(b)]^{r_0}} \{|\beta_0\omega|\, \nu\phi^{\nu-1}(b)\}^{-(\eta_0+1)/\delta_0} \lambda^{-(\nu[\eta_0+1]/\delta_0 + r_0)} \\ \times \Gamma\left(\frac{\eta_0+1}{\delta_0}\right) \exp\left\{i\omega\lambda^{\nu}\phi^{\nu}(b) + \frac{\pi i}{2}\left(\frac{\eta_0+1}{\delta_0}\right) \operatorname{sgn}(\beta_0\omega)\right\}. \end{aligned} \quad (6.3.38)$$

This completes our asymptotic analysis of (6.2.1) in the case where $\phi > 0$ in (a,b). The expansion of I itself is of course obtained by summing the expansions obtained for I_a and I_b. We wish to stress that care must always be taken in ordering terms. Indeed, it can happen that the first several terms in the expansion of I_a dominate the leading term in the expansion of I_b or vice versa.

There are several extensions of the above results that can be made. Of particular interest is the case where ϕ is negative in (a,b) and the case where either ϕ, ϕ' or both change sign in (a,b). We shall consider the first of these below, but shall delay consideration of the second until the following section where contributions from interior critical points will be discussed.

III. $\phi(t)<0$ IN (a,b)

At first glance this seems to be a trivial extension because all we need do is define

$$\hat{h}(t)=h(-t) \tag{6.3.39}$$

and observe that

$$I(\lambda)=\int_a^b h(\lambda\phi)f(t)\,dt=\int_a^b \hat{h}(\lambda\,|\phi|)f(t)\,dt \tag{6.3.40}$$

is an integral of the form already considered. We anticipate treating cases where $\phi(t)$ changes sign in (a,b) in which event we want to make explicit use of (6.3.39). For the most part, in this subsection we shall simply state results and emphasize that they are derived as those for the case $\phi>0$ in (a,b).

Because ϕ is negative in (a,b), we must now be concerned with the behavior of $h(t)$ as $t\to-\infty$. We require that h be oscillatory in this limit also and indeed we shall assume that as $t\to\infty$

$$\hat{h}(t)=h(-t)\sim\exp(i\hat{\omega}t^{\gamma})\sum_{m=0}^{\infty}\sum_{n=0}^{\hat{N}(m)}k_{mn}\,t^{-\rho_m}\,(\log t)^n\,,\gamma>0. \tag{6.3.41}$$

Here $\hat{\omega}$ is real, $\mathrm{Re}(\rho_m)\uparrow\infty$, $\hat{N}(m)$ is finite for each m.

If we assume that ϕ and f are both infinitely differentiable on (a,b), then again the only critical points are $t=a$ and $t=b$ which can be isolated by neutralization. Thus we have

$$I(\lambda)=I_a(\lambda)+I_b(\lambda), \tag{6.3.42}$$

where

$$I_a(\lambda)=\int_a^b \hat{h}(\lambda\,|\phi|)f_a\,dt, \tag{6.3.43}$$

$$I_b(\lambda)=\int_a^b \hat{h}(\lambda\,|\phi|)f_b\,dt. \tag{6.3.44}$$

Here f_a and f_b are as defined below (6.3.3).

We shall now exhibit asymptotic expansions for I_a and I_b along with the assumptions that distinguish the various cases.

$$I_a(\lambda)\sim\sum_{m=0}^{\infty}\lambda^{-(1+a_m)}\sum_{n=0}^{\bar{N}(m)}d_{mn}\binom{n}{j}(-\log\lambda)^j\,M^{(n-j)}[h(-t);z]\Big|_{z=1+a_m}; \tag{6.3.45}$$

$$\frac{f_a}{-\phi'}\bigg|_{t=\phi^{-1}(-u)}\sim\sum_{m=0}^{\infty}\sum_{n=0}^{\bar{N}(m)}d_{mn}\,u^{a_m}\,(\log u)^n,\qquad u\to0+,$$
$$\phi(t)<0,\quad \phi(a)=0. \tag{6.3.46}$$

If, in particular,

$$f(t) \sim \gamma_0 (t-a)^{\mu_0}, \qquad -\phi(t) \sim \alpha_0 (t-a)^{\nu_0}, \qquad \alpha_0, \nu_0 > 0, \qquad t \to a+, \quad (6.3.47)$$

then

$$I_a(\lambda) \sim (\alpha_0 \lambda)^{-(\mu_0+1)/\nu_0} \frac{\gamma_0}{\nu_0} M\left[h(-t); \frac{\mu_0+1}{\nu_0}\right]. \quad (6.3.48)$$

In the special case where $k_{mn} = 0$ for $n > 0$ in (6.3.41) and where, as $t \to a+$,

$$\phi(t) - \phi(a) \sim \alpha_0 (t-a)^{\nu_0}, \qquad f(t) \sim \gamma_0 (t-a)^{\mu_0}, \quad (6.3.49)$$

we have

$$I_a(\lambda) \sim \frac{k_{00}\,\gamma_0}{\nu_0\,[|\phi(a)|]^{\rho_0}} \{|\alpha_0 \hat{\omega}|\, \gamma\, |\phi(a)|^{\gamma-1}\}^{-(\mu_0+1)/\nu_0}\, \Gamma\left(\frac{\mu_0+1}{\nu_0}\right)$$

$$\times \lambda^{-(\gamma(\mu_0+1)/\nu_0 + \rho_0)} \exp\left\{i\hat{\omega}\lambda^\gamma |\phi(a)|^\gamma - \frac{\pi i}{2}\left(\frac{\mu_0+1}{\nu_0}\right) \operatorname{sgn}(\alpha_0 \hat{\omega})\right\},$$

$$\phi(a) < 0. \quad (6.3.50)$$

The analogous expansions for $I_b(\lambda)$ are the following:

$$I_b(\lambda) \sim \sum_{m=0}^{\infty} \lambda^{-(1+b_m)} \sum_{n=0}^{\tilde{N}(m)} p_{mn} \sum_{j=0}^{n} \binom{n}{j} (-\log \lambda)^j M^{(n-j)}[h(-t); z]\Big|_{z=1+b_m}; \quad (6.3.51)$$

$$\left.\frac{f_b}{\phi'}\right|_{t=\phi^{-1}(-u)} \sim \sum_{m=0}^{\infty} \sum_{n=0}^{\tilde{N}(m)} p_{mn}\, u^{b_m} (\log u)^n, \qquad u \to 0+, \qquad \phi(b) = 0; \quad (6.3.52)$$

$$I_b(\lambda) \sim (\lambda \beta_0)^{-(\eta_0+1)/\delta_0} \frac{\xi_0}{\delta_0} M\left[h(-t); \frac{\eta_0+1}{\delta_0}\right]; \quad (6.3.53)$$

$$-\phi(t) \sim \beta_0 (b-t)^{\delta_0}, \qquad \beta_0, \delta_0 > 0, \qquad f(t) \sim \xi_0 (b-t)^{\eta_0}. \quad (6.3.54)$$

Finally, if $k_{mn} = 0$, for $n > 0$ in (6.3.41) and if, as $t \to b-$,

$$\phi(t) - \phi(b) \sim \beta_0 (b-t)^{\delta_0}, \qquad \delta_0 > 0,$$
$$f \sim \xi_0 (b-t)^{\eta_0}, \quad (6.3.55)$$

then

$$I_b(\lambda) \sim \frac{k_{00}\,\xi_0}{\delta_0\,|\phi(b)|^{\rho_0}} \{|\beta_0 \hat{\omega}|\, \gamma\, |\phi(b)|^{\gamma-1}\}^{-(\eta_0+1)/\delta_0}\, \Gamma\left(\frac{\eta_0+1}{\delta_0}\right)$$

$$\times \lambda^{-(\gamma(\eta_0+1)/\delta_0 + \rho_0)} \exp\{i\hat{\omega}\lambda^\gamma |\phi(b)|^\gamma - \frac{\pi i}{2}\left(\frac{\eta_0+1}{\delta_0}\right) \operatorname{sgn}(\beta_0 \hat{\omega})\},$$

$$\phi(b) < 0. \quad (6.3.56)$$

Throughout this section we have assumed that the functions ϕ and f in (6.2.1) are infinitely differentiable in (a,b). This has enabled us to apply Corollary

6.2.3 and to ultimately derive an infinite asymptotic expansion of $I(\lambda)$ as $\lambda \to \infty$. If either ϕ, f, or both have only a finite number of continuous derivatives in (a,b), then only a finite asymptotic expansion of I can be obtained by the method of this section.

Let us suppose then that $\phi' > 0$ in (a,b) and that $\phi(a) = 0$. Let us further suppose that $F_a(u)$, defined by (6.3.6), satisfies the conditions placed on f in Theorem 6.2.4. We have from our discussion in Section 4.3 the estimate

$$M[h;z] = O(|y|^{[(x-r_0)/v]}), \qquad |y| \to \infty, \qquad x > r_0, \tag{6.3.57}$$

while by Theorem 6.2.4,

$$M[F_a;1-z] = O(|y|^{-n}), \qquad |y| \to \infty \tag{6.3.58}$$

holds for all x. Hence, we can displace the contour of integration in the Parseval formula (6.3.8) to the line $\operatorname{Re} z = R$ with the result

$$I_a(\lambda) = \sum_{m=0}^{M} \lambda^{-1-a_m} \sum_{n=0}^{\bar{N}(m)} \sum_{j=0}^{n} p_{mn} \binom{n}{j} (-\log \lambda)^j M^{(n-j)}[h;z]\Big|_{z=1+a_m}$$
$$+ \frac{\lambda^{-R}}{2\pi i} \int_{-\infty}^{\infty} \lambda^{-iy} M[h;R+iy]\, M[F_a;1-R-iy]\, dy. \tag{6.3.59}$$

Here M is the largest integer such that

$$1 + a_M < r_0 + v(n-1) \tag{6.3.60}$$

and R is such that

$$1 + a_M < R < r_0 + v(n-1). \tag{6.3.61}$$

It follows from (6.3.57), (6.3.58), and the Riemann-Lebesgue lemma 6.2.1 that the last term in (6.3.59) is $o(\lambda^{-R})$ and hence the sum represents a finite asymptotic expansion of $I_a(\lambda)$.

The conditions (6.3.60), (6.3.61) placed on M and R are not necessarily the sharpest possible and indeed can often be improved. We shall not discuss such improvements here, but rather remark that, in applications, the functions ϕ and f are most often piecewise infinitely differentiable, in which event, after appropriate neutralization, the results of Subsections I, II, and III become applicable.

6.4. Oscillatory Kernels: Continuation

In the previous section we considered the asymptotic behavior of

$$I(\lambda) = \textstyle\int_a^b h(\lambda\phi) f\, dt, \tag{6.4.1}$$

as $\lambda \to \infty$, in the case where $h(t)$ is oscillatory as $t \to \pm\infty$ and where the endpoints of integration $t = a$, $t = b$ are the only critical points in $[a,b]$. Indeed, it was assumed that both ϕ and f were infinitely differentiable in (a,b) and that neither ϕ nor ϕ' vanished there. We now wish to extend the results obtained in Section 6.3 by allowing critical points to occur in the interior of the interval of integration.

To begin, let us suppose that in (6.4.1) both ϕ and f are piecewise C^∞ functions and that there are a finite number of points in (a,b) at which either ϕ, ϕ' or both vanish. It then follows that there are a finite number of possible critical points in (a,b). Let us label these points $t = \alpha_j$, $j = 1, 2, \ldots, N$, where $\alpha_1 < \alpha_2 < \cdots < \alpha_N$. Thus, at $t = \alpha_j$ one or more of the following occur:

(1) ϕ vanishes.
(2) ϕ' vanishes.
(3) Some derivative of ϕ or f is discontinuous.

If we set $\alpha_0 = a$ and $\alpha_{N+1} = b$, then we can write

$$I(\lambda) = \sum_{j=0}^{N} I_j(\lambda), \tag{6.4.2}$$

$$I_j(\lambda) = \int_{\alpha_j}^{\alpha_{j+1}} h(\lambda\phi) f \, dt. \tag{6.4.3}$$

In this simple manner we have represented I as a sum of integrals, each of a form considered in Section 6.3. We note that, by assumption, both ϕ and f are infinitely differentiable in (α_j, α_{j+1}), $j = 0, 1, \ldots, N$.

It appears that having reduced $I(\lambda)$ to a sum of integrals previously treated, no further analysis is needed. We desire, however, to obtain an explicit formula for the contribution to the asymptotic expansion of I corresponding to each type of interior critical point. If we merely apply the results of Section 6.3 to each I_j and then sum the expansions thereby obtained, we shall not have accomplished our goal. We shall therefore employ a somewhat different procedure which is based on the total isolation of the interior critical points.

Suppose that we wish to study the contribution from the interior critical point $t = \alpha_j$, $0 < j < N + 1$. Suppose further that both ϕ and f are infinitely differentiable at $t = \alpha_j$ so that either $\phi(\alpha_j) = 0$ or $\phi'(\alpha_j) = 0$ (or both). (The contributions corresponding to jump discontinuities of ϕ, f or their derivatives are easier to obtain and will be considered in the exercises.)

The total isolation of $t = \alpha_j$ is readily accomplished by the neutralization process. Without going into detail we find that the determination of the contribution from $t = \alpha_j$ involves the asymptotic analysis of the integral

$$I_{\alpha_j}(\lambda) = \int_a^b f_{\alpha_j}(t)\, h(\lambda\phi)\, dt. \tag{6.4.4}$$

Here f_{α_j} is infinitely differentiable in (a,b), equals f identically throughout some neighborhood of $t = \alpha_j$, and vanishes for $a \le t \le \beta^*$ and $\beta^{**} \le t \le b$. Here

$$\alpha_{j-1} < \beta^* < \alpha_j, \qquad \alpha_j < \beta^{**} < \alpha_{j+1}. \tag{6.4.5}$$

We shall now consider separately the cases where $\phi(\alpha_j)=0$ and $\phi(\alpha_j)\neq 0$.

I. $\phi(\alpha_j)=0$. We first write

$$I_{\alpha_j}(\lambda)=\int_a^{\alpha_j} f_{\alpha_j}\, h(\lambda\phi)\,dt+\int_{\alpha_j}^b f_{\alpha_j}\, h(\lambda\phi)\,dt$$

$$=I_-(\lambda)+I_+(\lambda). \tag{6.4.6}$$

We note that the only critical point for I_- is the upper endpoint of integration, while the only critical point for I_+ is the lower endpoint of integration.

Because ϕ is infinitely differentiable at $t=\alpha_j$ we have

$$\begin{aligned}\phi(t)&\sim\frac{\phi^{(n)}(\alpha_j)}{n!}(t-\alpha_j)^n, && t\to\alpha_j+,\\ \phi(t)&\sim\frac{(-1)^n\phi^{(n)}(\alpha_j)}{n!}(\alpha_j-t)^n, && t\to\alpha_j-.\end{aligned}\tag{6.4.7}$$

Here n is an integer ≥ 1 and we assume that $\phi^{(n)}(\alpha_j)\neq 0$. If $n=1$, then $t=\alpha_j$ is a zero but not a stationary point of ϕ. If $n=2$, then $t=\alpha_j$ is a zero *and* a simple stationary point of ϕ. Finally, if $n>2$, then $t=\alpha_j$ is a zero and a stationary point of order $n-1$ of ϕ. For simplicity, we shall assume that $f(\alpha_j)\neq 0$.

To obtain the leading term in the expansion of I_{α_j}, we first suppose that in (6.4.7) n is even. Then ϕ has either a relative minimum or a relative maximum at $t=\alpha_j$, depending on the sign of $\phi^{(n)}(\alpha_j)$. The asymptotic expansions of $I_\pm(\lambda)$ can be obtained by using the results of Section 6.3. Indeed we find that the asymptotic expansions of I_+ and I_- agree to leading order and that

$$I_{\alpha_j}(\lambda)\sim\frac{2f(\alpha_j)}{n}\left(\frac{n!}{\lambda\,|\phi^{(n)}(\alpha_j)|}\right)^{1/n} M\left[h\,(\operatorname{sgn}\phi^{(n)}(\alpha_j)\,t);\frac{1}{n}\right],\qquad n\text{ even}. \tag{6.4.8}$$

Now suppose that n is odd. This means that ϕ must change sign at $t=\alpha_j$. The expansion of I_+ will now differ from that of I_- to leading order. We again use the results of Section 6.3 which yield

$$I_{\alpha_j}(\lambda)\sim\frac{f(\alpha_j)}{n}\left(\frac{n!}{\lambda\,|\phi^{(n)}(\alpha_j)|}\right)^{1/n}\left\{M\left[h(t);\frac{1}{n}\right]+M\left[h(-t);\frac{1}{n}\right]\right\},\qquad n\text{ odd}. \tag{6.4.9}$$

Of special interest to us is the case $n=1$. Indeed, in that event, because $t=\alpha_j$ is a zero but not a stationary point of ϕ, this case enables us to study the critical nature of the interior zeros of ϕ.

If $n=1$ in (6.4.7), then (6.4.9) becomes

$$I_{\alpha_j}(\lambda)\sim\frac{f(\alpha_j)}{\lambda\,|\phi'(\alpha_j)|}\{M[h(t);1]+M[h(-t);1]\}. \tag{6.4.10}$$

We can immediately conclude that if $\{M[h(t);1] + M[h(-t);1]\}$ is not zero, then $t = \alpha_j$ is a critical point. Even when this quantity is zero, $t = \alpha_j$ may still be a critical point because there may be nonzero contributions of lower order.

To examine this last point more closely, let us suppose that

$$\phi(t) \equiv (t - \alpha_j). \tag{6.4.11}$$

We also have

$$f(t) \sim \sum_{m=0}^{\infty} \frac{f^{(m)}(\alpha_j)}{m!}(t-\alpha_j)^m, \qquad t \to \alpha_j +,$$

$$f(t) \sim \sum_{m=0}^{\infty} \frac{(-1)^m f^{(m)}(\alpha_j)}{m!}(\alpha_j - t)^m, \qquad t \to \alpha_j -. \tag{6.4.12}$$

Now it is a simple matter to obtain an infinite asymptotic expansion of I_{α_j}. Indeed we find

$$I_{\alpha_j}(\lambda) \sim \sum_{m=0}^{\infty} \frac{\lambda^{-(m+1)} f^{(m)}(\alpha_j)}{m!}\{M[h(t);m+1] + (-1)^m M[h(-t);m+1]\}. \tag{6.4.13}$$

If any term in this sum is nonzero, then $t = \alpha_j$ is a critical point. Alternatively, if every term in the sum is zero, then $t = \alpha_j$ is not critical. The issue depends primarily on the kernel[1] and at that only through the quantities

$$M[h(t);m+1] + (-1)^m M[h(-t);m+1], \qquad m = 0, 1, 2, \ldots. \tag{6.4.14}$$

It can be shown that the above conclusions hold for general ϕ when $n = 1$. In particular, interior points at which ϕ vanishes are, in general, critical if the kernel is such that at least one of the quantities (6.4.14) is nonzero.

EXAMPLE 6.4.1. Let us suppose that $h(t) = \exp(it)$. We have

$$M[h(\pm t);z] = \Gamma(z)\exp\left\{\frac{\pm \pi i z}{2}\right\}. \tag{6.4.15}$$

Thus, whenever $t = \alpha_j$ is either a relative minimum or a relative maximum point of ϕ [n even in (6.4.7)], we find from (6.4.8) that

$$\int_a^b \exp\{i\lambda\phi\} f_{\alpha_j}\, dt \sim \frac{2f(\alpha_j)}{n}\Gamma\left(\frac{1}{n}\right)\left(\frac{n!}{\lambda\,|\phi^{(n)}(\alpha_j)|}\right)^{1/n}\exp\left\{\frac{\pi i}{2n}\operatorname{sgn}\phi^{(n)}(\alpha_j)\right\}. \tag{6.4.16}$$

When $t = \alpha_j$ is a simple stationary point, that is, when $n = 2$, (6.4.16) reduces to

$$\int_a^b \exp\{i\lambda\phi\} f_{\alpha_j}\, dt \sim \left(\frac{2\pi}{\lambda\,|\phi''(\alpha_j)|}\right)^{1/2} f(\alpha_j)\exp\left\{\frac{\pi i}{4}\operatorname{sgn}\phi''(\alpha_j)\right\}. \tag{6.4.17}$$

[1] If $f^{(m)}(\alpha_j) = 0$ for $m = 0, 1, 2, \ldots$, then $t = \alpha_j$ is not a critical point.

This will be recognized as the previously obtained "stationary phase" formula (6.1.5) in the case where ϕ vanishes at the stationary point.

If n is odd, then we have from (6.4.9) and (6.4.15)

$$\int_a^b \exp\{i\lambda\phi\} f_{\alpha_j}\, dt \sim \frac{2f(\alpha_j)}{n}\Gamma\left(\frac{1}{n}\right)\left(\frac{n!}{\lambda\,|\phi^{(n)}(\alpha_j)|}\right)^{1/n}\cos\left(\frac{\pi}{2n}\right). \tag{6.4.18}$$

Note that when $n = 1$ the right side vanishes. Moreover, because

$$M[e^{it};m+1] + (-1)^m M[e^{-it};m+1] = \Gamma(m+1)\,\{e^{(\pi i(m+1))/2} + (-1)^m e^{(-\pi i(m+1))/2}\} \tag{6.4.19}$$

vanishes for $m = 0, 1, 2, \ldots$, we can conclude that the interior zeros are not critical points for (6.4.1) when h is the Fourier kernel.

EXAMPLE 6.4.2. Now suppose that $h = J_\nu(t)$ where ν is a nonnegative integer. From the Appendix we have

$$M[J_\nu(t);z] = \frac{2^{z-1}\,\Gamma\left(\dfrac{z+\nu}{2}\right)}{\Gamma\left(\dfrac{\nu-z}{2}+1\right)}. \tag{6.4.20}$$

Because $J_\nu(t)$ is an even (odd) function about $t = 0$ when ν is even (odd), we have

$$M[J_\nu(-t);z] = (-1)^\nu M[J_\nu(t);z]. \tag{6.4.21}$$

Thus, it follows from (6.4.8) and (6.4.20) that when n is even in (6.4.7)

$$\int_a^b J_\nu(\lambda\phi) f_{\alpha_j}\, dt \sim \frac{2f(\alpha_j)}{n}\left(\frac{n!}{\lambda|\phi^{(n)}(\alpha_j)|}\right)^{1/n} 2^{1/n-1}\,\frac{\Gamma\left(\dfrac{1}{2n}+\dfrac{\nu}{2}\right)}{\Gamma\left(\dfrac{\nu}{2}-\dfrac{1}{2n}+1\right)} \times [\operatorname{sgn}(\phi^{(n)}(\alpha_j))]^\nu. \tag{6.4.22}$$

If n is odd in (6.4.7), then from (6.4.9) we have

$$\int_a^b J_\nu(\lambda\phi) f_{\alpha_j}\, dt \sim \frac{f(\alpha_j)}{n}\left(\frac{n!}{\lambda|\phi^{(n)}(\alpha_j)|}\right)^{1/n} 2^{1/n-1}\,\frac{\Gamma\left(\dfrac{1}{2n}+\dfrac{\nu}{2}\right)}{\Gamma\left(\dfrac{\nu}{2}-\dfrac{1}{2n}+1\right)}[1+(-1)^\nu]. \tag{6.4.23}$$

We note that

$$M[J_\nu(t);m+1] + (-1)^m M[J_\nu(-t);m+1] = \frac{2^m\,\Gamma(\frac{1}{2}(m+\nu+1))}{\Gamma(\frac{1}{2}(\nu+1-m))}(1+(-1)^{\nu+m}) \tag{6.4.24}$$

is not zero whenever $\nu + m$ is even. Thus, we have that interior zeros of ϕ are critical points for (6.4.1) whenever the kernel is an integer order Bessel function of the first kind.

It is of interest to investigate further just why the zeros of ϕ are critical for (6.4.1) when $h = J_\nu(t)$ but are not when $h = \exp\{it\}$. We can attempt to answer this by using a heuristic argument analogous to the one used in Section 5.4 for algebraic kernels. Indeed, we argued there that any point in (a,b) having a neighborhood throughout which $h(\lambda\phi)$ cannot be replaced by its asymptotic expansion for large argument would be critical. If we again use this argument we might conclude that any zero of ϕ is a critical point for (6.4.1). This, however, is contradicted by the result obtained in Example 6.4.1 for the Fourier kernel $h = \exp\{it\}$. In this case, however, $h(\lambda\phi) = \exp\{i\lambda\phi\}$ is its own asymptotic expansion which holds throughout $[a,b]$ regardless of whether or not ϕ vanishes in (a,b). For $h = J_\nu(t)$ this is not the case, because the asymptotic expansion of $J_\nu(\lambda\phi)$ for large argument cannot be used throughout any interval which contains a zero of ϕ.

II. $\phi(\alpha_j) \neq 0$. By assumption, because $\phi(t)$ cannot change sign in the effective range of integration in (6.4.4), we have that $\operatorname{sgn}\phi(t) = \operatorname{sgn}\phi(\alpha_j)$ in that region.

Let us first suppose that $\phi(\alpha_j)$ is positive and that (6.2.2) holds with $c_{0n} = 0$ for $n \geq 1$. We still have the decomposition (6.4.6) and the remarks following that equation remain valid. Now, however,

$$\begin{aligned} \phi(t) - \phi(\alpha_j) &\sim \frac{\phi^{(n)}(\alpha_j)}{n!}(t-\alpha_j)^n, \qquad t \to \alpha_j +, \\ \phi(t) - \phi(\alpha_j) &\sim \frac{(-1)^n \phi^{(n)}(\alpha_j)}{n!}(\alpha_j - t)^n, \qquad t \to \alpha_j - . \end{aligned} \tag{6.4.25}$$

Here n is an integer ≥ 2 and $\phi^{(n)}(\alpha_j) \neq 0$. We shall again assume that $f(\alpha_j) \neq 0$.

It follows from (6.3.28) and (6.4.25) that

$$I_+(\lambda) \sim \kappa(\alpha_j, \nu, n)\, \lambda^{-(\nu/n + r_0)} \exp\left\{i\omega\lambda^\nu\phi^\nu(\alpha_j) + \frac{i\pi}{2n}\operatorname{sgn}(\omega\phi^{(n)}(\alpha_j))\right\}, \tag{6.4.26}$$

where

$$\kappa(\alpha_j, \nu, n) = \frac{c_{00} f(\alpha_j)}{n[\phi(\alpha_j)]^{r_0}} \left\{\frac{|\phi^{(n)}(\alpha_j)\omega|}{n!}\, \nu\, \phi^{\nu-1}(\alpha_j)\right\}^{-1/n} \Gamma\left(\frac{1}{n}\right). \tag{6.4.27}$$

Moreover, from (6.3.38) and (6.4.25) we have

$$I_-(\lambda) \sim \kappa(\alpha_j, \nu, n)\, \lambda^{-(\nu/n + r_0)} \exp\left\{i\lambda^\nu\omega\phi^\nu(\alpha_j) + \frac{i\pi}{2n}(-1)^n \operatorname{sgn}(\omega\phi^{(n)}(\alpha_j))\right\}. \tag{6.4.28}$$

Thus

$$I_{\alpha_j}(\lambda) \sim \begin{cases} 2\kappa(\alpha_j, \nu, n)\, \lambda^{-(\nu/n + r_0)} \exp\{i\lambda^\nu \omega\phi^\nu(\alpha_j) + \dfrac{i\pi}{2n} \operatorname{sgn}(\omega\phi^{(n)}(\alpha_j))\}, & n \text{ even}, \\ 2\kappa(\alpha_j, \nu, n)\, \lambda^{-(\nu/n + r_0)} \exp\{i\lambda^\nu \omega\phi^\nu(\alpha_j)\} \cos \dfrac{\pi}{2n}, & n \text{ odd}. \end{cases} \tag{6.4.29}$$

The analysis when $\phi(\alpha_j)$ is negative proceeds exactly as above. Indeed, if we assume that (6.3.41) holds with $k_{0n} = 0$ for $n \geq 1$, then we find by using (6.3.50), (6.3.56), and (6.4.25) that

$$I_{\alpha_j}(\lambda) \sim \begin{cases} 2\tilde{\kappa}(\alpha_j, \gamma, n)\, \lambda^{-(\gamma/n + \rho_0)} \exp\{i\hat{\omega}\lambda^\gamma\, |\phi(\alpha_j)|^\gamma - \dfrac{\pi i}{2n} \operatorname{sgn}(\hat{\omega}\phi^{(n)}(\alpha_j))\}, & n \text{ even}, \\ 2\tilde{\kappa}(\alpha_j, \gamma, n)\, \lambda^{-(\gamma/n + \rho_0)} \exp\{i\hat{\omega}\lambda^\gamma\, |\phi(\alpha_j)|^\gamma\} \cos \dfrac{\pi}{2n}, & n \text{ odd}. \end{cases} \tag{6.4.30}$$

Here

$$\tilde{\kappa}(\alpha_j, \gamma, n) \sim \frac{k_{00} f(\alpha_j)}{n[|\phi(\alpha_j)|]^{\rho_0}} \left\{ \frac{|\phi^{(n)}(\alpha_j)\hat{\omega}|}{n!} |\phi(\alpha_j)|^{\gamma - 1} \gamma \right\}^{-1/n} \Gamma\left(\frac{1}{n}\right). \tag{6.4.31}$$

EXAMPLE 6.4.3. Suppose that $h(t) = \exp\{it\}$. Then in (6.2.2), $c_{00} = 1$, $r_0 = 0$, $\nu = 1$, and $\omega = 1$. Thus, it follows from (6.4.27) and (6.4.29) that

$$\int_a^b \exp\{i\lambda\phi\} f_{\alpha_j}\, dt \sim \begin{cases} \dfrac{2 f(\alpha_j)}{n} \left(\dfrac{n!}{|\phi^{(n)}(\alpha_j)|\lambda} \right)^{1/n} \Gamma\left(\dfrac{1}{n}\right) \\ \quad \times \exp\{i\lambda\phi(\alpha_j) + \dfrac{\pi i}{2n} \operatorname{sgn} \phi^{(n)}(\alpha_j)\}, & n \text{ even}, \quad \phi > 0, \\ \dfrac{2 f(\alpha_j)}{n} \left(\dfrac{n!}{|\phi^{(n)}(\alpha_j)|\lambda} \right)^{1/n} \Gamma\left(\dfrac{1}{n}\right) \\ \quad \times \exp\{i\lambda\phi(\alpha_j)\} \cos \dfrac{\pi}{2n}, & n \text{ odd}, \quad \phi > 0. \end{cases} \tag{6.4.32}$$

As, is readily seen, the same results hold for $\phi(\alpha_j) < 0$.

EXAMPLE 6.4.4. Now suppose that $h = J_\mu(t)$ where μ is any real number. We have from (6.1.17) that as $t \to \infty$

$$J_\mu(t) \sim \frac{1}{\sqrt{2\pi t}} \sum_{\pm} \exp\left\{ \pm i\left(\frac{\pi}{4} + \frac{\mu\pi}{2} - t \right) \right\}. \tag{6.4.33}$$

If we assume that $\phi(\alpha_j)$ is positive, then upon replacing $J_\mu(\lambda\phi)$ in (6.4.1) by the asymptotic expansion (6.4.33) we obtain two integrals of the form considered in this subsection. Each can be asymptotically evaluated by using formula (6.4.29). Indeed, after some calculation we find that, if in (6.4.25) n is even, then

$$\int_a^b J_\mu(\lambda\phi)\, f_{\alpha_j}\, dt \sim \frac{4f(\alpha_j)}{n[\phi(\alpha_j)]^{1/2}} \left(\frac{n!}{|\phi^{(n)}(\alpha_j)|}\right)^{1/n} \Gamma\left(\frac{1}{n}\right) \lambda^{-(1/n+1/2)}$$
$$\times \cos\left\{\lambda\phi(\alpha_j) + \frac{\pi}{2n} \operatorname{sgn} \phi^{(n)}(\alpha_j) - \frac{\pi}{4} - \frac{\mu\pi}{2}\right\}, \qquad (6.4.34)$$

while if n is odd, then

$$\int_a^b J_\mu(\lambda\phi)\, f_\alpha\, dt \sim \frac{4f(\alpha_j)}{n[\phi(\alpha_j)]^{1/2}} \left(\frac{n!}{|\phi^{(n)}(\alpha_j)|}\right)^{1/n} \Gamma\left(\frac{1}{n}\right) \lambda^{-(1/n+1/2)}$$
$$\times \cos\left\{\lambda\phi(\alpha_j) - \frac{\pi}{4} - \frac{\mu\pi}{2}\right\} \cos\frac{\pi}{2n}. \qquad (6.4.35)$$

To conclude this section we remark that, in principle, we can obtain as many terms in the above expansions as desired. To find any more than the leading term is quite tedious, however. Moreover, if either ϕ or f is not a piecewise C^∞ function in (a,b), then, as pointed out at the end of Section 6.3, at best a finite expansion can be obtained.

6.5. Exercises

6.1. Calculate the leading two terms in the asymptotic expansion, as $\lambda \to \infty$, of each of the following integrals:

(a) $\int_0^\pi e^{i\lambda(\sin t + t)} \cos(\sqrt{t+1})\, t^{3/2}\, dt$.

(b) $\int_0^1 e^{i\lambda t^3} \sin \pi t\, dt$.

(c) $\int_{-1}^1 e^{i\lambda \cosh t} \sqrt{1-t^2}\, dt$.

(d) $\int_{-1}^1 \sin(\lambda \log(1+t^2)) \sinh(1-t^2)\, dt$.

6.2. (a) Show that, as $\lambda \to \infty$,

$$\int_0^1 x^{-1/2} (1-x^2)^{-3/4} \sin\left(\frac{\pi x}{2}\right) \cos(\lambda \sqrt{1-x^2})\, dx$$
$$\sim \sqrt{\frac{\pi}{2\lambda}} + \frac{\pi}{4} \Gamma\left(\frac{3}{4}\right)\left(\frac{2}{\lambda}\right)^{3/4} \cos\left[\lambda - \frac{3\pi}{8}\right].$$

(b) Show that the error in (a) is $O(\lambda^{-7/4})$.

6.3. Show that the result (6.1.22) can also be written in the form

$$J_\lambda(\lambda r) \sim \sum_{\pm} \sqrt{\frac{\pi}{\lambda}} \frac{(1 \mp i)}{2(r^2-1)^{1/4}} \left(\frac{r\, e^{i\sqrt{r^2-1}}}{1+i\sqrt{r^2-1}} \right)^{\pm\lambda}, \qquad \lambda \to \infty, \qquad r > 1. \tag{6.5.1}$$

Here the notation $\sum_{\pm}$ means add the summand with upper sign to the summand with lower sign.

6.4. (a) Let

$$I(\lambda) = \int_{-1}^{1} f(p) \exp\{i\lambda [x\sqrt{1-p^2} + yp]\}\, dp$$

with $x > 0$, $f(p) \in C^\infty(-1,1)$. Suppose that, as $|p| \to 1$, $f(p) = 0((1-p^2)^\sigma)$ with $\sigma > -\frac{3}{4}$. Show that the leading term of the asymptotic expansion of $I(\lambda)$ arises from the stationary point $p = p_0$ where

$$\frac{p_0}{\sqrt{1-p_0^2}} = \frac{y}{x}.$$

(b) Show that

$$I(\lambda) \sim \sqrt{\frac{2\pi}{\lambda}}\, x f\left(\frac{y}{\sqrt{x^2+y^2}}\right) \frac{\exp\left\{i\lambda\sqrt{x^2+y^2} - \dfrac{i\pi}{4}\right\}}{(x^2+y^2)^{3/4}}. \tag{6.5.2}$$

(c) What role does the constraint on the behavior of f near $p = \pm 1$ play in the determination of (6.5.2)?

6.5. Let

$$I(\lambda) = \int_a^b (t-a)^\delta g(t) \exp\{i\lambda\phi(t)\}\, dt, \qquad \delta > -1. \tag{6.5.3}$$

Here $g \in C^\infty[a,b]$ vanishes infinitely smoothly at $t = c < b$ and is identically zero on $[c,b]$. Also, $\phi' = \alpha\nu(t-a)^{\nu-1} h(t)$ with h infinitely differentiable and positive on $[a,c]$ and $h(a) = 1$.

(a) Show that the change of variable from s to t defined by

$$\phi(t) - \phi(a) = \mu s^\nu, \qquad \mu = \operatorname{sgn}(\alpha) \tag{6.5.4}$$

is infinitely differentiable and one-to-one when we define $(s^\nu)^{1/\nu} = s$.

(b) Show that, under the transformation (6.5.4), (6.5.3) becomes

$$I(\lambda) = e^{i\lambda\phi(a)} \int_0^\infty s^\delta G(s) \exp\{i\lambda\mu s^\nu\}\, ds, \tag{6.5.5}$$

where

$$G(s) = \begin{cases} g(t(s)) \dfrac{dt}{ds} \left(\dfrac{t(s) - a}{s} \right)^{\delta}, & s^{\nu} < |\phi(c) - \phi(a)|, \\ 0, & s^{\nu} \geq |\phi(c) - \phi(a)|, \end{cases} \tag{6.5.6}$$

is infinitely differentiable on $[0, \infty)$.

(c) Set

$$G(s) \sim \sum_{n=0}^{\infty} G_n s^n \tag{6.5.7}$$

and apply the method of Section 3.4 to obtain

$$I(\lambda) \sim \exp\left\{ i\lambda \phi(a) + \frac{i\pi\mu(\delta + 1)}{\nu} \right\} \sum_{n=0}^{\infty} \frac{G_n \Gamma\left(\dfrac{\delta + n + 1}{\nu} \right)}{\nu \lambda^{(\delta + n + 1)/\nu}} e^{i\pi\mu n/2\nu} . \tag{6.5.8}$$

(d) Show that

$$G_0 = \frac{g(a)}{|\alpha|^{(\delta + 1)/\nu}}. \tag{6.5.9}$$

(e) How must (6.5.8) be modified if in (6.5.3) $(t - a)^{\delta}$ is replaced by $(t - a)^{\delta} \log(t - a)$?

6.6. With little modification the discussion of Section 2.7 can be applied to the case of scattering in two dimensions. In this case (2.7.20) is replaced by

$$I(\lambda) = \rho_0^2 \int_{n \cdot \mu_+ \leq 0} \mathbf{n}(\boldsymbol{\eta}) \cdot \boldsymbol{\mu}_{\pm} \exp\{i\lambda[\boldsymbol{\mu}_+ - \boldsymbol{\mu}_-] \cdot \boldsymbol{\eta}\}\, ds. \tag{6.5.10}$$

Here s is arc-length along the curve defined by $\boldsymbol{\eta} = \boldsymbol{\eta}(s)$. The unit vectors $\boldsymbol{\mu}_+$ and $\boldsymbol{\mu}_-$ are respectively in the directions of incidence and observation of the signal. Also $\mathbf{n} = \mathbf{n}(s)$ is the unit outward normal and the choice $\boldsymbol{\mu}_{\pm}$ in the amplitude depends on what boundary condition is used.

(a) Show that the phase function $\phi = [\boldsymbol{\mu}_+ - \boldsymbol{\mu}_-] \cdot \boldsymbol{\eta}$ is stationary at a point on the boundary curve $\boldsymbol{\eta} = \boldsymbol{\eta}(s)$ where $\boldsymbol{\mu}_+$ and $\boldsymbol{\mu}_-$ make equal angles with the normal $\mathbf{n}(s)$. (This point is called the point of *specular reflection* which suggests that the contribution corresponding to the stationary point represents the reflected wave.)

(b) Under the assumption that the stationary point is an interior point of the domain of integration, show that the asymptotic expansion of $I(\lambda)$ has, to leading order, the following parametric representation:

$$I(\lambda) \sim \rho_0^2 \sqrt{\frac{2\pi}{\lambda\, \kappa(s)\, |\boldsymbol{\mu}_+ - \boldsymbol{\mu}_-|}}\, \mathbf{n}(s) \cdot \boldsymbol{\mu}_{\pm} \exp\left\{ i\lambda[\boldsymbol{\mu}_+ - \boldsymbol{\mu}_-] \cdot \boldsymbol{\eta}(s) + \frac{i\pi}{4} \right\}, \tag{6.5.11}$$

$$\boldsymbol{\mu}_+ \cdot \mathbf{T}(s) = \boldsymbol{\mu}_- \cdot \mathbf{T}(s). \tag{6.5.12}$$

Here $\mathbf{T}(s)$ is the unit tangent vector to the curve $\boldsymbol{\eta} = \boldsymbol{\eta}(s)$ and $\kappa(s) = |\dot{\mathbf{T}}(s)|$ is the curvature.

(c) Show that in the special case of backscattering, that is, when $\boldsymbol{\mu}_+ = -\boldsymbol{\mu}_-$, the expansion becomes

$$I(\lambda) \sim \rho_0^2 \sqrt{\frac{\pi}{\lambda\kappa(s)}} \exp\left\{-2i\lambda\mathbf{n}(s)\cdot\boldsymbol{\eta}(s) + \frac{i\pi}{4}\right\}, \tag{6.5.13}$$

$$\boldsymbol{\mu}_+ \cdot \mathbf{T}(s) = 0. \tag{6.5.14}$$

6.7. Consider

$$I(\lambda) = \textstyle\int_a^b f(t) \exp\{i\lambda\phi(t)\}\, dt. \tag{6.5.15}$$

Suppose that ϕ has at least $k+4$ continuous derivatives on $[a,b]$ for some positive integer k and that ϕ has a stationary point at $t=a$ of order $k-1$. Suppose further that $f(t)$ has at least four continuous derivatives on $[a,b]$ and is identically zero in some left half-neighborhood of $t=b$.

(a) Show that

$$\begin{aligned} I(\lambda) = k^{-1} \exp\left\{i\lambda\phi(a) + \frac{i\mu\pi}{2k}\right\} &\left\{\Gamma\left(\frac{1}{k}\right)\left(\frac{k!}{\lambda\,|\phi^{(k)}(a)|}\right)^{1/k} f(a)\right. \\ &+ \Gamma\left(\frac{2}{k}\right)\left(\frac{k!}{\lambda|\phi^{(k)}(a)|}\right)^{2/k}\left[f^{(1)}(a) - \frac{2\,\phi^{(k+1)}(a)f(a)}{k(k+1)\,\phi^{(k)}(a)}\right] e^{i\mu\pi/2k} \\ &+ \Gamma\left(\frac{3}{k}\right)\left(\frac{k!}{\lambda|\phi^{(k)}(a)|}\right)^{3/k}\left[\frac{f^{(2)}(a)}{2} - \frac{3\,\phi^{(k+1)}(a)f^{(1)}(a)}{k(k+1)\,\phi^{(k)}(a)}\right. \\ &- \frac{3\,\phi^{(k+2)}(a)f(a)}{k(k+1)(k+2)\,\phi^{(k)}(a)} + \left(\frac{\phi^{(k+1)}(a)}{(k+1)|\phi^{(k)}(a)|}\right)^2 \frac{f(a)}{2} \\ &\left.\times \frac{3}{2k}\left(\frac{3}{k}+1\right)\right] e^{i\mu\pi/k} + O(\lambda^{-4/k})\bigg\}. \end{aligned} \tag{6.5.16}$$

Here $\mu = \operatorname{sgn} \phi^{(k)}(a)$.

(b) Similarly, show that if the roles of a and b are interchanged, then

$$\begin{aligned} I(\lambda) = k^{-1} \exp\left\{i\lambda\phi(b) + \frac{i\mu'\pi}{2k}\right\} &\left\{\Gamma\left(\frac{1}{k}\right)\left(\frac{k!}{\lambda\,|\phi^{(k)}(b)|}\right)^{1/k} f(b)\right. \\ &- \Gamma\left(\frac{2}{k}\right)\left(\frac{k!}{\lambda|\phi^{(k)}(b)|}\right)^{2/k}\left[f^{(1)}(b) - \frac{2\,\phi^{(k+1)}(b)f(b)}{k(k+1)\,\phi^{(k)}(b)}\right] e^{i\mu'\pi/2k} \\ &+ \Gamma\left(\frac{3}{k}\right)\left(\frac{k!}{\lambda|\phi^{(k)}(b)|}\right)^{3/k}\left[\frac{f^{(2)}(b)}{2} - \frac{3\,\phi^{(k+1)}(b)f^{(1)}(b)}{k(k+1)\,\phi^{(k)}(b)}\ \frac{f(b)}{2}\right. \\ &- \frac{3\,\phi^{(k+2)}(b)f(b)}{k(k+1)(k+2)\,\phi^{(k)}(b)} + \left(\frac{\phi^{(k+1)}(b)}{(k+1)\,|\phi^{(k)}(b)|}\right)^2 \\ &\left.\times \frac{3}{2k}\left(\frac{3}{k}+1\right)\right] e^{i\mu'\pi/k} + O(\lambda^{-4/k})\bigg\}. \end{aligned} \tag{6.5.17}$$

Here $\mu' = (-1)^k \operatorname{sgn} \phi^{(k)}(b)$.

(c) Show that if ϕ has only one interior stationary point at $t = c$, $a < c < b$, and if f vanishes sufficiently smoothly at $t = a$ and $t = b$, then

$$I(\lambda) = k^{-1} \exp\{i\lambda\phi(c)\} \left\{ \Gamma\left(\frac{1}{k}\right)\left(\frac{k!}{\lambda\,|\phi^{(k)}(c)|}\right)^{1/k} f(c) \left[2\cos\left(\frac{\mu\pi}{2k}\right) \right.\right.$$
$$\left. + [1 + (-1)^k]\, i \sin\left(\frac{\mu\pi}{2k}\right)\right] + \Gamma\left(\frac{2}{k}\right)\left(\frac{k!}{\lambda\,|\phi^{(k)}(c)|}\right)^{2/k} \left[f^{(1)}(c) \right.$$
$$\left.\left. - \frac{2\,\phi^{(k+1)}(c) f(c)}{k(k+1)\,\phi^{(k)}(c)}\right][1 - (-1)^k]\, i \sin\left(\frac{\mu\pi}{k}\right) + O(\lambda^{-3/k})\right\}. \tag{6.5.18}$$

Here $\mu = \operatorname{sgn} \phi^{(k)}(c)$.

(d) Modify (6.5.18) to account for a discontinuity in f at $t = c$.

(e) Modify (6.5.18) to account for a discontinuity of $\phi^{(k)}$ at $t = c$.

6.8. Let

$$I(\lambda) = \int_a^b f(t)\, h(\lambda\phi)\, dt \tag{6.5.19}$$

with $h(t)$ oscillatory in whichever of the limits $t \to \pm\infty$ is relevant. Assume that $\phi(a) = 0$ and that ϕ and f are otherwise as in Exercise 6.7(a). Show that we can obtain an asymptotic expansion of $I(\lambda)$ by modifying (6.5.16) in the following manner:

(i) Set $\phi(a) = 0$.

(ii) Replace $\Gamma\left(\dfrac{j}{k}\right) \exp \dfrac{i\mu j\pi}{2k} = M\left[e^{i\mu t}; \dfrac{j}{k}\right]$ by $M\left[h(\mu t); \dfrac{j}{k}\right]$.

6.9. Let

$$I(\lambda; a) = \int_0^{\pi/2} \frac{\operatorname{Ai}(-\lambda\sqrt{1 + a^2 - 2a\cos\theta})\, d\theta}{(1 + a^2 - 2a\cos\theta)^{1/4}},$$

where $\operatorname{Ai}(t)$ is the Airy function.

(a) Show that for $\lambda \to \infty$ with $0 < a < 1$

$$I(\lambda; a) \sim \frac{\cos\left[\frac{2}{3}\lambda^{3/2}(1 - a)^{3/2}\right]}{\lambda(1 - a)^{1/2}\sqrt{2a}}$$

$$- \frac{\cos\left[\frac{2}{3}\lambda^{3/2}(1 + a^2)^{3/4} + \frac{\pi}{4}\right]}{\lambda^{7/4}\sqrt{\pi}\, a\,(1 + a^2)^{1/8}}$$

(b) Show that

$$I(\lambda;1) = \sqrt{\frac{\pi}{\lambda}\frac{3^{-1/3}}{\Gamma(\frac{5}{6})}} - \frac{\cos\left[2^{7/4}\frac{\lambda^{3/2}}{3}+\frac{\pi}{4}\right]}{\lambda^{7/4}\sqrt{\pi}\,2^{1/8}} + O(\lambda^{-5/2}).$$

6.10. Consider

$$I(\lambda;a) = \int_0^{\pi/2} \frac{J_\nu\left(\lambda\sqrt{1+a^2-2a\cos\theta}\right)}{\sqrt{1+a^2-2a\cos\theta}}\,d\theta, \qquad \nu > 0, \qquad 0 < a \le 1.$$

Here J_ν is the Bessel function of the first kind.

(a) Show that

$$I(\lambda;a) \sim \frac{2\cos\left[\lambda(1-a)-\frac{\nu\pi}{2}\right]}{\lambda(1-a)\sqrt{a}}, \qquad 0 < a < 1, \qquad \lambda \to \infty.$$

(b) Show that

$$I(\lambda;1) \sim \frac{1}{\nu}, \qquad \lambda \to \infty.$$

6.11. Consider

$$I(\lambda) = \textstyle\int_a^b h(\lambda\,\phi(t))\,f_a(t)\,dt$$

under the following conditions:

(i) $h(t) \sim \exp\{i\omega t^\nu\} \sum_{m=0}^{\infty} C_m t^{-r_m}, \qquad t \to \infty, \qquad \omega \text{ real}, \qquad \nu > 0;$

(ii) $f_a(t) \in C^\infty(a,b]$ with f vanishing for $\gamma \le t, \qquad a < \gamma < b;$
(iii) $f_a(t) \sim \gamma_0(t-a)^{\mu_0}, \qquad t \to a+, \qquad -1 < \mathrm{Re}(\mu_0);$
(iv) h and $\phi \in C^\infty(a,\gamma]$;
(v) $\phi(t) - \phi(a) \sim \alpha_0(t-a)^{\nu_0}, \qquad t \to a+; \qquad \phi(a), \alpha_0, \nu_0 > 0;$
$\phi'(t) > 0$ on $(a,\gamma]; \qquad \phi'(t) \sim \nu_0\,\alpha_0(t-a)^{\nu_0-1}, \qquad t \to a+.$

Define for any real k

$$S_k = \exp\{i\omega t^\nu\} \sum_{\mathrm{Re}(r_m)<k} C_m t^{-r_m}$$

and

$$I_k(\lambda) = \textstyle\int_a^b S_k(\lambda\phi)\,f_a(t)\,dt.$$

show that, for any real k,

$$I(\lambda) = I_k(\lambda) + O(\lambda^{-k-\nu(\mu_0+1)/\nu_0}), \qquad \lambda \to \infty.$$

REFERENCES

A. ERDÉLYI. Asymptotic representations of Fourier integrals and the method of stationary phase. *SIAM J. Appl. Math.* **3,** 17–27, 1955.

A derivation of the method of stationary phase via integration by parts is presented. The amplitude is allowed to have an algebraic singularity at the stationary point.

A. ERDÉLYI. *Asymptotic Expansions*. Dover, New York, 1956.

In this reference also, the author derives the stationary phase results via integration by parts.

A. ERDÉLYI. Asymptotic expansions of Fourier integrals involving logarithmic singularities. *SIAM J. Appl. Math.* **4,** 38–47, 1956.

The method of stationary phase is extended to the case where the amplitude has a logarithmic singularity at the stationary point.

R. A. HANDELSMAN and N. BLEISTEIN. Asymptotic expansions of integral transforms with oscillatory kernels; a generalization of the method of stationary phase. *SIAM J. Math. Anal.* **4,** 3, 1973.

Sections 6.2 to 6.4 are based on this paper.

D. S. JONES. Fourier transforms and the method of stationary phase. *JIMA* **2,** 197–222, 1966.

In this paper the author develops the method of stationary phase in the context of generalized functions.

D. S. JONES. *Generalized Functions*. McGraw-Hill, London, 1966.

The author presents an extensive discussion of the asymptotic analysis of integrals of Fourier type.

LORD (W. THOMSON) KELVIN. On the wave produced by a single impulse in water of any depth or in a dispersive medium. *Philos. Mag.* **23,** 252–255, 1887. Also in *Math. and Phys. Papers* **4,** 303–306, 1910.

In this paper what is now called the general principle of stationary phase was first formulated.

G. G. STOKES. On the numerical calculation of a class of definite integrals and infinite series. *Camb. Philos. Trans.* **9,** 166–187, 1856. Also in *Math. and Phys. Papers* **2,** 329–357, 1883.

It appears that Stokes was the first to apply the principle of stationary phase in his investigations of the Airy functions.

G. N. WATSON. The limits of applicability of the principle of stationary phase. *Proc. Camb. Philos. Soc.* **19,** 49–55, 1918.

A rigorous justification of the method of stationary phase is presented.

7 The Method of Steepest Descents

7.1. Preliminary Results

Throughout this chapter we shall be concerned with integrals of the form

$$I(\lambda) = \int_C g(z) \exp\{\lambda w(z)\}\, dz. \tag{7.1.1}$$

Here C is a fixed contour in the complex z plane, while $g(z)$ and $w(z)$ are analytic functions in some region D that includes C.[1]

Our objective is to study $I(\lambda)$ in the limit $\lambda \to \infty$. We note that, with the integrand functions analytic, the Laplace-type integrals studied in Section 5.1 and the Fourier-type integrals studied in Section 6.1 are special cases of (7.1.1). As we shall find, complex function theory, whose use is made possible by the assumed analyticity of w and g, affords the additional machinery needed to handle the general class of integrals (7.1.1).

The results we have already obtained for Fourier- and Laplace-type integrals serve as partial motivation for the method of analysis to be employed here. Indeed, we have found when considering Fourier-type integrals (6.1.1), with f and g infinitely differentiable, that the critical points are the endpoints of integration and the stationary points of ϕ. Although the stationary points are, in general, the dominant critical points, we must consider the contribu-

[1] We can allow g and w to have singularities at the endpoints of C as long as $I(\lambda)$ is convergent.

tion from the endpoints if more than a leading term is desired. In contrast, we found that, for Laplace-type integrals (5.1.2), the entire asymptotic expansion depends only on the behavior of the integrand functions in a small neighborhood of the global minimum of ϕ along the interval of integration. As a result, the analysis of Laplace-type integrals is significantly simpler than that of Fourier-type integrals.

Complex function theory, and in particular Cauchy's integral theorem, tells us that we may deform the contour of integration C in (7.1.1) to a large extent without changing the value of $I(\lambda)$. This fact, coupled with the remarks of the preceding paragraph, suggest that we should seek to replace C by a contour or sum of contours in such a manner that not only is the value of $I(\lambda)$ unaltered, but also the resulting contour integrals are all of Laplace type. Our immediate task therefore is to develop the theory necessary to systematically carry out this deformation.

Let us first consider $w(z)$ which has already been assumed analytic in the region D. We shall further assume that $w \not\equiv$ const. because otherwise the

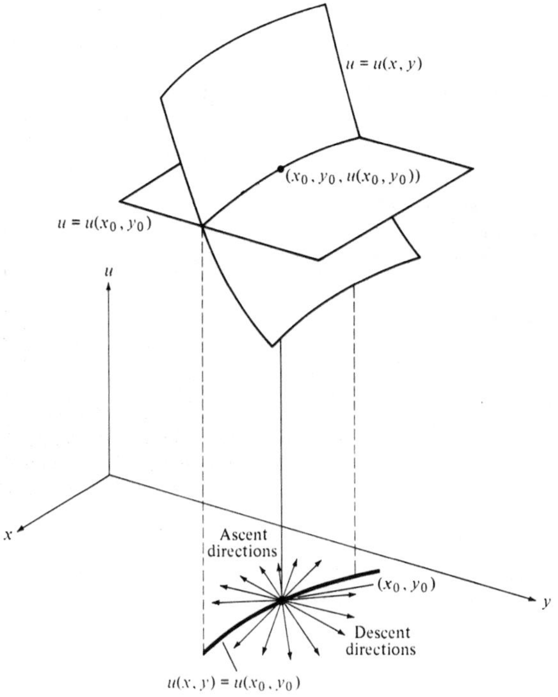

Figure 7.1.1.

asymptotic analysis of (7.1.1) is trivial. We now set

$$z = x + iy, \qquad w(z) = u(x,y) + iv(x,y) \tag{7.1.2}$$

with $u(x,y)$ and $v(x,y)$ real.

It will prove important for us to be able to identify curves in D along which $u(x,y)$ and hence $|\exp\{\lambda w(z)\}|$ is monotonic. To aid us in this task we shall now make several definitions.

DEFINITION 7.1.1. Let $z_0 = x_0 + iy_0$ be a point in D. Then a direction away from $z = z_0$ in which u decreases from the value $u(x_0, y_0)$ is called a *direction of descent* from $z = z_0$. (See Figure 7.1.1.) In an analogous way we define a *direction of ascent* as a direction away from z_0 in which $u(x,y)$ *increases* from the value $u(x_0, y_0)$.

The concepts of directions of descent and ascent lead us naturally to the following.

DEFINITION 7.1.2. If C is a directed curve from the point $z = z_0$ to the not necessarily finite point $z = z_1$ and is such that its tangent is always in a direction of descent (ascent), then C is called a *path of descent* (*ascent*).

Of course, emanating from $z = z_0$, there are many (in fact, a continuum of) directions of descent and hence many paths of descent. It is often possible and highly desirable to identify directions in which the rate[2] of descent (ascent) is maximal. Such directions are called *directions of steepest descent* (*ascent*). We might point out that at any point in D at which $\nabla u \neq 0$, the direction of steepest descent is unique and coincides with that of $(-\nabla u)$; while the direction of steepest ascent coincides with that of ∇u.

DEFINITION 7.1.3. A directed curve whose tangent at each point has a direction of steepest descent (ascent) is called a *path of steepest descent* (*ascent*).

The words "descent" and "ascent" have, of course, a topographical connotation. Indeed, their use in the present context is prompted by the graph of the surface $u = u(x,y)$ which is suggestive of a rolling countryside. We carry this analogy further in the following.

DEFINITION 7.1.4. A point $z_1 = x_1 + iy_1$ is said to lie in a *valley of* $w(z)$ with respect to $z_0 = x_0 + iy_0$ if $u(x_1, y_1) < u(x_0, y_0)$ and on a *hill of* $w(z)$ with respect to $z = z_0$ if $u(x_1, y_1) > u(x_0, y_0)$. If $u(x_1, y_1) = u(x_0, y_0)$, then $z = z_1$ is said to lie on the *boundary* of a hill and valley of w with respect to $z = z_0$.[3]

[2] Here, rate is with respect to arc-length or any other reasonable parameter.

[3] Note that for u harmonic or, equivalently, the real part of an analytic function, if $u = \text{const.}$ along a curve which separates either two hills or two valleys, then $u \equiv \text{const.}$ in D.

Thus, as we "move" along a path of descent from $z = z_0$, we are moving steadily into a valley, while as we "move" along a path of ascent, we are moving steadily up a hill. In order to penetrate a valley most "rapidly" (climb a hill most "rapidly") we would choose to move along a path of steepest descent (ascent).

To aid us in identifying the curves of steepest descent and steepest ascent we prove the following.

LEMMA 7.1. The curves of steepest descent and steepest ascent from any point $z = z_0 = (x_0 + iy_0)$ are those curves defined by

$$v(x_0, y_0) = \text{Im}(w) = v(x, y). \tag{7.1.3}$$

PROOF. Let δw denote the variation in w near $z = z_0$. It is defined by

$$\delta w = w(z) - w(z_0) = \delta u + i\delta v. \tag{7.1.4}$$

Clearly, $\delta v = 0$ if and only if z lies on the curve(s) $v(x,y) = v(x_0, y_0)$. In that event we have

$$\delta w = \delta u. \tag{7.1.5}$$

In general,

$$|\delta w|^2 = (\delta u)^2 + (\delta v)^2 = \kappa^2 \rho^{2n} [1 + 0(\rho)], \tag{7.1.6}$$

where κ is the magnitude of the first nonvanishing derivative, n is the order of that derivative, and $\rho = |z - z_0|$. Hence,

$$|\delta u| \leq |\delta w| \tag{7.1.7}$$

with equality holding only when $\delta v = 0$. But equality implies that $|\delta u|$ is maximal. If we now let $z \to z_0$ in such a manner that $\delta v = 0$, then $|\delta u|$ remains maximal. Hence, we can conclude that the directions of steepest descent and steepest ascent at $z = z_0$ coincide with the directions of the tangent to the curves defined by (7.1.3) at $z = z_0$. Finally, it follows from Definition 7.1.3 that (7.1.3) defines the curves of steepest descent and steepest ascent through $z = z_0$. This completes the proof.

A somewhat more difficult problem is the explicit determination of the directions of steepest descent and steepest ascent at points where dw/dz vanishes. The answer is given in the following.

THEOREM 7.1. Suppose

$$\left.\frac{d^q w}{dz^q}\right|_{z=z_0} = 0, \qquad q = 1, 2, \ldots, n-1, \qquad \left.\frac{d^n w}{dz^n}\right|_{z=z_0} = ae^{i\alpha}, \qquad a > 0. \tag{7.1.8}$$

If $z - z_0 = \rho e^{i\theta}$, then the directions of steepest descent, steepest ascent, and constant u at $z = z_0$ are as given in Table 7.1.

Table 7.1

Directions of	θ	
Steepest Descent	$-\frac{\alpha}{n}+(2p+1)\frac{\pi}{n}$	$p=0, 1, \ldots, n-1$
Steepest Ascent	$-\frac{\alpha}{n}+\frac{2p\pi}{n}$	$p=0, 1, \ldots, n-1$
Constant u	$-\frac{\alpha}{n}+\left(p+\frac{1}{2}\right)\frac{\pi}{n}$	$p=0, 1, \ldots, 2n-1$

PROOF. We introduce

$$\delta w = w(z) - w(z_0) = \frac{ae^{i\alpha}}{n!}\rho^n e^{in\theta}[1+O(\rho)], \qquad \rho \to 0, \tag{7.1.9}$$

which follows from (7.1.8). We now consider

$$\frac{\delta w}{\rho^n} = \frac{ae^{i[\alpha+n\theta]}}{n!}[1+O(\rho)] \tag{7.1.10}$$

and note that in a direction of steepest descent $\delta w/\rho^n$ is real and negative. Hence, the directions of steepest descent correspond to those values of θ which satisfy

$$(\alpha + n\theta) = (2p+1)\pi, \tag{7.1.11}$$

where p is an integer. Upon solving for θ we obtain the first entry of Table 7.1. Because we are only interested in distinct directions we require that $0 \le p \le n-1$.

In a direction of steepest ascent $\delta w/\rho^n$ is real and positive so that the corresponding values of θ must satisfy

$$\alpha + n\theta = 2p\pi, \qquad p = 0, 1, \ldots, n-1. \tag{7.1.12}$$

This yields entry 2 of Table 7.1.

Finally, in a direction of constant u, δw is purely imaginary which implies that

$$\alpha + n\theta = (p+\tfrac{1}{2})\pi, \qquad p = 0, 1, 2, \ldots, 2n-1. \tag{7.1.13}$$

This then yields entry 3 of Table 7.1 and the proof is complete.

Let us consider two boundary curves (curves of constant u) emanating from $z = z_0$, with z_0 as defined in the theorem above, with initial directions

$$-\frac{\alpha}{n} + \left(2k+1 \pm \frac{1}{2}\right)\frac{\pi}{n}, \qquad k \text{ an integer}. \tag{7.1.14}$$

These curves contain in their acute angle a valley of $w(z)$ with respect to $z = z_0$ and exactly one steepest descent path. This path, moreover, has the initial direction

$$-\frac{\alpha}{n} + (2k+1)\frac{\pi}{n}. \tag{7.1.15}$$

Note that in the range

$$-\frac{\alpha}{n} + \left(2k + \frac{1}{2}\right)\frac{\pi}{n} < \theta < -\frac{\alpha}{n} + \left(2k + \frac{3}{2}\right)\frac{\pi}{n} \tag{7.1.16}$$

δw has a negative real part which is consistent with our definition of a valley.

We also have that the two boundary curves emanating from $z = z_0$ with initial directions

$$-\frac{\alpha}{n} + \left(2k \pm \frac{1}{2}\right)\frac{\pi}{n}, \qquad k \text{ an integer} \tag{7.1.17}$$

contain in their acute angle a hill of w with respect to $z = z_0$ and exactly one path of steepest ascent having the initial direction

$$-\frac{\alpha}{n} + 2k\frac{\pi}{n}. \tag{7.1.18}$$

Thus we find that any neighborhood of $z = z_0$ is divided into $2n$ sectors[4] each having a vertex angle equal to π/n. These sectors are alternately valleys and hills. The bisectors of the vertex angles of these sectors define the directions of steepest descent and steepest ascent at $z = z_0$. Thus as we proceed around any circle enclosing $z = z_0$ (but no other point at which $dw/dz = 0$) we alternately intersect curves of steepest descent and steepest ascent, there being exactly n curves of each type in all.

We should point out that Theorem 7.1 is equivalent to a form of the maximum principle for harmonic functions and the maximum modulus theorem for analytic functions. Indeed, from our results, it follows that in the interior of D, $u = \text{Re } w$ can attain neither a maximum nor a minimum value so long as $w \not\equiv$ const. Naturally, similar statements hold for $v = \text{Im}(w)$ and for $|\exp \lambda w|$.

Let us return to Theorem 7.1 and, in particular, to the case where $n = 2$. Now there are two distinct directions of steepest descent away from $z = z_0$ defined by

$$\theta = -\frac{\alpha}{2} + \frac{\pi}{2}, \qquad -\frac{\alpha}{2} + \frac{3\pi}{2} \qquad \text{(descent)}. \tag{7.1.19}$$

[4] In actuality, except when $w(z) - w(z_0) = (z - z_0)^n$, the boundaries of these sectors are not straight lines but can be approximated by such whenever the neighborhood is small.

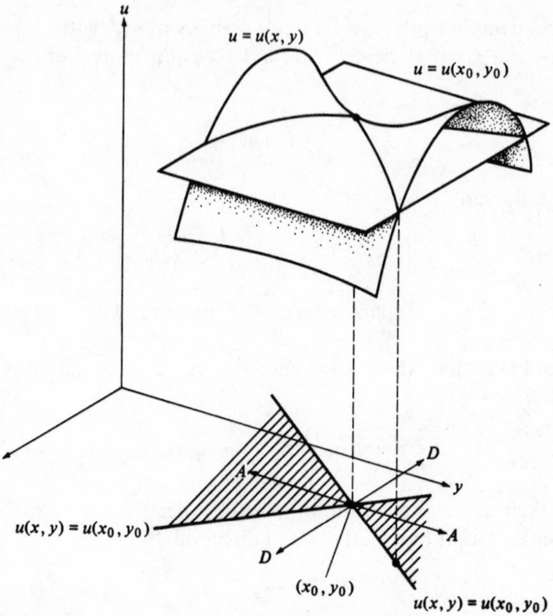

Figure 7.1.2. The Surface $u(x, y)$ near a (Simple) Saddle Point.

The two directions of steepest ascent are defined by

$$\theta = -\frac{\alpha}{2}, \qquad -\frac{\alpha}{2} + \pi \qquad \text{(ascent)}. \tag{7.1.20}$$

Thus we find that the two steepest descent directions are opposite as are the two steepest ascent directions. Figure 7.1.2 depicts a typical surface $u = u(x, y)$ about the point (x_0, y_0) in the case $n = 2$. In this figure the directions of steepest descent are labeled (D) while the directions of steepest ascent are labeled (A). We see that, locally, the surface is shaped like a saddle. For this reason a complex stationary point, that is, a point at which $dw/dz = 0$, is called a *saddle point* of w no matter how many derivatives of w vanish there.

In Figure 7.1.3 we depict a portion of the surface $u = u(x, y)$ near $z = z_0$ in the case where $n = 3$. It is convenient to designate as the order of the saddle point at $z = z_0$, the order of the last vanishing derivative of w at $z = z_0$. Thus, Figure 7.1.2 depicts a surface near a saddle point of order 1 (or simple saddle point), while Figure 7.1.3 depicts a surface near a saddle point of order 2, often referred to as a *monkey saddle*.

As we might anticipate, the explicit determination of steepest descent paths

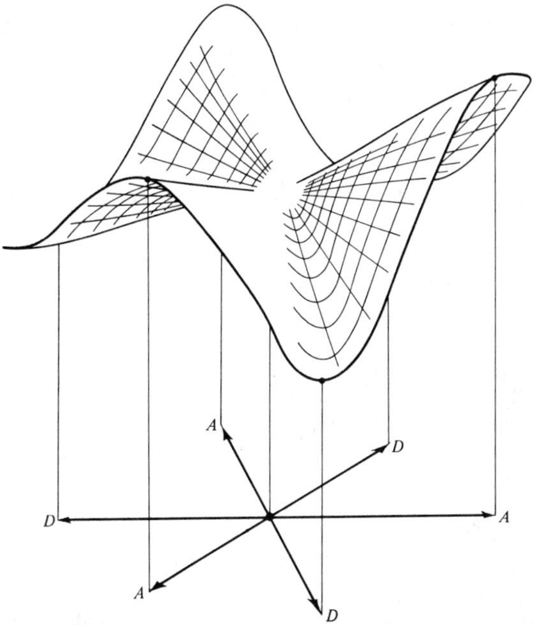

Figure 7.1.3. Monkey Saddle.

is rather difficult even when $w(z)$ is a fairly simple function. Fortunately, for our purposes it will not be necessary to determine these curves in great detail and moreover any features that will be required we shall find are readily obtainable. Nevertheless, it is instructive to consider an example for which the steepest descent paths from the saddle points can be obtained in as much detail as desired.

EXAMPLE 7.1. In Section 2.5 we considered the integral representation of the Airy function of positive argument and found it to be of the form (7.1.1) with, from (2.5.10),

$$w(z) = z - \frac{z^3}{3}. \tag{7.1.21}$$

Because

$$w'(z) = 1 - z^2, \qquad w''(z) = -2z, \tag{7.1.22}$$

we find that w has two simple saddle points at $z = \pm 1$. Moreover,

$$w(\pm 1) = \pm \tfrac{2}{3}, \qquad w''(\pm 1) = \mp 2. \tag{7.1.23}$$

We wish to study the hills, valleys, and paths of steepest descent and ascent corresponding to each of the two saddle points. By carrying along the $\pm$ signs as in (7.1.23) we can consider both saddle points simultaneously. We first note that

$$v(\pm 1,0) = \mathrm{Im}(w(\pm 1)) = 0 \tag{7.1.24}$$

and therefore the paths of steepest descent and ascent from $z = \pm 1$ must lie along the curves

$$v(x, y) = \mathrm{Im}(w) = -y\left(x^2 - \frac{y^2}{3} - 1\right) = 0. \tag{7.1.25}$$

Hence these paths, which are depicted in Figure 7.1.4 consist of the straight line $y = 0$ and the two branches of the hyperbola $x^2 - y^2/3 = 1$.

We naturally want to distinguish among the ascent and descent paths. To do this we need only use Table 7.1 with $n = 2$ to determine the steepest descent and ascent directions at the saddle points. From (7.1.23) we have that

$$\arg w''(\pm 1) = \alpha_\pm, \qquad \alpha_+ = \pi, \qquad \alpha_- = 0. \tag{7.1.26}$$

We can now fill in Table 7.1 for each of the saddle points. Indeed for the saddle point at $z = 1$ we find that the directions of steepest descent are given by $\arg(z - 1) = 0, \pi$ so that the paths of steepest descent lie along the line $y = 0$. We also find that the directions of steepest ascent at $z = 1$ are $\arg(z - 1) = \pm\pi/2$ and hence the branch of the hyperbola $x^2 - y^2/3 = 1$ through $z = 1$ is to be

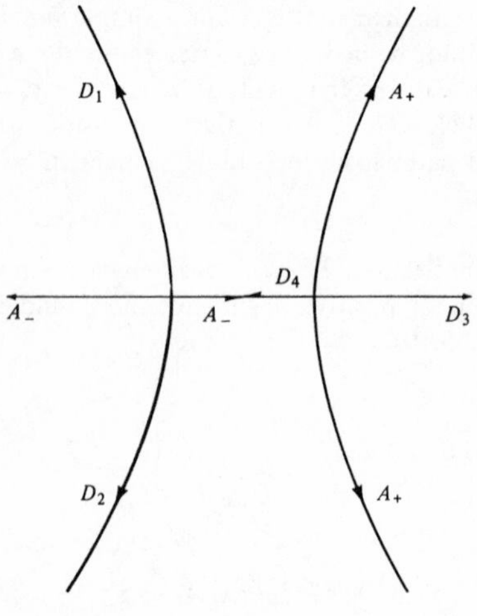

Figure 7.1.4.

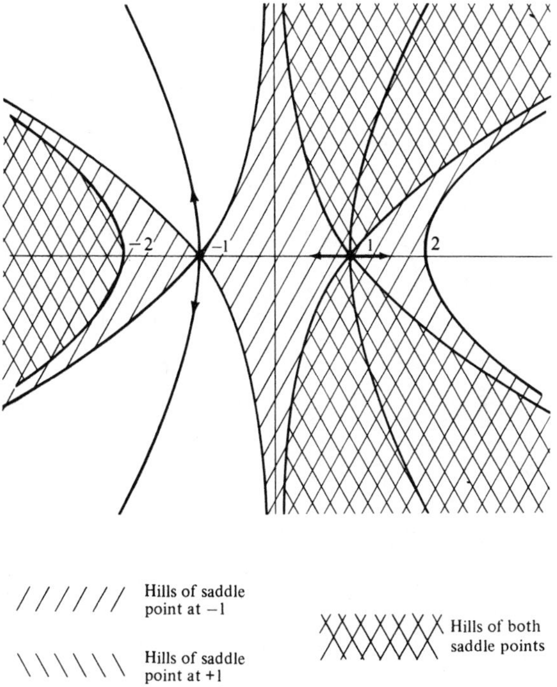

Figure 7.1.5. Hills and Valleys of $z - z^3/3$.

viewed as two semiinfinite paths of steepest ascent away from the saddle point.

For the saddle point at $z = -1$, $\alpha = 0$ and the directions of steepest descent are $\arg(z+1) = \pi/2,\ 3\pi/2$. Hence, the branch of the hyperbola $x^2 - y^2/3 = 1$ through $z = -1$ are two semi-infinite paths of steepest descent away from this saddle point. Also, the directions of steepest ascent in this case are along the line $y = 0$.

We note that the line segment $y = 0$, $|x| < 1$ is a path of steepest descent from the saddle point at $z = 1$ and a path of steepest ascent for the saddle point at $z = -1$. Upon reflection, we realize that this is no contradiction. Indeed, if we consider any two points along a path of constant v, then the connecting path will be a steepest descent curve from one point and a steepest ascent curve from the other. We also observe that the path of steepest descent from $z = 1$, directed toward $z = -1$ has a nonunique continuation away from $z = -1$. In fact, whenever a steepest descent path goes through a saddle point, it may continue along any of the steepest descent paths from that saddle

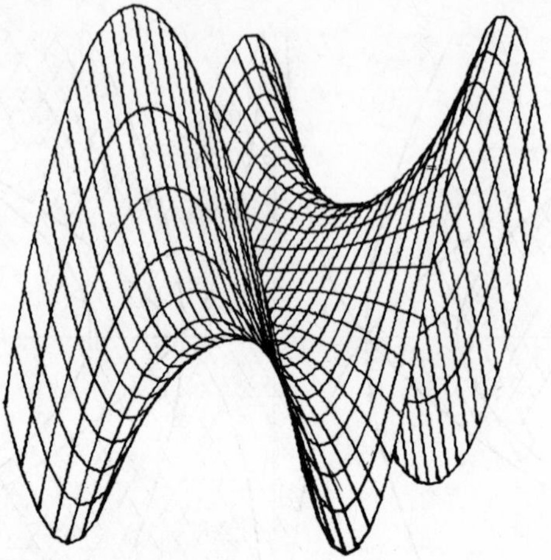

Figure 7.1.6. The Surface

$$u = Re\left(z - \frac{z^3}{3}\right) = x\left(y^2 - \frac{x^2}{3} + 1\right),$$

Viewed from a Point in the Third Quadrant of the (x, y) Plane. (Computer plot by PUREJOY, provided by ERDA Mathematics and Computing Laboratory at Courant Institute of Mathematical Sciences at New York University and modified for implementation at the University of Denver by Louis Krupp.)

point. Finally in Figure 7.1.5 we depict the hills and valleys of the function (7.1.21) with respect to the saddle points.

7.2. The Method of Steepest Descents

We wish now to study, in detail, the asymptotic behavior of (7.1.1) as $\lambda \to +\infty$. The technique to be employed is known as the *method of steepest descents* and consists of the following five basic steps:

(1) Identify the possible critical points of the integrand. These are the endpoints of integration, singular points of $g(z)$ or $w(z)$ and saddle points of $w(z)$.

(2) Determine the paths of steepest descent from each of the critical points.

(3) Justify, via Cauchy's integral theorem, the deformation of the original contour of integration C onto one or more of the paths of steepest descent found in (2).

(4) Determine the asymptotic expansions of the integrals that arise as a result of the deformation in (3). (Note that each integral is of Laplace type.)

(5) Sum the asymptotic expansions obtained to determine the asymptotic expansion of $I(\lambda)$.

We shall, of course, expand on the above outline and, in fact, shall analyze each of the steps in great detail. Before doing so however, we wish to make some general remarks. As we might expect, although step (2) offers no conceptual difficulties, it is often tedious to carry out. It may happen that we can judge in advance that certain critical points have steepest descent contours which will not be involved in the ultimate deformation of C described in step (3). In that event we need not identify these contours at all. Thus, whenever possible, steps (2) and (3) should be applied simultaneously rather than in strict sequence. We shall elaborate on this point below. (See, in particular, Example 7.2.1.)

The importance of step (3) cannot be overemphasized. Unfortunately, it is not only the pivotal step in the analysis, it is also often the most difficult to apply. This last statement might appear strange, because the actual asymptotic expansion is derived in steps (4) and (5). We consider these latter steps simple, however, because the theory for Laplace-type integrals has already been fully developed in Chapters 4 and 5. Also we have previously alluded to the computational difficulty of step (2). As we shall see, we do not need detailed information about the steepest descent contours in the large. In fact, quantitative information about these contours is needed only near the critical points themselves. Away from the critical points, qualitative information will prove sufficient.

Suppose then that the critical points identified in step (1) are $z_0, z_1, \ldots, z_n$. Then from our theory for Laplace-type integrals, we know that the contribution to the asymptotic expansion of $I(\lambda)$ corresponding to $z = z_i$ is $O(\exp\{\lambda \operatorname{Re}(w(z_i))\}\, \lambda^{\alpha_i})$ for some finite α_i. We might argue that only those terms should be retained which correspond to values of i for which $\operatorname{Re}(w(z_i))$ is maximal, because all others are asymptotically negligible compared to them. We recall, however, that in many applications, asymptotically negligible terms are associated with physical phenomena of interest and hence should be retained and studied. Furthermore, in the analysis of Stokes phenomenon, such terms often play a crucial role.

We now wish to study the implementation of our five basic steps. Steps (1) and (5) are trivial and need no special discussion. Step (2) has already been considered to a large extent in Section 7.1. This leaves steps (3) and (4). We shall begin with (4), the easier of the two. Thus suppose that in (7.1.1) C is already a path of steepest descent from a critical point $z = \hat{z}_0$ to some other point $z = \hat{z}$. In most cases of practical interest $\hat{z}$ is the point at ∞ and $u = \operatorname{Re}(w)$ has the limit $-\infty$ as $z \to \infty$ along the path of steepest descent.

Let us for the moment suppose that $z = z_0$ is a saddle point of order $n - 1$. Hence (7.1.8) holds and the directions of steepest descent are determined, as indicated in Table 7.1, by the n distinct angles

$$\theta = -\frac{\alpha}{n} + (2p+1)\frac{\pi}{n}, \qquad p = 0, 1, \ldots, n-1. \tag{7.2.1}$$

If we make the change of variable in (7.1.1) defined by

$$t = -[w(z) - w(z_0)], \tag{7.2.2}$$

then we find that the path of steepest descent C is mapped onto the positive real axis. Indeed we have

$$I(\lambda) = \exp(\lambda w(z_0)) \int_0^\infty G(t) \exp(-\lambda t)\, dt, \tag{7.2.3}$$

where

$$G(t) = g(z)\frac{dz}{dt} = -\frac{g(z)}{w'(z)}\bigg|_{z = w^{-1}(w(z_0) - t)} \tag{7.2.4}$$

We see from (7.2.3) that our problem has been reduced to the analysis of an integral of Laplace type. Hence we may use the results of Chapter 4. Indeed, if we assume that as $t \to 0+$,

$$G(t) \sim \sum_{m=0}^{\infty} \sum_{k=0}^{\overline{N}(m)} p_{mk}\, t^{a_m} (\log t)^k \tag{7.2.5}$$

subject to the conditions following (4.3.15), then it follows from (4.4.24) that

$$I(\lambda) \sim \exp(\lambda\, w(z_0)) \sum_{m=0}^{\infty} \lambda^{-(a_m+1)} \sum_{k=0}^{\overline{N}(m)} p_{mk} \sum_{j=0}^{k} \binom{k}{j} (-\log \lambda)^j \times \left(\frac{d}{dz}\right)^{k-j} \Gamma(z)\bigg|_{z=a_m+1} \tag{7.2.6}$$

As always, it is desirable to relate the constants in (7.2.6) to the original integrand functions $w(z)$ and $g(z)$. Let us do this now for the leading term in the case where, as $z \to z_0$ along C

$$g(z) = g_0(z-z_0)^{\beta-1} + o(z-z_0)^{\beta-1}, \qquad \operatorname{Re}(\beta) > 0. \tag{7.2.7}$$

We then find after some computation that, in (7.2.5),

$$a_0 = \frac{\beta}{n} - 1, \qquad p_{0_k} = 0, \qquad n \geq 1, \qquad p_{00} = \frac{g_0}{n}\left(\frac{n!}{|w^{(n)}(z_0)|}\right)^{\beta/n} \exp(i\beta\theta) \tag{7.2.8}$$

with θ given by (7.2.1). In determining p_{00} we have used the relations

$$t = \frac{1}{n!}|w^{(n)}(z_0)|\,|z-z_0|^n\,[1 + 0(|z-z_0|)],$$

$$z - z_0 = t^{1/n}\left(\frac{n!}{|w^{(n)}(z_0)|}\right)^{1/n} \exp(i\theta)\,[1 + O(t^{1/n})],$$

which follow from (7.1.8) and (7.2.2). We have taken care to define the phase angle of -1 so that $\arg(t) = 0$ on the path of integration as required.

It now follows from (7.2.6) and (7.2.8) that to leading order

$$I(\lambda) \sim \frac{g_0}{n}\left(\frac{n!}{\lambda|w^{(n)}(z_0)|}\right)^{\beta/n} \Gamma\left(\frac{\beta}{n}\right)\exp\left[\lambda\, w(z_0) + i\beta\left([2p+1]\frac{\pi}{n} - \frac{\alpha}{n}\right)\right].^{5} \quad (7.2.9)$$

We might point out that (7.2.9) remains valid even when n is not an integer (that is, when w has a branch point at $z = z_0$) if $n!/|w^{(n)}(z_0)|$ and α are replaced by the appropriate constants in the expansion of δw along the contour C.

Two special cases arise so often that it is worthwhile to exhibit separate formulas for them. Thus, if at $z = z_0$, w has a simple saddle point, that is, $n = 2$ and $g(z)$ is regular, then (7.2.9) becomes

$$I(\lambda) \sim g(z_0)\sqrt{\frac{\pi}{2\lambda|w''(z_0)|}}\exp\left[\lambda\, w(z_0) + i\left((2p+1)\frac{\pi}{2} - \frac{\alpha}{2}\right)\right],^{5} \qquad p = 0, 1; \quad (7.2.10)$$

while, if at $z = z_0$, $w(z)$ has no saddle point and is regular, that is, $n = 1$, and g has an algebraic branch point, then

$$I(\lambda) \sim \frac{g_0\,\Gamma(\beta)}{(|w'(z_0)|\lambda)^{\beta}}\exp[\lambda\, w(z_0) + i\beta(\pi - \alpha)].^{5} \quad (7.2.11)$$

We note that the various asymptotic expansions just obtained depend only on the local properties of $G(t)$ near $t = 0+$ or, equivalently, the local properties of w and g near $z = z_0$ in a sector containing the particular direction of steepest descent. This observation has certain important implications. Firstly, we need only know detailed information about the path of steepest descent near the critical point in question. In fact, suppose that C_1 is a contour that coincides with the original steepest descent contour C for some finite length starting from $z = z_0$, but then continues merely as a descent contour. Then the integral along C_1 would have the same asymptotic expansion as that derived above for the integral along C, differing at most from the latter by an exponentially smaller quantity.[6] We recall, however, that such "small" terms may have a physical significance and hence may be of interest. Modulo such cases we shall say that C_1 and C are *asymptotically equivalent contours.*

Suppose now that C_2 is *any* descent contour from $z = z_0$ and initially in the same valley as C. Suppose further that the sectors of validity of the approximations for w and g used above contain a finite length of C_2 from $z = z_0$. Then we can conclude that C and C_2 are asymptotically equivalent contours because, by Cauchy's integral theorem, C_2 can be deformed onto C near $z = z_0$ thereby arriving at a contour of the type C_1 described above.

Again we must stress the fact that we have not, as yet, considered the most difficult task in the analysis. Indeed, although it is relatively simple to deter-

[5] We point out that the asymptotic validity of these results depends on (7.2.6).

[6] This is the basis of the so-called *saddle point method.*

mine directions of steepest descent from various critical points and to determine valleys of $w(z)$ with respect to these points, it is, in general, not so simple to justify the replacement of an original contour C by one or more paths, each being asymptotically equivalent to a steepest descent contour from a critical point. In fact, the only reason we need to know anything about the steepest descent contours away from the critical points is to accomplish this justification. These latter issues shall be clarified in the examples of the present and subsequent sections. It is our belief that the finer points of the method of steepest descents can best be learned by careful study of examples such as those to be presented here.

EXAMPLE 7.2.1. Here we shall consider the Airy functions originally introduced in Section 7.1. Much of the groundwork for our analysis has been laid in that section. We begin by repeating the integral representations (2.5.9):

$$f_n(s) = \frac{1}{2\pi i} \int_{C_n} \exp\left(-\frac{z^3}{3} + sz\right) dz, \qquad n = 1, 2, 3, \tag{7.2.12}$$

which are related through (2.5.11), (2.5.12) to the Airy functions Ai and Bi. In (7.2.12) the contours C_n are as shown in Figure 2.5. We seek the asymptotic expansion of $f_n(s)$ as $s \to +\infty$. We observe that the exponent in (7.2.12) is not in the "canonical" form $\lambda w(z)$. To place it in that form, however, we need only introduce the "stretched" variable of integration $\sqrt{s}\, z$ which yields

$$f_n(s) = I_n(\lambda) = \frac{\lambda^{1/3}}{2\pi i} \int_{C_n} \exp[\lambda\, w(z)]\, dz. \tag{7.2.13}$$

Here

$$w(z) = z - \frac{z^3}{3}, \qquad \lambda = s^{3/2} \tag{7.2.14}$$

and the contours C_n can, by Cauchy's theorem, remain unchanged.

In Section 7.1 we have developed all of the information necessary for the implementation of steps (1) and (2) of our basic procedure. In particular, we call attention to Figure 7.1.4. We shall also require the information contained in Table 7.2 which, in turn, is obtained from (7.1.23) and the discussion following (7.1.25).

We now wish to replace the contours C_n by combinations of steepest descent contours from the two saddle points $z = \pm 1$. [Note that because the contours C_n are infinite in extent and because $w(z)$ is an entire function, there are no other critical points.] We can use Table 7.2 to aid us in this task. For example, we know that along C_1 (or equivalently along the imaginary axis) $\mathrm{Re}(w(z)) \le 0$, while $w(+1) = \frac{2}{3}$. It then follows that the saddle point $z = 1$ *cannot* contribute to the asymptotic expansion of $I_1(\lambda)$ because any such contribution would perforce be exponentially *larger* than the integrand all along C_1. From this we can conclude that the replacement of C_1 should not involve any contours

Table 7.2

Saddle Point	*Descent Path*	$w(z)$	$w''(z)$	θ
$z = +1$	D_3	$\frac{2}{3}$	-2	0
$z = +1$	D_4	$\frac{2}{3}$	-2	π
$z = -1$	D_1	$-\frac{2}{3}$	2	$\frac{\pi}{2}$
$z = -1$	D_2	$-\frac{2}{3}$	2	$-\frac{\pi}{2}$

emanating from $z = 1$. In general, given an original contour C along which $\text{Re}(w) \le a$, then no critical point $z = z_0$ at which $\text{Re}(w(z_0)) > a$ can contribute to the asymptotic expansion of the integral along C. We say that such a critical point is *inadmissible.*

The utility of these last observations will not be apparent in the present example because, as we shall soon see, step (3) of our procedure is readily carried out. However, in cases where there are many critical points and, in particular, saddle points of higher order than the first, the concept of inadmissibility is quite useful in that it eliminates certain steepest descent contours from consideration.

Upon considering Figures 7.1.4 and 7.1.5, we find that we can replace C_1 by $D_1 - D_2$, C_2 by $D_3 - D_1 - D_4$, and C_3 by $D_2 + D_4 - D_3$. In arriving at these conclusions we have, of course, used Cauchy's integral theorem and the fact that, in each case, the contours of steepest descent go to infinity in the same valley as does the original contour C_i. Although this completes step (3), we must warn the reader that the implementation of this step is not always so simple.

We are now prepared to apply steps (4) and (5). We shall be content here with obtaining leading terms only. Let us first introduce

$$I^j(\lambda) = \frac{\lambda^{1/3}}{2\pi i} \int_{D_j} \exp[\lambda w(z)]\, dz \tag{7.2.15}$$

so that

$$\begin{aligned} I_1(\lambda) &= I^1(\lambda) - I^2(\lambda), \\ I_2(\lambda) &= I^3(\lambda) - I^1(\lambda) - I^4(\lambda), \\ I_3(\lambda) &= I^2(\lambda) + I^4(\lambda) - I^3(\lambda . \end{aligned} \tag{7.2.16}$$

Because each of the saddle points is simple and because $g(z) \equiv 1$ we may use (7.2.10) and Table 7.2 to conclude that, as $\lambda \to \infty$,

$$I^1(\lambda) \sim \frac{\lambda^{-1/6}}{4\sqrt{\pi}} \exp\left[-\frac{2\lambda}{3}\right]; \qquad I^2(\lambda) \sim -\frac{\lambda^{-1/6}}{4\sqrt{\pi}} \exp\left[-\frac{2\lambda}{3}\right];$$

$$I^3(\lambda) \sim \frac{\lambda^{-1/6}}{4i\sqrt{\pi}} \exp\left[\frac{2\lambda}{3}\right]; \qquad I^4(\lambda) \sim -\frac{\lambda^{-1/6}}{4i\sqrt{\pi}} \exp\left[\frac{2\lambda}{3}\right]. \tag{7.2.17}$$

It then follows from (7.2.16) and (7.2.17) that

$$\begin{aligned} I_1(\lambda) &\sim \frac{\lambda^{-1/6}}{2\sqrt{\pi}} \exp\left[-\frac{2\lambda}{3}\right], \\ I_2(\lambda) &\sim \frac{\lambda^{-1/6}}{2i\sqrt{\pi}} \exp\left[\frac{2\lambda}{3}\right], \\ I_3(\lambda) &\sim -\frac{\lambda^{-1/6}}{2i\sqrt{\pi}} \exp\left[\frac{2\lambda}{3}\right]. \end{aligned} \tag{7.2.18}$$

Finally, the asymptotic expansions of the Airy functions themselves are obtained from (2.5.10), (2.5.12), (7.2.13), and (7.2.18) which yield

$$\begin{aligned} \mathrm{Ai}(s) &\sim \frac{s^{-1/4}}{2\sqrt{\pi}} \exp\left[-\frac{2s^{3/2}}{3}\right], \\ \mathrm{Bi}(s) &\sim \frac{s^{-1/4}}{\sqrt{\pi}} \exp\left[\frac{2s^{3/2}}{3}\right], \end{aligned} \tag{7.2.19}$$

as $s \to +\infty$.

EXAMPLE 7.2.2. The Hankel function of type j, of argument kr, and of order ka, is denoted by $H_{ka}^{(j)}(kr)$ and has the integral representation

$$H_{ka}^{(j)}(kr) = \frac{1}{\pi} \int_{C_j} \exp\left\{ik\left[r\cos z + a\left(z - \frac{\pi}{2}\right)\right]\right\} dz, \qquad j = 1, 2. \tag{7.2.20}$$

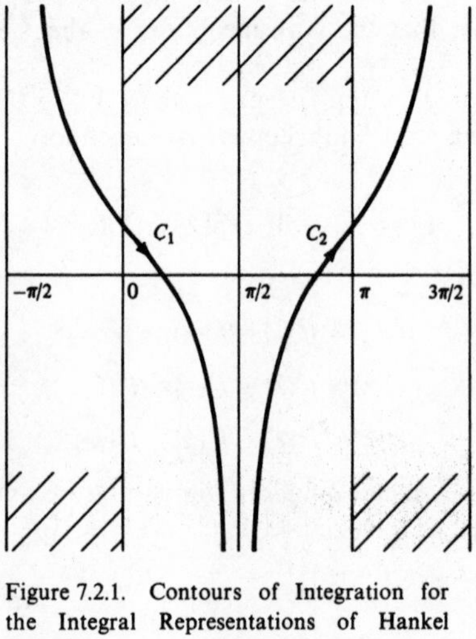

Figure 7.2.1. Contours of Integration for the Integral Representations of Hankel Functions.

Here the contours C_j are as depicted in Figure 7.2.1 and, in particular, extend to ∞ in the unshaded regions of that diagram. We have expressed the argument and order by kr and ka respectively because such forms naturally arise in applications.

The main purpose of this example is to illustrate how asymptotic expansions can be obtained via the method of steepest descents when quantitative information about the paths of steepest descent is known only near the critical points. Here we shall seek the asymptotic expansions of the two Hankel functions as $k \to \infty$. In the important application of wave propagation, this corresponds to what is called the "high-frequency limit."

The integrals (7.2.20) can be cast in the form (7.1.1) by introducing

$$\lambda = kr, \qquad \beta = \frac{a}{r}, \tag{7.2.21}$$

in which event we have

$$H_{ka}^{(j)}(kr) = I_j(\lambda;\beta) = \frac{1}{\pi}\int_{C_j} \exp[\lambda\, w(z;\beta)]\, dz, \qquad j = 1, 2. \tag{7.2.22}$$

Here

$$w(z;\beta) = i\left[\cos z + \beta\left(z - \frac{\pi}{2}\right)\right]. \tag{7.2.23}$$

In most problems a and r have the dimension of length, while k has the dimension of $(\text{length})^{-1}$. Thus, λ and β are both dimensionless. In this example we shall assume that $\lambda \to \infty$ while

$$0 < \beta < 1. \tag{7.2.24}$$

Furthermore, for convenience we shall henceforth suppress the explicit dependence of I_j and w on β.

It is clear from (7.2.23) and the definitions of the contours C_j that the only critical points are the saddle points of w. Because

$$w'(z) = i[-\sin z + \beta]; \qquad w''(z) = -i\cos z \tag{7.2.25}$$

we have that, in the strip $-\pi/2 < \mathrm{Re}(z) < 3\pi/2$, w has two simple saddle points, denoted by $z = z_+, z_-$ and defined by

$$\sin z_\pm = \beta, \qquad 0 < z_+ < \frac{\pi}{2}, \qquad \frac{\pi}{2} < z_- = \pi - z_+ < \pi. \tag{7.2.26}$$

There are additional saddle points on the real axis located at the points $z = z_\pm \pm 2n\pi$, $n = 1, 2, 3, \ldots$. We claim, however, that they can be ignored, but shall delay establishing this claim until later on in the analysis.

If we define the inverse function $\sin^{-1}\beta$ so that

$$0 < \sin^{-1}\beta < \frac{\pi}{2} \qquad \text{or} \qquad z_+ = \sin^{-1}\beta, \tag{7.2.27}$$

then it follows that

$$w(z_\pm) = \pm i\left[\sqrt{1-\beta^2} + \beta\left(\sin^{-1}\beta - \frac{\pi}{2}\right)\right]$$
$$= \pm i[\sqrt{1-\beta^2} - \beta\cos^{-1}\beta];$$
$$w''(z_\pm) = \mp i\sqrt{1-\beta^2}. \tag{7.2.28}$$

Here we have taken

$$0 < \cos^{-1}\beta < \frac{\pi}{2}. \tag{7.2.29}$$

Now upon applying Theorem 7.1 we find from (7.2.28) and Table 7.1 that the directions of steepest descent at the two saddle points are

$$\theta(z_+) = -\frac{\pi}{4}, \frac{3\pi}{4}, \qquad \theta(z_-) = \frac{\pi}{4}, -\frac{3\pi}{4}. \tag{7.2.30}$$

The paths of steepest descent from the saddle points are those curves defined by

$$v(x,y) = \text{Im}(w) = \cos x \cosh y + \beta\left(x - \frac{\pi}{2}\right)$$
$$= \pm(\sqrt{1-\beta^2} - \beta\cos^{-1}\beta) = \text{Im}(w(z_\pm)) \tag{7.2.31}$$

along which

$$u(x,y) = \text{Re}(w) = \sin x \sinh y - \beta y \tag{7.2.32}$$

decreases monotonically.

Although we cannot draw the steepest descent paths here as precisely as we could the cubics that arose in the analyses of the Airy functions, we can nevertheless obtain enough qualitative information to carry out our procedure. We observe from (7.2.32) that u continually decreases as $|y| \to \infty$ so long as

$$\left|x - \frac{\pi}{2}\right| < \frac{\pi}{2}; \qquad y \to -\infty,$$
$$\frac{3\pi}{2} \geq \left|x - \frac{\pi}{2}\right| \geq \frac{\pi}{2}; \qquad y \to +\infty. \tag{7.2.33}$$

Also, in order that the left-hand side of (7.2.31) remain finite as $|y| \to \infty$, it is necessary that x approach an odd multiple of $\pi/2$. Hence, the vertical lines $x = \pm\pi/2$, $x = 3\pi/2$ are the asymptotes of the paths of steepest descent (and of the paths of steepest ascent as well).

The information we have obtained is sufficient to complete Figure 7.2.2 which depicts the saddle points under consideration and the paths of steepest descent and steepest ascent from them. Also shown are the hills and valleys of u with

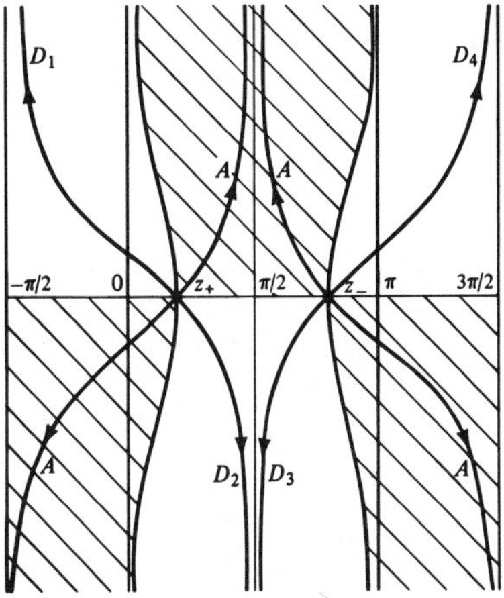

Figure 7.2.2. Hills (Shaded) and Valleys for $w(z) = i[\cos z + \beta(z - \pi/2)]$, $\beta < 1$.

respect to these saddle points. They, of course, are bounded by the curves $u(x, y) = \text{Re}(w(z_{\pm}))$. The qualitative information necessary to approximate these boundary curves is obtained in a manner similar to the analysis above.

Our next task is to replace the contours C_1 and C_2 by linear combinations of steepest descent paths from the saddle points. Upon referring to Figures 7.2.1 and 7.2.2 we readily find that we should replace C_1 by $D_2 - D_1$ and C_2 by $D_4 - D_3$. Aside from their local behavior near the saddle points, we need no further information about the paths of steepest descent to obtain the complete asymptotic expansions of the integrals I_j.

All of the information needed to find the leading terms of the desired expansions has already been obtained. Moreover, the computations are rather straightforward, so that we shall only present the results here. Indeed we have

$$I_j(\lambda) \sim \sqrt{\frac{2}{\pi\lambda}} \frac{\exp\left\{(-1)^{j+1} i\left[\lambda(\sqrt{1-\beta^2} - \beta\cos^{-1}\beta) - \dfrac{\pi}{4}\right]\right\}}{(1-\beta^2)^{1/4}}, \qquad j = 1, 2. \tag{7.2.34}$$

In terms of the original variables, (7.2.34) becomes

$$H_{ka}^{(j)}(kr) \sim \sqrt{\frac{2}{\pi k}} \frac{\exp\left\{(-1)^{j+1} i\left[k\sqrt{r^2 - a^2} - ka\cos^{-1}\left(\dfrac{a}{r}\right) - \dfrac{\pi}{4}\right]\right\}}{(r^2 - a^2)^{1/4}}, \frac{a}{r} < 1, j = 1, 2. \tag{7.2.35}$$

We note that we can also write

$$H_\nu^{(j)}(\rho) \sim \sqrt{\frac{2}{\pi}} \frac{\exp\left\{(-1)^{j+1} i\left[\sqrt{\rho^2 - \nu^2} - \nu \cos^{-1}\left(\frac{\nu}{\rho}\right) - \frac{\pi}{4}\right]\right\}}{(\rho^2 - \nu^2)^{1/4}}; \quad \frac{\nu}{\rho} < 1, \; j = 1, 2, \tag{7.2.36}$$

which is the leading term of the asymptotic expansion of $H_\nu^{(j)}(\rho)$ as *both* ν and $\rho \to \infty$ with $\nu < \rho$. We might further point out that, to leading order, the asymptotic expansion of $H_\nu^{(j)}(\rho)$ for finite order and large argument can be recovered from (7.2.36) by assuming ν fixed and ρ large, so long as we set

$$\lim_{\rho \to \infty} \cos^{-1}\left(\frac{\nu}{\rho}\right) = \frac{\pi}{2}. \tag{7.2.37}$$

From (7.2.34)–(7.2.36) we see that the results we have obtained are invalid when the order and argument of the Hankel functions are equal, that is, when $r/a = 1$ or $\nu = \rho$. In that event, we have $z_+ = z_- = \pi/2$, while $w''(\pi/2) = 0$. Thus, if $r/a = 1$, instead of having two distinct saddle points of first order, we have one saddle point of higher order. All of the above calculations, however, have been based on the assumption that there are two distinct saddle points. This explains the breakdown of our results when $r = a$. We naturally want to treat this anomalous case and we shall do so in the following example.

There is still one point that remains to be discussed and that is the question of the ignored saddle points $z = z_\pm \pm 2n\pi$, $n = 1, 2, 3, \ldots$. To understand why the ignoring of these saddle points was not only valid but indeed necessary, we first point out that the configuration of hills, valleys, and paths of steepest descent and ascent corresponding to the saddle points $z_\pm + 2n\pi$ for any fixed integer n, is precisely that given in Figure 7.2.2 with the values of x shifted by $2n\pi$. The original contours C_1 and C_2, however, are confined to the strip $-\pi/2 < x < 3\pi/2$. It should be clear that it is impossible to replace either C_1 or C_2 by a combination of steepest descent contours which involves any emanating from the saddle points $z_+ \pm 2n\pi$, $n = 1, 2, \ldots$ except in a trivial manner. This is so because to accomplish this at least one hill of (7.2.32) would have to be crossed "at infinity"; an operation which cannot be justified by Cauchy's theorem. Indeed, we can say that with C_1 and C_2 as originally described, all saddle points other than $z = z_\pm$ are inadmissible and must be ignored.

EXAMPLE 7.2.3. We shall now consider the Hankel functions when the order and argument are equal and large. Hence we set $\beta = 1$ in (7.2.21) so that

$$w(z) = i\left[\cos z + \left(z - \frac{\pi}{2}\right)\right], \qquad w'(z) = i[-\sin z + 1],$$

$$w''(z) = -i \cos z, \qquad w'''(z) = i \sin z. \tag{7.2.38}$$

We now find that there are no longer two distinct saddle points, but rather a single saddle point of order 2 located at $z = \pi/2$. Indeed we have

$$w\left(\frac{\pi}{2}\right) = w'\left(\frac{\pi}{2}\right) = w''\left(\frac{\pi}{2}\right) = 0, \qquad w'''\left(\frac{\pi}{2}\right) = i \tag{7.2.39}$$

so that in (7.1.8), $n = 3$, and $\alpha = \pi/2$. It then follows from Table 7.1 that the directions of steepest descent are

$$\theta = \frac{\pi}{6}, \frac{5\pi}{6}, -\frac{\pi}{2}. \tag{7.2.40}$$

The steepest descent paths are as depicted in Figure 7.2.3 from which we readily conclude that we may replace C_1 by $D_2 - D_1$ and C_2 by $D_3 - D_2$. Then a simple calculation yields

$$I_j(\lambda; 1) \sim -\frac{\Gamma(\frac{1}{3})}{\pi\, \lambda^{1/3}}\left(\frac{4}{3}\right)^{1/6} \exp\left[(-1)^{j+1}\frac{2\pi i}{3}\right]; \qquad j = 1, 2 \tag{7.2.41}$$

or, equivalently,

$$H_{ka}^{(j)}(ka) \sim -\frac{\Gamma(\frac{1}{3})}{\pi\, (ka)^{1/3}}\left(\frac{4}{3}\right)^{1/6} \exp\left[(-1)^{j+1}\frac{2\pi i}{3}\right]; \qquad j = 1, 2. \tag{7.2.42}$$

From (7.2.34), we see that $I_j(\lambda;\beta) = O(\lambda^{-1/2})$ when $0 < \beta < 1$, while from (7.2.41) it follows that $I_j(\lambda;1) = O(\lambda^{-1/3})$. Evidently there is a transition in order as $\beta \to 1$ which is not completely described by our results. It would be desirable to have a single expansion that remains uniformly valid for all β in $[0, 1]$. Such an expansion will indeed be obtained in Chapter 9 where the general subject of uniform asymptotic expansions is discussed.

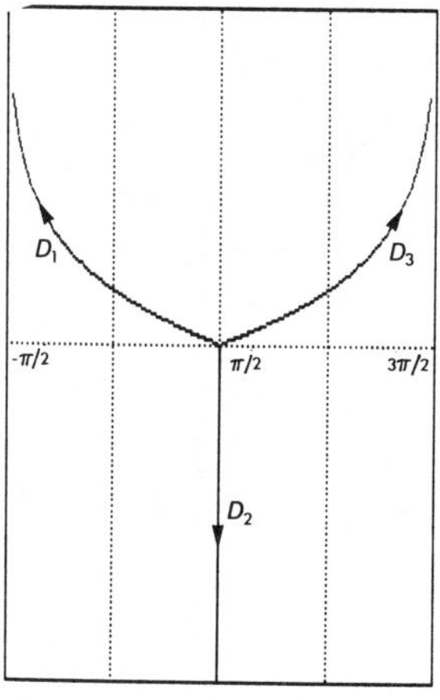

Figure 7.2.3. Paths of Steepest Descent for $w(z) = i[\cos z + z - \pi/2]$, $\beta = 1$.

To complete our analysis of the Hankel function, let us consider the following.

EXAMPLE 7.2.4. If in (7.2.21) we take $\beta > 1$, so that the order of the Hankel function is greater than the argument, then we find that there are again two simple saddle points $z = z_{\pm}$ in the strip $-\pi/2 < \mathrm{Re}(z) < 3\pi/2$. These are defined by

$$z_{\pm} = \frac{\pi}{2} \pm i \cosh^{-1} \beta. \tag{7.2.43}$$

The steepest descent paths from the saddle points are shown in Figure 7.2.4. As is readily seen, we may replace C_1 by $D_3 - D_2 - D_1$ and C_2 by $D_4 + D_2 - D_3$. We also note that the contributions from $z = z_+$ are exponentially smaller than those from $z = z_-$ and hence can be ignored. Indeed, after some calculation we find that to leading order

$$I_j(\lambda;\beta) \sim \sqrt{\frac{2}{\pi\lambda}} \frac{\exp\left\{\lambda\left[\beta \cosh^{-1}\beta - \sqrt{\beta^2 - 1}\right] + (-1)^j \frac{i\pi}{2}\right\}}{(\beta^2 - 1)^{1/4}}; \qquad j = 1, 2, \qquad \beta > 1. \tag{7.2.44}$$

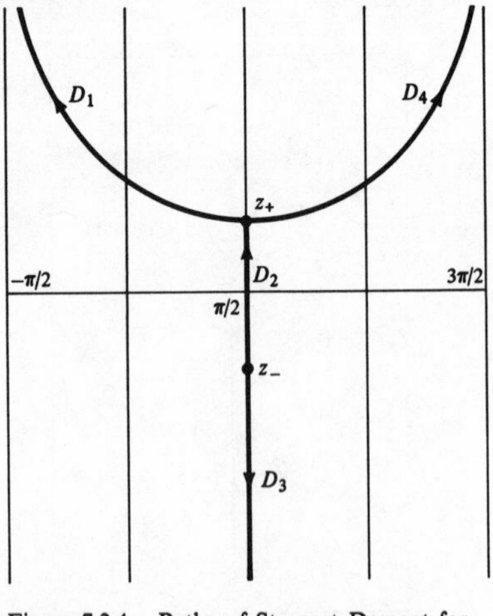

Figure 7.2.4. Paths of Steepest Descent for $w(z) = i[\cos z + \beta(z - \pi/2)]$, $\beta > 1$.

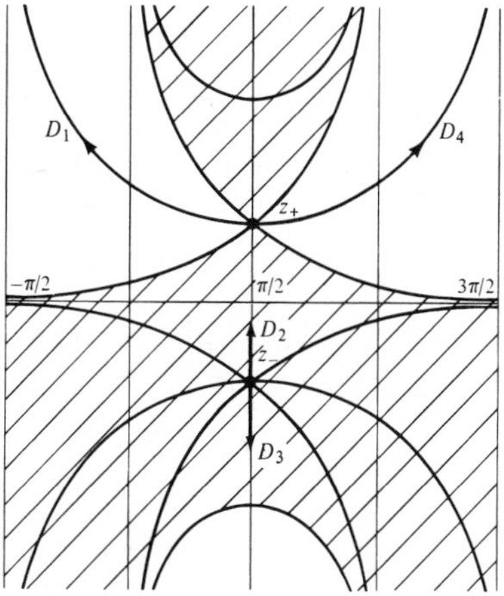

Figure 7.2.4.(a) Hills (Shaded) and Valleys with Respect to the Saddle Point z_+ for $w(z) = i[\cos z + \beta(z - \pi/2)]$, $\beta > 1$.

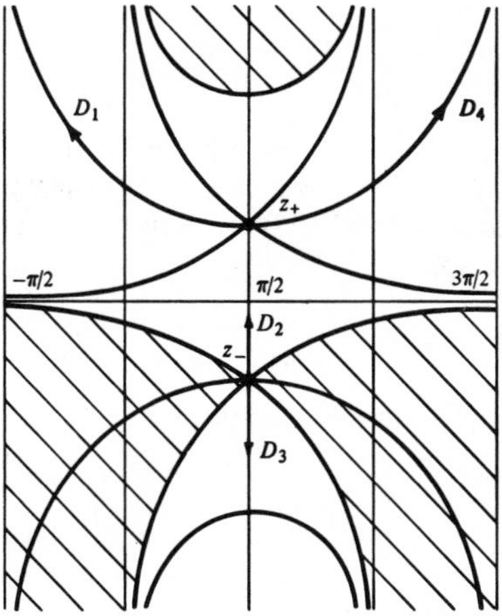

Figure 7.2.4.(b) Hills (Shaded) and Valleys with Respect to the Saddle Point at z_- for $w(z) = i[\cos z + \beta(z - \pi/2)]$, $\beta > 1$.

From (7.2.44), it follows that

$$H_{ka}^{(j)}(kr) \sim \sqrt{\frac{2}{\pi k}} \frac{\exp\left\{ka \cosh^{-1}\left(\frac{a}{r}\right) - k\sqrt{a^2 - r^2} + (-1)^j \frac{\pi i}{2}\right\}}{(a^2 - r^2)^{1/4}},$$

$$j = 1, 2, \qquad \frac{a}{r} > 1 \tag{7.2.45}$$

and

$$H_\nu^{(j)}(\rho) \sim \sqrt{\frac{2}{\pi}} \frac{\exp\left\{\nu \cosh^{-1}\left(\frac{\nu}{\rho}\right) - \sqrt{\nu^2 - \rho^2} + (-1)^j \frac{\pi i}{2}\right\}}{(\nu^2 - \rho^2)^{1/4}};$$

$$j = 1, 2, \qquad \frac{\nu}{\rho} > 1. \tag{7.2.46}$$

To illustrate how the method of steepest descents is applied to integrals over contours of finite extent let us consider the following.

EXAMPLE 7.2.5. Here we shall show that, in certain instances, the stationary phase results obtained in Section 6.1 can be derived via the method of steepest descents. Thus, let us consider the Fourier-type integral

$$I(\lambda) = \int_a^b g(t) \exp(i\lambda\, f(t))\, dt. \tag{7.2.47}$$

We shall assume that a and b are real and finite while both f and g are *entire* functions of the complex variable t. Furthermore, we shall assume that $f(t)$ is real for t in $[a,b]$.

Let us suppose for simplicity that

$$f'(t_0) = 0, \qquad f''(t_0) \neq 0, \qquad a < t_0 < b \tag{7.2.48}$$

and that $f'(t)$ does not vanish at any other point in the complex t plane. In the notation of Chapter 6, $t = t_0$ is a simple stationary point of f. Here, however, we choose to look upon (7.2.47) as a contour integral so that the results of this section can be applied.

If we set

$$i\, f(t) = w(t), \tag{7.2.49}$$

then we have that $w(t)$ has a simple saddle point at $t = t_0$. Because there are no other saddle points, the set of critical points consists of $t = t_0$, $t = a$, and $t = b$.

Our first task is to determine the directions of steepest descent from each of the critical points along with the configuration of the hills and valleys of $u = \mathrm{Re}(w)$ at infinity. From Table 7.1 we find that the directions of steepest descent at $t = t_0$ are given by

$$\arg(t - t_0) = -\frac{\mu\pi}{4} + \frac{\pi}{2},\; -\frac{\mu\pi}{4} + \frac{3\pi}{2}. \tag{7.2.50}$$

Here

$$\mu = \operatorname{sgn} f''(t_0) = \operatorname{sgn} f'(b) = -\operatorname{sgn} f'(a). \tag{7.2.51}$$

The directions of steepest descent from the two endpoints of integration a, b are respectively

$$\arg(t-a) = \pi + \frac{\pi}{2}\mu, \tag{7.2.52}$$

$$\arg(t-b) = \pi - \frac{\pi}{2}\mu. \tag{7.2.53}$$

The configuration of the hills and valleys of $u = \operatorname{Re}(w)$ with respect to the saddle point clearly depends on the value of μ. In each of the two possible cases it is readily determined. The hills and valleys at infinity are, of course, independent of the particular finite point under consideration. In Figures 7.2.5 and 7.2.6 we have depicted typical configurations corresponding to $\mu = -1$ and $\mu = +1$, respectively. Also, paths asymptotically equivalent to the steepest descent paths from the critical points are depicted. In both cases we find that the original contour C can be replaced by the combination of contours $D_a + \mu(D_1 - D_2) - D_b$. Hence, we can write

$$\begin{aligned} I(\lambda) &= \int_{D_a} + \int_{\mu D_1} - \int_{\mu D_2} - \int_{D_b} g \exp(i\lambda f)\, dt \\ &= I_a - I_b + \mu(I_1 - I_2). \end{aligned} \tag{7.2.54}$$

The leading terms in the asymptotic expansions of I_1 and I_2 are obtained

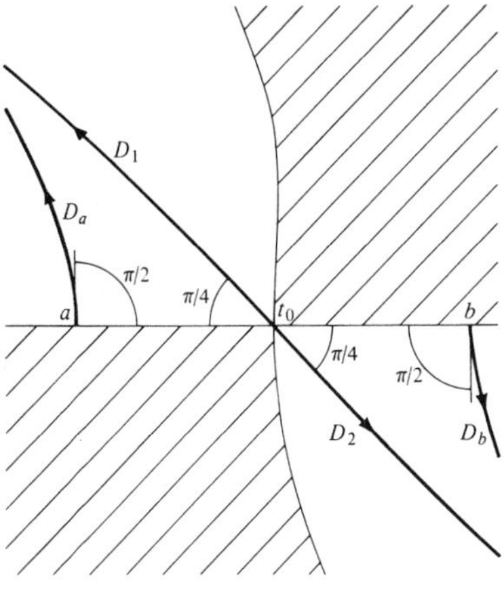

Figure 7.2.5. Descent Paths for $w = if$, with $\mu = -1$.

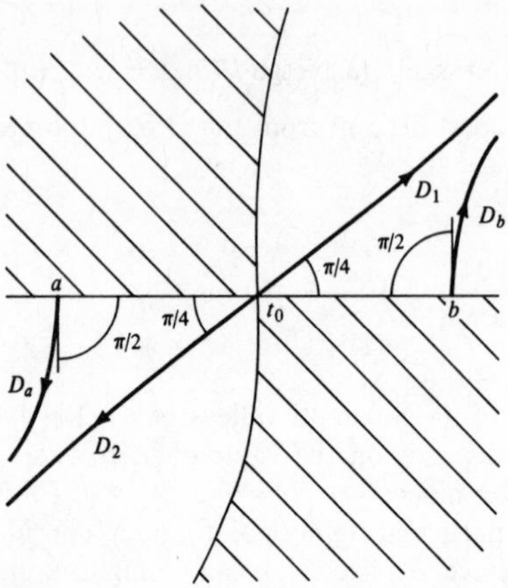

Figure 7.2.6. Descent Paths for $w = if$, with $\mu = 1$.

by applying (7.2.10). Indeed we find upon using the definitions of D_1 and D_2 that

$$I_1(\lambda) \sim \mu\, g(t_0) \sqrt{\frac{\pi}{2\lambda\,|f''(t_0)|}} \exp\left(i\lambda f(t_0) + \frac{\mu\pi i}{4}\right),$$

$$I_2(\lambda) \sim -\,\mu\, g(t_0) \sqrt{\frac{\pi}{2\lambda\,|f''(t_0)|}} \exp\left(i\lambda f(t_0) + \frac{\mu\pi i}{4}\right). \tag{7.2.55}$$

From this it follows that

$$\mu(I_1 - I_2) \sim g(t_0) \sqrt{\frac{2\pi}{\lambda\,|f''(t_0)|}} \exp\left(i\lambda f(t_0) + \frac{\mu\pi i}{4}\right). \tag{7.2.56}$$

To obtain the asymptotic expansions of I_a and I_b to leading order, we apply (7.2.11) with $\beta = 1$ and $\pi - \alpha$ given by (7.2.52) and (7.2.53), respectively. In this manner we find

$$I_a \sim \frac{g(a)}{\lambda\,|f'(a)|} \exp[i\lambda f(a) + i(\pi + \frac{\mu}{2}\pi)],$$

$$I_b \sim \frac{g(b)}{\lambda\,|f'(b)|} \exp[i\lambda f(b) + i(\pi - \frac{\mu}{2}\pi)], \tag{7.2.57}$$

and hence

$$I_a - I_b \sim \frac{1}{i\lambda}\left[\frac{g(b)}{f'(b)} \exp(i\lambda f(b)) - \frac{g(a)}{f'(a)} \exp(i\lambda f(a))\right]. \tag{7.2.58}$$

As we might have anticipated from the fact that $f(a)$, $f(b)$, and $f(t_0)$ are all real, the endpoint contributions are comparable to the saddle point contribution at least to exponential order. The leading term of the asymptotic expansion of $I(\lambda)$ comes from the saddle point contribution and, indeed, is given by (7.2.56). This is in agreement with the stationary phase formula (6.1.5). If any more than the leading term is required, then the contribution from the endpoints must also be included.

From this example we might argue that we should have delayed considering Fourier-type integrals until after the method of steepest descents had been developed. This would imply, however, that the method of stationary phase is a special case of the method of steepest descents. That this implication is false is seen from the fact that the derivation of the stationary phase formula does not require the strong analyticity assumptions made in the present section. Indeed, although the two methods are applicable to overlapping classes of integrals, neither is by any means a special case of the other.

To conclude this section, we wish to consider a fundamental question. In step (1) of our basic procedure, we indicated that the set of possible critical points for (7.1.1) includes the endpoints of integration, points of nonanalyticity of either w or g, and saddle points of w. Except perhaps for the saddle points themselves, the critical nature of this set has been motivated by previous discussions. We now shall investigate the critical nature of the saddle points to better understand why such points play a vital role in the method of steepest descents.

Suppose that in (7.1.1), C is a contour of finite extent and that w and g are entire functions. If, in particular, $w \equiv z$, then there are no saddle points of w so that the only possible critical points are the endpoints of integration. The configuration of the hills and valleys of $u = \text{Re}(w)$ at infinity is of course quite simple. Indeed, there is just one hill and one valley. The steepest descent contours from the endpoints must perforce go to infinity in the same valley, namely the left half-plane. As a result, C can *always* be replaced by a combination of these two contours.

Now suppose that w has a single saddle point in the finite plane located at $z = z_0$. Then the configuration of the hills and valleys at infinity is more complicated than above. In the simplest case where $z = z_0$ is a simple saddle point, there are two hills which alternate between two valleys. Now the steepest descent contours from the endpoints need not go to infinity in the same valley. Indeed let us suppose that they do not. Then it is impossible to replace C by a combination of these two contours alone. The reason for this is that, in order to connect them a hill at infinity must be crossed which is not permissible. Thus, if only steepest descent contours are to be used, then an additional critical point must come into play. As the results of Example 7.2.5 indicate, this point is the saddle point $z = z_0$.

Thus we might say that the saddle point and the descent paths emanating from it are the means by which distinct valleys can be connected at infinity.

Hence, the saddle point is, or is not, critical according as to whether or not the location of the original contour C necessitates such a connection.

In the following sections we shall, via four examples worked out in detail, extend and apply the results already obtained. In the first two of these examples we shall consider the determination of sectors of validity for asymptotic expansions obtained by the method of steepest descents. This will involve the occurrence of the Stokes' phenomenon discussed in Section 1.6. In the last two examples we shall illustrate how the method of steepest descents arises in applications by considering a partial differential equation that models certain wave propagation problems and by establishing a simple version of the central limit theorem of probability theory.

In the previous chapters, methods originally developed for integrals with exponential kernels were extended to integrals with more general kernel functions. In a completely analogous manner, the method of steepest descents can be generalized to integrals of the form

$$I(\lambda) = \int_C h(\lambda w(z))\, g(z)\, dz. \tag{7.2.59}$$

Here $h(t)$ is assumed to be a transcendental function of t, having asymptotic expansions as $t \to \infty$ of the form (4.4.7) valid in various sectors of the complex plane. It is further assumed that these sectors cover enough of the plane to allow for any required deformation of the contour C.

In contrast with the method of steepest descents already developed, there are two new features with which we must now contend. Firstly, zeros of $w(z)$ are possible critical points of the integrand. Because this was also true in our extensions of Laplace's method and the method of stationary phase, this should not be surprising. Secondly, there now can exist boundary curves which separate *two hills*, two regions of exponential growth, of the kernel function. (We recall that for the exponential function a boundary always separates a hill and a valley.[7])

The following functions illustrate this phenomenon: $\sin t$, $J_\nu(t)$, and $\mathrm{Ai}(t)$. For the first two, the entire real axis is a boundary curve which separates two hills with respect to the origin, while for $\mathrm{Ai}(t)$, the negative real axis is a boundary between two hills with respect to the origin.

As a consequence of this latter feature, there need not always exist a path of descent from a critical point, that is, a path along which $|h(\lambda w)|$ decreases exponentially and monotonically. Thus, after deformation, we find that, in general, C has been replaced by a combination of descent paths and boundary paths. On the former, the extension of Laplace's method developed in Sections 5.2 and 5.3 can be applied, while on the latter, either the theory of Section 5.4 or that of Sections 6.2 to 6.4 is applicable because the kernel is either algebraic or oscillatory. The actual procedure is outlined in Exercises 7.27 and 7.28.

[7]See the discussion on page 256.

7.3. The Airy Function for Complex Argument

Let us again consider the function $f_1(s) = \mathrm{Ai}(s)$ defined by (7.2.12). We shall now, however, assume that s is a complex variable and write

$$s = |s|\, e^{i\theta}. \tag{7.3.1}$$

Upon introducing the "stretched" variable of integration $\sqrt{|s|}\, z$ into (7.2.12), we obtain

$$f_1(s) = I_1(\lambda;\theta) = \frac{\lambda^{1/3}}{2\pi i}\int_{C_1} \exp[\lambda w(z;\theta)]\, dz, \qquad \lambda = |s|^{3/2}. \tag{7.3.2}$$

Here C_1 is as in Figure 2.5 and

$$w(z;\theta) = -\frac{z^3}{3} + e^{i\theta} z. \tag{7.3.3}$$

From (7.3.3) we easily find that

$$u(x,y;\theta) = \mathrm{Re}[w(z;\theta)] = -\left(\frac{x^3}{3} - xy^2\right) + x\cos\theta - y\sin\theta \tag{7.3.4}$$

and

$$v(x,y;\theta) = \mathrm{Im}[w(z;\theta)] = -\left(x^2 y - \frac{y^3}{3}\right) + x\sin\theta + y\cos\theta. \tag{7.3.5}$$

Because

$$w'(z;\theta) = -z^2 + e^{i\theta}, \qquad w''(z;\theta) = -2z, \tag{7.3.6}$$

there are simple points at

$$z = z_\pm = \pm e^{i\theta/2}. \tag{7.3.7}$$

Also

$$w(z_\pm;\theta) = \pm\tfrac{2}{3}\, e^{3i\theta/2}, \qquad w''(z_\pm;\theta) = \mp 2e^{i\theta/2}. \tag{7.3.8}$$

It follows from (7.1.8), (7.3.8), and Table 7.1 that the directions of steepest descent from the two saddle points are

$$\arg(z - z_+) = -\frac{\theta}{4},\ -\frac{\theta}{4} + \pi, \qquad \arg(z - z_-) = \pm\frac{\pi}{2} - \frac{\theta}{4}. \tag{7.3.9}$$

As usual, our major problem is to justify the deformation of the original contour of integration C_1 onto a combination of contours, each asymptotically equivalent to a steepest descent path from one of the two saddle points.

The curves of steepest descent from the saddle points are defined by

$$-\left(x^2 y - \frac{y^3}{3}\right) + x\sin\theta + y\cos\theta = \pm\frac{2}{3}\sin\left(\frac{3\theta}{2}\right). \tag{7.3.10}$$

For $\theta \neq 0$, these are rather difficult to determine explicitly. Since, as we have pointed out, away from the saddle points the curves used to replace C_1 need only be descent paths, we proceed in a more qualitative way.

We know that any path having a directed tangent vector whose nonzero components are always of opposite sign to the corresponding components of ∇u is a path of descent. Thus we can determine paths of descent merely by considering the vector field defined by

$$\nabla u = (u_x, u_y) = (y^2 - x^2 + \cos\theta,\, 2xy - \sin\theta) \tag{7.3.11}$$

and, in particular, the signs of the components of ∇u.

Suppose first that $0 < \theta < \pi/2$. Then u_x is negative "outside" the two branches of the hyperbola

$$x^2 - y^2 = \cos\theta \tag{7.3.12}$$

and positive between them. Also, u_y is positive outside the two branches of the hyperbola

$$2xy = \sin\theta \tag{7.3.13}$$

and negative between them. These hyperbolas clearly intersect at the saddle points themselves. The situation is schematically depicted in Figure 7.3.1. The directions of the horizontal and vertical arrows reflect the signs of u_x

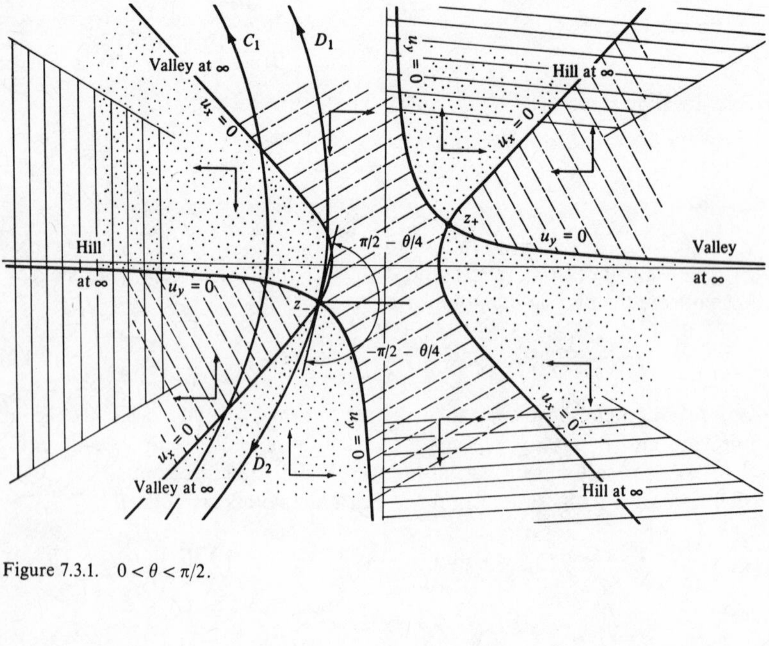

Figure 7.3.1. $0 < \theta < \pi/2$.

and u_y respectively in the various regions shown. In addition to the original contour C_1 we have drawn D_1 and D_2 which, from the above discussions, are seen to be descent paths from $z = z_-$. Moreover, they are assumed to be coincident with the steepest descent paths near that saddle point and hence have directions at $z = z_-$ given by the latter of (7.3.9). It is readily seen that C_1 is deformable onto the contour $D_1 - D_2$.

Now suppose that $\theta = \pi/2$. In this event, the hyperbola (7.3.12) degenerates into the 45-degree lines $x = \pm y$. The resulting vector field is qualitatively described in Figure 7.3.2. We again find that C_1 can be replaced by $D_1 - D_2$ with D_1 and D_2 asymptotically equivalent to the steepest descent paths from the saddle point $z = z_-$.

For $\pi/2 < \theta < \pi$, $u_x = 0$ on a hyperbola having a vertical focal axis. Between the two branches of the hyperbola, u_x is negative, while outside of the branches, u_x is positive. The sign of u_y remains as in the cases already considered, thereby producing the situation depicted in Figure 7.3.3. Once again we find that C_1 can be replaced by $D_1 - D_2$ and hence only the saddle point at $z = z_-$ is involved.

Finally, when $\theta = \pi$, the sign of u_x is as in the immediately preceding case. Now, however, u_y is zero on the coordinate axes $x = 0$, $y = 0$. The vector field ∇u is described schematically in Figure 7.3.4. We now find that we can no longer replace C_1 by descent contours from $z = z_- = -i$ alone. Any deforma-

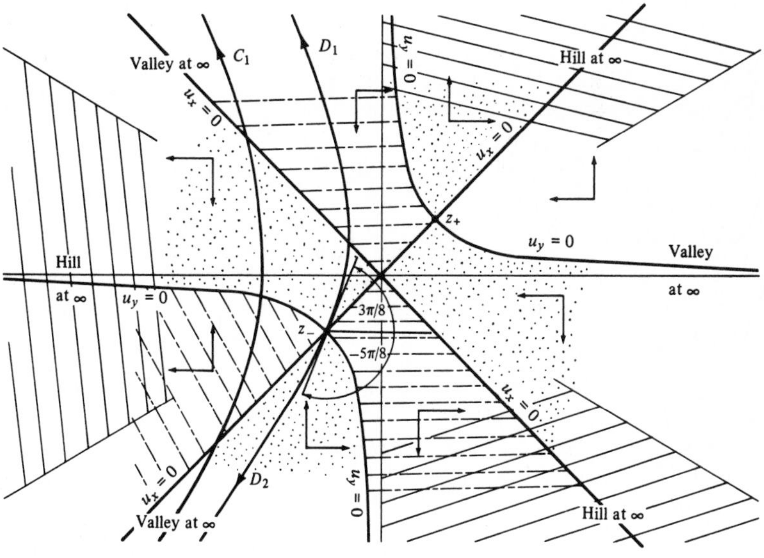

Figure 7.3.2. $\theta = \pi/2$.

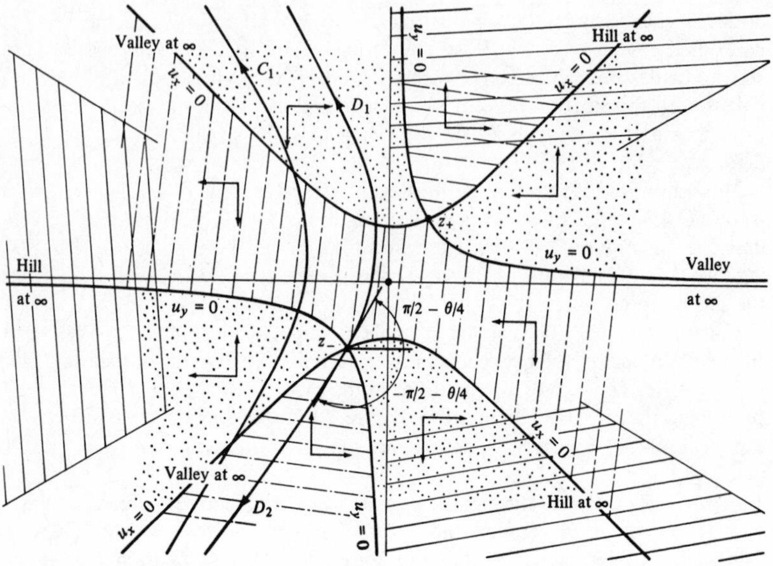

Figure 7.3.3. $\pi/2 < \theta < \pi$.

tion of C_1 onto descent paths from the saddle points must include paths from $z = z_+ = i$ as well. In Figure 7.3.4, D_1 and D_2 are seen to be descent paths from $z = z_-$ while D_3 and D_4 are descent paths from $z = z_+$. Furthermore, we find that C_1 can be replaced by the composite contour $D_4 + D_1 - D_2 - D_3$.

From the above discussion we can conclude that, for $0 \le \arg(s) < \pi$, the asymptotic expansion of Ai(s) involves a dominant contribution from the saddle point at $z = z_-$ only,[8] whereas, for $\theta = \pi$, it involves contributions from both saddle points. Moreover, because $w(z_\pm, \pi)$ is purely imaginary, the two contributions in the latter case are of equal order. Because completely analogous statements hold for the range $-\pi \le \arg(s) < 0$, it follows that the asymptotic expansion of the Airy function Ai(s), as determined by the method of steepest descent, and which is valid in the sector $|\arg(s)| < \pi$, has a Stokes line along the negative real axis. In other words, the analytic continuation of the expansion valid in $|\arg(s)| < \pi$ is *not* the asymptotic expansion of Ai(s) outside of this sector.

Let us now calculate the leading term of the expansion itself. If $|\arg(s)| < \pi$, then C_1 can always be replaced by $D_1 - D_2$ as shown above. Because D_1 is

[8] However, in the range $\pi/3 < \theta < \pi$, z_+ is in the valley of w with respect to z_-. Now we can find descent paths through z_+ as well as z_-. Thus we would include a term which remains subdominant in $\pi/3 < \theta < \pi$ but becomes dominant in $\pi < \theta < 5\pi/3$. See Exercise 7.26.

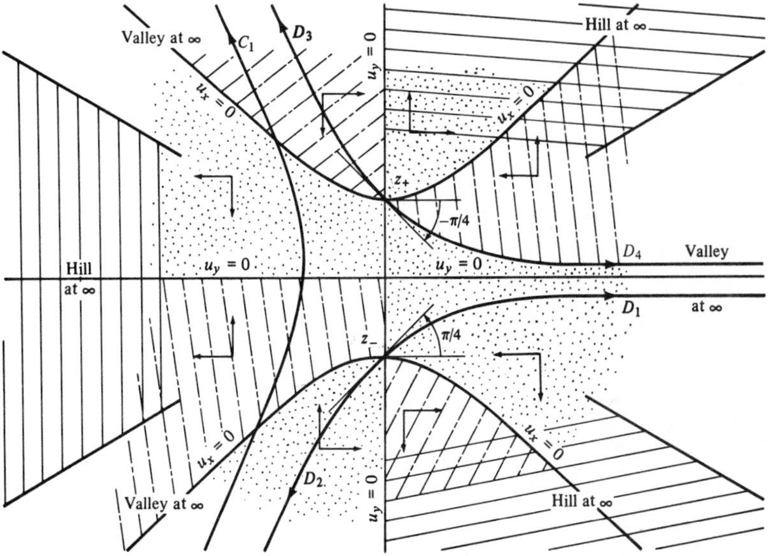

Figure 7.3.4. $\theta = \pi$.

asymptotically equivalent to the steepest descent path from $z = z_-$ with initial direction $\arg(z - z_-) = \pi/2 - \theta/4$, while D_2 is asymptotically equivalent to the steepest descent path from $z = z_-$ with initial direction $\arg(z - z_-) = -\pi/2 - \theta/4$, it follows from (7.2.10) and (7.3.8) that

$$I_1(\lambda;\theta) \sim \frac{\lambda^{-1/6}}{2\sqrt{\pi}} \exp\left\{-\frac{2}{3}\lambda\, e^{3i\theta/2} - \frac{i\theta}{4}\right\}, \qquad |\theta| < \pi. \tag{7.3.14}$$

Upon using the relations

$$\lambda\, e^{3i\theta/2} = s^{3/2}, \qquad \lambda^{-1/6}\, e^{-i\theta/4} = s^{-1/4}, \tag{7.3.15}$$

we finally obtain

$$\mathrm{Ai}(s) \sim \frac{s^{-1/4}}{2\sqrt{\pi}} \exp\left\{-\frac{2\, s^{3/2}}{3}\right\}, \qquad |s| \to \infty, \qquad |\arg(s)| < \pi. \tag{7.3.16}$$

An expansion of Ai(s), valid as $s \to -\infty$, can be found by using the information obtained above. Indeed, we have

$$\mathrm{Ai}(s) \sim \frac{|s|^{-1/4}}{\sqrt{\pi}} \sin\left(\frac{2\,|s|^{3/2}}{3} + \frac{\pi}{4}\right), \qquad \arg(s) = \pi. \tag{7.3.17}$$

We leave the details of this last derivation to the exercises.

7.4. The Gamma Function for Complex Argument

Although we have previously derived the asymptotic expansion of the gamma function for complex argument in Section 3.2, we wish to do so again via the method of steepest descents. We start from the integral representation of $\Gamma(s+1)$ given by (2.2.7) which states

$$\Gamma(s+1) = \frac{\exp\{-\pi i s\}}{2i \sin \pi s} \lambda^{s+1} I(\lambda;\theta), \qquad s = \lambda e^{i\theta}. \tag{7.4.1}$$

Here

$$I(\lambda;\theta) = \int_C \exp[\lambda w(z;\theta)]\, dz \tag{7.4.2}$$

with C the contour depicted in Figure 2.2 and

$$w(z;\theta) = e^{i\theta} \log z - z. \tag{7.4.3}$$

Because

$$w'(z;\theta) = \frac{e^{i\theta}}{z} - 1, \qquad w''(z;\theta) = -\frac{e^{i\theta}}{z^2}, \tag{7.4.4}$$

we find that w has one simple saddle point on each of the sheets of the infinite sheeted Riemann surface of $\log z$. These are defined by

$$z = z_n = \exp\{i\theta + 2n\pi i\}, \qquad n = 0, \pm 1, \pm 2, \ldots . \tag{7.4.5}$$

The saddle points, together with the origin, which is a branch point for w, constitute the set of possible critical points of I. Also, in what follows, we shall refer to the region defined by $0 \le \arg(z) < 2\pi$ as the "principal sheet" of the Riemann surface of $\log z$.

It follows from (7.4.4) and Table 7.1 that the directions of steepest descent from $z = z_n$ are given by

$$\arg(z - z_n) = \frac{\theta}{2}, \frac{\theta}{2} + \pi. \tag{7.4.6}$$

The paths of steepest descent from the saddle points are not as readily determined. We can, however, resort to the mode of analysis employed in the previous example. Thus, if we set

$$w = u(x,y) + i\, v(x,y), \qquad z = x + iy = re^{i\phi}, \tag{7.4.7}$$

then it follows from (7.4.3) that

$$\begin{aligned} u &= \cos\theta \log r - \phi \sin\theta - x, \\ v &= \sin\theta \log r + \phi \cos\theta - y. \end{aligned} \tag{7.4.8}$$

Because

$$\nabla u = (u_x, u_y) = \left(\frac{\cos(\phi - \theta)}{r} - 1, \frac{\sin(\phi - \theta)}{r} \right), \tag{7.4.9}$$

we have that, on each sheet of the Riemann surface of $\log z$, $u_x = 0$ along the circle through the origin and the saddle point on that sheet. The curves along which $u_y = 0$ are simply the rays defined by

$$\phi = \theta + n\pi, \qquad n = \pm 1, \pm 2, \ldots . \tag{7.4.10}$$

In Figures 7.4.1, 7.4.2, and 7.4.3, we show schematically the gradient field of u in the cases $0 < \theta < \pi/2$, $\theta = \pi/2$, and $\pi/2 < \theta < \pi$, respectively. In each instance we can verify that the contours designated by D_1 and D_2 are asymptotically equivalent to paths of steepest descent from $z = z_0$, the saddle point on the principal sheet. Furthermore, we find that the original contour C can, in each case, be replaced by the combination of contours $D_2 - D_1$. As a result, $z = z_0$ is the only contributing critical point.

The asymptotic expansion of I, to leading order, is obtained by applying (7.2.10) to each of the descent integrals and thus forming the indicated combination. In this manner we find

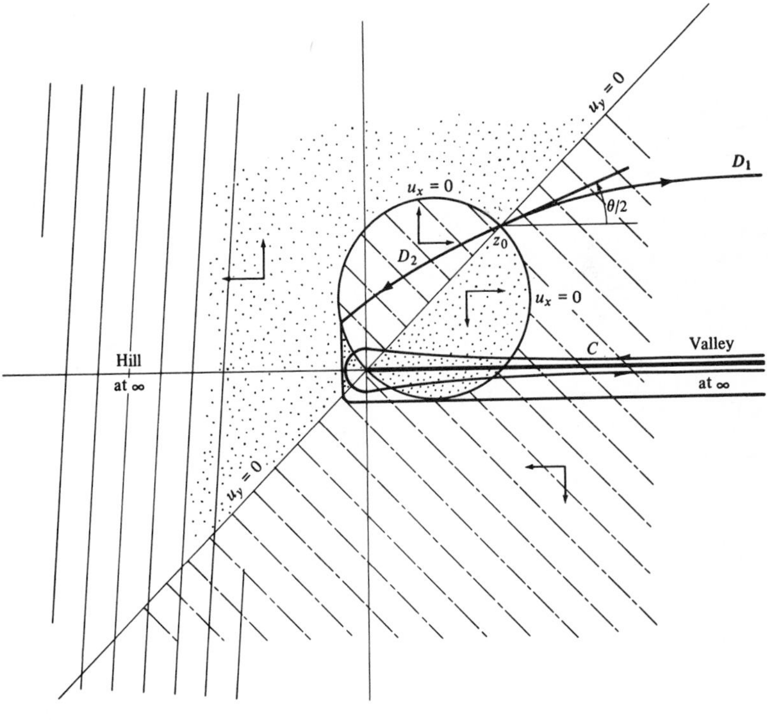

Figure 7.4.1. $0 < \theta < \pi/2$.

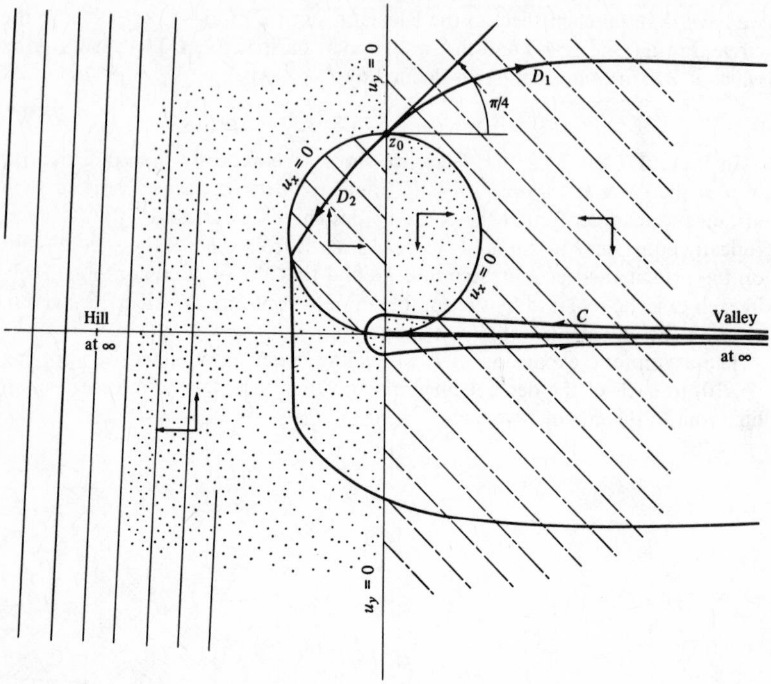

Figure 7.4.2. $\theta = \pi/2$.

$$I(\lambda;\theta) \sim -\sqrt{\frac{2\pi}{\lambda}} \exp\left\{\lambda[i\theta e^{i\theta} - e^{i\theta}] + \frac{i\theta}{2}\right\}, \qquad 0 < \theta < \pi. \tag{7.4.11}$$

We shall delay writing the corresponding expansion of $\Gamma(s+1)$ itself until a wider range of values of $\arg(s)$ has been considered.

The case $\theta = 0$ is somewhat more difficult to treat. Indeed, although (7.4.1) is correct for s real, it is quite awkward due to the presence of removable singularities at the positive integers. Furthermore, if we were interested in real positive s only, then we would naturally use the simpler representation (5.1.22) and apply Laplace's method as in Example 5.1.1. Nevertheless we shall find it instructive to pursue the asymptotic analysis of $I(\lambda;0)$.

In Figure 7.4.4 we depict the gradient field of u when $\theta = 0$. There is no apparent replacement of C by a combination of descent paths from $z = z_0$ alone. Suppose then we attempt to make use of the saddle point $z = z_1$. It can be shown that it is possible to replace C by a combination of descent paths from $z = z_0$ *and* $z = z_1$. Two of these contours, however, are spirals along which $|\arg(z)| \to \infty$ while $|z| \to 0$. Fortunately, a detailed knowledge of these

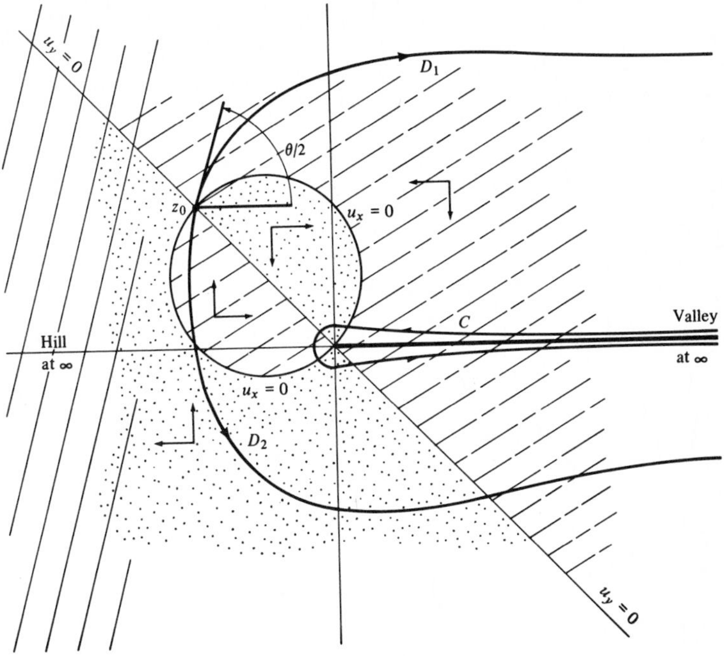

Figure 7.4.3. $\pi/2 < \theta < \pi$.

fairly complicated curves is not necessary. In fact, we need not consider them at all.

From Figure 7.4.5, we see that C can be replaced by the combination $D_2 - D_1 - D_3 + D_4 + \bar{C}$. The contours D_i, $i = 1, 2, 3, 4$, are each asymptotically equivalent to a steepest descent contour from one of the two saddle points under consideration. Although $\bar{C}$ is not even a descent contour it has the property that along it

$$u(x,y) < \log r - x \ll 0. \tag{7.4.12}$$

This last inequality assures us that the integral along $\bar{C}$ is asymptotically negligible compared to the integrals along the paths D_i, $i = 1, 2, 3, 4$. Heuristically, we might say that $\bar{C}$ connects the contour $D_2 - D_1$ to the contour $D_4 - D_3$ "deep" in a valley of both saddle points.

It is now a simple matter to obtain the leading term of the expansion of $I(\lambda;0)$. Indeed, upon applying (7.2.10) to each of the relevant integrals and forming the appropriate combination, we have

$$I(\lambda;0) \sim \sqrt{\frac{2\pi}{\lambda}}\, e^{-\lambda}\, (e^{2\pi i\lambda} - 1). \tag{7.4.13}$$

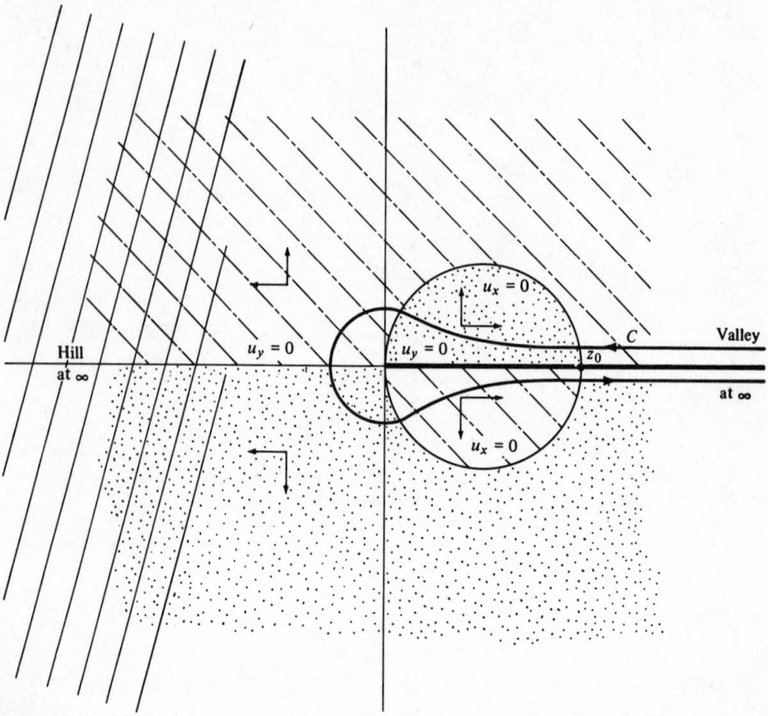

Figure 7.4.4. $\theta = 0$.

In the range $-\pi < \theta < 0$, the saddle point at $z = z_1$ lies on the principal sheet, while the one at $z = z_0$ lies on the next "lower" sheet. The analysis in this case is completely analogous to that given above for $0 < \theta < \pi$. Indeed, we find that

$$I(\lambda;\theta) \sim \sqrt{\frac{2\pi}{\lambda}} \exp\left\{\lambda[i(\theta + 2\pi)\, e^{i\theta} - e^{i\theta}] + \frac{i\theta}{2}\right\}, \qquad -\pi < \theta < 0. \quad (7.4.14)$$

Our original goal was, of course, to obtain an asymptotic expansion of $\Gamma(s + 1)$. To this end, we note that

$$\lim_{|s| \to \infty} \frac{\exp(-\pi is)}{2i \sin \pi s} = -1, \qquad 0 < \theta < \pi,$$

$$\lim_{|s| \to \infty} \frac{\exp(\pi is)}{2i \sin \pi s} = 1, \qquad -\pi < \theta < 0. \qquad (7.4.15)$$

It then follows from (7.4.1), (7.4.11), and (7.4.13) to (7.4.15) that

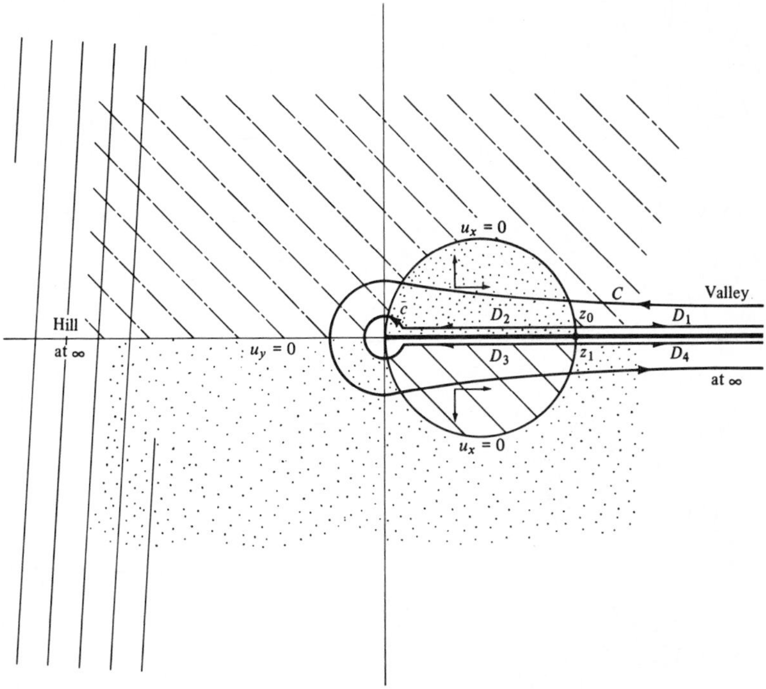

Figure 7.4.5. $\theta = 0$.

$$\Gamma(s+1) \sim \sqrt{2\pi s}\left(\frac{s}{e}\right)^s, \qquad |\arg(s)| < \pi.^{9} \tag{7.4.16}$$

When $|\arg(s)| = \pi$, $\Gamma(s+1)$ has poles at the negative integers. In the representation (7.4.1), these poles are due to the factor $(\sin \pi s)^{-1}$. Because $I(\lambda;\theta)$ is perfectly well behaved for θ near π, we could obtain the asymptotic expansion of $I(\lambda;\pi)$ and use it in (7.4.1) to find a meaningful expansion of $\Gamma(s+1)$ along the negative real axis. In Figure 7.4.6 we show ∇u along with the descent paths from the relevant saddle point. A simpler procedure, however, would be to use the relation

$$\Gamma(s+1) = -\frac{\pi}{\sin \pi s\, \Gamma(-s)} \tag{7.4.17}$$

and then replace $\Gamma(-s)$ by its already determined asymptotic expansion for $\arg(-s) = 0$.

[9] If we set $x = s + 1$ in (3.2.40) and use the result $\lim_{s\to\infty}(1 + 1/s)^s = e$, then we can show that (7.4.16) agrees with the former result.

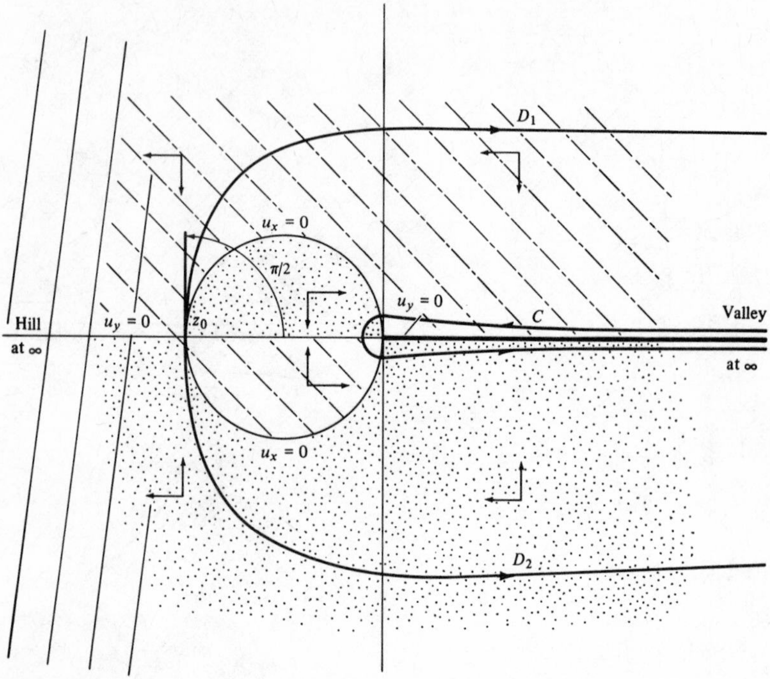

Figure 7.4.6. $\theta = \pi$.

7.5. The Klein-Gordon Equation

Here we shall consider the following "source" problem for the Klein-Gordon equation:

$$c^2 u_{xx} - u_{tt} - b^2 u = \delta(x) \exp\{-i\omega_0 t\}, \quad -\infty < x < \infty, \quad t > 0, \quad \omega_0 > b, \quad (7.5.1)$$

$$u(x,0) = u_t(x,0) = 0, \qquad u(x,t) \equiv 0, \qquad t < 0. \qquad (7.5.2)$$

From the discussion of Section 2.6, we find that the solution $u(x,t)$ represents a wave propagating in a one-dimensional dispersive medium. The wave itself is produced by a stationary disturbance or source located at the origin.

We imagine that the source is "turned on" at time $t = 0$ and subsequently oscillates with frequency ω_0. Because the medium is source-free for $t < 0$, the last of conditions (7.5.2) must be satisfied.

Our first objective is to obtain an integral representation of $u(x,t)$. To accomplish this we introduce

$$\hat{u}(x,\omega) = \int_0^\infty u(x,t)\, e^{i\omega t}\, dt, \qquad (7.5.3)$$

the Fourier transform of $u(x,t)$ with respect to time. From (7.5.1), we readily find that $\hat{u}$ satisfies the ordinary differential equation

$$c^2\,\hat{u}_{xx} + (\omega^2 - b^2)\,\hat{u} = \frac{i\,\delta(x)}{\omega - \omega_0}. \tag{7.5.4}$$

The solution to (7.5.4) which, when inserted into the Fourier inversion formula, results in the satisfaction of the "outgoing" condition is

$$\hat{u}(x,\omega) = \frac{\exp\{i(\omega^2 - b^2)^{1/2}\,|x|/c\}}{2c(\omega-\omega_0)\,(\omega^2 - b^2)^{1/2}}. \tag{7.5.5}$$

Thus the desired integral representation is given by

$$u(x,t) = \frac{1}{2\pi}\int_\Gamma \frac{\exp\{i[(\omega^2 - b^2)^{1/2}\,\frac{|x|}{c} - \omega t]\}\,d\omega}{2c(\omega-\omega_0)\,(\omega^2 - b^2)^{1/2}}. \tag{7.5.6}$$

Here on the contour Γ, $\mathrm{Re}(\omega)$ goes from $-\infty$ to ∞ and Γ passes *above* the three singularities of the integrand, each of which lies on the real axis. Furthermore, we shall take the lines drawn from the points $\omega = \pm b$ vertically downward to infinity, as the branch cuts for $(\omega^2 - b^2)^{1/2}$. (See Figure 7.5.1.)
If we introduce the dimensionless quantities

$$\lambda = bt, \qquad \nu = \frac{\omega}{b}, \qquad \theta = \frac{|x|}{ct}, \tag{7.5.7}$$

then

$$u(x,t) = U(\lambda;\theta) = \frac{1}{4\pi bc}\int_{\Gamma'} \frac{\exp[\lambda\phi(\nu;\theta)]}{(\nu-\nu_0)\,(\nu^2-1)^{1/2}}\,d\nu. \tag{7.5.8}$$

Here

$$\phi(\nu;\theta) = i[(\nu^2-1)^{1/2}\,\theta - \nu], \qquad \nu_0 = \frac{\omega_0}{b}. \tag{7.5.9}$$

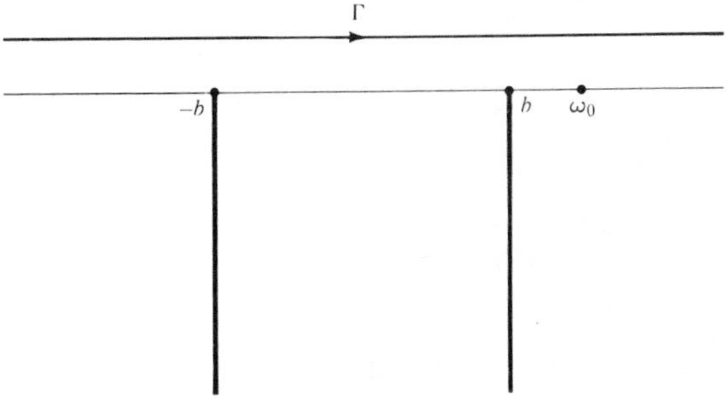

Figure 7.5.1. ω Plane, Contour Γ, Branch Points $\pm b$, Branch Cuts Extending from $\pm b$ to $\pm b - i\infty$ Pole at ω_0.

Because the new contour Γ' is simply Γ stretched by the positive number b, we have Figure 7.5.2.

For $\theta > 1$, that is, for $|x| > ct$, we can close the contour of integration in the upper half-plane. It then follows by Cauchy's integral theorem that

$$u(x,t) \equiv 0, \qquad |x| > ct. \tag{7.5.10}$$

This last result is simply a reflection of the fact that the signal travels at a finite speed. Hence, if we remain ahead of the wave fronts $|x| = ct$, then we do not observe any disturbance at all. The fronts themselves travel with the "characteristic speed" c. We might also mention that the portion of the wave at and just behind the fronts is called the "precursor" because it is the first signal to reach any particular point x.

We wish to study $U(\lambda;\theta)$ in the limit $\lambda \to \infty$, that is, as time gets large compared to b^{-1}. To accomplish this, we shall apply the method of steepest descents to (7.5.8). From (7.5.9) we have

$$\phi' = i\left[\frac{\nu\theta}{(\nu^2-1)^{1/2}} - 1\right], \qquad \phi'' = -i\left[\frac{\theta}{(\nu^2-1)^{3/2}}\right]. \tag{7.5.11}$$

Thus, ϕ has a simple saddle point at each ν satisfying

$$\theta = \frac{(\nu^2-1)^{1/2}}{\nu}. \tag{7.5.12}$$

If $0 < \theta < 1$, then there is a solution $\hat{\nu}(\theta)$ to (7.5.12) such that $1 < \hat{\nu}(\theta) < \infty$ and given by

$$\hat{\nu}(\theta) = \frac{1}{\sqrt{1-\theta^2}}. \tag{7.5.13}$$

Moreover, $-\hat{\nu}(\theta)$ also satisfies (7.5.12) because, with the branch cuts as shown

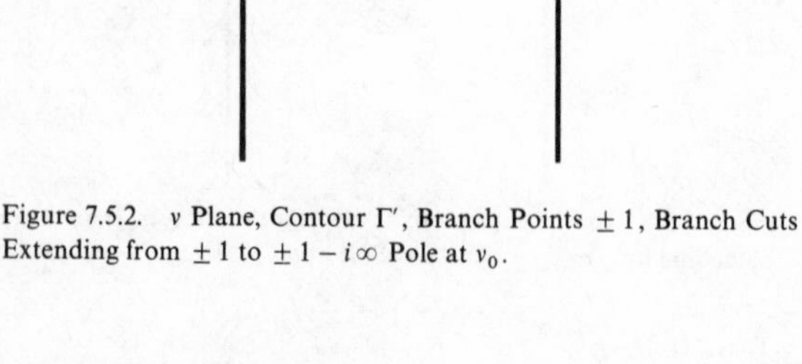

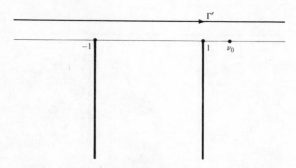

Figure 7.5.2. ν Plane, Contour Γ', Branch Points ± 1, Branch Cuts Extending from ± 1 to $\pm 1 - i\infty$ Pole at ν_0.

in Figure 7.5.2, $(v^2-1)^{1/2}/v$ is even and positive for v real and outside of the interval $(-1,1)$. Clearly $v=\pm\hat{v}(\theta)$ are the only saddle points.

At the saddle points we have

$$\phi(\pm\hat{v}(\theta)) = \mp i\sqrt{1-\theta^2},$$
$$\phi''(\pm\hat{v}(\theta)) = \mp i\frac{(1-\theta^2)^{3/2}}{\theta^2}. \tag{7.5.14}$$

As $\theta\to 1-$, $\hat{v}(\theta)$ goes to $+\infty$ and we must expect difficulties to arise. Also, when $\theta=0$ the saddle points coincide with the branch points and again we might anticipate complications. We shall investigate these issues in subsequent chapters but now insist that $0<\theta<1$.

From Table 7.1 and (7.5.14) we obtain

Saddle Point	*Directions of Steepest Descent*
$\hat{v}(\theta)$	$-\frac{\pi}{4}, \frac{3\pi}{4}$
$-\hat{v}(\theta)$	$\frac{\pi}{4}, -\frac{3\pi}{4}$

To obtain sufficient qualitative information about the paths of steepest descent from the saddle points, we first note that for $|v|\gg 1$

$$\phi = -iv \pm i\theta v + 0(|v|^{-1}). \tag{7.5.15}$$

Here the plus sign is taken when $|v|\to\infty$ outside of the branch cuts, while the minus sign is taken when $|v|\to\infty$ in between the branch cuts.

Because $|\theta|<1$, we have from (7.5.15) that, on a path of steepest descent, $\text{Im}(v)\to-\infty$. Furthermore, the fact that $\text{Im}(\phi)$ is constant on paths of steepest descent implies that $\text{Re}(v)$ must remain finite on such paths. Finally, because $\text{Re}(\pm v(\theta))=0$, we must have that $\text{Re}(\phi)<0$ on all descent paths away from the saddle points.

From the information just obtained, we can conclude that the steepest descent paths are qualitatively as shown in Figure 7.5.3. This diagram is symmetric about the vertical, and although this information is not needed, we note that the asymptote for D_1 is

$$\text{Re}(v) = \sqrt{\frac{1+\theta}{1-\theta}}, \tag{7.5.16}$$

while that for D_2 is

$$\text{Re}(v) = \sqrt{\frac{1-\theta}{1+\theta}}. \tag{7.5.17}$$

If v_0 is located as in Figure 7.5.3, then we have by Cauchy's

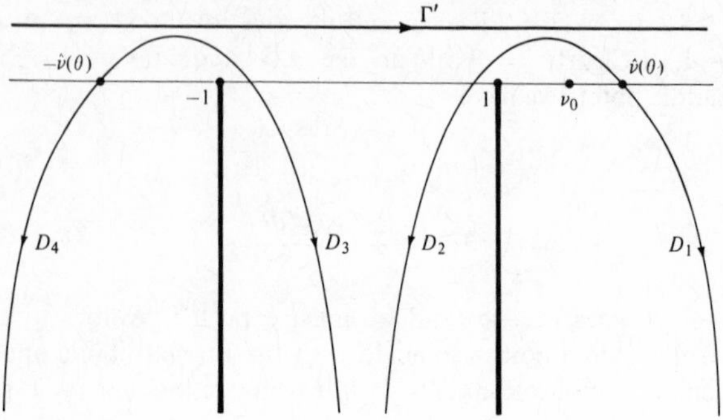

Figure 7.5.3. Paths of Descent $\hat{v}(\theta) > v_0$.

theorem

$$U(\lambda;\theta) = \frac{1}{4\pi bc} \int_{D_1 - D_2 + D_3 - D_4} \frac{\exp\{\lambda\phi(v;\theta)\}}{(v - v_0)(v^2 - 1)^{1/2}}\, dv. \tag{7.5.18}$$

Alternatively, if $|v_0| > \hat{v}(\theta)$, then a residue term must also be included. See Figure 7.5.4. Indeed, we have

$$U(\lambda;\theta) = \frac{1}{4\pi bc} \int_{D_1 - D_2 + D_3 - D_4} \frac{\exp\{\lambda\phi(v;\theta)\}}{(v - v_0)(v^2 - 1)^{1/2}}\, dv$$

$$+ \frac{\exp\{i\lambda\,[\sqrt{v_0^2 - 1}\,\theta - v_0]\}}{2i\,bc\,(v_0^2 - 1)^{1/2}}, \qquad v_0 > \hat{v}(\theta). \tag{7.5.19}$$

The leading terms of the saddle point contributions are readily obtained. We need only apply (7.2.10) which yields

$$\frac{1}{4\pi bc} \int_{D_1 - D_2} \frac{\exp\{\lambda\phi(v;\theta)\}}{(v - v_0)(v^2 - 1)^{1/2}} dv \sim \frac{\exp\left\{-i\lambda\sqrt{1 - \theta^2} - \dfrac{\pi i}{4}\right\}}{\sqrt{2\pi\lambda}\,2bc(1 - \theta^2)^{1/4}[(1 - \theta^2)^{-1/2} - v_0]} \tag{7.5.20}$$

$$\frac{1}{4\pi bc} \int_{D_3 - D_4} \frac{\exp\{\lambda\phi(v;\theta)\}}{(v - v_0)(v^2 - 1)^{1/2}} dv \sim \frac{\exp\left\{i\lambda\sqrt{1 - \theta^2} + \dfrac{\pi i}{4}\right\}}{\sqrt{2\pi\lambda}\,2bc(1 - \theta^2)^{1/4}[(1 - \theta^2)^{-1/2} + v_0]} \tag{7.5.21}$$

Upon combining (7.5.18)–(7.5.21) we find that

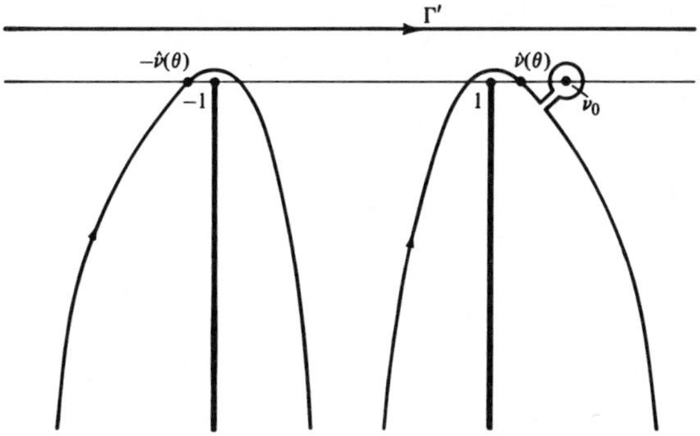

Figure 7.5.4. Path of Integration Replacing Γ' When $\hat{\nu}(\theta) < \nu_0$.

$$U(\lambda;\theta) \sim \frac{1}{2ibc\sqrt{2\pi\lambda}\,(1-\theta^2)^{1/4}}$$

$$\times \left\{ \frac{\exp\left[-i\lambda\sqrt{1-\theta^2} - \frac{3\pi i}{4}\right]}{[\nu_0 - (1-\theta^2)^{-1/2}]} + \frac{\exp\left[i\lambda\sqrt{1-\theta^2} + \frac{3\pi i}{4}\right]}{[\nu_0 + (1-\theta^2)^{-1/2}]} \right\},$$

$$1 > \theta > \frac{(\nu_0^2 - 1)^{1/2}}{\nu_0} \tag{7.5.22}$$

and

$$U(\lambda;\theta) \sim \frac{\exp\{i\lambda\,[\sqrt{\nu_0^2 - 1}\,\theta - \nu_0]\}}{2ibc\,(\nu_0^2 - 1)^{1/2}}, \qquad 0 < \theta < \frac{(\nu_0^2 - 1)^{1/2}}{\nu_0}. \tag{7.5.23}$$

Let us compare (7.5.22) with (7.5.23). We first note that the latter being 0(1) as $\lambda \to \infty$ is a stronger signal than the former which is $0(\lambda^{-1/2})$. The right side of (7.5.23) is clearly a wave which oscillates at the source frequency ν_0. It is the main signal, but is observed continuously only if we move at a sufficiently slow speed. Indeed, we must remain in the region $|x| < (\nu_0^2 - 1)^{1/2}\, ct/\nu_0$ to "see" a sustained signal from the source. Outside of this region, we do not observe the main signal but rather see the algebraically damped wave (7.5.22) whenever $ct > |x| > (\nu_0^2 - 1)^{1/2}\, ct/\nu_0$. Because the region near the wave front has been called the precursor region, we have that (7.5.22) represents the asymptotic expansion to leading order of the precursor wave.

If $\theta = (\nu_0^2 - 1)^{1/2}/\nu_0$, $\nu_0 > 1$, then the saddle point $\hat{\nu}(\theta)$ and the pole of the integrand at $\nu = \nu_0$ coincide. This case is not covered by the above analysis because the integrand is not regular on the steepest descent paths from $\nu = \nu_0$. Nevertheless, with minor modifications, the method of steepest descents can be applied. Indeed, we find that

$$U(\lambda;\theta) \sim \frac{\exp\left\{i\lambda\sqrt{1-\theta^2}+\dfrac{\pi i}{4}\right\}}{2bc\sqrt{2\pi\lambda}\,[v_0+(1-\theta^2)^{-1/2}](1-\theta^2)^{1/4}}$$
$$+\frac{\exp\{-i\lambda\sqrt{1-\theta^2}\}}{4ibc\theta}\left[\sqrt{1-\theta^2}-\frac{e^{i\pi/4}}{\sqrt{2\pi\lambda}}\frac{(1-\theta^2)^{1/4}}{\theta}\right], \qquad v_0=\hat{v}(\theta) \tag{7.5.24}$$

but shall leave the details of the derivation to the exercises.

We note that neither (7.5.22) nor (7.5.23) has (7.5.24) as its limit when $\theta \to (v_0^2-1)^{1/2}/v_0$. Indeed, in both cases $|U| \to \infty$ in this limit. A single expansion which goes smoothly from (7.5.22) to (7.5.24) as θ increases throughout the range $0<\theta<1$ is said to be uniformly valid as the saddle point $\hat{v}(\theta)$ passes through the pole at $v=v_0$. Such an expansion would be useful, in that it would enable us to study the interaction of the precursor with the main signal. We shall consider this situation in Chapter 9.

7.6. The Central Limit Theorem for Identically Distributed Random Variables

Let $X_1, X_2, \ldots, X_N$ be independent random variables having the common probability density function f. We shall assume here that $f(x)=0(e^{-\delta|x|})$, $\delta>0$, $|x|\to\infty$, so that all moments of f exist. In particular, we have that the mean μ and the variance σ^2 of each X_i are given by

$$\mu=\int_{-\infty}^{\infty} x f(x)\,dx \qquad \text{and} \qquad \sigma^2=\int_{-\infty}^{\infty} x^2 f(x)\,dx-\mu^2,$$

respectively.

In Section 2.3, we introduced the characteristic function of a random variable as the Fourier transform of its probability density function. Thus, the characteristic function of each X_i is

$$\phi(\alpha)=\int_{-\infty}^{\infty} e^{i\alpha x} f(x)\,dx. \tag{7.6.1}$$

Let us now suppose that Y is a random variable with probability density function $g(x)$. Suppose further that

$$Z=aY+b, \qquad a,b \text{ constant}, \ a>0. \tag{7.6.2}$$

If we denote the probability density function of Z by $h(x)$, then we clearly have

$$h(x)=\frac{1}{a}g\left(\frac{x-b}{a}\right). \tag{7.6.3}$$

Hence $\hat{\phi}(\alpha)$, the characteristic function of Z, is related to $\bar{\phi}(\alpha)$, the characteristic function of Y, in the following way:

$$\hat{\phi}(\alpha) = \frac{1}{a}\int_{-\infty}^{\infty} e^{i\alpha x}\, g\left(\frac{x-b}{a}\right) dx$$
$$= e^{ib\alpha}\, \bar{\phi}(\alpha a). \tag{7.6.4}$$

If we now set

$$Y = \sum_{i=1}^{N} X_i \tag{7.6.5}$$

and

$$Z = \frac{Y - N\mu}{\sigma\sqrt{N}}, \tag{7.6.6}$$

then we have

$$\hat{\phi}(\alpha) = e^{-i\sqrt{N}\mu\alpha/\sigma}\, \bar{\phi}\left(\frac{\alpha}{\sigma\sqrt{N}}\right). \tag{7.6.7}$$

But by the convolution theorem for Fourier transforms [see (2.3.13) and (2.3.14)] it follows that

$$\bar{\phi}(\alpha) = [\phi(\alpha)]^N \tag{7.6.8}$$

and hence

$$\hat{\phi}(\alpha) = e^{-i\sqrt{N}\mu\alpha/\sigma}\left[\phi\left(\frac{\alpha}{\sigma\sqrt{N}}\right)\right]^N. \tag{7.6.9}$$

Our objective is to study the limiting behavior, as $N \to \infty$, of the density function $h(x;N)$ of the standardized random variable Z defined by (7.6.5) and (7.6.6). We have from (7.6.9) and the Fourier inversion formula that

$$h(x;N) = \frac{1}{2\pi}\int_{-\infty}^{\infty} \exp\left\{-i\sqrt{N}\frac{\mu\alpha}{\sigma} - i\alpha x\right\}\left[\phi\left(\frac{\alpha}{\sigma\sqrt{N}}\right)\right]^N d\alpha. \tag{7.6.10}$$

If we introduce the quantities

$$z = \frac{\alpha}{\sigma\sqrt{N}}, \qquad \beta = \frac{\sigma x}{\sqrt{N}}, \tag{7.6.11}$$

then (7.6.10) becomes

$$h(x;N) = \frac{\sigma\sqrt{N}}{2\pi}\int_{-\infty}^{\infty} \exp\{N\, w(z;\beta)\}\, dz. \tag{7.6.12}$$

Here

$$w(z;\beta) = \log\phi - i(\mu+\beta)\, z. \tag{7.6.13}$$

We note that the assumed decay of f guarantees that $\phi(z)$ is analytic in the strip $|\text{Im}(z)| < \delta$. Also, it is readily seen that $\phi(z) \to 0$ as $|z| \to \infty$ in this strip. Hence, as $|z| \to \infty$ in $|\text{Im}(z)| < \delta$, $\text{Re}(\log \phi) \to -\infty$, and therefore so does $\text{Re}(w)$.

We now want to apply the method of steepest descents to study (7.6.12) as $N \to \infty$ with $\beta \ll 1$. This last restriction, of course, implies that $|x| \ll \sqrt{N}/\sigma$. Before proceeding, we note that

$$\phi(0) = 1, \qquad \phi'(0) = i\mu,$$
$$\phi''(0) = -(\sigma^2 + \mu^2), \qquad \phi'''(0) = -i\mu_3, \tag{7.6.14}$$

where μ_3 is the third moment of f.

We seek the saddle points of w, which satisfy

$$w'(z;\beta) = \frac{\phi'(z)}{\phi(z)} - i(\beta + \mu) = 0. \tag{7.6.15}$$

We also have

$$w''(z;\beta) = \frac{\phi''(z)}{\phi(z)} - \left[\frac{\phi'(z)}{\phi(z)}\right]^2. \tag{7.6.16}$$

Certain saddle points of w are clearly not admissible. To see why, we note that, for z real, $|\phi| \le 1$ and hence $\text{Re}(w) \le 0$. Therefore, the entire real axis lies in the valley of w with respect to any saddle point at which $\text{Re}(w) > 0$. Thus, such saddle points are inadmissible. Also, $|\phi| < 1$ for z real and nonzero, so that all points on the real axis lie in the valley of w with respect to the origin.

The above discussion suggests that we should seek a saddle point near $z = 0$ and that, of the admissible saddle points, this will be the dominant one. We cannot hope to obtain the saddle point near $z = 0$ precisely. However, by taking advantage of the smallness of β, we can obtain a useful approximation. From (7.6.14) we find that (7.6.15) takes the approximate form

$$i\mu - (\sigma^2 + \mu^2)\, z - \frac{i\,\mu_3\, z^2}{2} + O(z^3) = i(\beta + \mu)\left[1 + i\mu z - \frac{(\sigma^2 + \mu^2)\, z^2}{2} + O(z^3)\right]. \tag{7.6.17}$$

For small β we can solve (7.6.17) by iteration. Indeed, after two iterations, we find that the desired saddle point $z = z_0$ is given approximately by

$$z_0 = -\frac{i\beta}{\sigma^2} - \frac{i\beta^2}{\sigma^4}\left\{\mu + \frac{(\mu\,[\sigma^2 + \mu^2] - \mu_3)}{2\,\sigma^2}\right\} + O(\beta^3). \tag{7.6.18}$$

We must now calculate w and w'' at $z = z_0$. From (7.6.14) and (7.6.16) we have

$$w''(z_0;\beta) = \left\{\frac{\phi''(0)}{\phi(0)} - \left(\frac{\phi'(0)}{\phi(0)}\right)^2\right\}\{1 + O(\beta)\}$$
$$= -\sigma^2\,\{1 + 0(\beta)\}. \tag{7.6.19}$$

Hence the paths of steepest descent from $z = z_0$ are approximately horizontal and $-w''(z_0;\beta)$ is approximately the variance σ^2. The above discussion shows that the deformation of the original contour of integration, namely the real axis, onto paths of steepest descent through $z = z_0$ is easily justified.

To find $w(z_0;\beta)$ we first write

$$w(z;\beta) = w(0;\beta) + w'(0;\beta)z + w''(0;\beta)\frac{z^2}{2} + O(z^3). \tag{7.6.20}$$

We then use (7.6.13)–(7.6.16) to find

$$\begin{aligned} w(z_0;\beta) &= -i\beta\left[-\frac{i\beta}{\sigma^2} + O(\beta^2)\right] - \frac{\sigma^2}{2}[1 + O(\beta)]\left[-\frac{\beta^2}{\sigma^4} + O(\beta^3)\right] \\ &= -\frac{\beta^2}{2\sigma^2} + O(\beta^3). \end{aligned} \tag{7.6.21}$$

Upon applying the steepest descent formula (7.2.10) to each of the two steepest descent integrals we obtain

$$h \sim \frac{1}{\sqrt{2\pi}}\exp\left\{-\frac{N\beta^2}{2\sigma^2} + O(N\beta^3)\right\}[1 + O(\beta) + O(N^{-1})]. \tag{7.6.22}$$

In terms of the original variables x and N this becomes

$$h(x;N) \sim \frac{1}{\sqrt{2\pi}}\exp\left\{-\frac{x^2}{2}\left(1 + O\left(\frac{\sigma^3 x}{\sqrt{N}}\right)\right)\right\}\left\{1 + O\left(\frac{\sigma x}{\sqrt{N}}\right) + O(N^{-1})\right\}. \tag{7.6.23}$$

A random variable X having the probability density function

$$h(x) = \frac{1}{\sqrt{2\pi}}e^{-x^2/2}, \qquad -\infty < x < \infty,$$

is said to be a *normal random variable* with mean zero and variance one. From (7.6.23) we can conclude that Z defined by (7.6.5) and (7.6.6) approaches such a random variable as $N \to \infty$. More precisely, there is a region centered about zero, the mean of Z, whose width is $o(\gamma)$, $\gamma = N^{1/6}/\sigma$, and throughout which the probability density function of Z approaches, as $N \to \infty$, that of a normal random variable with mean zero and variance one. This result is known as the *central limit theorem* of probability theory.

The central limit theorem can be established under conditions on f much less restrictive than those assumed here. In particular the assumption that f decreases exponentially as $|x| \to \infty$ can be replaced by the assumption that the first three moments of f are finite. Furthermore, we need not insist that the X_i's are identically distributed. We have considered the special case above because it most readily lends itself to analysis by the method of steepest descents.

7.7. Exercises

7.1. Let $w(z) = u(x,y) + i\,v(x,y)$ with $u_0 = u(x_0, y_0)$ and $v_0 = v(x_0, y_0)$.

(a) Use (7.1.9) to show that if $z_0 = x_0 + i\,y_0$ is a point of analyticity for w and if (7.1.8) holds, then

$$u^2 + v^2 = u_0^2 + v_0^2 + \frac{2a\rho^n}{n!}\left[u_0 \cos(n\theta + \alpha) + v_0 \sin(n\theta + \alpha)\right] + O(\rho^{n+1}),$$

$$\rho \to 0+. \qquad (7.7.1)$$

Here, $z - z_0 = \rho e^{i\theta}$.

(b) Use the result in (a) to prove the *maximum modulus theorem*: If z_0 is an interior point of the domain of analyticity of w, then $|w(z)|$ cannot attain a local maximum there unless $w(z)$ is identically a constant.

7.2. Let $w(z) = z^4/4 - z$.

(a) Find the saddle points of $w(z)$ and fill out Table 7.1 for each of them.

(b) Find the "valleys at infinity" for $w(z)$, that is, the regions at ∞ in which $\exp\{\lambda\, w(z)\}$ decays exponentially to zero.

(c) Draw approximate paths of steepest descent of w from the saddle points.

7.3. Repeat Exercise 7.2 for the function $w = -z^7/7 + \frac{3}{5} z^5 - z^3 + z - tz$:

(a) $t > 1$.
(b) $0 < t < 1$.
(c) $t < 0$.
(d) $t = 0$.
(e) $t = 1$.

7.4. Repeat Exercise 7.2 for the function

$$w = -\frac{z^2}{2} - e^{i\theta} z. \qquad (7.7.2)$$

7.5. Consider w as given by (7.7.2).

(a) For each of the following choices of θ, show that the path of steepest descent from the origin is a segment of a hyperbola:

(i) $\theta = 0, \pi$.

(ii) $\theta = \pm\theta_0, \qquad 0 < \theta_0 < \frac{\pi}{4}$.

(iii) $\theta = \pm\frac{\pi}{4}$.

(iv) $\theta = \pm\theta_0, \quad \frac{\pi}{4} < \theta_0 < \frac{3\pi}{4}.$

(v) $\theta = \pm\frac{3\pi}{4}.$

(vi) $\theta = \pm\theta_0, \quad \frac{3\pi}{4} < \theta_0 < \frac{5\pi}{4}.$

(b) What is the significance of the angles $\theta = \pm\pi/2$?

(c) Discuss the family of descent paths for each choice of θ in (a).

(d) Show that for $\pi/4 < \theta_0 < 3\pi/4$, there are descent paths from the origin which end at $+\infty$ and descent paths which end at $-\infty$.

(e) Suppose that $e^{i\theta} = i$. Suppose further that in (7.1.1) $g(z)$ has an algebraic branch point at $z = 0$ with the branch cut being the positive real axis. Then show that the contour C of Figure 7.7.1 can be deformed onto the contour D, consisting of a circle around the origin and descent contours. In Figure 7.7.1, the dashed segments of D depict paths on the next lower sheet of the Riemann surface of $g(z)$.

(f) Show that C is also equivalent to the set of contours in Figure 7.7.2.

(g) Corresponding to the two deformations of contour depicted in Figures 7.7.1 and 7.7.2 are two distinct asymptotic expansions of the underlying integral (7.1.1). Explain how this can be possible.

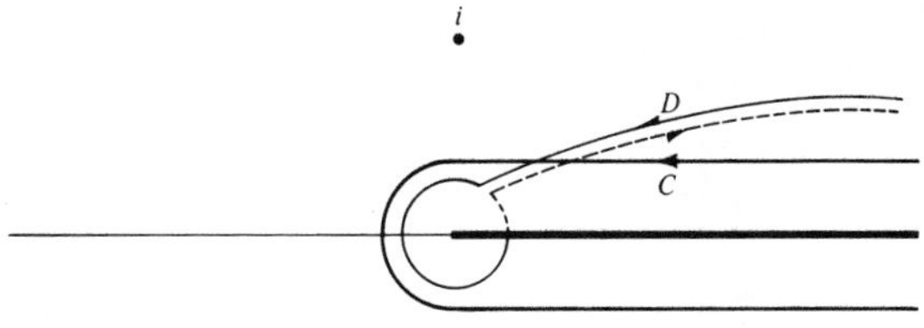

Figure 7.7.1.

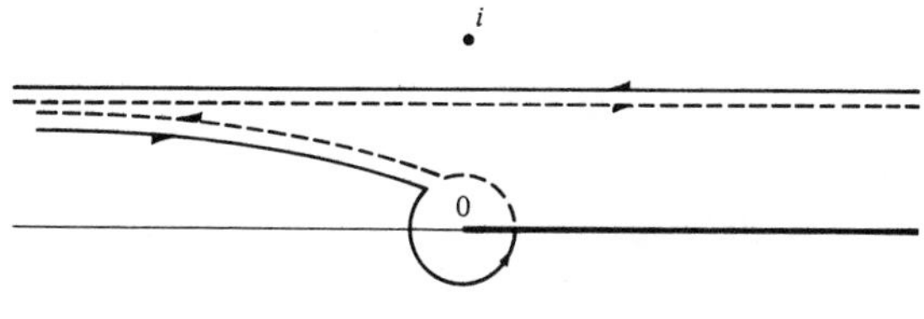

Figure 7.7.2.

7.6. Draw a diagram such as Figure 7.1.5 for each of the following functions:

(a) $w(z) = z^3$.

(b) $w(z) = z - \dfrac{z^2}{2}$.

(c) $w(z) = \cosh z$.

7.7. Use the method of steepest descents to calculate the asymptotic expansion of the following integrals as $\lambda \to \infty$. In each case the contour C is an infinite straight line parallel to but above the real axis. Also in parts (a) through (d) compute an expansion correct to $0(\lambda^{-1/2})$ while in parts (e) and (f) compute the leading term only.

(a) $$\int_C \frac{\exp\left\{i\lambda\left(\dfrac{z^3}{3} - z\right)\right\} dz}{z-2}.$$

(b) $$\int_C \frac{\exp\left\{i\lambda\left(\dfrac{z^3}{3} - z\right)\right\} dz}{z}.$$

(c) $$\int_C \frac{\exp\left\{i\lambda\left(\dfrac{z^2}{2} - z\right)\right\} dz}{\sqrt{z-2}}.$$

(d) $$\int_C \frac{\exp\left\{i\lambda\left(\dfrac{z^2}{2} - z\right)\right\} dz}{\sqrt{z}}.$$

(e) $$\int_C \sin \lambda z \exp\left\{\frac{iz^3}{3}\right\} dz.$$

(f) $$\int_C \exp\left\{i\lambda\left[\frac{z^3}{3} + \cos z\right]\right\} dz.$$

7.8. Let $w(z) = i\left[z \cos\theta + \sqrt{1 - z^2} \sin\theta\right]$. Here the square root is taken to be positive for $-1 < \mathrm{Re}(z) < 1$ and the branch cuts are taken from ± 1 to $\pm(1 + i\infty)$, respectively.

(a) Show that for $0 < \theta < \pi/2$, the location of the saddle point and the paths of steepest descent from it are as depicted in Figure 7.7.3.

(b) Show that for $\pi/2 < \theta < \pi$, the location of the saddle point and the steepest descent paths from it are as depicted in Figure 7.7.4.

(c) Discuss the case $\theta = \pi/2$.

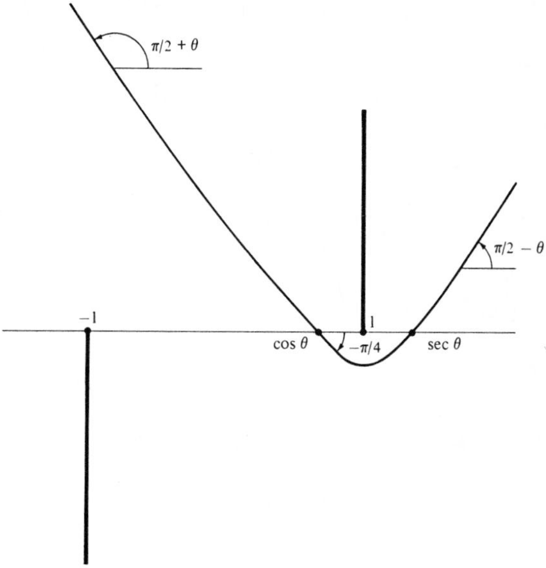

Figure 7.7.3. $0 < \theta < \pi/2$.

7.9. Let

$$I(\lambda) = \int_C (z - z_0)^\nu g(z) \exp\{\lambda w(z)\} \, dz. \tag{7.7.3}$$

Here $g(z)$ and $w(z)$ are analytic near $z = z_0$ and $w'(z_0) \neq 0$. For ν not an integer, we suppose that the branch cut for $(z - z_0)^\nu$ from z_0 to ∞ is a descent path for $w(z)$. Finally, let C be a "loop" contour counterclockwise about the branch cut with endpoints at ∞.

(a) Show that if

$$w'(z_0) = -\left|w'(z_0)\right| e^{-i\theta}, \tag{7.7.4}$$

then the direction of steepest descent at $z = z_0$ is $\arg(z - z_0) = \theta$.

(b) Show that the change of variable of integration

$$-t = w(z) - w(z_0) \tag{7.7.5}$$

leads to the integral

$$I(\lambda) = \exp\{\lambda \, w(z_0)\} \int_{0+} t^\nu \, G(t) \, e^{-\lambda t} \, dt, \tag{7.7.6}$$

where

$$G(t) = \left(\frac{z - z_0}{t}\right)^\nu g(z) \frac{dz}{dt}$$

is an analytic function in some neighborhood of $t = 0$.

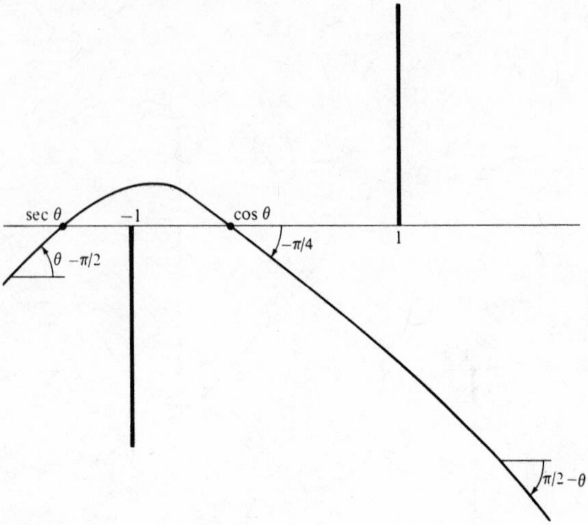

Figure 7.7.4. $\pi/2 < \theta < \pi$.

(c) Assume

$$G(t) = \sum_{m=0}^{\infty} c_m t^m \tag{7.7.7}$$

near $t = 0$. Apply Exercise 4.8 to show that

$$I(\lambda) \sim \exp\{\lambda w(z_0)\} \sum_{m=0}^{\infty} \frac{-(2\pi i) c_m e^{i\pi(m+\nu)}}{\Gamma(-m-\nu) \lambda^{m+\nu+1}}, \qquad \lambda \to \infty. \tag{7.7.8}$$

Note that this result holds even when ν is an integer.

(d) Show that in (7.7.8)

$$c_0 = \frac{g(z_0) e^{i(\nu+1)\theta}}{|w'(z_0)|^{\nu+1}}. \tag{7.7.9}$$

7.10. (a) Find an asymptotic expansion, as $\lambda \to \infty$, of the integral

$$I(\lambda, \theta; r) = \int_C z^{-r-1} \exp\left\{-\lambda\left[\frac{z^2}{2} + e^{i\theta} z\right]\right\} dz \tag{7.7.10}$$

in the case $\theta = 0$. Here C is the $0+$ contour of Figure 7.7.1.

(b) Show that the result obtained in part (a) is valid for $|\theta| < 3\pi/4$. [*Caution*: Different arguments are required for the ranges $|\theta| < \pi/2$ and $|\theta| \geq \pi/2$.]

(c) The Weber function $D_r(\zeta)$ is given by

$$D_r(\zeta) = \frac{\Gamma(r+1)}{2\pi i\, \lambda^{r/2}} \exp\left[-\frac{\zeta^2}{4} + r\pi i\right] I(\lambda, \theta; r) \tag{7.7.11}$$

with I defined by (7.7.10), r not a negative integer, and $\zeta = \sqrt{\lambda}\, e^{i\theta}$. Show that

$$D_r(\zeta) \sim \zeta^r \exp\left\{-\frac{\zeta^2}{4}\right\}, \qquad |\zeta| \to \infty, \qquad |\arg(\zeta)| < \frac{3\pi}{4}. \tag{7.7.12}$$

(d) Show that

$$D_r(\zeta) \sim -\frac{\sqrt{2\pi}}{\Gamma(-r)\zeta^{r+1}} \exp\left\{\frac{\zeta^2}{4} + r\pi i\right\} [1 + 0(|\zeta|^{-2})]$$

$$+ \zeta^r \exp\left\{-\frac{\zeta^2}{4}\right\} [1 + 0(|\zeta|^{-2})], \qquad |\zeta| \to \infty, \qquad \frac{\pi}{4} < \arg(\zeta) < \frac{5\pi}{4}. \tag{7.7.13}$$

[*Hint*: Deform the contour of integration as described in Exercise 7.5.]

(e) Explain how both results (7.7.12) and (7.7.13) can be valid for $\pi/4 < \arg(\zeta) < 3\pi/4$.

(f) Formally apply L'Hospital's rule to the representation (7.7.11) to obtain the result

$$D_{-n}(\zeta) = \frac{\exp\left\{-\frac{\zeta^2}{4}\right\}}{\Gamma(n)} \int_0^\infty z^{n-1} \exp\left\{-\left[\frac{z^2}{2} + \zeta z\right]\right\} dz, \qquad n = 1, 2, \ldots. \tag{7.7.14}$$

(g) Derive an asymptotic expansion of the integral (7.7.14) and show that the result agrees with (7.7.12) for $r = -n$.

7.11. Let

$$I(\lambda) = \int_{-\infty}^{\infty} \frac{\exp\{i\lambda[z \cos\theta + \sqrt{1-z^2}\sin\theta]\}\, dz}{(z - \cos\alpha)\sqrt{1-z^2}}. \tag{7.7.15}$$

Here the contour of integration passes above the branch point $z = -1$, below the branch point $z = 1$, and above the pole $z = \cos\alpha$. Furthermore, $\sqrt{1-z^2}$ is defined to be positive for $-1 < \text{Re}(z) < 1$ and $\sin\theta \geq 0$.

(a) Show that for $\cos\theta > \cos\alpha$

$$I(\lambda) \sim \sqrt{\frac{2\pi}{\lambda}} \frac{\exp\left\{i\lambda - \frac{i\pi}{4}\right\}}{\cos\theta - \cos\alpha} [1 + 0(\lambda^{-1})]. \tag{7.7.16}$$

(b) Show that for $\cos\theta < \cos\alpha$ and $\sin\alpha > 0$

$$I(\lambda) \sim -\frac{2\pi i \exp\{i\lambda\cos(\theta-\alpha)\}}{\sin\alpha} + \sqrt{\frac{2\pi}{\lambda}}\frac{\exp\left\{i\lambda - \dfrac{i\pi}{4}\right\}}{\cos\theta - \cos\alpha}[1 + 0(\lambda^{-1})]. \quad (7.7.17)$$

7.12. Let

$$I(\lambda) = \int_C \frac{g(z)}{z - z_0}\exp\{\lambda\, w(z)\}\, dz. \quad (7.7.18)$$

Here $g(z)$ and $w(z)$ are analytic near $z = z_0$ and the contour C, $w(z)$ has a simple saddle point at $z = z_0$, and C passes from one valley of w with respect to z_0 to another, avoiding z_0 in a counterclockwise manner.

(a) Introduce a new variable of integration t by the equation

$$w(z) - w(z_0) = -t$$

and then use the result of Exercise 4.8 to show that

$$I(\lambda) \sim \pi i\, g(z_0)\exp\{\lambda\, w(z_0)\}\,[1 - 0(\lambda^{-1/2})], \quad (7.7.19)$$

that is, the leading term of the asymptotic expansion is given by the "half-residue" corresponding to the pole.

(b) Show that if C avoids z_0 in a clockwise manner, then the result is the negative of (7.7.18).

(c) In Exercise 7.11, show that

$$I(\lambda) \sim -\frac{\pi i \exp\{i\lambda\}}{\sin\alpha}, \qquad \sin\theta = \sin\alpha > 0. \quad (7.7.20)$$

(d) For the integral (7.5.8) show that when $\theta = (v_0^2 - 1)^{1/2}/v_0$,

$$U(\lambda;\theta) \sim -\frac{i\exp\{-i\lambda v_0^{-1}\}}{4bc\,(v_0^2 - 1)^{1/2}}. \quad (7.7.21)$$

7.13. Let

$$I(\lambda) = \int_C (z - z_0)^\beta g(z)\exp[\lambda\, w(z)]\, dz. \quad (7.7.22)$$

Here $w(z)$ and $g(z)$ are analytic near z_0 and C, while $w(z)$ has a saddle point at $z = z_0$ with $w''(z_0) = ae^{i\alpha}$, $a > 0$. The contour C passes from one valley of $w(z)$ with respect to z_0 to another, avoiding z_0 in a clockwise manner and directed into the valley containing the direction of steepest descent given by $\arg(z - z_0) = \pi/2 - \alpha/2$. Also $(z - z_0)^\beta$ is defined on C by its principal value. Then show that

$$I(\lambda) \sim g(z_0)e^{\lambda w(z_0)}\left[\frac{2e^{i(\pi-\alpha)}}{\lambda a}\right]^{(\beta+1)/2}\cos\frac{\pi\beta}{2}\, e^{i\pi\beta/2}\,\Gamma\left[\frac{\beta+1}{2}\right] \quad (7.7.23)$$

[*Hint*: Introduce t as a new variable of integration as in Exercise 7.12(a) and reduce $I(\lambda)$ to a loop integral to which Exercise 4.8 can be applied.]

7.14. Let

$$I(\lambda) = \int_C e^{-\lambda z^3} \frac{g(z)}{z}\, dz, \tag{7.7.24}$$

where $g(z)$ is analytic in a domain containing the origin and the contour C.

(a) Let C be the contour C_1 of Figure 2.5. Obtain the asymptotic expansion of I as $\lambda \to \infty$. [*Hint*: Introduce the new variable of integration t defined by $t^{1/3} e^{2\pi i/3} = z$.]

(b) Repeat part (a) with C the contour C_2 in Figure 2.5. [*Hint*: Here set $t^{1/3} = z$.]

(c) Repeat part (a) with C being the contour C_3 in Figure 2.5. [*Hint*: Here set $t^{1/3} e^{\pi i/3} = z$.]

(d) Let C be a contour from $\infty e^{4\pi i/3}$ to ∞ which avoids the origin in a counterclockwise manner. Obtain the asymptotic expansion of $I(\lambda)$.

(e) Let C be the contour $0+$ and again obtain the asymptotic expansion of $I(\lambda)$.

7.15. Let $I(\lambda)$ be given by (7.7.18). Suppose that $g(z)$ and $w(z)$ are analytic in a domain including the point $z = z_0$ and the contour of integration C. Further, suppose that $w(z)$ has a saddle point of order $n - 1$ at $z = z_0$ and that C passes from one valley of w with respect to z_0 to any other. Derive the leading term of the asymptotic expansion of I in this case.

7.16. Let $I(\lambda)$ be given by (7.7.3). Suppose that w, g, and C are as in Exercise 7.15. Find the leading term of the asymptotic expansion of I in this case.

7.17. The Sommerfeld representation for the Bessel function $J_\nu(\lambda)$ is given by

$$J_\nu(\lambda) = \frac{1}{2\pi} \int_{C_3} \exp\{i\lambda\, w(z;\beta)\}\, dz,$$

$$w(z;\beta) = i\left[\cos z + \beta\left(z - \frac{\pi}{2}\right)\right], \qquad \beta = \frac{\nu}{\lambda}. \tag{7.7.25}$$

Here C_3 is the contour shown in Figure 7.7.5.

(a) By comparing (7.7.25) with (7.2.22) and C_3 with the contours C_1 and C_2 of Figure 7.2.1, show that

$$J_\nu(\lambda) = \tfrac{1}{2}\left[H_\nu^{(1)}(\lambda) + H_\nu^{(2)}(\lambda)\right].$$

(b) Use the method of steepest descents to show that

$$J_\nu(\lambda) \sim \frac{\exp\left\{-\nu \cosh^{-1}\dfrac{\nu}{\lambda} + \sqrt{\nu^2 - \lambda^2}\right\}}{\sqrt{2\pi}\,(\nu^2 - \lambda^2)^{1/4}}, \qquad \lambda, \nu \to \infty, \qquad \nu > \lambda. \tag{7.7.26}$$

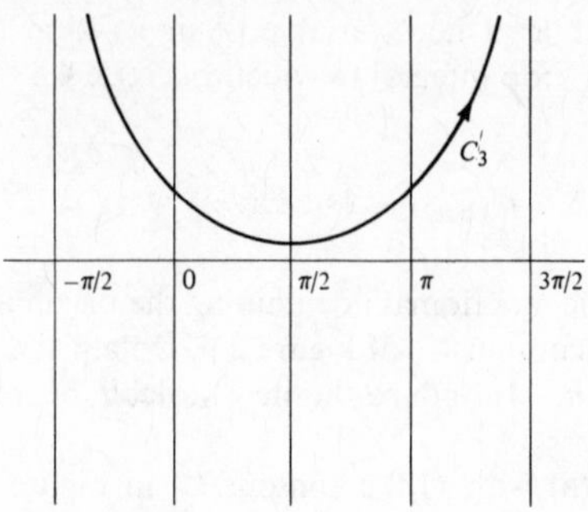

Figure 7.7.5.

(c) Show that this result can be rewritten as

$$J_\nu(\lambda) \sim \frac{1}{\sqrt{2\pi}\,(\nu^2-\lambda^2)^{1/4}} \frac{e^{\sqrt{\nu^2-\lambda^2}}}{\left(\dfrac{\nu}{\lambda}+\sqrt{\dfrac{\nu^2}{\lambda^2}-1}\right)^{\nu}}, \qquad \lambda,\ \nu \to \infty, \qquad \nu > \lambda. \tag{7.7.27}$$

(d) From (c), deduce that for fixed λ,

$$J_\nu(\lambda) \sim \frac{1}{\sqrt{2\pi\nu}} \left(\frac{e\lambda}{2\nu}\right)^{\nu}, \qquad \nu \to \infty. \tag{7.7.28}$$

7.18. (a) Calculate the leading term of the asymptotic expansion of $J_\nu(\nu)$ as $\nu \to \infty$.

(b) Calculate the leading term of the asymptotic expansion of $J_\nu(\lambda)$, $\nu,\ \lambda \to \infty$, $\nu < \lambda$.

7.19. Let

$$S(\lambda;\phi) = \sum_{n=0}^{\infty} J_n(\lambda)\, e^{in\phi}. \tag{7.7.29}$$

(a) Show that the series converges. [*Hint*: Use (7.7.28).]
(b) Show that

$$S(\lambda;\phi) = \sum_{n=0}^{\infty} \frac{e^{in\phi}}{2\pi} \int_{C_3} \exp\left\{i\lambda \cos z + in\left(z - \frac{\pi}{2}\right)\right\} dz. \tag{7.7.30}$$

Here C_3 is the contour shown in Figure 7.7.5.

(c) Show that because $\operatorname{Im}(z) \geq y_0 > 0$ on C_3, the summation and integration processes can be interchanged in (7.7.30) and obtain

$$S(\lambda;\phi) = \frac{1}{2\pi} \int_{C_3} \frac{\exp\{i\lambda \cos z\}}{1 - e^{i[\phi + (z - \pi/2)]}}\, dz. \tag{7.7.31}$$

(d) Verify that, as $\lambda \to \infty$,

$$S(\lambda;\phi) \sim S_1(\lambda;\phi) = \frac{\exp\{\pm i\lambda \mp i\pi/4\}}{\sqrt{2\pi\lambda}\,(1 \pm ie^{i\phi})}, \qquad \frac{\pi}{2} < |\phi| < \frac{3\pi}{2}. \tag{7.7.32}$$

(e) Verify that, as $\lambda \to \infty$,

$$S(\lambda;\phi) \sim \exp\{i\lambda \sin\phi\} + S_1(\lambda;\phi), \qquad |\phi| < \frac{\pi}{2}.$$

(f) Verify that, as $\lambda \to \infty$,

$$S\left(\lambda;\frac{\pi}{2}\right) \sim \frac{1}{2}\exp\{i\lambda\}.$$

7.20. Define

$$S(t) = \sum_{n=0}^{\infty} \frac{(-t^2)^n}{n!\,(2n)!}. \tag{7.7.33}$$

(a) Show that

$$S(t) = \frac{1}{2\pi i}\oint \frac{\exp\left\{z - \dfrac{t^2}{z^2}\right\}dz}{z}, \tag{7.7.34}$$

where the contour of integration is any circle about the origin.

(b) Show that

$$S(t) = \frac{1}{2\pi i}\oint \frac{\exp\left\{\lambda\left[z - \dfrac{1}{z^2}\right]\right\}dz}{z}, \qquad \lambda = t^{2/3}. \tag{7.7.35}$$

(c) Use the method of steepest descents to show that

$$S(t) \sim \frac{2^{1/3}}{t^{1/3}\sqrt{3\pi}}\exp\{2^{-5/3}\,3t^{2/3}\}\sin\left(\frac{3^{3/2}\,t^{2/3}}{2^{5/3}} + \frac{\pi}{3}\right). \tag{7.7.36}$$

7.21. Consider the differential equation

$$(D^2 - \lambda^2)^3\,u + \lambda^6\,tu = 0, \qquad D = \frac{d}{dt}. \tag{7.7.37}$$

(a) Show that seven solutions to (7.7.37) are given by

$$u_j = \int_{\Gamma_j} \exp\left\{-\lambda\left[\frac{z^7}{7} - \frac{3}{5}z^5 + z^3 - z + tz\right]\right\}dz, \qquad j = 1, \ldots, 7, \tag{7.3.38}$$

with contours Γ_j as shown in Figure 7.7.6.

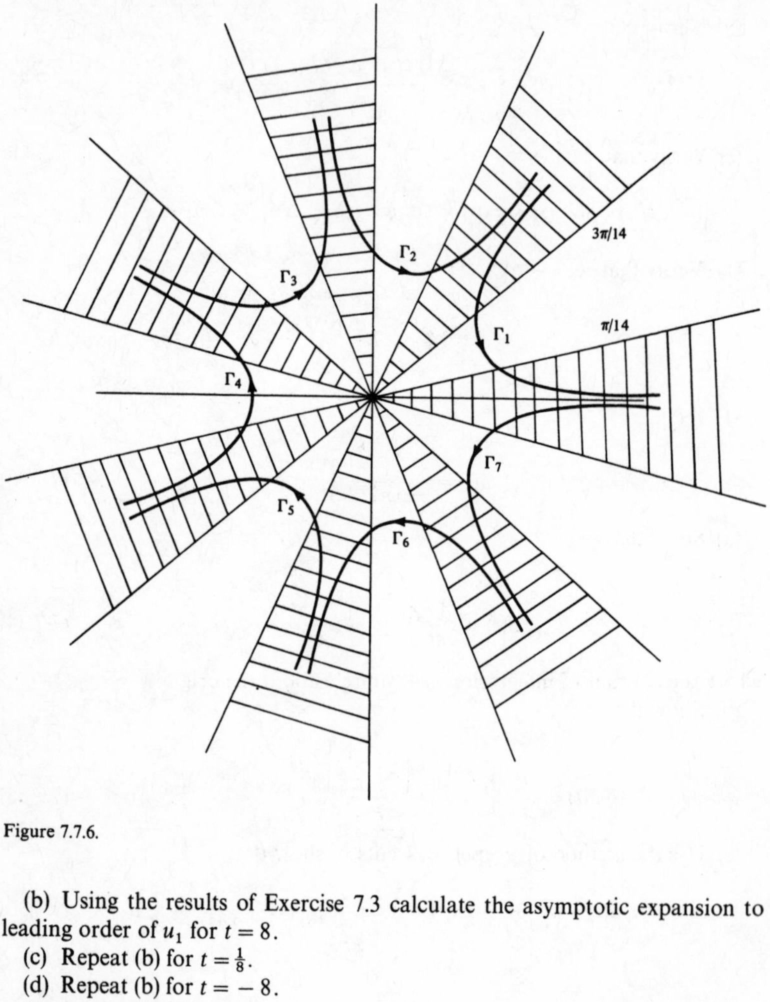

Figure 7.7.6.

(b) Using the results of Exercise 7.3 calculate the asymptotic expansion to leading order of u_1 for $t = 8$.

(c) Repeat (b) for $t = \frac{1}{8}$.

(d) Repeat (b) for $t = -8$.

(e) Repeat (b) for $t = 0$.

(f) Repeat (b) for $t = 1$.

7.22. Let

$$I_n(\omega) = \int_0^\infty \exp\left\{ -t^{-n} + i\omega t - i\alpha \frac{t^\nu}{\nu} \right\} t^{-r}\, dt. \tag{7.7.39}$$

(a) If $1 < \nu$ and $\alpha > 0$, then introduce the new variable of integration

$$\tau = t\,\omega^{-1/(\nu-1)} \tag{7.7.40}$$

and show that

$$I_n(\omega) \sim \frac{2\pi}{\nu-1} \alpha^{\gamma} \omega^{-\beta} \exp\left\{-\left(\frac{\omega}{\alpha}\right)^{-n/(\nu-1)} + i\frac{\nu-1}{\nu} \omega^{\nu/(\nu-1)}\alpha^{-1/(\nu-1)} - i\pi/4\right\}. \tag{7.7.41}$$

Here,

$$\beta = \frac{\nu/2 + r - 1}{\nu - 1}, \qquad \gamma = \frac{r - 1/2}{\nu - 1}. \tag{7.7.42}$$

(b) Show that $I_n(\omega) \to 0$ as $\omega \to \infty$ if either of the following conditions are true

(i) $0 \le r, \quad \nu > 2;$

(ii) $0 < r, \quad 1 < \nu \le 2, \quad r > 1 - \frac{\nu}{2}.$

(c) Show that $I_n(\omega) \to 0$ as $\omega \to \infty$ if $\alpha \le 0$, $r > 0$.

7.23. Let

$$I(\omega) = \int_0^\infty f(t) \exp\{i\omega t\}\, dt. \tag{7.7.43}$$

Suppose that $f(t)$ is locally integrable and

$$f(t) \sim \exp\{-i\alpha t^\nu\} \sum_{m=0}^{\infty} C_m t^{-r_m}, \qquad t \to \infty, \tag{7.7.44}$$

with $\mathrm{Re}(r_m) \uparrow \infty$. Set $f(t) = s(t) + g(t)$ where

$$s(t) = \exp\{-t^{-2} - i\alpha t^\nu\} \sum_{\substack{m \\ \mathrm{Re}(r_m) \le 1}} C_m t^{-r_m} \tag{7.7.45}$$

and

$$g(t) = f(t) - s(t).$$

(a) Show that $g(t)$ is absolutely integrable on $[0, \infty)$.

(b) Use the Riemann-Lebesgue lemma and Exercise 7.22 to show that $I(\omega) \to 0$ as $\omega \to \infty$ whenever the conditions of either 7.22(b) or 7.22(c) are satisfied. In these conditions $\mathrm{Re}(r_0) = r$.

7.24. Let

$$I_k(z) = \int_0^\infty \exp\{-t^{-(k+1)} + i\omega t^\nu\}\, t^{z-r-1}\, dt, \qquad z = x + iy, \qquad x < r. \tag{7.7.46}$$

(a) Show that $I_k(z)$ can be analytically continued to the entire z plane as a holomorphic function by deforming the contour of integration in an appropriate way.

(b) Introduce the new variable of integration τ defined by

$$t = |y|^{1/\nu} \tau \tag{7.7.47}$$

and use the method of steepest descent to show that as $|y| \to \infty$,

$$I_k = \begin{cases} 0(|y|^{(x-r)/v-\frac{1}{2}}), & \operatorname{sgn}(\omega) = -\operatorname{sgn}(y), \\ 0(|y|^{-N}), & \text{any } N, \quad \operatorname{sgn}(\omega) = \operatorname{sgn}(y). \end{cases} \tag{7.7.48}$$

7.25. Let

$$M[f;z] = \int_0^\infty f(t)\, t^{z-1}\, dt, \qquad z = x + iy. \tag{7.7.49}$$

Suppose that

$$f(t) \sim \exp\{i\omega t^v\} \sum_{m=0}^{\infty} C_m\, t^{-r_m}, \qquad t \to \infty,$$

$$f(t) = 0(t^c), \qquad t \to 0+, \qquad -c < \operatorname{Re}(r_0). \tag{7.7.50}$$

Introduce

$$s_k(t) = \exp\{i\omega t^v - t^{-(k+1)}\} \sum_{\substack{m \\ \operatorname{Re}(r_m) \le \operatorname{Re}(r_0)+k}} C_m\, t^{-r_m} \tag{7.7.51}$$

and

$$f_k = f - s_k. \tag{7.7.52}$$

(a) Show that $M[f_k;z]$ is analytic for $-c < x < \operatorname{Re}(r_0) + k$ and $M[f_k;z] \to 0$ as $|y| \to \infty$ in this strip.

(b) Use Exercise 7.24 to show that

(i) $M[s_k;z]$ can be analytically continued to the entire z plane as a holomorphic function and

(ii) $M[s_k;z] = 0(|y|^{(x-\operatorname{Re}(r_0))/v-\frac{1}{2}})$.

7.26. (a) Show that the line connecting z_+ and z_- in Figure 7.3.1 is a path of ascent with respect to z_- for $0 < \theta < \pi/3$, a boundary for $\theta = \pi/3$, and a path of descent with respect to z_- for $\pi/3 < \theta < \pi$.

(b) For $\pi/3 < \theta < \pi$, justify deforming the contour C in Figure 7.3.1 onto a sum of descent contours from z_- and from z_+.

(c) Show that (7.3.17) is actually valid for $\pi/3 < \theta \le \pi$ if we replace $|s|$ by $se^{i\pi}$.

(d) Show that, with $|s|$ replaced by $se^{i\pi}$, (7.3.17) is valid for $\pi/3 < \theta < 5\pi/3$.

7.27. Let

$$I(\lambda) = \int_\Gamma h(\lambda\, w(z))\, g(z)\, dz. \tag{7.7.53}$$

Here Γ is a path of steepest descent from z_0 to ∞ for $h(\lambda\, w(z))$ and

$$w(z_0) = w'(z_0) = \cdots = w^{(n-1)}(z_0) = 0, \qquad w^{(n)}(z_0) = a\, e^{i\alpha} \ne 0. \tag{7.7.54}$$

(a) Let

$$t = w(z). \tag{7.7.55}$$

If Γ is an image of the ray of descent,

$$\arg t = \beta \tag{7.7.56}$$

in the t plane. Then show that locally

$$z - z_0 = \left(\frac{n!}{a}\right)^{1/n} t^{1/n} e^{-i\alpha/n} [1 + O(|t|^{1/n})] \tag{7.7.57}$$

and the direction of Γ at z_0 must be of the form

$$\arg(z - z_0) = \frac{\beta - \alpha}{n} + \frac{2\pi p}{n} \tag{7.7.58}$$

for some fixed integer $p, 0 \le p \le n-1$.

(b) Let

$$t = se^{i\beta} \tag{7.7.59}$$

and

$$h_\beta(\lambda s) = h(\lambda s e^{i\beta}). \tag{7.7.60}$$

Then show that

$$I(\lambda) = \int_0^\infty h_\beta(\lambda s)\, G(s)\, ds, \tag{7.7.61}$$

where

$$G(s) = g(z(s)) \frac{dz}{ds}. \tag{7.7.62}$$

(c) If

$$G(s) \sim \sum_{m=0}^{\infty} p_m s^{r_m}, \qquad s \to 0+, \tag{7.7.63}$$

with $\operatorname{Re} r_m \uparrow +\infty$, show that

$$I(\lambda) \sim \sum_{m=0}^{\infty} p_m \lambda^{-r_m - 1} M[h_\beta; r_m + 1]. \tag{7.7.64}$$

(d) If

$$g(z) = g_0(z - z_0)^{\gamma - 1} + o(|z - z_0|^{\gamma - 1}) \tag{7.7.65}$$

in some sector containing the descent direction at z_0, then show that

$$I(\lambda) \sim \frac{g_0}{n} \left(\frac{n!}{a\lambda}\right)^{\gamma/n} \exp\left\{\frac{i\gamma(\beta - \alpha + 2\pi p)}{n}\right\} M\left[h_\beta; \frac{\gamma}{n}\right]. \tag{7.7.66}$$

(e) Specialize these results to the case

$$h(t) = e^t \tag{7.7.67}$$

and show that (7.7.66) agrees with (7.2.9).

(f) Suppose that Γ is a boundary contour between two hills. Show that the above results hold and are derived in exactly the same way.

7.28. (a) Show that the results for the Airy function $\mathrm{Ai}(t)$ in Section 7.3 can be summarized as shown in Figure 7.7.7. [The point $t = -\frac{1}{2}$ and the capital letters are related to part (b).]

(b) For the function

$$\mathrm{Ai}(\lambda\, w(z)); \qquad w(z) = \frac{z^2}{2} - z, \tag{7.7.68}$$

show that the hills and valleys of the exponent with respect to its zeros are as shown in Figure 7.7.8. Here $z = 0$ and $z = 2$ are the points where $w = 0$ and $z = 1$ is the saddle point. [*Hint*: Find the images of the relevant contours and points of Figure 7.7.8 under the mapping $t = w(z)$; verify that $t = -\frac{1}{2}$ is the preimage of the saddle point.]

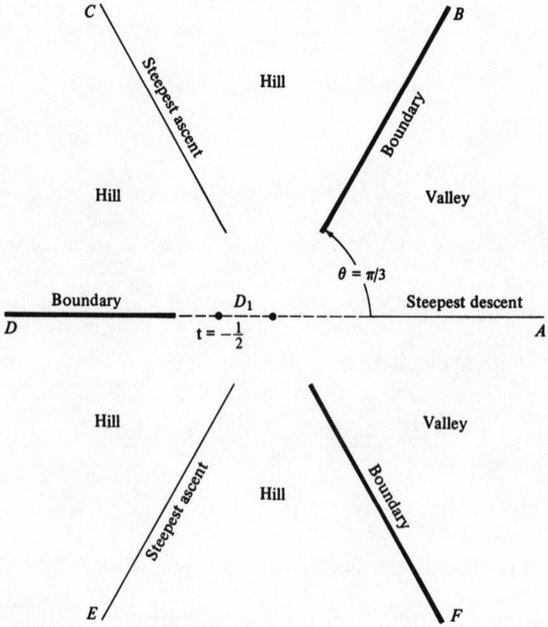

Figure 7.7.7.

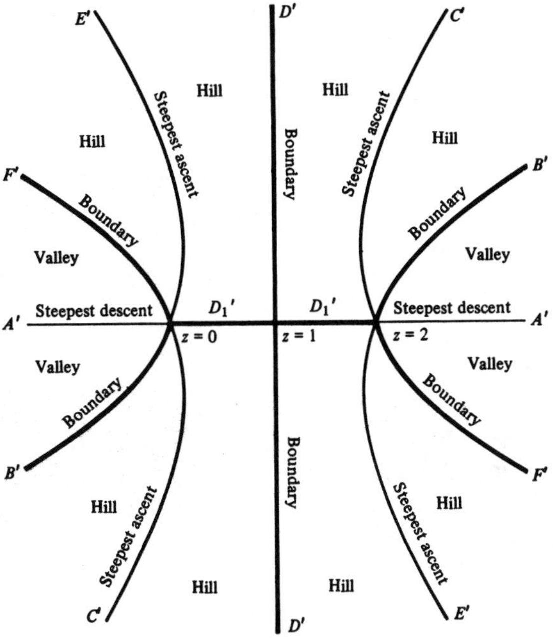

Figure 7.7.8.

7.29. Let

$$I(\lambda;\Gamma)=\int_{\Gamma}\operatorname{Ai}(\lambda\, w(z))\, g(z)\, dz. \tag{7.7.69}$$

Here $w(z)$ is as defined in (7.7.68) above; $g(z)$ is assumed to be analytic whenever necessary to perform the deformations of contour required below. With the contours described in Figure 7.7.9 show that

(a)
$$I(\lambda;B_2)\sim\lambda^{-1}\left\{g(1)\sin\left[\frac{2}{3}\left(\frac{\lambda}{2}\right)^{3/2}\right]\right\}+\frac{2}{3}\frac{g(0)}{\lambda}. \tag{7.7.70}$$

(b)
$$I(\lambda;D_1)\sim\frac{g(0)}{3\lambda}. \tag{7.7.71}$$

(c)
$$I(\lambda;B_1)\sim\frac{i\,g(1)}{\lambda}\cos\left\{\frac{2}{3}\left(\frac{\lambda}{2}\right)^{3/2}\right\}. \tag{7.7.72}$$

(d)
$$I(\lambda;D_5)\sim\frac{g(z_4)\exp\{-\frac{2}{3}[\lambda\, w(z_4)]^{3/2}\}}{2\sqrt{\pi}\,\lambda^{7/4}\,[w(z_4)]^{3/4}\,(z_4-1)}. \tag{7.7.73}$$

(e)
$$I(\lambda;B_4)\sim -I(\lambda;D_1). \tag{7.7.74}$$

(f) Find the leading term of the asymptotic expansion of the integral from z_2 to $i\infty$.

(g) Find the leading term of the asymptotic expansion of the integral from z_3 to $-\infty$.

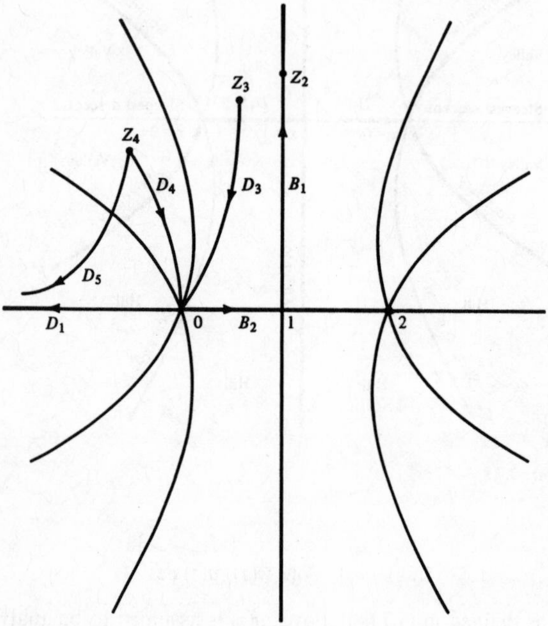

Figure 7.7.9.

REFERENCES

V. Barcilon and N. Bleistein. Scattering of inertial waves in a rotating fluid. *Stud. Appl. Math.* **48,** 91–104, 1969.
Exercise 7.19 is based on the results of this paper.

N. Bleistein and R. A. Handelsman. A generalization of the method of steepest descent. *JIMA* **10,** 211–230, 1972.
The generalization mentioned at the end of Section 7.2 is developed in this paper.

N. Bleistein, R. A. Handelsman, and J. S. Lew. Functions whose Fourier transforms decay at infinity: An extension of the Riemann-Lebesgue lemma. *SIAM J. Math. Anal.* **3,** 485–495, 1972.
Exercises 7.24 and 7.25 are based on this paper.

N. G. de Bruijn. *Asymptotic Methods in Analysis.* North-Holland Publishing Co., Amsterdam, 1958.
Detailed discussions of the saddle point method and the method of steepest descents are given along with several examples.

G. F. Carrier, M. Krook, and C. E. Pearson. *Functions of a Complex Variable.* McGraw-Hill, New York, 1966.
The authors discuss the general problem of contour integration and in particular the method of steepest descents.

E. T. Copson. *Asymptotic Expansions.* University Press, Cambridge, 1965.
In this monograph an extensive discussion of the method of steepest descents is presented.

P. Debye. Näherungsformeln für die Zylinderfunktionen für grosse Werte des Arguments und unbeschränkt veränderliche Werte des Index. *Math. Annal.* **67,** 535–558, 1909.
The method of steepest descents as introduced by Riemann is further developed here.

K. O. Friedrichs. *Special Topics in Analysis.* Lecture notes. Courant Inst. Math. Sci., New York, 1953–1954.
The asymptotic expansion of the Airy function for complex argument is developed here via the method of steepest descents.

B. Granoff and N. Bleistein. Asymptotic solutions of a 6th order equation with two turning points. Part I. Derivation by the method of steepest descent. *SIAM J. Math. Anal.* **3,** 45–57, 1972.
Exercise 7.21 is based on results obtained in this paper.

H. Jeffreys. *Asymptotic Approximations.* University Press, Oxford, 1962.
This book offers a detailed description of the method of steepest descents.

D. S. Jones. High-frequency refraction and diffraction in general media. *Phil. Trans. Roy. Soc. London* **A 255,** 341–387, 1963.
In this paper the zeros of the argument of the kernel function play a crucial role as critical points for a contour integral.

H. M. Nussenzveig. High-frequency scattering by an impenetrable sphere. *Anals of Phys.* **34,** 23–95, 1965.
In this paper the application of the method of steepest descents leads to a residue series of subdominant terms having an important physical interpretation. Also, the forms of the asymptotic expansions (7.7.27), (7.7.28) are introduced.

F. W. J. Olver. Why steepest descents? *SIAM Review* **12,** 228–247, 1970.
The author makes use of steepest descent paths to obtain numerical error estimates in an asymptotic expansion.

B. RIEMANN. Sullo suolgimento del quotiente di due serie pergeometriche in fraziune contriva infinita. *Collected Works*. 2nd Ed. Dover, New York, 1953.
The basic idea of the method of steepest descents is presented here.

L. SIROVICH. *Techniques of Asymptotic Analysis*. Springer-Verlag, New York, 1971.
The author presents comprehensive discussions of the method of steepest descents and of the saddle point method.

E. C. TITCHMARSH. *The Theory of Functions*. 2nd Ed. University Press, Oxford, 1952.
This is an excellent reference for the subject of contour integration in the complex plane.

B. L. VAN DER WAERDEN. On the method of saddle points. *Appl. Sci. Res.* **B2,** 33–45, 1951.
The method of steepest descents is developed by a different technique than in the text.

G. N. WATSON. *A Treatise on the Theory of Bessel Functions*. University Press, Cambridge, Paperback, 1966.
The method of steepest descents is applied to obtain asymptotic expansions for the various Bessel functions in the complex plane.

8 Asymptotic Expansion of Multiple Integrals

8.1. Introduction

A natural extension to the problems treated in Chapters 4 through 6 would be to consider the asymptotic behavior of functions defined by multiple integrals of the form

$$I(\lambda) = \int_{\mathscr{D}} h(\lambda\phi(\mathbf{x}))\, g(\mathbf{x})\, d\mathbf{x}, \qquad \mathbf{x} = (x_1, x_2, \ldots, x_n). \tag{8.1.1}$$

Here $\mathscr{D}$ is some (not necessarily bounded) domain in n-dimensional x space.

Unfortunately, the techniques developed for the one-dimensional analog of (8.1.1) cannot be directly extended to the general multidimensional case. There is a class of multidimensional integrals, however, for which we can develop an asymptotic theory. Indeed, throughout the remainder of the chapter, we shall assume that the kernel function $h(t) \equiv \exp(t)$ and hence we shall restrict our consideration to the study of integrals of the form

$$I(\lambda) = \int_{\mathscr{D}} \exp\{\lambda\phi(\mathbf{x})\}\, g(\mathbf{x})\, d\mathbf{x}. \tag{8.1.2}$$

We shall further assume that ϕ is real, and λ is either real, in which event $I(\lambda)$ is said to be of Laplace type, or purely imaginary, in which event $I(\lambda)$ is said to be of Fourier type. Although we could consider (8.1.2) in each of the limits $|\lambda| \to \infty$ and $|\lambda| \to 0$, we shall only investigate the former here because it is of greater interest both mathematically and physically.

There are many problems of mathematical physics in which integrals of the form (8.1.2) arise. Of special interest are those of electromagnetic theory which involve the diffraction or scattering of radiation by obstacles. Indeed, in Section 2.7 we formulated such a problem. There we considered only the scalar or "acoustic" case, but we could treat the vector case as well.

8.2. Asymptotic Expansion of Double Integrals of Laplace Type

Here we shall consider the behavior, as $\lambda \to \infty$, of functions $I(\lambda)$ defined by integrals of the form

$$I(\lambda) = \int_{\mathscr{D}} \exp\{\lambda\phi(\mathbf{x})\}\, g_0(\mathbf{x})\, d\mathbf{x}, \qquad \mathbf{x} = (x_1, x_2). \tag{8.2.1}$$

For ease of discussion we shall assume that the domain $\mathscr{D}$ is finite, simply connected, and bounded by a smooth curve Γ. More precisely, if Γ is parametrized by the relations

$$x_1 = x_1(s), \qquad x_2 = x_2(s), \qquad 0 \le s \le L, \tag{8.2.2}$$

where s is arc-length, then both $x_1(s)$ and $x_2(s)$ are to be continuously differentiable on $[0, L]$. In addition, we suppose that, as s increases, Γ is described in a counterclockwise sense. With regard to the functions $\phi(\mathbf{x})$ and $g_0(\mathbf{x})$ we assume both are sufficiently differentiable for the operations below.

As we have indicated, our objective is to study $I(\lambda)$ as $\lambda \to \infty$. From our analysis of one-dimensional integrals of Laplace type, we might anticipate that the results here will depend heavily on the behavior of ϕ. In particular, it is reasonable to expect that the critical points for $I(\lambda)$ are those points in $\bar{\mathscr{D}}$, the closure of $\mathscr{D}$, at which ϕ achieves its absolute maximum. As we shall show, this is indeed the case.

Let us first suppose that $\nabla\phi \neq \mathbf{0}$ in $\bar{\mathscr{D}}$. This of course implies that the absolute maximum of ϕ in $\bar{\mathscr{D}}$ must occur on the boundary Γ. For simplicity we assume that the point at which this occurs is unique and we label it $\mathbf{x}_0 = (x_{10}, x_{20})$. We have that the first directional derivative of ϕ in the direction of $\mathbf{T}$, the unit tangent vector to Γ, must vanish at $\mathbf{x} = \mathbf{x}_0$. If we let $s = 0$ correspond to $\mathbf{x} = \mathbf{x}_0$, so that

$$x_1(0) = x_{10}, \qquad x_2(0) = x_{20}, \tag{8.2.3}$$

then

$$\nabla\phi \cdot \mathbf{T}\big|_{s=0} = \phi_{x_1}(\mathbf{x}_0)\, \dot{x}_1(0) + \phi_{x_2}(\mathbf{x}_0)\, \dot{x}_2(0) = 0, \qquad (\dot{\ }) = \frac{d}{ds}. \tag{8.2.4}$$

It follows from (8.2.4) that $\nabla\phi$ is normal to Γ at $\mathbf{x} = \mathbf{x}_0$. If we introduce

$$\mathbf{N}(s) = (\dot{x}_2(s), -\dot{x}_1(s)), \tag{8.2.5}$$

the unit outward normal vector to Γ, then we find that

$$\nabla\phi(\mathbf{x}_0) = |\nabla\phi(\mathbf{x}_0)|\, \mathbf{N}(0). \tag{8.2.6}$$

Actually we can only conclude from (8.2.4) that $\nabla\phi(\mathbf{x}_0)$ is either parallel or antiparallel to $\mathbf{N}(0)$. That (8.2.6) is correct follows from the fact that the absolute maximum of ϕ in $\bar{\mathscr{D}}$ occurs at $\mathbf{x}=\mathbf{x}_0$ which, in turn, implies that $\nabla\phi$ must point out from $\mathscr{D}$ at that point.

In the treatment of one-dimensional integrals of Laplace type we found that if ϕ' is nonzero throughout the domain of integration, then the asymptotic expansion can be generated via integration by parts. The same is true here except now the analog of integration by parts is the divergence theorem of vector calculus.

We begin by introducing the vector

$$\mathbf{H}_0 = g_0 \frac{\nabla\phi}{|\nabla\phi|^2}. \tag{8.2.7}$$

Because

$$e^{\lambda\phi}\, g_0 = -\frac{e^{\lambda\phi}}{\lambda}\nabla\cdot\mathbf{H}_0 + \frac{1}{\lambda}\nabla\cdot(\mathbf{H}_0\, e^{\lambda\phi}) \tag{8.2.8}$$

we have by the divergence theorem

$$I(\lambda) = \frac{1}{\lambda}\oint_\Gamma e^{\lambda\phi}\, \mathbf{H}_0\cdot\mathbf{N}\, ds - \frac{1}{\lambda}\int_{\mathscr{D}} e^{\lambda\phi}\, g_1\, d\mathbf{x}. \tag{8.2.9}$$

Here we have set

$$g_1 = \nabla\cdot\mathbf{H}_0. \tag{8.2.10}$$

We claim that, to leading order, our problem has been reduced to the study of the boundary integral in (8.2.9) which, of course, is one-dimensional and of Laplace type. Indeed the area integral in (8.2.9) is of the same form as $I(\lambda)$ itself and has the multiplicative factor λ^{-1}. Hence it is reasonable to expect that it is of lower order than $I(\lambda)$ and that the leading term in the expansion comes from the boundary integral. We shall elaborate on this point below.

Let us now consider

$$J(\lambda) = \frac{1}{\lambda}\oint_\Gamma e^{\lambda\phi}\, \mathbf{H}_0\cdot\mathbf{N}\, ds \tag{8.2.11}$$

to which we can apply the theory developed in Chapter 5. Indeed, if we set

$$\psi(s) = \phi(\mathbf{x}(s)), \tag{8.2.12}$$

then it follows by Laplace's formula (5.1.21), with ϕ in that formula set equal to $-\psi(s)$, that

$$J(\lambda) \sim e^{\lambda\phi(\mathbf{x}_0)}\sqrt{\frac{2\pi}{\lambda^3|\ddot{\psi}(0)|}}\,(\mathbf{H}_0\cdot\mathbf{N})\Big|_{s=0}. \tag{8.2.13}$$

Here we have assumed that the maximum of ψ at $s=0$ is simple so that $\ddot{\psi}(0)<0$.

Upon applying the divergence theorem to

$$I_1(\lambda) = \frac{1}{\lambda}\int_{\mathscr{D}} e^{\lambda\phi}\, g_1\, d\mathbf{x} \tag{8.2.14}$$

we obtain

$$I_1(\lambda) = \frac{1}{\lambda^2}\oint_{\Gamma} e^{\lambda\psi}\, \mathbf{H}_1\cdot\mathbf{N}\, ds - \frac{1}{\lambda^2}\int_{\mathscr{D}} e^{\lambda\phi}\, g_2\, d\mathbf{x}. \tag{8.2.15}$$

Here

$$\mathbf{H}_1 = g_1\frac{\nabla\phi}{|\nabla\phi|^2} \quad \text{and} \quad g_2 = \nabla\cdot\mathbf{H}_1. \tag{8.2.16}$$

Laplace's method can be applied to the boundary integral in (8.2.15) and the area integral can be easily estimated. Indeed we find that

$$\frac{1}{\lambda^2}\oint_{\Gamma} e^{\lambda\psi}\, \mathbf{H}_1\cdot\mathbf{N}\, ds = O(e^{\lambda\phi(\mathbf{x}_0)}\,\lambda^{-5/2}), \tag{8.2.17}$$

$$\frac{1}{\lambda^2}\int_{\mathscr{D}} e^{\lambda\phi}\, g_2\, d\mathbf{x} = O(e^{\lambda\phi(\mathbf{x}_0)}\,\lambda^{-2}).^{1} \tag{8.2.18}$$

It therefore follows that $I_1 = O(e^{\lambda\phi(\mathbf{x}_0)}\,\lambda^{-2})$ and hence $I(\lambda) \sim J(\lambda)$. Thus from (8.2.13) we find that

$$I(\lambda) \sim e^{\lambda\phi(\mathbf{x}_0)}\sqrt{\frac{2\pi}{\lambda^3|\ddot{\psi}(0)|}}\,(\mathbf{H}_0\cdot\mathbf{N})\Big|_{s=0} \tag{8.2.19}$$

We of course wish to express (8.2.19) in terms of the original functions ϕ and g_0. From (8.2.6) and (8.2.7) we have

$$(\mathbf{H}_0\cdot\mathbf{N})\Big|_{s=0} = \frac{g_0(\mathbf{x}_0)}{|\nabla\phi(\mathbf{x}_0)|}. \tag{8.2.20}$$

Also

$$\begin{aligned}\ddot{\psi}(0) &= \frac{d}{ds}\nabla\phi\Big|_{s=0}\cdot\mathbf{T}(0) + \nabla\phi(\mathbf{x}_0)\left(\frac{d\mathbf{T}}{ds}\right)\Big|_{s=0} \\ &= \phi_{x_1x_1}(\mathbf{x}_0)\,\dot{x}_1^2(0) + 2\,\phi_{x_1x_2}(\mathbf{x}_0)\,\dot{x}_1(0)\,\dot{x}_2(0) \\ &\quad + \phi_{x_2x_2}(\mathbf{x}_0)\,\dot{x}_2^2(0) + \kappa(0)\,\nabla\phi(\mathbf{x}_0)\cdot\mathbf{n}(0).\end{aligned} \tag{8.2.21}$$

Here $\kappa(0)$ is the curvature of Γ at $\mathbf{x} = \mathbf{x}_0$ and $\mathbf{n}(0)$ is the principal normal vector

[1] Actually the estimate (8.2.18) is extremely crude. It is, however, sufficient for our present purposes.

to Γ at that point. Because $\mathbf{n}(0) = \mp \mathbf{N}(0)$ according to whether Γ is convex or concave at $\mathbf{x} = \mathbf{x}_0$, we have

$$\ddot{\psi}(0) = \phi_{x_1x_1}(\mathbf{x}_0)\,\dot{x}_1^2(0) + 2\,\phi_{x_1x_2}(\mathbf{x}_0)\,\dot{x}_1(0)\,\dot{x}_2(0) + \phi_{x_2x_2}(\mathbf{x}_0)\,\dot{x}_2^2(0) \mp \kappa(0)\,|\nabla\phi(\mathbf{x}_0)|. \tag{8.2.22}$$

Clearly the minus (plus) sign holds when Γ is convex (concave) at $\mathbf{x} = \mathbf{x}_0$.

We can further reduce our expression for $\ddot{\psi}(0)$ by noting that (8.2.6) implies

$$\dot{x}_1(0) = \frac{-\phi_{x_2}(\mathbf{x}_0)}{|\nabla\phi(\mathbf{x}_0)|}, \qquad \dot{x}_2(0) = \frac{\phi_{x_1}(\mathbf{x}_0)}{|\nabla\phi(\mathbf{x}_0)|}. \tag{8.2.23}$$

Upon combining (8.2.19), (8.2.20), (8.2.22), and (8.2.23) we finally obtain

$$I(\lambda) \sim \exp\{\lambda\phi(\mathbf{x}_0)\}\, g_0(\mathbf{x}_0)\sqrt{\frac{2\pi}{\lambda^3}} \times \left[\left|\phi_{x_1x_1}\,\phi_{x_2}^2 - 2\,\phi_{x_1x_2}\,\phi_{x_1}\,\phi_{x_2} + \phi_{x_2x_2}\,\phi_{x_1}^2 \mp \kappa|\nabla\phi|^3\right|\right]_{\mathbf{x}=\mathbf{x}_0}^{-1/2}. \tag{8.2.24}$$

If g_0 and ϕ are both infinitely differentiable in $\bar{\mathscr{D}}$, then the above process can be repeated indefinitely and an infinite expansion can be obtained. In this section, however, we shall be content with obtaining leading terms only, while in the following section, where Laplace integrals of higher dimension are considered, further terms will be found.

We now consider $I(\lambda)$ in the case where the absolute maximum of ϕ in $\bar{\mathscr{D}}$ occurs only at the *interior* point $\mathbf{x}_0 = (x_{10}, x_{20})$. In particular we assume that

$$\nabla\phi(\mathbf{x}_0) = 0, \qquad \phi_{x_1x_1}(\mathbf{x}_0)\,\phi_{x_2x_2}(\mathbf{x}_0) - \phi_{x_1x_2}^2(\mathbf{x}_0) > 0, \qquad \phi_{x_1x_1}(\mathbf{x}_0) < 0 \tag{8.2.25}$$

and that $\nabla\phi$ is nonzero at all other points of $\bar{\mathscr{D}}$. If, in a given problem, this last restriction is not satisfied, then $\mathscr{D}$ can always be subdivided into regions so that, in each, $\nabla\phi$ vanishes at only one point.

Because we anticipate that $\mathbf{x} = \mathbf{x}_0$ is the dominant critical point, we must be concerned with the local behavior of ϕ near this point. By Taylor's theorem we have

$$\phi(\mathbf{x}) = \phi(\mathbf{x}_0) + \phi_{x_1x_1}(\mathbf{x}_0)\frac{(x_1 - x_{10})^2}{2} + \phi_{x_1x_2}(\mathbf{x}_0)\,(x_1 - x_{10})\,(x_2 - x_{20}) + \phi_{x_2x_2}(\mathbf{x}_0)\frac{(x_2 - x_{20})^2}{2} + \cdots. \tag{8.2.26}$$

Our method of analysis involves the reduction of $I(\lambda)$ to an integral of canonical form in which the exponent ϕ is replaced by a quadratic function. Once this is accomplished, the asymptotic expansion, at least to leading order, is readily obtained as we shall show.

It is well known from linear algebra theory that there exists an orthogonal[2] matrix Q which diagonalizes the symmetric matrix

$$A = (\phi_{x_i x_j}(\mathbf{x}_0)), \qquad i, j = 1, 2. \tag{8.2.27}$$

Indeed

$$Q^T A Q = \begin{pmatrix} \lambda_1 & 0 \\ 0 & \lambda_2 \end{pmatrix}, \tag{8.2.28}$$

where λ_1 and λ_2 are the eigenvalues of A. These eigenvalues are both negative and are such that

$$\lambda_1 \lambda_2 = \det A = \phi_{x_1 x_1}(\mathbf{x}_0)\, \phi_{x_2 x_2}(\mathbf{x}_0) - \phi^2_{x_1 x_2}(\mathbf{x}_0). \tag{8.2.29}$$

Let us introduce the vector $\mathbf{z} = (z_1, z_2)$ and the function $f(\mathbf{z})$ defined by

$$(\mathbf{x} - \mathbf{x}_0)^T = QR\mathbf{z}^T, \qquad R = \begin{pmatrix} |\lambda_1|^{-1/2} & 0 \\ 0 & |\lambda_2|^{-1/2} \end{pmatrix}, \qquad f(\mathbf{z}) = \phi(\mathbf{x}_0) - \phi(\mathbf{x}(\mathbf{z})). \tag{8.2.30}$$

We note that $\mathbf{x} = \mathbf{x}_0$ corresponds to $\mathbf{z} = \mathbf{0}$ and that near $\mathbf{z} = \mathbf{0}$,

$$f(\mathbf{z}) \sim \tfrac{1}{2}(z_1^2 + z_2^2). \tag{8.2.31}$$

We now seek a second change of variables so that this local behavior will hold in the large.

We can readily show that because $\nabla\phi$ vanishes in $\overline{\mathscr{D}}$ only at $\mathbf{x} = \mathbf{x}_0$, there exist functions

$$\xi_i = h_i(\mathbf{z}), \qquad i = 1, 2 \tag{8.2.32}$$

such that

$$|\boldsymbol{\xi}|^2 = h_1^2 + h_2^2 = 2f(\mathbf{z}), \qquad \boldsymbol{\xi} = (\xi_1, \xi_2) \tag{8.2.33}$$

and

$$h_i = z_i + o(|\mathbf{z}|), \qquad i = 1, 2, \tag{8.2.34}$$

as $|\mathbf{z}| \to 0$. If we denote by D the image of $\mathscr{D}$ under the two transformations (8.2.30) and (8.2.32), then we further have that the Jacobian

$$J(\boldsymbol{\xi}) = \left| \det\!\left(\frac{\partial x_i}{\partial \xi_j} \right) \right| \tag{8.2.35}$$

of the composite transformation is nonzero and finite for $\boldsymbol{\xi}$ in $\overline{D}$. In other

---

[2] An orthogonal matrix $Q = (q_{ij})$ is such that $\sum_{i=1}^{n} q_{ij}\, q_{ik} = \delta_{jk}$, $j, k = 1, 2$. Here δ_{jk} is the Kronecker delta. We note that $Q^{-1} = Q^T$.

words, the mapping from $\mathscr{D}$ to D is one-to-one. We also note that

$$J(\mathbf{0}) = (\lambda_1\lambda_2)^{-1/2} = (\phi_{x_1x_1}(\mathbf{x}_0)\,\phi_{x_2x_2}(\mathbf{x}_0) - \phi^2_{x_1x_2}(\mathbf{x}_0))^{-1/2} \tag{8.2.36}$$

Although only the existence of the functions h_1 and h_2 is needed for the actual determination of the asymptotic expansion, it is instructive to indicate their construction. Thus let us introduce the "polar" coordinates consisting of the "radial" variable

$$\rho^2 = f(\mathbf{z}) \tag{8.2.37}$$

and the "angular" variable θ defined by

$$\frac{\nabla_z\rho\cdot\nabla_z\theta}{f^{1/2}} = 0. \tag{8.2.38}$$

From (8.2.38) we see that $\theta(\mathbf{z})$ is constant along a "ray" or orthogonal trajectory of f. That constant, moreover, is determined by

$$\lim_{\mathbf{z}\to\mathbf{0}} \frac{\nabla_z f}{|\nabla_z f|} = (\cos\theta, \sin\theta), \tag{8.2.39}$$

where the limit is taken along a fixed orthogonal trajectory. We now set

$$\xi_1 = h_1 = \sqrt{2}\rho\cos\theta, \qquad \xi_2 = h_2 = \sqrt{2}\rho\sin\theta \tag{8.2.40}$$

and clearly (8.2.34) is satisfied. That the Jacobian $J(\xi)$ has the stated properties follows from the simple geometrical fact that the level curves of ϕ and their orthogonal trajectories form a simple covering of $\overline{\mathscr{D}} - \mathbf{x}_0$. This in turn relies on the fact that $\nabla\phi \neq 0$ in $\overline{\mathscr{D}} - \mathbf{x}_0$.

In terms of the new variables ξ_1, ξ_2, the integral $\mathbf{I}(\lambda)$ becomes

$$I(\lambda) = \exp\{\lambda\phi(\mathbf{x}_0)\}\,K(\lambda), \tag{8.2.41}$$

where

$$K(\lambda) = \int_D G_0(\xi)\exp\left\{-\frac{\lambda}{2}\,\xi\cdot\xi\right\}d\xi. \tag{8.2.42}$$

Here

$$G_0(\xi) = J(\xi)\,g_0(\mathbf{x}(\xi)). \tag{8.2.43}$$

Our problem now is to obtain an asymptotic expansion of $K(\lambda)$. This integral is significantly simpler than (8.2.1) due to the simple nature of the exponent in (8.2.42). As the first step we set

$$G_0(\xi) = G_0(\mathbf{0}) + \xi\cdot\mathbf{H}_0(\xi), \qquad \mathbf{H}_0 = (H_{01}, H_{02}). \tag{8.2.44}$$

Clearly $\mathbf{H}_0$ is not uniquely determined by (8.2.44). Our results however will be independent of the particular $\mathbf{H}_0$ chosen as long as it is well behaved throughout $\overline{D}$. We can select, for example,

$$H_{01} = \frac{1}{\xi_1}[G_0(\xi_1, \xi_2) - G_0(0, \xi_2)],$$

$$H_{02} = \frac{1}{\xi_2}[G_0(0, \xi_2) - G_0(0, 0)], \tag{8.2.45}$$

which we note are well behaved, even at the origin, as long as G_0 is differentiable with respect to both of its arguments there.

Perhaps it is worthwhile to motivate the representation (8.2.44). We anticipate that the only critical point for $K(\lambda)$ is the origin $\boldsymbol{\xi} = \mathbf{0}$. Because $\boldsymbol{\xi} \cdot \mathbf{H}_0$ vanishes at $\boldsymbol{\xi} = \mathbf{0}$, we might expect that the contribution corresponding to $G_0(\mathbf{0})$ will dominate that corresponding to $\boldsymbol{\xi} \cdot \mathbf{H}_0$. Of course this will only be so if $G_0(\mathbf{0})$ is not zero which we shall assume.

To establish the validity of the preceding remark we first note that

$$\exp\left\{-\frac{\lambda}{2}\boldsymbol{\xi}\cdot\boldsymbol{\xi}\right\}(\boldsymbol{\xi}\cdot\mathbf{H}_0) = -\frac{1}{\lambda}\left[\nabla\cdot\left(\mathbf{H}_0 \exp\left\{-\frac{\lambda}{2}\boldsymbol{\xi}\cdot\boldsymbol{\xi}\right\}\right) - \left(\nabla\cdot\mathbf{H}_0 \exp\left\{-\frac{\lambda}{2}\boldsymbol{\xi}\cdot\boldsymbol{\xi}\right\}\right)\right] \tag{8.2.46}$$

so that by the divergence theorem we have

$$K(\lambda) = G_0(\mathbf{0}) \int_D \exp\left\{-\frac{\lambda}{2}\boldsymbol{\xi}\cdot\boldsymbol{\xi}\right\} d\boldsymbol{\xi} - \frac{1}{\lambda}\oint_{\hat{\Gamma}} (\mathbf{H}_0 \cdot \hat{\mathbf{N}}) \exp\left\{-\frac{\lambda}{2}\boldsymbol{\xi}\cdot\boldsymbol{\xi}\right\} ds$$

$$+ \frac{1}{\lambda}\int_D G_1(\boldsymbol{\xi}) \exp\left\{-\frac{\lambda}{2}\boldsymbol{\xi}\cdot\boldsymbol{\xi}\right\} d\boldsymbol{\xi}. \tag{8.2.47}$$

Here $\hat{\mathbf{N}}$ is the unit outward normal to $\hat{\Gamma}$, the boundary of D, and

$$G_1 = \nabla\cdot\mathbf{H}_0. \tag{8.2.48}$$

Because $\boldsymbol{\xi} \neq \mathbf{0}$ on $\hat{\Gamma}$, the boundary integral in (8.2.47) is exponentially small as $\lambda \to \infty$. Also due to the factor λ^{-1}, the integral involving G_1 is dominated by

$$K_1(\lambda) = G_0(\mathbf{0}) \int_D \exp\left\{-\frac{\lambda}{2}\boldsymbol{\xi}\cdot\boldsymbol{\xi}\right\} d\boldsymbol{\xi}. \tag{8.2.49}$$

We now prove the following.

LEMMA 8.2.1. If $\boldsymbol{\xi} = \mathbf{0}$ is an interior point of D, then, as $\lambda \to \infty$,

$$\hat{K}(\lambda) = \int_D \exp\left\{-\frac{\lambda}{2}\boldsymbol{\xi}\cdot\boldsymbol{\xi}\right\} d\boldsymbol{\xi} = \frac{2\pi}{\lambda} + o(\lambda^{-m}) \tag{8.2.50}$$

for any m.

PROOF. Let R_1 be the radius of a circular disk C_1, centered at $\boldsymbol{\xi} = \mathbf{0}$ and contained in D. Let R_2 be the radius of a circular disk C_2, centered at $\boldsymbol{\xi} = \mathbf{0}$ and containing D. (See Figure 8.2.1.) Then clearly

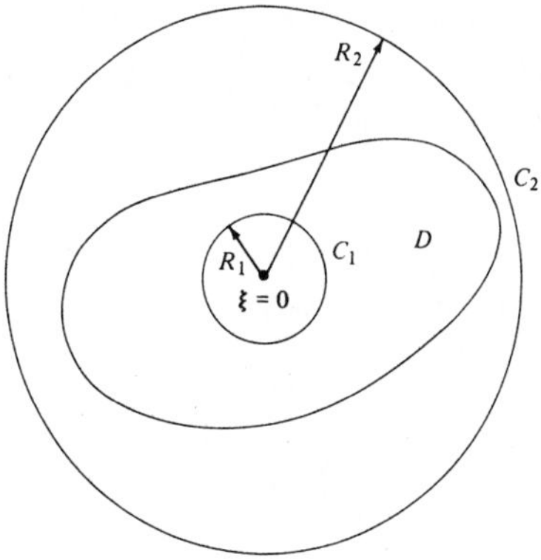

Figure 8.2.1.

$$\int_{C_1} \exp\left\{-\frac{\lambda}{2}\xi\cdot\xi\right\} d\xi \le K(\lambda) \le \int_{C_2} \exp\left\{-\frac{\lambda}{2}\xi\cdot\xi\right\} d\xi. \tag{8.2.51}$$

Simple calculations yield

$$\frac{2\pi}{\lambda}\left(1-\exp\left\{-\frac{\lambda}{2}R_1^2\right\}\right) \le \hat{K}(\lambda) \le \frac{2\pi}{\lambda}\left(1-\exp\left\{-\frac{\lambda}{2}R_2^2\right\}\right), \tag{8.2.52}$$

which proves the lemma if we simply let $\lambda \to \infty$.

It now follows from (8.2.41), (8.2.47), and the above lemma that

$$I(\lambda) \sim \frac{2\pi}{\lambda} G_0(\mathbf{0}) \exp\{\lambda\phi(\mathbf{x}_0)\}. \tag{8.2.53}$$

But, as is easily seen from (8.2.36) and (8.2.43)

$$G_0(\mathbf{0}) = \frac{1}{\sqrt{\lambda_1\lambda_2}} g_0(\mathbf{x}_0) = \frac{g_0(\mathbf{x}_0)}{(\phi_{x_1x_1}(\mathbf{x}_0)\,\phi_{x_2x_2}(\mathbf{x}_0) - \phi_{x_1x_2}^2(\mathbf{x}_0))^{1/2}} \tag{8.2.54}$$

so that we finally have

$$\mathrm{I}(\lambda) \sim \frac{2\pi}{\lambda} \frac{g_0(\mathbf{x}_0)\exp\{\lambda\phi(\mathbf{x}_0)\}}{(\phi_{x_1x_1}(\mathbf{x}_0)\,\phi_{x_2x_2}(\mathbf{x}_0) - \phi_{x_1x_2}^2(\mathbf{x}_0))^{1/2}}. \tag{8.2.55}$$

From this last result we see that the asymptotic expansion of I, at least to leading order, is indeed independent of the selection of the functions h_1, h_2, and $\mathbf{H}_0$. Of course we must have (8.2.34) and (8.2.44) satisfied and $\mathbf{H}_0$ must be well behaved throughout D. We might then argue that the above procedure is not necessary and could be avoided. This is in fact true, there being several distinct methods for arriving at (8.2.55). The advantages of our procedure will not be apparent until the following section where infinite expansions are obtained. There we shall find that the method we have described yields a concise representation of the complete asymptotic expansion of I. It appears, moreover, that the other methods we have alluded to can recover this representation only with some difficulty.

From (8.2.55) we see that when the absolute maximum of ϕ occurs at $\mathbf{x} = \mathbf{x}_0$, an interior point of $\mathscr{D}$,

$$\exp\{-\lambda\phi(\mathbf{x}_0)\}\, I(\lambda) = O(\lambda^{-1}), \tag{8.2.56}$$

as $\lambda \to \infty$. This is to be compared with the previous result which showed that when the absolute maximum of ϕ occurs at the boundary point $\mathbf{x} = \mathbf{x}_0$ and $\nabla\phi(\mathbf{x}_0) \neq 0$,

$$\exp\{-\lambda\phi(\mathbf{x}_0)\}\, I(\lambda) = O(\lambda^{-3/2}), \tag{8.2.57}$$

as $\lambda \to \infty$.

It still remains to consider the case where absolute maximum occurs at the boundary point $\mathbf{x} = \mathbf{x}_0$ with $\nabla\phi(\mathbf{x}_0) = 0$. We shall still assume that (8.2.25) holds so that the analysis for this case proceeds precisely as that for an interior maximum, through Equation (8.2.47). Now, however, $\boldsymbol{\xi} = \mathbf{0}$ lies on the boundary of D and hence Lemma 8.2.1 does not apply. Furthermore, we can no longer conclude that, as $\lambda \to \infty$, the boundary integral in (8.2.47) is exponentially smaller than the area integrals.

To handle this case, we now prove the following.

LEMMA 8.2.2. Let $\hat{K}(\lambda)$ be defined by (8.2.50) and suppose that $\boldsymbol{\xi} = \mathbf{0}$ lies on the boundary of D. Then, as $\lambda \to \infty$,

$$K(\lambda) \sim \frac{\pi}{\lambda}. \tag{8.2.58}$$

PROOF. There is no loss of generality in assuming that the tangent to $\hat{\Gamma}$ at $\boldsymbol{\xi} = \mathbf{0}$ is horizontal. We let C_a and C_1 be respectively the circular disk of radius a and the semicircular region of radius R_1 depicted in Figure 8.2.2.

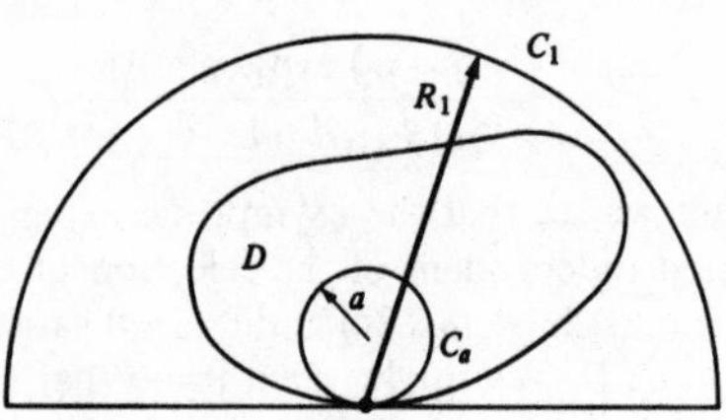

Figure 8.2.2.

(Note that C_a is tangent to D at $\boldsymbol{\xi} = \mathbf{0}$.) As is readily seen

$$\int_{C_a} \exp\left\{-\frac{\lambda}{2}\xi\cdot\xi\right\} d\xi \le \hat{K}(\lambda) \le \int_{C_1} \exp\left\{-\frac{\lambda}{2}\xi\cdot\xi\right\} d\xi. \tag{8.2.59}$$

We now have

$$\int_{C_1} \exp\left\{-\frac{\lambda}{2}\xi\cdot\xi\right\} d\xi = \frac{\pi}{\lambda}\left(1 - \exp\left\{-\frac{\lambda}{2}R_1^2\right\}\right), \tag{8.2.60}$$

$$\int_{C_a} \exp\left\{-\frac{\lambda}{2}\xi\cdot\xi\right\} d\xi = \frac{\pi}{\lambda}\left(1 - \frac{1}{\pi}\int_0^{\pi} \exp\left\{-2\lambda a^2 \sin^2\theta\right\} d\theta\right). \tag{8.2.61}$$

The integral on the right side of (8.2.61) can be asymptotically evaluated via Laplace's method. In this manner we find that

$$\int_{C_a} \exp\left\{-\frac{\lambda}{2}\xi\cdot\xi\right\} d\xi \sim \frac{\pi}{\lambda} - \frac{\sqrt{2\pi}}{2a\,\lambda^{3/2}}. \tag{8.2.62}$$

The lemma is finally established upon combining (8.2.59), (8.2.60), and (8.2.62).

Returning to the boundary integral in (8.2.47) we find by applying Laplace's method that it is $O(\lambda^{-3/2})$ and hence does not affect the expansion of $I(\lambda)$ to leading order. Indeed, it follows from Lemma 8.2.2 that, when $\nabla\phi$ vanishes at the boundary point $\mathbf{x} = \mathbf{x}_0$, which, in turn, is the point where ϕ achieves its absolute maximum in $\bar{\mathscr{D}}$,

$$I(\lambda) \sim \frac{\pi}{\lambda}\,\frac{\exp\{\lambda\phi(\mathbf{x}_0)\}\, g_0(\mathbf{x}_0)}{(\phi_{x_1x_1}(\mathbf{x}_0)\,\phi_{x_2x_2}(\mathbf{x}_0) - \phi_{x_1x_2}^2(\mathbf{x}_0))^{1/2}}, \tag{8.2.63}$$

as $\lambda \to \infty$. We observe that (8.2.63) differs from (8.2.55), the corresponding result for an interior maximum, only by the multiplicative factor $\frac{1}{2}$. This, of course, is not surprising, especially in the light of the analogous result for one-dimensional integrals.

8.3. Higher-Dimensional Integrals of Laplace Type

We shall now consider the asymptotic behavior, as $\lambda \to \infty$, of integrals of the form

$$I(\lambda) = \int_{\mathscr{D}} \exp\{\lambda\phi(\mathbf{x})\}\, g_0(\mathbf{x})\, d\mathbf{x}, \qquad \mathbf{x} = (x_1, \ldots, x_n). \tag{8.3.1}$$

Here n is any positive integer and $\mathscr{D}$ is a bounded simply connected domain in n-dimensional Euclidean space. The boundary of $\mathscr{D}$, which we denote by Γ, is an $(n-1)$-dimensional hypersurface. We assume that it can be represented parametrically by

$$\mathbf{x} = \mathbf{x}(\boldsymbol{\sigma}), \qquad \boldsymbol{\sigma} = (\sigma_1, \ldots, \sigma_{n-1})\ \varepsilon\ \mathscr{P}. \tag{8.3.2}$$

As in Section 8.2, we suppose that ϕ and g_0 are sufficiently differentiable functions of their arguments throughout $\bar{\mathscr{D}}$ for the operations to follow. Also the same is to be true of the functions $x_i(\boldsymbol{\sigma})$ for $\boldsymbol{\sigma}$ in $\bar{\mathscr{P}}$.

To obtain an asymptotic expansion of $I(\lambda)$, as $\lambda \to \infty$, we shall apply techniques analogous to those described in Section 8.2 for the case $n=2$. Indeed, let us first assume that $\nabla\phi$ does not vanish in $\bar{\mathscr{D}}$ so that the absolute maximum of ϕ in $\bar{\mathscr{D}}$ is only achieved on Γ. We shall further assume that $\mathbf{x}=\mathbf{x}_0$ is the only point on Γ where this maximum occurs. Because the parametrization in (8.3.2) can always be selected so that $\mathbf{x}_0=\mathbf{x}_0(\boldsymbol{\sigma}_0)$ with $\boldsymbol{\sigma}_0=(\sigma_{0,1}, \ldots, \sigma_{0,n-1})$ an interior point of $\mathscr{P}$, we must have

$$\nabla_x\phi \cdot \left.\frac{\partial \mathbf{x}}{\partial \sigma_i}\right|_{\sigma=\sigma_0}=0, \qquad i=1,2,\ldots,n-1, \qquad \nabla_x=\left(\frac{\partial}{\partial x_1}, \ldots, \frac{\partial}{\partial x_n}\right). \tag{8.3.3}$$

If we set

$$\psi(\boldsymbol{\sigma})=\phi(\mathbf{x}(\boldsymbol{\sigma})), \tag{8.3.4}$$

then (8.3.3) is equivalent to

$$\left.\nabla_\sigma \psi\right|_{\sigma=\sigma_0}=\mathbf{0}, \qquad \nabla_\sigma=\left(\frac{\partial}{\partial \sigma_1}, \ldots, \frac{\partial}{\partial \sigma_{n-1}}\right). \tag{8.3.5}$$

To ensure that $\boldsymbol{\sigma}=\boldsymbol{\sigma}_0$ is a maximum point of ψ, we assume that the quadratic form

$$(\boldsymbol{\sigma}-\boldsymbol{\sigma}_0)\,B(\boldsymbol{\sigma}-\boldsymbol{\sigma}_0)^T; \qquad B=(\psi_{\sigma_i\sigma_j}(\boldsymbol{\sigma}_0)), \qquad i,j=1,2,\ldots,n-1 \tag{8.3.6}$$

is negative definite.

As in the two-dimensional case, we now set

$$\mathbf{H}_0=g_0\frac{\nabla\phi}{|\nabla\phi|^2} \tag{8.3.7}$$

so that by the divergence theorem

$$I(\lambda)=\frac{1}{\lambda}\int_\Gamma (\mathbf{H}_0\cdot\mathbf{N})\exp\{\lambda\phi\}\,d\Sigma-\frac{1}{\lambda}\int_{\mathscr{D}} g_1\exp\{\lambda\phi\}\,d\mathbf{x}. \tag{8.3.8}$$

Here $\mathbf{N}$ is the unit outward normal vector to Γ, $d\Sigma$ is the differential element of "surface area" on the $(n-1)$-dimensional hypersurface Γ, and

$$g_1=\nabla\cdot\mathbf{H}_0. \tag{8.3.9}$$

On the right-hand side of (8.3.8), because the last term is an integral of the same form as $I(\lambda)$ multiplied by λ^{-1}, it is reasonable to suppose that the leading term in the expansion of I comes from the boundary integral. Thus we can look upon the integral over $\mathscr{D}$ on the right side of (8.3.8) as an error term. Furthermore, we can apply the same integration by parts procedure to this error term. Indeed, this process can be repeatedly applied with the result that after M steps we have the exact representation

$$I(\lambda)=-\sum_{j=0}^{M-1}(-\lambda)^{-(j+1)}\int_\Gamma (\mathbf{H}_j\cdot\mathbf{N})\exp\{\lambda\phi\}\,d\Sigma+\frac{(-1)^M}{\lambda^M}\int_{\mathscr{D}} g_M\exp\{\lambda\phi\}\,d\mathbf{x}. \tag{8.3.10}$$

Here the quantities g_j and $\mathbf{H}_j$ are defined recursively by

$$\mathbf{H}_j = g_j \frac{\nabla \phi}{|\nabla \phi|^2}, \qquad j = 0, 1, 2, \ldots, M-1, \qquad \nabla = \nabla_x.$$
$$g_{j+1} = \nabla \cdot \mathbf{H}_j, \tag{8.3.11}$$

In terms of the parametrization (8.3.2), each term of the sum in (8.3.10) is an $(n-1)$-dimensional integral of the form (8.3.1) with $\mathbf{x}$ replaced by $\boldsymbol{\sigma}$. Furthermore, we know that the absolute maximum of ψ in $\mathscr{P}$ occurs in the interior of $\mathscr{P}$. To obtain a useful asymptotic expansion of I, each integral in the sum must itself be expanded and the resulting terms then appropriately combined. In addition, it must be shown that the last term on the right in (8.3.10) is asymptotically small compared to each term in the sum and hence constitutes the error made in approximating $I(\lambda)$ by the sum. If this is so, then our problem has been reduced to the study of integrals of the form (8.3.1) in the case where the absolute maximum of ϕ occurs in the interior of $\mathscr{D}$. Therefore, we shall suspend the discussion of this case until after the analysis of an interior maximum.

Let us now suppose that the absolute maximum of ϕ in $\overline{\mathscr{D}}$ is achieved only at the interior point $\mathbf{x} = \mathbf{x}_0$. Then we must have

$$\nabla\phi\big|_{\mathbf{x}=\mathbf{x}_0} = \mathbf{0}. \tag{8.3.12}$$

Also, near $\mathbf{x} = \mathbf{x}_0$,

$$\phi(\mathbf{x}) - \phi(\mathbf{x}_0) \approx \tfrac{1}{2}(\mathbf{x} - \mathbf{x}_0)\, A(\mathbf{x} - \mathbf{x}_0)^T, \tag{8.3.13}$$

where

$$A = (\phi_{x_i x_j}(\mathbf{x}_0)), \qquad i, j = 1, 2, \ldots, n. \tag{8.3.14}$$

We shall assume that the quadratic form in (8.3.13) is negative definite and that $\nabla\phi$ does not vanish at any other point in $\overline{\mathscr{D}}$.

Following the discussion in Section 8.2, we let Q be an orthogonal matrix which diagonalizes A so that

$$Q^T A Q = \Lambda. \tag{8.3.15}$$

Here

$$\Lambda = \begin{pmatrix} \lambda_1 & & 0 \\ & \ddots & \\ 0 & & \lambda_n \end{pmatrix}, \tag{8.3.16}$$

where $\lambda_1, \ldots, \lambda_n$ are the not necessarily distinct eigenvalues of A. We, of course, have that each λ_i is negative and

$$\det A = \prod_{i=1}^{n} \lambda_i. \tag{8.3.17}$$

If we now set

$$(\mathbf{x}-\mathbf{x}_0)^T = QR\mathbf{z}^T, \qquad R = \begin{pmatrix} |\lambda_1|^{-1/2} & & 0 \\ & \ddots & \\ 0 & & |\lambda_n|^{-1/2} \end{pmatrix}, \tag{8.3.18}$$

then near $\mathbf{z} = \mathbf{0}$,

$$f(\mathbf{z}) = \phi(\mathbf{x}_0) - \phi(\mathbf{x}(\mathbf{z})) \approx \tfrac{1}{2}\,\mathbf{z}\cdot\mathbf{z}. \tag{8.3.19}$$

To make this approximation hold in the large, we introduce the quantities

$$\xi_i = h_i(\mathbf{z}), \qquad i = 1, 2, \ldots, n. \tag{8.3.20}$$

These are to be such that

$$h_i = z_i + o(|\mathbf{z}|), \qquad |\mathbf{z}| \to 0, \qquad i = 1, 2, \ldots, n \tag{8.3.21}$$

and

$$\sum_{i=1}^{n} h_i^2 = 2f. \tag{8.3.22}$$

Because $\nabla\phi$ vanishes in $\bar{\mathscr{D}}$ only at $\mathbf{x} = \mathbf{x}_0$, it is known that such quantities exist for which the Jacobian

$$J(\boldsymbol{\xi}) = \frac{\partial(x_1, \ldots, x_n)}{\partial(\xi_1, \ldots, \xi_n)}, \qquad \boldsymbol{\xi} = (\xi_1, \ldots, \xi_n) \tag{8.3.23}$$

is positive and finite throughout D, the image of $\mathscr{D}$ under the two changes of variables (8.3.18) and (8.3.20). It follows from (8.3.18) and (8.3.21) that

$$J(\mathbf{0}) = \prod_{i=1}^{n} |\lambda_i|^{-1/2} = (|\det(\phi_{x_i x_j}(\mathbf{x}_0))|)^{-1/2}. \tag{8.3.24}$$

The functions h_i are obviously not uniquely determined by (8.3.21) and (8.3.22). Nevertheless, it turns out that, in this case, we need no further information about these functions to derive the complete asymptotic expansion of I.

Upon introducing $\xi_1, \ldots, \xi_n$ as the variables of integration in (8.3.1) we obtain

$$I(\lambda) = \exp\{\lambda\phi(\mathbf{x}_0)\} \int_D G_0(\boldsymbol{\xi}) \exp\left\{-\frac{\lambda}{2}\,\boldsymbol{\xi}\cdot\boldsymbol{\xi}\right\} d\boldsymbol{\xi}. \tag{8.3.25}$$

Here

$$G_0(\boldsymbol{\xi}) = g_0(\mathbf{x}(\boldsymbol{\xi}))\, J(\boldsymbol{\xi}). \tag{8.3.26}$$

We now set

$$G_0(\boldsymbol{\xi}) = G_0(\mathbf{0}) + \boldsymbol{\xi}\cdot\mathbf{H}_0. \tag{8.3.27}$$

As in Section 8.2, $\mathbf{H}_0$ is not uniquely determined. Nevertheless we need only require that $\mathbf{H}_0$ be well behaved throughout D. One possible choice is given by

$$H_1 = \frac{1}{\xi_1}[G_0(\xi_1, \ldots, \xi_n) - G_0(0, \xi_2, \ldots, \xi_n)],$$
$$H_2 = \frac{1}{\xi_2}[G_0(0, \xi_2, \ldots, \xi_n) - G_0(0, 0, \xi_3, \ldots, \xi_n)],$$
$$\vdots$$
$$H_n = \frac{1}{\xi_n}[G_0(0, \ldots, \xi_n) - G_0(0, \ldots, 0)]. \tag{8.3.28}$$

The ambiguity in $\mathbf{H}_0$ will not affect our determination of the asymptotic expansion.

It follows by the divergence theorem that

$$I(\lambda) = \exp\{\lambda\phi(\mathbf{x}_0)\}\left[G_0(\mathbf{0})\int_D \exp\left\{-\frac{\lambda}{2}\boldsymbol{\xi}\cdot\boldsymbol{\xi}\right\}d\boldsymbol{\xi}\right.$$
$$\left. - \frac{1}{\lambda}\int_{\hat{\Gamma}}(\mathbf{H}_0\cdot\hat{\mathbf{N}})\exp\left\{-\frac{\lambda}{2}\boldsymbol{\xi}\cdot\boldsymbol{\xi}\right\}d\hat{\Sigma} + \frac{1}{\lambda}\int_D G_1(\boldsymbol{\xi})\exp\left\{-\frac{\lambda}{2}\boldsymbol{\xi}\cdot\boldsymbol{\xi}\right\}d\boldsymbol{\xi}\right]. \tag{8.3.29}$$

Here $\hat{\Gamma}$ denotes the boundary of D, $d\hat{\Sigma}$ is the differential element of surface area on $\hat{\Gamma}$, and $\hat{\mathbf{N}}$ is the unit outward normal to $\hat{\Gamma}$. Clearly, the boundary integral is exponentially small as $\lambda \to \infty$. Anticipating that the integrals over D are not exponentially small, we shall neglect the boundary integral in our future considerations.

Upon repeating the process M times we obtain

$$I(\lambda) \sim \exp\{\lambda\phi(\mathbf{x}_0)\}\left[\sum_{j=0}^{M-1}\lambda^{-j}G_j(\mathbf{0})\int_D \exp\left\{-\frac{\lambda}{2}\boldsymbol{\xi}\cdot\boldsymbol{\xi}\right\}d\boldsymbol{\xi}\right.$$
$$\left. + \lambda^{-M}\int_D G_M(\boldsymbol{\xi})\exp\left\{-\frac{\lambda}{2}\boldsymbol{\xi}\cdot\boldsymbol{\xi}\right\}d\boldsymbol{\xi}\right]. \tag{8.3.30}$$

The functions G_j are defined recursively by

$$\begin{aligned} G_j(\boldsymbol{\xi}) &= G_j(\mathbf{0}) + \boldsymbol{\xi}\cdot\mathbf{H}_j(\boldsymbol{\xi}), \\ G_{j+1}(\boldsymbol{\xi}) &= \nabla\cdot\mathbf{H}_j(\boldsymbol{\xi}). \end{aligned} \tag{8.3.31}$$

It should be noted that (8.3.30) is *not* an exact representation because we have omitted the M exponentially small boundary integrals.

Because

$$\left|\lambda^{-M}\int_D \exp\left\{-\frac{\lambda}{2}\boldsymbol{\xi}\cdot\boldsymbol{\xi}\right\}G_M(\boldsymbol{\xi})\,d\boldsymbol{\xi}\right| < \frac{K}{\lambda^M}\int_D \exp\left\{-\frac{\lambda}{2}\boldsymbol{\xi}\cdot\boldsymbol{\xi}\right\}d\boldsymbol{\xi} \tag{8.3.32}$$

for some constant K, we immediately have that the sum in (8.3.30) represents an asymptotic expansion of $I(\lambda)$ to M terms. More precisely

$$I(\lambda) \sim \exp\{\lambda\phi(\mathbf{x}_0)\} \sum_{j=0}^{M-1} \lambda^{-j} G_j(\mathbf{0}) \int_D \exp\left\{-\frac{\lambda}{2}\xi\cdot\xi\right\} d\xi \tag{8.3.33}$$

is an asymptotic expansion of I to M terms with respect to the asymptotic sequence

$$\left(\lambda^{-j} \exp\{\lambda\phi(\mathbf{x}_0)\} \int_D \exp\left\{-\frac{\lambda}{2}\xi\cdot\xi\right\} d\xi\right), \qquad j = 0, 1, 2, \ldots. \tag{8.3.34}$$

We can improve on this result however by simplifying the asymptotic sequence. Indeed we now prove the following.

LEMMA 8.3.1. Let $\xi = \mathbf{0}$ lie in the interior of D. Then, as $\lambda \to \infty$,

$$\hat{K}(\lambda) = \int_D \exp\left\{-\frac{\lambda}{2}\xi\cdot\xi\right\} d\xi = \left(\frac{2\pi}{\lambda}\right)^{n/2} + o(\lambda^{-m}) \tag{8.3.35}$$

for all m.

PROOF. The proof follows closely that of Lemma 8.2.1. Indeed, let S_{R_1} be a sphere of radius R_1, centered at $\xi = \mathbf{0}$, and contained in D. Let S_{R_2} be a sphere of radius R_2, centered at $\xi = \mathbf{0}$ and containing D. Then

$$\int_{S_{R_1}} \exp\left\{-\frac{\lambda}{2}\xi\cdot\xi\right\} d\xi \le K(\lambda) \le \int_{S_{R_2}} \exp\left\{-\frac{\lambda}{2}\xi\cdot\xi\right\} d\xi. \tag{8.3.36}$$

We have

$$\int_{S_{R_2}} \exp\left\{-\frac{\lambda}{2}\xi\cdot\xi\right\} d\xi = \left(\frac{2}{\lambda}\right)^{n/2} \omega_n \int_0^{\sqrt{(\lambda/2)}R_2} e^{-r^2} r^{n-1}\, dr, \tag{8.3.37}$$

where

$$\omega_n = \frac{2(\pi)^{n/2}}{\Gamma(n/2)} \tag{8.3.38}$$

is the surface area of the unit sphere in n space. Thus we obtain the bound

$$\int_{S_{R_2}} \exp\left\{-\frac{\lambda}{2}\xi\cdot\xi\right\} d\xi \le \frac{2\left(\dfrac{2\pi}{\lambda}\right)^{n/2}}{\Gamma(n/2)} \int_0^\infty e^{-r^2} r^{n-1}\, dr = \left(\frac{2\pi}{\lambda}\right)^{n/2} \tag{8.3.39}$$

To determine a lower bound we write

$$\int_{S_{R_1}} \exp\left\{-\frac{\lambda}{2}\xi\cdot\xi\right\} d\xi = \left(\frac{2\pi}{\lambda}\right)^{n/2} \frac{2}{\Gamma(n/2)}$$

$$\times\left[\int_0^\infty e^{-r^2} r^{n-1}\, dr - \int_{\sqrt{(\lambda/2)}R_1}^\infty e^{-r^2} r^{n-1}\, dr\right]. \tag{8.3.40}$$

If $\sqrt{\lambda/2}\,R_1$ is greater than 1 and n is greater than or equal to 2, then we have

$$\int_{S_{R_1}} \exp\left\{-\frac{\lambda}{2}\boldsymbol{\xi}\cdot\boldsymbol{\xi}\right\} d\boldsymbol{\xi} \geq \left(\frac{2\pi}{\lambda}\right)^{n/2}\left[1-\frac{e^{(-\lambda/2)R_1^2}}{\Gamma(n/2)}\right]. \tag{8.3.41}$$

Upon combining (8.3.36), (8.3.39), and (8.3.41), we obtain finally

$$\left(\frac{2\pi}{\lambda}\right)^{n/2}\left[1-\frac{\exp\left\{-\dfrac{\lambda R_1^2}{2}\right\}}{\Gamma(n/2)}\right] \leq \hat{K}(\lambda) \leq \left(\frac{2\pi}{\lambda}\right)^{n/2} \tag{8.3.42}$$

for λ sufficiently large. This completes the proof.

By using the result of this last lemma in (8.3.33) we find that

$$I(\lambda) \sim \exp\{\lambda\phi(\mathbf{x}_0)\} \sum_{j=0}^{M-1} (2\pi)^{n/2}\,\lambda^{-((n/2)+j)}\,G_j(\mathbf{0}) \tag{8.3.43}$$

is an asymptotic expansion of $I(\lambda)$ to M terms with respect to the asymptotic sequence

$$(\lambda^{-((n/2)+j)}\exp(\lambda\phi(\mathbf{x}_0)\}), \qquad j = 0, 1, 2, \ldots . \tag{8.3.44}$$

As usual, we wish to express $G_j(\mathbf{0})$ in terms of the original functions ϕ and g_0. Except for the leading term this is extremely difficult to accomplish. One significant simplification is achieved, however, through the following.

LEMMA 8.3.2. Let the functions $G_j(\boldsymbol{\xi})$ be defined by (8.3.31). Then

$$G_j(\mathbf{0}) = \frac{1}{2^j j!}\Delta_\xi^j\, G_0\Big|_{\xi=0}, \qquad \Delta_\xi = \left(\frac{\partial^2}{\partial\xi_1^2}+\cdots+\frac{\partial^2}{\partial\xi_n^2}\right). \tag{8.3.45}$$

PROOF. Consider any n vector $\mathbf{F}(\boldsymbol{\xi})$. Then, because $\Delta_\xi\boldsymbol{\xi} = \mathbf{0}$, we have

$$\Delta_\xi(\boldsymbol{\xi}\cdot\mathbf{F}) = \boldsymbol{\xi}\cdot\Delta_\xi\mathbf{F} + 2\nabla_\xi\cdot\mathbf{F}. \tag{8.3.46}$$

Repeated applications of this result yield

$$\Delta_\xi^j(\boldsymbol{\xi}\cdot\mathbf{F}) = \boldsymbol{\xi}\cdot\Delta_\xi^j\,\mathbf{F} + 2j\,\Delta_\xi^{j-1}(\nabla_\xi\cdot\mathbf{F}), \tag{8.3.47}$$

which, when evaluated at $\boldsymbol{\xi} = \mathbf{0}$, becomes

$$\Delta_\xi^j(\boldsymbol{\xi}\cdot\mathbf{F})\big|_{\xi=0} = 2j\,\Delta_\xi^{j-1}(\nabla_\xi\cdot\mathbf{F})\big|_{\xi=0}. \tag{8.3.48}$$

Now for any ℓ and j we have

$$\begin{aligned}\Delta_\xi^j\, G_\ell\big|_{\xi=0} &= \Delta_\xi^j(\boldsymbol{\xi}\cdot\mathbf{H}_\ell)\big|_{\xi=0} = 2j\,\Delta_\xi^{j-1}(\nabla_\xi\cdot\mathbf{H}_\ell)\\ &= 2j\,\Delta_\xi^{j-1}\,G_{\ell+1}\big|_{\xi=0}.\end{aligned} \tag{8.3.49}$$

From this last result we readily find that

$$\begin{aligned}\Delta_\xi^j G_0\big|_{\xi=0} &= 2j\,\Delta_\xi^{j-1} G_1\big|_{\xi=0} \\ &= 2^2 j(j-1)\,\Delta_\xi^{j-2} G_2\big|_{\xi=0} \\ &\;\;\vdots \\ &= 2^j j!\, G_j\big|_{\xi=0}\end{aligned}$$

and the lemma is proved.

If we use (8.3.45) in (8.3.43), then we obtain

$$I(\lambda) \sim \exp\{\lambda\phi(\mathbf{x}_0)\}\left(\frac{2\pi}{\lambda}\right)^{n/2} \sum_{j=0}^{M-1} \frac{\Delta_\xi^j G_0\big|_{\xi=0}}{(2\lambda)^j j!}. \tag{8.3.50}$$

Thus we have established our previous claim that our expansion is independent of the particular vectors $\mathbf{H}_j$ selected in the integration by parts procedure. Although it still remains to express the quantities $\Delta_\xi^j G_0\big|_{\xi=0}$ in terms of ϕ and g_0, we wish to point out that (8.3.50) is an extremely concise representation of the asymptotic expansion. In fact, whenever $G_0(\xi)$ is explicitly determined, the complete asymptotic expansion is readily obtained. Unfortunately, we rarely can determine $G_0(\xi)$ explicitly in which event the computation of $\Delta_\xi^j G_0\big|_{\xi=0}$ is extremely awkward for $j \geq 1$.

From (8.3.24) and (8.3.26) we have that

$$G_0(\mathbf{0}) = \frac{g_0(\mathbf{x}_0)}{(|\det(\phi_{x_i x_j}(\mathbf{x}_0))|)^{1/2}} \tag{8.3.51}$$

and hence to leading order

$$I(\lambda) \sim \frac{\exp\{\lambda\phi(\mathbf{x}_0)\}}{(|\det(\phi_{x_i x_j}(\mathbf{x}_0))|)^{1/2}}\left(\frac{2\pi}{\lambda}\right)^{n/2} g_0(\mathbf{x}_0). \tag{8.3.52}$$

The quantity

$$\Delta_\xi G_0\big|_{\xi=0} = \Delta_\xi(g_0 J)\big|_{\xi=0}$$

is quite complicated when expressed in terms of g_0 and ϕ and we shall merely quote the result.

$$\begin{aligned}\Delta G_0\big|_{\xi=0} = &-(|\det(\phi_{x_i x_j}(\mathbf{x}_0)|)^{-1/2}\,[\div \phi_{x_s x_r x_q}\, B_{sq} B_{rp} (g_0)_{x_p} + \mathrm{Tr}(CB) \\ &+ g_0\{\phi_{x_p x_q x_r}\,\phi_{x_s x_t x_u}\,(\tfrac{1}{4} B_{ps} B_{qr} B_{tu} + \tfrac{1}{6} B_{ps} B_{qt} B_{ru}) \\ &- \tfrac{1}{4}\,\phi_{x_p x_q x_r x_s}\, B_{pr} B_{qs}\}]_{\mathbf{x}=\mathbf{x}_0}\end{aligned} \tag{8.3.53}$$

Here we have used the summation convention where repeated indices are to be summed from 1 to n. The matrices B and C are defined by

$$B = (B_{pq}), \qquad B_{pq}\,\phi_{x_q x_r}(\mathbf{x}_0) = \delta_{pr}, \tag{8.3.54}$$

$$C = ((g_0)_{x_p x_q}), \tag{8.3.55}$$

and Tr(CB) denotes the trace of the matrix CB.

Although we have not exhibited the calculations leading to (8.3.53), we wish to emphasize that (8.3.22) is the only information concerning the functions H_i used in these calculations. This presumably would be the case for the higher coefficients as well, a result that as yet has not been established.

Let us now suppose that the absolute maximum of ϕ in $\bar{\mathscr{D}}$ occurs at the boundary point $\mathbf{x} = \mathbf{x}_0$ with $\Delta\phi(\mathbf{x}_0) = 0$. The analysis for this case is more complicated than that for an interior maximum because the boundary integrals produced in the integration by parts procedure are *not* asymptotically negligible, as $\lambda \to \infty$. Indeed we now have

$$I(\lambda) \sim \exp\{\lambda\phi(\mathbf{x}_0)\}\left[\sum_{j=0}^{M-1}\left(\lambda^{-j} G_j(\mathbf{0}) \int_D \exp\left\{-\frac{\lambda}{2}\xi\cdot\xi\right\} d\xi \right.\right.$$
$$\left. - \lambda^{-(j+1)} \int_{\hat{\Gamma}} (\mathbf{H}_j\cdot\hat{\mathbf{N}}) \exp\left\{-\frac{\lambda}{2}\xi\cdot\xi\right\} d\hat{\Sigma}\right)$$
$$\left. + \lambda^{-M}\int_D \exp\left\{-\frac{\lambda}{2}\xi\cdot\xi\right\} G_M(\xi)\, d\xi\right]. \tag{8.3.56}$$

Here the quantities $\mathbf{H}_j(\xi)$ and $G_j(\xi)$ are still difined by (8.3.31).

All of the integrals in (8.3.56) can be asymptotically evaluated by the method of this section. The leading term arises from the first integral in the first sum on the right side of (8.3.56). Indeed it dominates the contribution from the first integral in the second sum by the factor $\lambda^{1/2}$.

In order to obtain an explicit representation of the leading term we must use the following lemma whose proof is left to the exercises.

LEMMA 8.3.3. Let $\xi = \mathbf{0}$ lie on a smooth portion of the boundary of D. Then as $\lambda \to \infty$

$$\hat{K}(\lambda) = \int_D \exp\left\{-\frac{\lambda}{2}\xi\cdot\xi\right\} d\xi \sim \frac{1}{2}\left(\frac{2\pi}{\lambda}\right)^{n/2}. \tag{8.3.57}$$

From this result we find that in the present case

$$I(\lambda) \sim \frac{\exp\{\lambda\phi(\mathbf{x}_0)\}}{2\sqrt{|\det(\phi_{x_i x_j}(\mathbf{x}_0))|}}\, g_0(\mathbf{x}_0)\left(\frac{2\pi}{\lambda}\right)^{n/2}. \tag{8.3.58}$$

To conclude this section we return to the case where $\nabla\phi \neq 0$ in $\bar{\mathscr{D}}$. We have from (8.3.10) that now

$$I(\lambda) \sim \frac{1}{\lambda}\int_\Gamma g_0 \frac{(\nabla\phi\cdot\mathbf{N})}{|\nabla\phi|^2} \exp\{\lambda\phi\}\, d\Sigma. \tag{8.3.59}$$

In terms of the parametrization of the boundary (8.3.2), (8.3.59) becomes

$$I(\lambda) \sim \frac{1}{\lambda} \int_{\mathscr{P}} g_0 \frac{W \, \nabla\phi \cdot \mathbf{N}}{|\nabla\phi|^2} \exp\{\lambda\psi\} \, d\boldsymbol{\sigma}. \tag{8.3.60}$$

To define W, we first introduce the $n \times (n-1)$ Jacobian matrix of the transformation from $\mathbf{x}$ to $\boldsymbol{\sigma}$,

$$\mathscr{J} = \left(\frac{\partial x_i}{\partial \sigma_j}\right) \tag{8.3.61}$$

and then set

$$W = |\det \mathscr{J}^T \mathscr{J}|^{1/2}. \tag{8.3.62}$$

We have agreed to assume that the maximum of ϕ in $\bar{\mathscr{D}}$ occurs, in this case, only at the boundary point $\mathbf{x} = \mathbf{x}_0$ and that $\mathbf{x}(\boldsymbol{\sigma}_0) = \mathbf{x}_0$ with $\boldsymbol{\sigma}_0$ an interior point of $\mathscr{P}$. If we assume further that the maximum of ψ at $\boldsymbol{\sigma} = \boldsymbol{\sigma}_0$ is simple, then we can apply the results obtained above for an interior maximum. Indeed, it follows from (8.3.52) that, to leading order,

$$I(\lambda) \sim \frac{(2\pi)^{(n-1)/2} \, g_0(\mathbf{x}_0) \, W \exp\{\lambda\phi(\mathbf{x}_0)\}}{\lambda^{(n+1)/2} \, |\det(\psi_{\sigma_i \sigma_j}(\boldsymbol{\sigma}_0))|^{1/2} \, |\nabla\phi(\mathbf{x}_0)|}. \tag{8.3.63}$$

We remark that this result can be expressed in terms of the original variables. The result is

$$I(\lambda) \sim \frac{(2\pi)^{(n-1)/2} \, g_0(\mathbf{x}_0) \exp\{\lambda\phi(\mathbf{x}_0)\}}{\lambda^{(n+1)/2} \, |J|^{1/2}}. \tag{8.3.64}$$

Here

$$J = \phi_{x_p} \phi_{x_q} \operatorname{cof}[\phi_{x_p x_q} - k h_{x_p x_q}].^3 \tag{8.3.65}$$

In this equation subscripts denote differentiation with respect to the x_j's and we must sum from 1 to n over repeated subscripts. We use $h(\mathbf{x}) = 0$ to represent the $(n-1)$-dimensional boundary "surface" in the neighborhood of $\mathbf{x}_0$ and the factor k is determined by

$$\nabla\phi = k \nabla h \tag{8.3.66}$$

This result is derived by Jones. The derivation is outlined in the exercises.

8.4. Multiple Integrals of Fourier Type

Throughout this section we shall consider multiple integrals of the form

$$I(\lambda) = \int_{\mathscr{D}} g_0(\mathbf{x}) \exp\{i\lambda\phi(\mathbf{x})\} \, d\mathbf{x}, \qquad \mathbf{x} = (x_1, x_2, \ldots, x_n). \tag{8.4.1}$$

[3] We use the symbol $\operatorname{cof}[A_{pq}]$ to denote the cofactor of the element A_{pq} in the matrix $[A_{pq}]$.

Here ϕ is real and $\mathscr{D}$ is a bounded connected domain in Euclidean n space. We wish to study $I(\lambda)$ as $\lambda \to \infty$. From our experience with one-dimensional integrals of Fourier type, we should expect that the analysis here will be more complicated than that of the previous section where multidimensional integrals of Laplace type were considered. Nevertheless, we shall find that much of the discussion of the last section will be useful here.

The treatment of Laplace-type integrals was simplified by the fact that only those points in $\bar{\mathscr{D}}$ at which ϕ achieved its absolute maximum were critical. In analogy with one-dimensional integrals of Fourier type, we might anticipate that now the set of *possible* critical points will be more extensive. Indeed we claim that this set consists of

(1) points in $\mathscr{D}$ at which $\nabla\phi = 0$;
(2) all points on Γ, the boundary of $\mathscr{D}$;
(3) points in $\mathscr{D}$ where either ϕ or g_0 fails to be infinitely differentiable with respect to any of their arguments. These are clearly analogs of the critical points found for the one-dimensional case.

Indeed, points at which $\nabla\phi$ vanishes are *stationary points* of ϕ and points on Γ correspond to the endpoints of integration in one dimension. The critical nature of these points can, of course, only be established by determining the associated contributions to the asymptotic expansion of $I(\lambda)$.

To facilitate our discussion, we shall assume below that both ϕ and g_0 are infinitely differentiable in $\mathscr{D}$. This will enable us, in many cases, to derive infinite expansions. However, if there are critical points of type (3) present, then only a finite expansion can be obtained by the method of this section.

Let us first suppose that $\nabla\phi \neq 0$ in $\bar{\mathscr{D}}$. Then there are no stationary points and, as in the derivation of (8.3.10), we can repeatedly apply the divergence theorem to obtain

$$I(\lambda) = -\sum_{j=0}^{M-1} (-i\lambda)^{-(j+1)} \int_{\mathscr{P}} (\mathbf{H}_j \cdot \mathbf{N})\, W \exp\{i\lambda\psi\}\, d\boldsymbol{\sigma}$$
$$+ \frac{(-1)^M}{(i\lambda)^M} \int_{\mathscr{D}} g_M(\mathbf{x}) \exp\{i\lambda\phi(\mathbf{x})\}\, d\mathbf{x}; \qquad \boldsymbol{\sigma} = (\sigma_1, \sigma_2, \ldots, \sigma_{n-1}). \tag{8.4.2}$$

Here

$$\mathbf{x} = \mathbf{x}(\boldsymbol{\sigma}), \qquad \boldsymbol{\sigma}\, \varepsilon\, \mathscr{P} \tag{8.4.3}$$

is a parametric representation of the boundary Γ, W is defined by (8.3.61) and (8.3.62), and

$$\psi(\boldsymbol{\sigma}) = \phi(\mathbf{x}(\boldsymbol{\sigma})). \tag{8.4.4}$$

The quantities g_j and $\mathbf{H}_j$ are defined recursively by

$$\left.\begin{aligned} \mathbf{H}_j &= g_j \frac{\nabla\phi}{|\nabla\phi|^2} \\ g_{j+1} &= \nabla\cdot\mathbf{H}_j \end{aligned}\right\}, \qquad j = 0, 1, 2, \ldots. \tag{8.4.5}$$

Thus, in this case, our problem has been reduced to the study of a sequence of $(n-1)$-dimensional integrals of the form (8.4.1). Because ϕ must achieve maximum and minimum values on Γ, the function ψ must have stationary points in the domain $\mathscr{P}$. For this reason we shall now consider (8.4.1) in the case where $\nabla\phi$ vanishes at one or more interior points of $\mathscr{D}$. We shall return to the present case below, where boundary critical points are studied in detail.

Just as in the treatment of one-dimensional Fourier-type integrals, we shall find here that the isolation of critical points via the process of neutralization is extremely useful. Thus, let us pause to introduce the concept of multidimensional neutralizers. Let $\mathbf{x}_0$ be an arbitrary but fixed point in $\mathbf{x}$ space. Then a function $v(\mathbf{x};\mathbf{x}_0)$ is said to be a *neutralizer about* $\mathbf{x} = \mathbf{x}_0$ if it satisfies the following conditions:

(1) There exists a neighborhood N_0 of $\mathbf{x}_0$ throughout which $v(\mathbf{x}, \mathbf{x}_0) \equiv 1$.
(2) There exists a neighborhood N_1 of $\mathbf{x}_0$, which contains $\mathbf{N}_0$, and outside of which $v(\mathbf{x};\mathbf{x}_0) \equiv 0$.
(3) In all of $\mathbf{x}$ space, v is infinitely differentiable with respect to each of its arguments, and $0 \le v \le 1$.

The actual construction of a multidimensional neutralizer is unimportant because, in our analysis, we shall need only the above defining properties. The reader, however, must convince himself that such functions exist. This easily follows from the one-dimensional case for which we have explicit examples. Indeed, let $\bar{v}(x)$ be a neutralizer about $x = 0$. In particular, let δ_1 and δ_2 be constants with $0 < \delta_1 < \delta_2$ and such that $\bar{v} \equiv 1$ for $0 \le |x| < \delta_1$ and $\bar{v} \equiv 0$ for $|x| > \delta_2$. Then clearly

$$v(\mathbf{x};\mathbf{x}_0) = \bar{v}(|\mathbf{x} - \mathbf{x}_0|^2) \tag{8.4.6}$$

is a neutralizer about $\mathbf{x} = \mathbf{x}_0$. Here N_0 and N_1 are respectively the spherical regions $|\mathbf{x} - \mathbf{x}_0| < \delta_1$ and $|\mathbf{x} - \mathbf{x}_0| < \delta_2$. That the properties in (3) above are satisfied follows from the corresponding properties of $\bar{v}(x)$.

Before proceeding with the asymptotic analysis of $I(\lambda)$, we wish to prove two lemmas that we shall use throughout the remainder of this section.

LEMMA 8.4.1. Consider

$$J(\lambda) = \int_{\mathscr{D}} \hat{g}(\mathbf{x}) \exp\{i\lambda\phi\}\, d\mathbf{x}. \tag{8.4.7}$$

Suppose that $\nabla\phi \ne \mathbf{0}$ in $\bar{\mathscr{D}}$ while $\hat{g}$ belongs to $C^\infty(\bar{\mathscr{D}})$ and vanishes on Γ.[4]

[4] Note that $\hat{g}$ could be the function g_0 in (8.4.1) neutralized about any interior point x_0.

Then, as $\lambda \to \infty$,

$$J(\lambda) = o(\lambda^{-R}) \tag{8.4.8}$$

for all R.

PROOF. Because $\nabla\phi \neq \mathbf{0}$ in $\overline{\mathscr{D}}$, (8.4.2) holds with $I(\lambda) = J(\lambda)$ and $g_0 = \hat{g}$. Due to the stated properties of $\hat{g}$, each boundary integral in the sum vanishes, and hence we find that $J(\lambda) = 0(\lambda^{-M})$. If we now let M get arbitrarily large, then the lemma follows.

LEMMA 8.4.2. Suppose that in (8.4.7) $\nabla\phi$ vanishes in $\overline{\mathscr{D}}$ only at the interior point $\mathbf{x} = \mathbf{x}_0$. Suppose further that $\hat{g}$ satisfies the hypothesis of Lemma 8.4.1 and, in addition, vanishes C^∞ smoothly at $\mathbf{x} = \mathbf{x}_0$. Then, as $\lambda \to \infty$,

$$J(\lambda) = o(\lambda^{-R}) \tag{8.4.9}$$

for all R.

PROOF. Each step in the proof of Lemma 8.4.1 remains valid and hence so does the conclusion. The only possible source of difficulty is the vanishing of $\nabla\phi$ at $\mathbf{x} = \mathbf{x}_0$ because each integration by parts introduces the factor $\nabla\phi/|\nabla\phi|^2$. However, the fact that $\hat{g}$ belongs to $C^\infty(\overline{\mathscr{D}})$ and vanishes at $\mathbf{x} = \mathbf{x}_0$ implies that the singularities introduced are all removable.

The two preceding lemmas show that, in (8.4.1), whenever the function g_0 vanishes infinitely smoothly at each critical point, $I(\lambda)$ goes to zero faster than any power of λ^{-1} as $\lambda \to \infty$. This is actually a principle that we have alluded to in the past and that will be of future use, especially in Chapter 9.

Let us now suppose that ϕ has one or more stationary points in the interior of $\mathscr{D}$. We wish to determine the contribution to the asymptotic expansion of I corresponding to one such point. If $\nabla\phi$ vanishes at $\mathbf{x} = \mathbf{x}_0 \,\varepsilon\, \mathscr{D}$, then the desired contribution is given by the asymptotic expansion of

$$I_0(\lambda) = \int_{\mathscr{D}} g_0(\mathbf{x})\, v(\mathbf{x}; \mathbf{x}_0) \exp\{i\lambda\phi(\mathbf{x})\}\, d\mathbf{x}. \tag{8.4.10}$$

Here, in the definition of the neutralizer $v(\mathbf{x}; \mathbf{x}_0)$, the neighborhood N_1 of $\mathbf{x}_0$ should be small enough so that it contains no stationary points of ϕ other than $\mathbf{x} = \mathbf{x}_0$ and is itself contained in $\mathscr{D}$.

We shall assume that the stationary point $\mathbf{x} = \mathbf{x}_0$ is simple so that is

$$A = (\phi_{x_i x_j}(\mathbf{x}_0)), \qquad i, j = i, 2, \ldots, n, \tag{8.4.11}$$

then

$$\det A \neq 0. \tag{8.4.12}$$

Let us denote the positive eigenvalues of A by $\lambda_1, \lambda_2, \ldots, \lambda_r$, $r \leq n$, and the negative eigenvalues of A by $\lambda_{r+1}, \lambda_{r+2}, \ldots, \lambda_n$. The signature of A, which we shall denote by sig A, and which is the number of positive eigenvalues minus the number of negative eigenvalues, is clearly given by

$$\operatorname{sig} A = 2r - n. \tag{8.4.13}$$

We know that there exists an orthogonal matrix Q that diagonalizes A. Furthermore, Q can be chosen so that when we set

$$(\mathbf{x} - \mathbf{x}_0)^T = QR\mathbf{z}^T, \qquad R = \begin{bmatrix} \lambda_1^{-1/2} & & & & 0 \\ & \ddots & & & \\ & & \lambda_r^{-1/2} & & \\ & & & |\lambda_{r+1}|^{-1/2} & \\ & & & & \ddots \\ 0 & & & & |\lambda_n|^{-1/2} \end{bmatrix}, \tag{8.4.14}$$

then

$$f(\mathbf{z}) = \phi(\mathbf{x}(\mathbf{z})) - \phi(\mathbf{x}_0) \sim \frac{1}{2}\left(\sum_{i=1}^{r} z_i^2 - \sum_{i=r+1}^{n} z_i^2\right), \tag{8.4.15}$$

as $|\mathbf{z}| \to 0$.

To make this approximate behavior hold throughout the effective domain of integration in (8.4.10), we introduce the second change of variables defined by

$$\xi_i = h_i(\mathbf{z}), \qquad i = 1, \ldots, n. \tag{8.4.16}$$

Here

$$h_i = z_i + o(|\mathbf{z}|) \tag{8.4.17}$$

as $|\mathbf{z}| \to 0$, and

$$\sum_{i=1}^{r} h_i^2 - \sum_{i=r+1}^{n} h_i^2 = 2f. \tag{8.4.18}$$

In N_1, because $\nabla\phi$ vanishes only at $\mathbf{x} = \mathbf{x}_0$, we have that the functions h_i can be chosen so that the Jacobian

$$J(\xi) = \left|\det\left(\frac{\partial x_i}{\partial \xi_j}\right)\right|, \qquad \xi = (\xi_1, \ldots, \xi_n) \tag{8.4.19}$$

is finite and nonzero throughout $\hat{N}_1$, the image of $\hat{N}_1$ under (8.4.14) and (8.4.16). Also, it readily follows from (8.4.14) and (8.4.17) that

$$J(\mathbf{0}) = |\det A|^{-1/2}. \tag{8.4.20}$$

In terms of ξ, (8.4.10) becomes

$$I_0(\lambda) = \exp\{i\lambda\phi(\mathbf{x}_0)\} \int_{\hat{N}_1} G_0(\xi)\, v(\xi; \mathbf{0}) \exp\left\{\frac{i\lambda}{2}\, \rho\cdot\xi\right\} d\xi. \tag{8.4.21}$$

Here

$$\boldsymbol{\rho} = (\xi_1, \ldots, \xi_r, -\xi_{r+1}, \ldots, -\xi_n) \tag{8.4.22}$$

and

$$G_0(\boldsymbol{\xi}) = g_0(x(\boldsymbol{\xi}))\, J(\boldsymbol{\xi}). \tag{8.4.23}$$

With regard to $v(\boldsymbol{\xi};\mathbf{0})$, we need only know that it is a neutralizer about $\boldsymbol{\xi} = \mathbf{0}$ whose support is the region $\hat{N}_1$.

If we set

$$G_0(\boldsymbol{\xi}) = G_0(\mathbf{0}) + \boldsymbol{\rho}\cdot\mathbf{H}_0, \tag{8.4.24}$$

with $\mathbf{H}_0$ chosen so that it is well behaved throughout $\hat{N}_1$, then we can write

$$I_0(\lambda) = \exp\{i\lambda\phi(\mathbf{x}_0)\}\,[I_0^{(1)}(\lambda) + I_0^{(2)}(\lambda)]. \tag{8.4.25}$$

Here

$$I_0^{(1)}(\lambda) = G_0(\mathbf{0}) \int_{\hat{N}_1} v(\boldsymbol{\xi};\mathbf{0}) \exp\left\{\frac{i\lambda}{2}\,\boldsymbol{\rho}\cdot\boldsymbol{\xi}\right\} d\boldsymbol{\xi} \tag{8.4.26}$$

and

$$I_0^{(2)}(\lambda) = \int_{\hat{N}_1} (\boldsymbol{\rho}\cdot\mathbf{H}_0)\, v(\boldsymbol{\xi};\mathbf{0}) \exp\left\{\frac{i\lambda}{2}\,\boldsymbol{\rho}\cdot\boldsymbol{\xi}\right\} d\boldsymbol{\xi}. \tag{8.4.27}$$

Let us first consider $I_0^{(2)}(\lambda)$. Upon applying the divergence theorem, we obtain

$$I_0^{(2)}(\lambda) = -\frac{1}{i\lambda} \int_{\hat{N}_1} [v\nabla_\xi\cdot\mathbf{H}_0 + \mathbf{H}_0\cdot\nabla_\xi v] \exp\left\{\frac{i\lambda}{2}\,\boldsymbol{\rho}\cdot\boldsymbol{\xi}\right\} d\boldsymbol{\xi}. \tag{8.4.28}$$

Because $\nabla_\xi v$ is infinitely differentiable, vanishes on the boundary of $\hat{N}_1$ and at the stationary point $\boldsymbol{\xi} = \mathbf{0}$, it follows from Lemma 8.4.2 that

$$I_0^{(2)}(\lambda) = -\frac{1}{i\lambda} \int_{\hat{N}_1} v(\boldsymbol{\xi};\mathbf{0})\, G_1(\boldsymbol{\xi}) \exp\left\{\frac{i\lambda}{2}\,\boldsymbol{\rho}\cdot\boldsymbol{\xi}\right\} d\boldsymbol{\xi} + o(\lambda^{-R}) \tag{8.4.29}$$

for all R. Here

$$G_1(\boldsymbol{\xi}) = \nabla_\xi\cdot\mathbf{H}_0. \tag{8.4.30}$$

Upon applying the above procedure M times we obtain the finite expansion

$$I_0(\lambda) \sim \exp\{i\lambda\,\phi(\mathbf{x}_0)\}\left[\sum_{m=0}^{M-1} \left(-\frac{1}{i\lambda}\right)^m G_m(\mathbf{0}) \int_{\hat{N}_1} v(\boldsymbol{\xi};\mathbf{0}) \exp\left\{\frac{i\lambda}{2}\,\boldsymbol{\rho}\cdot\boldsymbol{\xi}\right\} d\boldsymbol{\xi}\right.$$
$$\left. + \left(-\frac{1}{i\lambda}\right)^M \int_{\hat{N}_1} G_M(\boldsymbol{\xi})\, v(\boldsymbol{\xi};\mathbf{0}) \exp\left\{\frac{i\lambda}{2}\,\boldsymbol{\rho}\cdot\boldsymbol{\xi}\right\} d\boldsymbol{\xi}\right]. \tag{8.4.31}$$

The functions $G_m(\boldsymbol{\xi})$ are defined recursively by

$$\left.\begin{aligned} G_m(\xi) &= G_m(\mathbf{0}) + \boldsymbol{\rho}\cdot\dot{\mathbf{H}}_m(\xi) \\ G_{m+1}(\xi) &= \nabla_\xi\cdot\mathbf{H}_m(\xi) \end{aligned}\right\}, \qquad m = 0, 1, 2, \ldots, \tag{8.4.32}$$

with the vectors $\mathbf{H}_m$ well behaved throughout $\hat{N}_1$.

In order to obtain a more useful expansion, we consider the following.

LEMMA 8.4.3. Let

$$K(\lambda) = \int_D v(\xi;\mathbf{0}) \exp\left\{\frac{i\lambda}{2}\,\boldsymbol{\rho}\cdot\xi\right\} d\xi, \tag{8.4.33}$$

where $\boldsymbol{\rho}$ is defined by (8.4.22) and $v(\xi;0)$ is any neutralizer about $\xi = \mathbf{0}$. If the support of v is contained in D, then, as $\lambda \to \infty$,

$$K(\lambda) = \left(\frac{2\pi}{\lambda}\right)^{n/2} \exp\left\{\frac{\pi i}{4}(2r-n)\right\} + o(\lambda^{-R}) \tag{8.4.34}$$

for all R.

PROOF. The proof, which consists of reducing (8.4.33) to a product of one-dimensional integrals and then applying to each the ordinary method of stationary phase, is left to the exercises.

Upon using (8.4.34) in (8.4.31) we obtain

$$I_0(\lambda) \sim (2\pi)^{n/2} \exp\left\{\frac{\pi i}{4}(2r-n) + i\lambda\phi(\mathbf{x}_0)\right\} \sum_{m=0}^{\infty} (i)^m \lambda^{-(m+n/2)}\, G_m(\mathbf{0}). \tag{8.4.35}$$

Here we have let M go to infinity in (8.4.31) and have made the obvious estimate of the "error" integral in that equation.

We can further simplify our results by proving the following.

LEMMA 8.4.4. Let the functions $G_m(\xi)$ be defined by (8.4.32). Then

$$G_m(\mathbf{0}) = \frac{1}{2^m\, m!}\,\bar{\Delta}^m\, G_0(\mathbf{0}). \tag{8.4.36}$$

Here the operator $\bar{\Delta}$ is defined by

$$\bar{\Delta} = \frac{\partial^2}{\partial\xi_1^2} + \cdots + \frac{\partial^2}{\partial^2\xi_r^2} - \frac{\partial^2}{\partial\xi_{r-1}^2} + \cdots - \frac{\partial^2}{\partial\xi_n^2}. \tag{8.4.37}$$

PROOF. The proof follows step-by-step that of Lemma 8.3.2. Indeed, we have for any n vector $\mathbf{F}(\xi)$

$$\bar{\Delta}(\boldsymbol{\rho}\cdot\mathbf{F}) = \boldsymbol{\rho}\cdot\bar{\Delta}\mathbf{F} + 2\nabla\cdot\mathbf{F}. \tag{8.4.38}$$

From this it follows that

$$\bar{\Delta}^m(\rho\cdot\mathbf{F})\big|_{\xi=0} = 2m\bar{\Delta}^{m-1}(\nabla\cdot\mathbf{F})\big|_{\xi=0}\,. \tag{8.4.39}$$

We thus find that for any ℓ

$$\bar{\Delta}^m\, G_\ell(\mathbf{0}) = \bar{\Delta}^m(\rho\cdot\mathbf{H}_\ell)\big|_{\xi=0} = 2m\,\bar{\Delta}^{m-1}(\nabla\cdot\mathbf{H}_\ell)\big|_{\xi=0}\,. \tag{8.4.40}$$

Finally, (8.4.32) yields

$$\begin{aligned}\bar{\Delta}^m\, G_0(0) &= 2m\,\bar{\Delta}^{m-1}\; G_1(0)\\ &= 2^2 m(m-1)\,\bar{\Delta}^{m-2}\; G_2(0)\\ &\;\;\vdots\\ &= 2^m\; m!\; G_m(\mathbf{0})\end{aligned} \tag{8.4.41}$$

and the lemma is proved.

It follows from (8.4.35) and (8.4.36) that

$$I_0(\lambda) \sim (2\pi)^{n/2} \exp\left\{\frac{\pi i}{4}(2r-n) + i\lambda\phi(\mathbf{x}_0)\right\}$$

$$\times \sum_{m=0}^{\infty} \frac{(i)^m}{2^m\; m!}\,\lambda^{-(m+n/2)}\,\bar{\Delta}^m\; G_0(\mathbf{0}). \tag{8.4.42}$$

It now remains to express the quantities $\bar{\Delta}^m\, G_0(0)$ in terms of the original functions g_0 and ϕ. We shall do this for $m=0$ and $m=1$ only because, as in the previous section, the expressions are exceedingly complicated for higher values of m. From (8.4.20) and (8.4.23) we have

$$G_0(\mathbf{0}) = \frac{g_0(\mathbf{x}_0)}{\sqrt{|\det(A)|}} = \frac{g_0(\mathbf{x}_0)}{\sqrt{|\det(\phi_{x_ix_j}(\mathbf{x}_0))|}}\cdot \tag{8.4.43}$$

Hence, to leading order, the expansion of I is given by

$$I_0(\lambda) \sim \left(\frac{2\pi}{\lambda}\right)^{n/2} \frac{g_0(\mathbf{x}_0)\exp\left\{i\lambda\phi(\mathbf{x}_0) + \dfrac{\pi i}{4}\operatorname{sig}(\phi_{x_ix_j}(\mathbf{x}_0))\right\}}{\sqrt{|\det(\phi_{x_ix_j}(\mathbf{x}_0))|}}, \tag{8.4.44}$$

where we have used the relation (8.4.13).

After a great deal of computation we find that $-\bar{\Delta}G_0$ is given by the right-hand side of (8.3.53)[5] with the quantities that appear there defined by (8.3.54) and (8.3.55). When this result is inserted into (8.4.42) an explicit expansion of I_0 to second order is obtained.

This completes our derivation of the contribution to the asymptotic expansion of I corresponding to a simple interior stationary point of ϕ. We note

[5] In deriving (8.3.53), explicit use is made of the fact that all of the eigenvalues of the matrix (ϕ_{pq}) are negative. Thus, $-\bar{\Delta}G_0 = \Delta G_0$ for that case.

that the analysis presented here is by no means the only procedure that could be followed. Indeed, we might argue that the entire integration by parts process could be eliminated by analyzing (8.4.10) in a manner similar to the treatment of $K(\lambda)$ in the proof of Lemma 8.4.3 outlined in Exercise 8.18. This is in fact true. However, had we followed this latter course, it would have been difficult to arrive at (8.4.42) which is a rather useful representation of the asymptotic expansion of I.

Let us now return to the consideration of boundary critical points. To isolate the boundary from the interior critical points we introduce the neutralizer function ν_B. The support of ν_B is to lie in $\mathscr{D}$ and all of the stationary points of ϕ in $\mathscr{D}$ are to lie in the region where $\nu_B \equiv 1$. Then the contribution from the boundary is given by the asymptotic expansion of

$$I_B(\lambda) = \int_{\mathscr{D}} g_0(\mathbf{x})(1-\nu_B)\exp\{i\lambda\phi\}\,d\mathbf{x}. \tag{8.4.45}$$

If $\nabla\phi$ vanishes on Γ, that is, if there are one or more stationary points of ϕ on the boundary, then their contributions to the asymptotic expansion of I_B are readily found, at least to leading order. Indeed, suppose that $\mathbf{x}_0 \in \Gamma$ and $\nabla\phi(\mathbf{x}_0) = \mathbf{0}$. Also suppose that $\det(\phi_{x_i x_j}(\mathbf{x}_0)) \neq 0$. Then the contribution from this point to leading order is given by

$$I_0(\lambda) \sim \frac{\frac{1}{2}\exp\left\{i\lambda\phi(\mathbf{x}_0) + \frac{\pi i}{4}\operatorname{sig}(\phi_{x_i x_j}(\mathbf{x}_0))\right\} g_0(\mathbf{x}_0)\left(\frac{2\pi}{\lambda}\right)^{n/2}}{\sqrt{|\det(\phi_{x_i x_j}(\mathbf{x}_0))|}}. \tag{8.4.46}$$

This result requires the following lemma whose proof is left to the exercises.

LEMMA 8.4.5. Consider

$$K(\lambda) = \int_D \nu(\boldsymbol{\xi};\mathbf{0})\exp\{i\lambda\boldsymbol{\rho}\cdot\boldsymbol{\xi}/2\}\,d\boldsymbol{\xi} \tag{8.4.47}$$

with $\boldsymbol{\xi} = \mathbf{0}$ on the boundary of D and ρ given by (8.4.22). Then, as $\lambda \to \infty$,

$$K(\lambda) \sim \frac{1}{2}\exp\left\{\frac{\pi i}{4}(2r-n)\right\}\left(\frac{2\pi}{\lambda}\right)^{n/2}. \tag{8.4.48}$$

Further terms in the expansion of I_0 are fairly difficult to obtain.

If there are no stationary points of ϕ on Γ, then $\nabla\phi \neq \mathbf{0}$ throughout the effective domain of integration in (8.4.45). Thus (8.4.2) holds with g_0 replaced by $g_0(1-\nu_B)$ and I replaced by I_B. However, $\nu_B \equiv 0$ on Γ so that upon letting $M \to \infty$ in (8.4.2), we obtain

$$I_B(\lambda) \sim -\sum_{j=0}^{\infty}(-i\lambda)^{-(j+1)}\int_{\mathscr{P}}(\mathbf{H}_j\cdot\mathbf{N})\,W\exp\{i\lambda\psi\}\,d\sigma. \tag{8.4.49}$$

The critical points for the boundary integrals in (8.4.49) are, of course, determined by the behavior of ψ and by the configuration of the boundary itself. The points on Γ at which

$$\nabla\phi\cdot\frac{\partial \mathbf{x}}{\partial \sigma_i}=0, \qquad i=1,2,\ldots,n-1 \tag{8.4.50}$$

are the stationary points of $\psi(\sigma)$. If these stationary points are simple, then, for each, we can apply the analysis leading to (8.4.42) to the integrals in (8.4.49). In this manner we could obtain the complete expansion corresponding to each of the stationary points of ψ. The leading terms are readily found and their derivations are left to the exercises. We do wish to note however that the contribution corresponding to a simple stationary point of ψ is $0(\lambda^{(-n-1)/2})$. This is to be compared with the contribution from a simple stationary point of ϕ which is $0(\lambda^{-n/2})$.

Even if all of the stationary points of ϕ are simple, the same need not be true of the stationary points of ψ. Indeed consider a point on Γ at which ψ is stationary. From (8.4.50), we see that the level surface of ϕ passing through this point is tangent to Γ there. Clearly the order of contact between this level surface of ϕ and Γ increases as the order of the stationary point increases and vice versa. In an extreme case, Γ is itself a level surface of ϕ. Then the order of contact is infinite all along Γ and the entire boundary is critical. Indeed, in that case

$$I_B(\lambda)\sim\frac{1}{i\lambda}\exp\{i\lambda\phi_B\}\int_{\mathscr{P}}(\mathbf{H}_0\cdot\mathbf{N})\,W\,d\sigma, \tag{8.4.51}$$

where ϕ_B is the constant value of ϕ on Γ.

We note that when n is greater than two, the contribution from a stationary point of ψ of infinite order dominates the contribution from any simple interior stationary point of ϕ. This may or may not be so for stationary points of ψ of finite order.

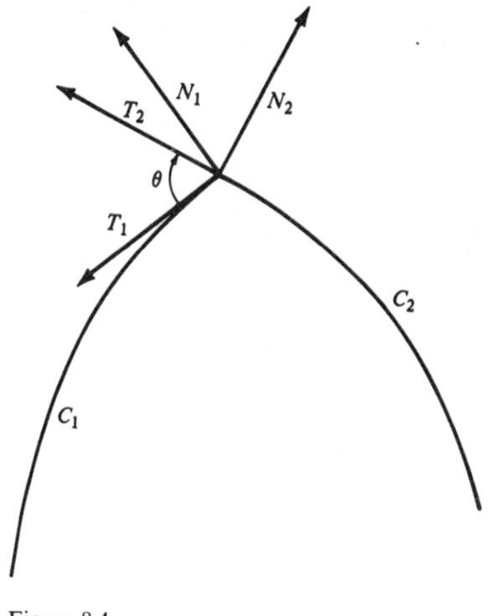

Figure 8.4.

Another interesting class of boundary critical points consists of those points at which Γ fails to have a continuously turning tangent plane (tangent line in two dimensions). In two dimensions this type of critical point is called a "corner." In three dimensions we could have isolated corner points or "edges" which are curves of critical points. Obviously the possibilities increase as the dimension increases. We cannot hope to be exhaustive here. Indeed, we shall be content with deriving the leading term of the contribution corresponding to a corner point in two dimensions.

Let us suppose that Γ is as depicted in Figure 8.4. Suppose further that the directed curves C_1 and C_2 have parametric representations given by

$$\mathbf{x} = \mathbf{x}_1(\sigma) = (x_1^{(1)}(\sigma), x_2^{(1)}(\sigma)), \qquad 0 \le \sigma \le \ell_1 \tag{8.4.52}$$

and

$$\mathbf{x} = \mathbf{x}_2(\sigma) = (x_1^{(2)}(\sigma), x_2^{(2)}(\sigma)), \qquad -\ell_2 \le \sigma \le 0, \tag{8.4.53}$$

respectively. In both cases the parameter σ is such that

$$\mathbf{T}_i = (\dot{x}_1^{(i)}, \dot{x}_2^{(i)}), \qquad (\dot{\ }) = \frac{d}{d\sigma} \tag{8.4.54}$$

is the unit tangent vector to C_i.

For continuity, we have

$$\mathbf{x}_1(0) = \mathbf{x}_2(0) = \mathbf{x}_0, \qquad \mathbf{x}_1(\ell_1) = \mathbf{x}_2(-\ell_2). \tag{8.4.55}$$

We shall assume that there is a corner at $\mathbf{x}_0$ so that

$$\mathbf{T}_1(0) \ne \mathbf{T}_2(0). \tag{8.4.56}$$

Although it is not necessary, we shall suppose that $\nabla\phi \ne \mathbf{0}$ in $\bar{\mathscr{D}}$. It then follows from (8.4.2) that

$$I(\lambda) \sim \frac{1}{i\lambda}\left[\int_0^{\ell_1} g_0(\mathbf{x}_1(\sigma))\frac{\nabla\phi\cdot\mathbf{N}_1}{|\nabla\phi|^2}\exp\{i\lambda\psi_1(\sigma)\}\,d\sigma \right.$$
$$\left. + \int_{-\ell_2}^{0} g_0(\mathbf{x}_2(\sigma))\frac{\nabla\phi\cdot\mathbf{N}_2}{|\nabla\phi|^2}\exp\{i\lambda\psi_2(\sigma)\}\,d\sigma\right]. \tag{8.4.57}$$

Here

$$\psi_i = \phi(\mathbf{x}_i(\sigma)) \tag{8.4.58}$$

and $\mathbf{N}_i$ is the unit outward normal vector to C_i, $i = 1, 2$.

Because we are only interested in the contribution from the corner, we can imagine that the amplitude functions in the two one-dimensional integrals on the right side of (8.4.57) are both neutralized about $\sigma = 0$. We shall further assume that neither ψ_1 nor ψ_2 is stationary at $\sigma = 0$. Now if I_c represents the contribution from the corner, then we find upon integrating by parts once that

$$I_c(\lambda) \sim \frac{1}{\lambda^2} \frac{g_0(\mathbf{x}_0) \exp\{i\lambda\, \phi(\mathbf{x}_0)\}}{|\nabla\phi(\mathbf{x}_0)|^2} \left[\frac{\nabla\phi \cdot \mathbf{N}_1}{\dot{\psi}_1} - \frac{\nabla\phi \cdot \mathbf{N}_2}{\dot{\psi}_2} \right]_{\sigma=0}. \tag{8.4.59}$$

We have, for $i = 1, 2$,

$$\dot{\psi}_i = \nabla\phi \cdot \mathbf{T}_i \tag{8.4.60}$$

and

$$\mathbf{N}_i = (\dot{x}_2^{(i)}, -\dot{x}_1^{(i)}). \tag{8.4.61}$$

Also, if θ denotes the angle between $\mathbf{T}_1(0)$ and $\mathbf{T}_2(0)$, then

$$\sin\theta = \dot{x}_1^{(2)}(0)\, \dot{x}_2^{(1)}(0) - \dot{x}_1^{(1)}(0)\, \dot{x}_2^{(2)}(0). \tag{8.4.62}$$

Upon using (8.4.60) to (8.4.62) in (8.4.59) we finally obtain

$$I_c(\lambda) \sim \frac{1}{\lambda^2} \frac{g_0(\mathbf{x}_0) \exp\{i\lambda\phi(\mathbf{x}_0)\} \sin\theta}{(\nabla\phi \cdot \mathbf{T}_1(0))\, (\nabla\phi \cdot \mathbf{T}_2(0))}. \tag{8.4.63}$$

Thus, to leading order, the contribution from a corner point is proportional to the sine of the opening angle of the corner itself. Furthermore, $I_c = 0(\lambda^{-2})$ and hence is *less* dominant than the contributions corresponding to either stationary points of ϕ or stationary points of ψ.

To conclude this section we wish to consider the contribution from a nonsimple interior stationary point of ϕ. We shall do this for the case $n = 2$ only and remark that the extension to higher dimensions will be apparent. Thus let us suppose that in (8.4.1) $n = 2$ and

$$\nabla\phi(\mathbf{x}_0) = \mathbf{0}, \qquad \mathbf{x}_0 \in \mathscr{D}, \tag{8.4.64}$$

$$\det A = \det(\phi_{x_i x_j}(\mathbf{x}_0)) = 0; \qquad \nabla(\det A)|_{x=x_0} \neq \mathbf{0}. \tag{8.4.65}$$

The results obtained above for a simple interior stationary point are clearly invalid in this case. Our objective is to obtain the correct contribution from $\mathbf{x}_0$. This contribution is given by the asymptotic expansion of

$$I_0(\lambda) = \int_{\mathscr{D}} \nu(\mathbf{x}; \mathbf{x}_0) \exp\{i\lambda\phi(\mathbf{x})\}\, g_0(\mathbf{x})\, d\mathbf{x}, \tag{8.4.66}$$

where ν is an appropriately constructed neutralizer about $\mathbf{x} = \mathbf{x}_0$.

It follows from (8.4.65) that the two eigenvalues of A are given by

$$\lambda_1 = \phi_{x_1x_1}(\mathbf{x}_0) + \phi_{x_2x_2}(\mathbf{x}_0), \qquad \lambda_2 = 0. \tag{8.4.67}$$

We shall assume that $\lambda_1 \neq 0$ and set

$$\mu = \operatorname{sgn}(\phi_{x_1x_1}(\mathbf{x}_0) + \phi_{x_2x_2}(\mathbf{x}_0)). \tag{8.4.68}$$

We can readily show that there exists a nonsingular linear transformation defined by

$$(\mathbf{x}-\mathbf{x}_0)^T = B\,\mathbf{z}^T, \qquad B=(B_{ij}), \qquad i,j=1,2, \tag{8.4.69}$$

which is such that, as $|z|\to 0$,

$$f(\mathbf{z}) = \phi(\mathbf{x}(\mathbf{z})) - \phi(\mathbf{x}_0) = \tfrac{1}{2}\,\mu z_1^2 + \tfrac{1}{3} z_2^3 + O([z_1^2+z_2^3]\,[z_1+z_2]). \tag{8.4.70}$$

The constants B_{ij} involve second- and third-order derivatives of ϕ at $\mathbf{x}=\mathbf{x}_0$. (See Exercise 8.21.)

The approximation (8.4.70) can be made to hold exactly throughout the support of v, which presumably has been chosen sufficiently small. This is accomplished by setting

$$\xi_i = h_i(\mathbf{z}), \qquad i=1,2. \tag{8.4.71}$$

Here

$$\frac{\mu}{2}h_1^2 + \frac{h_2^3}{3} = f(\mathbf{z}) \tag{8.4.72}$$

and

$$h_i = z_i + o(|\mathbf{z}|) \tag{8.4.73}$$

as $|z|\to 0$. Furthermore, the Jacobian

$$J(\xi) = \frac{\partial(\mathbf{x})}{\partial(\xi)} \tag{8.4.74}$$

is such that

$$J(\mathbf{0}) = \det B. \tag{8.4.75}$$

In terms of the variables ξ_1 and ξ_2, (8.4.66) becomes

$$I_0(\lambda) = \exp\{i\lambda\phi(\mathbf{x}_0)\}\int_{\hat{N}_1} v(\xi;\mathbf{0})\,G_0(\xi)\exp\left\{i\lambda\left(\frac{\mu\,\xi_1^2}{2}+\frac{\xi_2^3}{3}\right)\right\}d\xi. \tag{8.4.76}$$

Here $v(\xi;\mathbf{0})$ is a neutralizer about $\xi=\mathbf{0}$ with support $\hat{N}_1$ and

$$G_0(\xi) = g_0(\mathbf{x}(\xi))\,J(\xi). \tag{8.4.77}$$

We now write

$$G_0(\xi) = G_0(\mathbf{0}) + \xi_2\frac{\partial G_0(\mathbf{0})}{\partial \xi_2} + \hat{\rho}\cdot\mathbf{H}_0, \tag{8.4.78}$$

where

$$\hat{\rho} = (\mu\xi_1, \xi_2^2). \tag{8.4.79}$$

We note that $\hat{\rho}$ is the gradient of the phase function $(\mu\xi_1^2/2 + \xi_2^3/3)$. The reason for this expansion is that when (8.4.78) is inserted into (8.4.76), the last term can be integrated by parts. The presence of the second term in the expansion is necessary if $\mathbf{H}_0$ is to be well behaved in $\hat{N}_1$.

Upon inserting (8.4.78) into (8.4.76) and integrating the last term by parts, we obtain

$$I_0(\lambda) \sim \int_{\hat{N}_1} v(\xi;0)\left[G_0(\mathbf{0}) + \xi_2 \frac{\partial G_0(\mathbf{0})}{\partial \xi_2} - \frac{1}{i\lambda} G_1(\xi)\right] \times \exp\left\{i\lambda\left(\frac{\mu}{2}\xi_1^2 + \frac{1}{3}\xi_2^3\right)\right\} d\xi. \tag{8.4.80}$$

Here

$$G_1(\xi) = \nabla \cdot \mathbf{H}_0. \tag{8.4.81}$$

In deriving (8.4.80) we have used Lemma 8.4.2 to eliminate the term involving $(\nabla v \cdot \mathbf{H}_0)$.

Arguments similar to those used in the proof of Lemma 8.4.3 (see Exercise 8.18) show that, for all R,

$$\int_{\hat{N}_1} \exp\left\{i\lambda\left(\frac{\mu\xi_1^2}{2} + \frac{1}{3}\xi_2^3\right)\right\} v(\xi;\mathbf{0})\, d\xi = \sqrt{8\pi}\left(\frac{1}{3}\right)^{2/3} \Gamma\left(\frac{1}{3}\right) \cos\left(\frac{\pi}{6}\right) e^{\mu\pi i/4}\lambda^{-5/6} + 0(\lambda^{-R}), \tag{8.4.82}$$

$$\int_{\hat{N}_1} \exp\left\{i\lambda\left(\frac{\mu\xi_1^2}{2} + \frac{1}{3}\xi_2^3\right)\right\} v(\xi;\mathbf{0})\, \xi_2\, d\xi = i\sqrt{8\pi}\left(\frac{1}{3}\right)^{1/3} \Gamma\left(\frac{2}{3}\right) \sin\left(\frac{\pi}{3}\right) e^{\mu\pi i/4}\lambda^{-7/6} + o(\lambda^{-R}). \tag{8.4.83}$$

Also, by repeating the process to the integral involving G_1 in (8.4.80) we find

$$-\frac{1}{i\lambda}\int_{\hat{N}_1} v(\xi;\mathbf{0})\, G_1(\xi) \exp\left\{i\lambda\left(\frac{\mu}{2}\xi^2 + \frac{1}{3}\xi_2^3\right)\right\} d\xi = o(\lambda^{-7/6}). \tag{8.4.84}$$

Finally by combining (8.4.81) to (8.4.84) we obtain the following two-term expansion of I_0:

$$I_0(\lambda) \sim g_0(\mathbf{x}_0) \exp\{i\lambda\phi(\mathbf{x}_0)\} \det B \sqrt{8\pi}\left(\frac{1}{3}\right)^{2/3} \Gamma\left(\frac{1}{3}\right) \cos\left(\frac{\pi}{6}\right) e^{\pi i\mu/4}\lambda^{-5/6} + \frac{\partial G_0(\mathbf{0})}{\partial \xi_2}\sqrt{8\pi}\exp\{i\lambda\phi(\mathbf{x}_0)\}\left(\frac{1}{3}\right)^{1/3} \Gamma\left(\frac{2}{3}\right) e^{\mu i\pi/4} \sin\left(\frac{\pi}{3}\right)\lambda^{-7/6}. \tag{8.4.85}$$

Upon comparing (8.4.44) (with $n = 2$) and (8.4.85), we see that the contribution from a nonsimple stationary point is more dominant than that from a simple stationary point. Moreover, we can show that as the order of the stationary point increases, that is, the order of vanishing of $\det(\phi_{x_i x_j}(\mathbf{x}))$ as $\mathbf{x} \to \mathbf{x}_0$ increases, this dominance becomes more pronounced.

8.5. Parametric Expansions

In the asymptotic analysis of multidimensional integrals of form (8.4.1), the major step is most often the determination of the stationary points of ϕ. This requires finding those points in $\bar{\mathscr{D}}$ at which $\nabla\phi(\mathbf{x}) = 0$. In general, because this condition involves simultaneous transcendental equations, an explicit determination of the stationary points is difficult if not impossible. To avoid this complication we often take a parametric point of view. We shall discuss this point of view in the context of two examples; wave propagation in dispersive media and acoustic scattering by convex bodies.

As we have indicated in Section 2.6, the solution to the initial-value problem for any energy-conserving dispersive hyperbolic equation of second order, can be represented as a sum of integrals of the form

$$u(\mathbf{x},t) = \frac{1}{(2\pi)^{3/2}} \int_{-\infty}^{\infty} A(\mathbf{k}) \exp\{i[\mathbf{k}\cdot\mathbf{x} - \omega(\mathbf{k})t]\}\, d\mathbf{k}, \qquad \mathbf{k} = (k_1, k_2, k_3). \quad (8.5.1)$$

This integral is in *dimensional* form so that $\mathbf{x} = (x_1, x_2, x_3)$ represents a point in physical space and $t > 0$ represents time. In the specific example of Section 2.6, namely the Klein-Gordon equation, $\omega(\mathbf{k}) = \pm\sqrt{c^2 k^2 + b^2}$. In general, the *dispersion relation* $\omega = \omega(\mathbf{k})$, which relates the frequency ω to the wave vector $\mathbf{k}$, is determined by the particular differential equation under consideration.

In Section 2.6 we found that when dimensionless variables were introduced, a large parameter λ appeared which had the interpretation of "large time" compared to some fundamental time unit of the problem. Here, however, we shall leave the integral in dimensional form but in our subsequent analysis we shall interpret our asymptotic result as a large time expansion.

Let us now formally[6] apply the results of the previous section to (8.5.1). Because there are no boundaries, the stationary points are the dominant critical points. Hence if we set $\lambda = 1$, (8.4.44) yields

$$u(\mathbf{x}, t) \sim \frac{A(\mathbf{k}) \exp\left\{i[\mathbf{k}\cdot\mathbf{x} - \omega(\mathbf{k})t] - \dfrac{\pi i}{4} \operatorname{sig}\left(\dfrac{\partial^2 \omega(\mathbf{k})}{\partial k_i\, \partial k_j}\right)\right\}}{t^{3/2} \left|\det\left(\dfrac{\partial^2 \omega(\mathbf{k})}{\partial k_i\, \partial k_j}\right)\right|^{1/2}}, \quad (8.5.2)$$

$$\mathbf{x} = \mathbf{V}_g(\mathbf{k})t; \qquad \mathbf{V}_g(\mathbf{k}) = \nabla_k \omega. \quad (8.5.3)$$

Here (8.5.3) is simply the condition that the phase function

$$\phi = \mathbf{k}\cdot\mathbf{x} - \omega(\mathbf{k})t \quad (8.5.4)$$

be stationary.

[6] The results of the previous section remain valid for integrals over infinite domains under quite reasonable conditions on the integrand, but we have not shown this. In this sense, our results here are formal.

To obtain an explicit asymptotic expansion of $u(\mathbf{x}, t)$ it is necessary to invert (8.5.3). If this inversion could be accomplished, then we would obtain stationary points of the form $\mathbf{k} = \boldsymbol{\kappa}(\mathbf{x}, t)$. Upon setting $\mathbf{k} = \boldsymbol{\kappa}$ in (8.5.2) the desired expansion would be obtained. Because this procedure cannot be carried out in most cases, we content ourselves here with an implicit or parametric representation of the expansion. In this, we look upon the stationary condition (8.5.3) as defining for each fixed $\mathbf{k}$, a *ray* in space time. In turn, this ray defines a point moving in space with the *group velocity* $\mathbf{V}_g(\mathbf{k})$. Along any ray the asymptotic expansion of $u(\mathbf{x}, t)$ has the form (8.5.2) which is a plane wave with constant wave vector $\mathbf{k}$ and frequency $\omega(\mathbf{k})$. Furthermore, the amplitude of the plane wave decays as time progresses like $t^{-3/2}$. We note that this decay is the only variation of the amplitude along a ray.

As $\mathbf{k}$ ranges over all possible values, the corresponding rays fill out a certain region in space time. This region of course is determined solely by the group velocity vector $\mathbf{V}_g(\mathbf{k})$. To obtain the asymptotic expansion of $u(\mathbf{x}, t)$ to leading order at any fixed space time point $(\mathbf{x}, t)$ we need only sum (8.5.2) over all values of $\mathbf{k}$ corresponding to the rays passing through that point. In general, only one ray passes through a given point. An interesting situation arises when more than one ray passes through $(\mathbf{x}, t)$. Indeed, we can show that then there exist curves in space time called *caustics* along which $\det(\partial^2\omega/\partial k_i \partial k_j)$ evaluated at the stationary point $\mathbf{k} = \boldsymbol{\kappa}(\mathbf{x}, t)$ vanishes. These caustic curves, moreover, can be shown to be envelopes of the ray system itself. (See Figure 8.5.) Clearly (8.5.2) is invalid along a caustic. The analysis required to obtain a valid expansion is equivalent to studying (8.5.1) in the case where ϕ has stationary points of higher order and will not be discussed here.

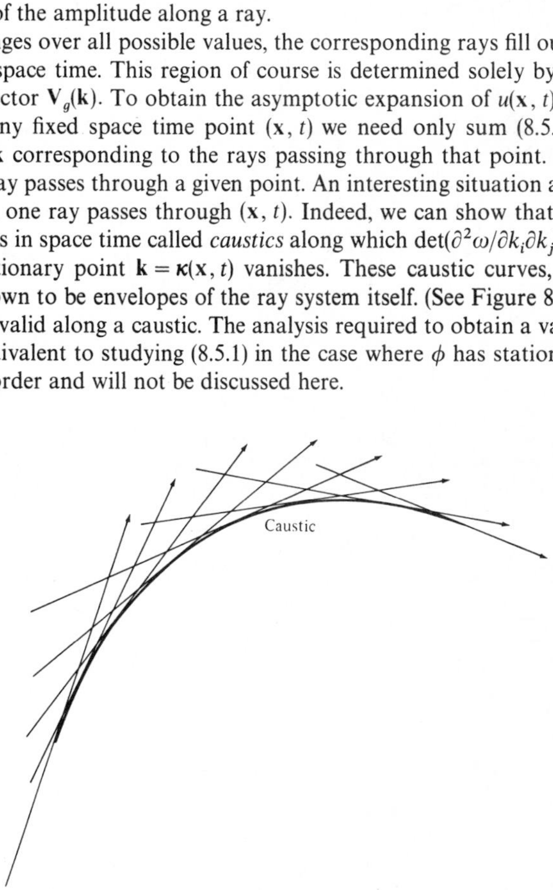

Figure 8.5. Caustic Curve of a Two-Dimensional Ray Field.

A further interpretation of our parametric result is perhaps in order. We define the energy associated with $u(\mathbf{x}, t)$ to be

$$E = \frac{1}{2}\int_{-\infty}^{\infty} |u(\mathbf{x}, t)|^2 \, d\mathbf{x}. \tag{8.5.5}$$

From (8.5.1) we find that we also have

$$E = \frac{1}{2}\int_{-\infty}^{\infty} |A(\mathbf{k})|^2 \, d\mathbf{k}$$

so long as $\omega(\mathbf{k})$ is real for real $\mathbf{k}$. (We remark that this is indeed the case when u represents the solution to an energy-conserving equation.)

Let us now consider the amount of energy in a volume element $\Delta\mathbf{k}$ centered around a given wave vector $\mathbf{k}$. This energy is constant in time and is given approximately by $\frac{1}{2}|A(\mathbf{k})|^2 \, \Delta\mathbf{k}$. As (8.5.3) indicates, the signal u eventually decomposes in such a manner that each wave vector $\mathbf{k}$, and hence any function of $\mathbf{k}$, propagates at its own velocity $\mathbf{V}_g(\mathbf{k})$. In particular, the energy packet $\frac{1}{2}|A|^2 \, \Delta\mathbf{k}$ travels at the group velocity $\mathbf{V}_g(\mathbf{k})$.

Let us now suppose that at a given time t, the wave vector $\mathbf{k}$ is located at the point $\mathbf{x}$ and that the packet of wave vectors in the volume $\Delta\mathbf{k}$ around $\mathbf{k}$ is located in a volume $\Delta\mathbf{x}$ about $\mathbf{x}$. Then we must have

$$\tfrac{1}{2}|u|^2 \, \Delta\mathbf{x} = \tfrac{1}{2}|A|^2 \, \Delta\mathbf{k}$$

so that

$$|u| = \frac{|A|}{(\Delta\mathbf{x}/\Delta\mathbf{k})^{1/2}}. \tag{8.5.6}$$

We fix the volume element $\Delta\mathbf{k}$. Then as time increases, the element $\Delta\mathbf{x}$, containing this packet of wave vectors, varies in size. This follows from (8.5.3), which shows that nearby wave vectors propagate at slightly different speeds. In any case, for time large, the ratio $(\Delta\mathbf{x}/\Delta\mathbf{k})^{1/2}$ is simply the Jacobian of the transformation from $\mathbf{k}$ space to $\mathbf{x}$ space defined by (8.5.3). This yields

$$|u| = \frac{A}{J}; \qquad J = \left(\frac{\Delta\mathbf{x}}{\Delta\mathbf{k}}\right)^{1/2} = \left|\det\left(\frac{\partial x_i}{\partial k_j}\right)\right|^{1/2} = t^{3/2}\left|\det\left(\frac{\partial^2 \omega}{\partial k_i \partial k_j}\right)\right|^{1/2} \tag{8.5.7}$$

which agrees with (8.5.2).

In summary, (8.5.2) and (8.5.3) is a parametric representation of a wave train which is locally planar but which actually has slowly varying wave vector and frequency. The energy of the prescribed initial state associated with a given wave vector $\mathbf{k}$ ultimately propagates with the group velocity $V_g(\mathbf{k})$ defined by (8.5.3). Thus we find that, in addition to computational advantages, the parametric representation affords certain physical insights not immediately apparent from an explicit representation.

As a second example let us consider the problem considered in Section 2.7, namely that of acoustic scattering by convex bodies. There we found that when the incident field is a plane wave, the Kirchhoff approximation for the scattered field is given by

$$u_s(\mathbf{x};k) \sim \frac{-ik\exp[ikr]}{2\pi r} I; \qquad r = (x_1^2 + x_2^2 + x_3^2)^{1/2}, \tag{8.5.8}$$

$$I = \int_L (\mathbf{n}(\xi)\cdot\boldsymbol{\mu}_\pm)\exp\{ik[\boldsymbol{\mu}_+ - \boldsymbol{\mu}_-]\cdot\xi\}\,d\sigma. \tag{8.5.9}$$

Here L is the "lit" region on the scattering body, $\boldsymbol{\mu}_+$ and $\boldsymbol{\mu}_-$ are unit vectors in the direction of incidence and observation respectively, ξ is the position vector of a point on L, and $d\sigma$ is the differential element of surface area on L. Also, $\mathbf{n}$ is the unit outward normal to L and the $\pm$ sign refers to the particular boundary condition assumed in deriving (8.5.9). That $\boldsymbol{\mu}_-$ is independent of ξ is a result of the assumption that the point of observation is "far" from the scattering body.

We shall seek an asymptotic expansion of I for "large" k. Because k is a dimensional quantity, we remark that we must interpret our results as being correct for wavelengths small compared to the fundamental body dimensions. Here, as in the previous problem, the major contribution to the asymptotic expansion of I comes from the interior stationary points. This contribution represents the far reflected field.

If we introduce σ_1 and σ_2 as coordinates on L, then the conditions for stationarity of the phase $\phi = (\boldsymbol{\mu}_+ - \boldsymbol{\mu}_-)\cdot\xi$ are

$$(\boldsymbol{\mu}_+ - \boldsymbol{\mu}_-)\cdot\xi_{\sigma_1} = 0, \qquad (\boldsymbol{\mu}_+ - \boldsymbol{\mu}_-)\cdot\xi_{\sigma_2} = 0. \tag{8.5.10}$$

These conditions imply that, at the stationary points, $(\boldsymbol{\mu}_+ - \boldsymbol{\mu}_-)$ is parallel to the normal $\mathbf{n}$. It is easy to see then that due to the convexity of L, there can be only one stationary point for a given $\boldsymbol{\mu}_-$. It also follows from (8.5.10) that $\boldsymbol{\mu}_+$ and $\boldsymbol{\mu}_-$ make the same angle with $\mathbf{n}$ at the stationary point. In other words, we have *Snell's law* which states that the angle of incidence equals the angle of reflection.

To first order the asymptotic expansion of I is given parametrically by (8.5.10) and

$$I \sim \frac{2\pi}{k}\,\frac{(\mathbf{n}\cdot\boldsymbol{\mu}_\pm)\,D\exp\left\{ik(\boldsymbol{\mu}_+ - \boldsymbol{\mu}_-)\cdot\xi + \dfrac{\pi i}{2}\right\}}{[|(\boldsymbol{\mu}_+ - \boldsymbol{\mu}_-)\cdot\xi_{\sigma_1\sigma_1}(\boldsymbol{\mu}_+ - \boldsymbol{\mu}_-)\cdot\xi_{\sigma_2\sigma_2} - [(\boldsymbol{\mu}_+ - \boldsymbol{\mu}_-)\cdot\xi_{\sigma_1\sigma_2}]^2|]^{1/2}}.^{7} \tag{8.5.11}$$

Here D^2 is the discriminant of the first fundamental form of differential geometry, that is,

$$D^2 = [(\xi_{\sigma_1}\cdot\xi_{\sigma_1})(\xi_{\sigma_2}\cdot\xi_{\sigma_2}) - (\xi_{\sigma_1}\cdot\xi_{\sigma_2})^2]. \tag{8.5.12}$$

The above parametric representation is to be understood as follows. Because

[7] That the signature of the matrix $(\phi_{\sigma_i\sigma_j})$ at the stationary point is 2 follows from the convexity of L.

of the convexity of L there is a one-to-one correspondence between stationary points on L and directions of reflection $\boldsymbol{\mu}_-$. Thus for each point on L we first determine the direction $\boldsymbol{\mu}_-$ so that the stationary conditions (8.5.10) hold. With $\boldsymbol{\mu}_-$ so determined (8.5.11) and (8.5.8) then represent the far reflected field in that direction from the scattering body.

We can express our result more compactly without altering its parametric nature. Indeed we note that, at the stationary point,

$$(\boldsymbol{\mu}_+ - \boldsymbol{\mu}_-) = \frac{(\boldsymbol{\xi}_{\sigma_1} \times \boldsymbol{\xi}_{\sigma_2})\, 2\cos\theta}{D}, \tag{8.5.13}$$

where θ is the angle of incidence (reflection). The *Gaussian curvature* κ of the surface L is given by

$$\kappa = \frac{d^2}{D^2}. \tag{8.5.14}$$

Here d^2 is the discriminant of the second fundamental form of differential geometry, that is,

$$d^2 = \frac{1}{D^2}\left[\boldsymbol{\xi}_{\sigma_1\sigma_1}\cdot(\boldsymbol{\xi}_{\sigma_1} \times \boldsymbol{\xi}_{\sigma_2})\, \boldsymbol{\xi}_{\sigma_2\sigma_2}\cdot(\boldsymbol{\xi}_{\sigma_1} \times \boldsymbol{\xi}_{\sigma_2}) - (\boldsymbol{\xi}_{\sigma_1\sigma_2}\cdot(\boldsymbol{\xi}_{\sigma_1} \times \boldsymbol{\xi}_{\sigma_2}))^2\right]. \tag{8.5.15}$$

Upon using (8.5.13) to (8.5.15) we find that (8.5.11) becomes

$$I \sim \pm\frac{\pi}{k}\frac{\exp\left\{ik[(\boldsymbol{\mu}_+ - \boldsymbol{\mu}_-)\cdot\boldsymbol{\xi}] + \dfrac{\pi i}{2}\right\}}{\sqrt{\kappa}}. \tag{8.5.16}$$

Finally, it follows from (8.5.8) and (8.5.16) that the scattered field is given approximately by

$$u_s \approx \pm\frac{\exp\left\{ik[r + (\boldsymbol{\mu}_+ - \boldsymbol{\mu}_-)\cdot\boldsymbol{\xi}]\right\}}{2r\sqrt{\kappa}} \tag{8.5.17}$$

It will be recalled that in Section 2.7 we pointed out that the Kirchhoff approximation involved certain inaccuracies near the boundary of L. Thus, when the stationary point lies on or near that boundary, we must expect that (8.5.17) is a poor representation of the scattered field. We also know from the previous section that, for the stationary point on the boundary, (8.5.16) is in error by a factor of $\frac{1}{2}$. An interesting mathematical question is, can an asymptotic expansion of I be obtained which remains uniformly valid as the stationary point approaches the boundary of L? In Section 9.6 we shall derive such an expansion.

8.6. Exercises

8.1. Consider

$$I(\lambda) = \int_{\mathscr{D}} \left(x^2 + \frac{y^2}{2}\right) \exp\left\{\lambda\left[y \sin\frac{\pi x}{2} - \frac{y^2}{2}\right]\right\} dx\, dy. \tag{8.6.1}$$

Calculate the leading term of the asymptotic expansion of $I(\lambda)$, as $\lambda \to \infty$, for each of the domains defined below:

(a) $|x+y| \leq 2, \quad |x-y| \leq 2.$

(b) $x^2 + y^2 \leq 4.$

(c) $x^2 + y^2 \leq 2.$

8.2. Consider

$$I(\lambda;\alpha) = \int_{\mathscr{D}} \exp\{\lambda\,[\sin x \cos y \sin\alpha + \cos x \cos\alpha]\}\, dx\, dy \tag{8.6.2}$$

with $\mathscr{D}$ defined by $0 \leq x \leq \pi/4$, $0 \leq y \leq 2\pi$. Calculate the leading term of the asymptotic expansion, as $\lambda \to \infty$, for each of the following integrals:

(a) $I(\lambda;0)$.

(b) $I\left(\lambda;\dfrac{\pi}{8}\right)$.

(c) $I\left(\lambda;\dfrac{\pi}{4}\right)$.

(d) $I\left(\lambda;\dfrac{3\pi}{8}\right)$.

(e) $I(-\lambda;0)$.

8.3. (a) Explain why, in the derivation of (8.3.57), D is confined to one side of a hyperplane through $\boldsymbol{\xi} = \mathbf{0}$.

(b) Use part (a) and the result (8.3.39) to show that

$$\hat{K}(\lambda) = \frac{1}{2}\left(\frac{2\pi}{\lambda}\right)^{n/2} [1 + O(\lambda^{-1/2})]. \tag{8.6.3}$$

Here $\hat{K}(\lambda)$ is defined by (8.3.57).

(c) Choose a number r such that a hypersphere of radius r tangent to D at $\boldsymbol{\xi} = \mathbf{0}$ is totally contained in D. (The number r is less than the minimal *principal radius* of the boundary surface of D at $\boldsymbol{\xi} = \mathbf{0}$.) Show that

$$J(\lambda) = \int_0^{2r} \rho^{n-1} F_n(\rho)\, e^{-\lambda\rho^2/2}\, d\rho < \hat{K}(\lambda). \tag{8.6.4}$$

Here

$$F_n(\rho) = \omega_{n-1} \int_0^{\cos^{-1}(\rho/2r)} (\sin \phi)^{n-2} \, d\phi \tag{8.6.5}$$

with ω_{n-1} the area of the surface of a unit sphere in $n-1$ dimensions. [ω_n is given by (8.3.38).]

(d) Use Laplace's method to show that

$$J(\lambda) = \frac{1}{2} F_n(0) \left[\frac{2}{\lambda}\right]^{n/2} \Gamma\left[\frac{n}{2}\right] [1 + O(\lambda^{-1/2})], \qquad \lambda \to \infty, \tag{8.6.6}$$

and calculate $F_n(0)$.

(e) Verify (8.3.57).

8.4. Explain why the results of Sections 8.2 and 8.3 are valid for $|\arg(\lambda)| < \pi/2$.

8.5. (a) Show that when $n = 2$, $\phi_{x_1x_1}(\mathbf{x}_0) = \phi_{x_2x_2}(\mathbf{x}_0) = 1$, and $\phi_{x_1x_2}(\mathbf{x}_0) = 0$. (8.3.53) reduces to

$$\begin{aligned}
\Delta_{\xi} G_0 \Big|_{\xi=0} &= \Delta_x g_0(\mathbf{x}_0) - \frac{\partial g_0}{\partial x_1}(\mathbf{x}_0) \left\{\phi_{x_1x_2x_2}(\mathbf{x}_0) + \phi_{x_1x_1x_1}(\mathbf{x}_0)\right\} \\
&\quad - \frac{\partial g_0(\mathbf{x}_0)}{\partial x_2} \left\{\phi_{x_1x_1x_2}(\mathbf{x}_0) + \phi_{x_2x_2x_2}(\mathbf{x}_0)\right\} \\
&\quad + g_0(\mathbf{x}_0) \left[\frac{5}{12} \phi^2_{x_1x_1x_1}(\mathbf{x}_0) + \frac{5}{12} \phi^2_{x_2x_2x_2}(\mathbf{x}_0)\right. \\
&\quad + \frac{3}{4} \phi^2_{x_1x_2x_2}(\mathbf{x}_0) + \frac{3}{4} \phi^2_{x_1x_1x_2}(\mathbf{x}_0) + \frac{\phi_{x_1x_1x_1}(\mathbf{x}_0)\, \phi_{x_2x_2x_1}(\mathbf{x}_0)}{2} \\
&\quad + \frac{\phi_{x_2x_2x_2}(\mathbf{x}_0)\, \phi_{x_1x_1x_2}(\mathbf{x}_0)}{2} - \frac{1}{4} \phi_{x_1x_1x_1x_1}(\mathbf{x}_0) \\
&\quad \left. - \frac{1}{4} \phi_{x_2x_2x_2x_2}(\mathbf{x}_0) - \frac{1}{2} \phi_{x_1x_1x_2x_2}(\mathbf{x}_0)\right].
\end{aligned} \tag{8.6.7}$$

(b) Calculate the result in (a) directly from the definition of G_1, (8.3.45), and the definition of the change of variables (8.3.19) to (8.3.22).

8.6. In (8.3.60), choose the parameters σ_j, $j = 1, \ldots, n-1$, so that $\boldsymbol{\sigma} = \mathbf{0}$ at $\mathbf{x} = \mathbf{x}_0$ and the σ_j's are arc-lengths in the principal directions of curvature of the boundary at $\mathbf{x}_0$. Then show that

$$I(\lambda) \sim \frac{1}{\lambda} \left(\frac{2\pi}{\lambda}\right)^{(n-1)/2} \frac{g_0(\mathbf{x}_0)}{|\nabla\phi|} \frac{\exp\{\lambda\phi(\mathbf{x}_0)\}}{|J|^{1/2}}. \tag{8.6.8}$$

Here

$$J = \det\left(\frac{\partial^2 \psi}{\partial \sigma_i \, \partial \sigma_j}\right)\Bigg|_{\sigma=0} \tag{8.6.9}$$

[Exercises 8.7 to 8.10 establish the validity of (8.3.64) in the text.]

8.7. Consider the quadratic form

$$f(\mathbf{x}) = \Phi_{pq} \, x_p \, x_q.$$

Here $\Phi = (\Phi_{pq})$ is a real symmetric matrix and repeated indices are to be summed from 1 to n.

(a) Show that the values of $\mathbf{x}$ for which $f(\mathbf{x})$ attains an extremal value subject to the constraints

$$x_p \, x_p = 1, \qquad x_p \, N_p = 0$$

are found by solving the equations

$$\Phi_{pq} \, x_q = \alpha x_p + \beta N_p, \qquad p = 1, 2, \ldots, n \tag{8.6.10}$$

subject to the two constraints. Also show that there are $n-1$ extrema and that to each extremal point, say $\mathbf{x} = \mathbf{x}^r$, the corresponding value of α, say α_r is such that $\alpha_r = f(\mathbf{x}^r)$.

(b) Let Ψ be an orthogonal matrix whose last column is comprised of the elements of the vector $\mathbf{N} = (\mathbf{N}_1, \ldots, \mathbf{N}_n)$ in part (a). Define $\boldsymbol{\eta}$ by

$$\mathbf{x}^T = \Psi \boldsymbol{\eta}^T \tag{8.6.11}$$

and set

$$F(\eta_1, \ldots, \eta_n) = \Phi_{pq} \, \Psi_{pi} \, \Psi_{qj} \, \eta_i \, \eta_j. \tag{8.6.12}$$

Then show that the problem corresponding to that of part (a) is to find those $\boldsymbol{\eta}$ which extremize $F(\eta_1, \ldots, \eta_n)$ subject to the constraints

$$\eta_i \, \eta_i = 1, \qquad \eta_n = 0.$$

(c) Show that in (b) we can as well delete the last component of $\boldsymbol{\eta}$ and the last column of Ψ. If we call the resulting $n \times (n-1)$ dimensional matrix $\mathcal{J}$, then show that the problem now is to extremize

$$\begin{aligned} F(\eta_1, \eta_2, \ldots, \eta_{n-1}, 0) &= \Phi_{pq} \, \mathcal{J}_{pi} \, \mathcal{J}_{qj} \, \eta_i \, \eta_j \\ &= L_{ij} \, \eta_i \, \eta_j \end{aligned}$$

subject to the constraint

$$\eta_i \, \eta_i = 1. \tag{8.6.13}$$

Here i and j are to be summed from 1 to $n-1$.

(d) Finally show that the problem in part (c) is equivalent to solving

$$L_{ij} \, \eta_j = \alpha \eta_i, \qquad i = 1, 2, \ldots, n-1$$

subject to (8.6.13) and hence the α_i's are the eigenvalues of the matrix $L = (L_{ij})$. [Note that $\det(L_{ij}) = \prod_{i=1}^{n-1} \alpha_i$.]

8.8. Let the $n \times (n-1)$ matrix $\mathscr{J}$ of Exercise 8.7(c) be given by

$$\mathscr{J} = \left(\frac{\partial x_i}{\partial \sigma_j}\right)$$

with the parameters $\sigma_1, \sigma_2, \ldots, \sigma_{n-1}$ chosen as in Exercise 8.6. Let the surface $\mathscr{P}$ in (8.3.60) also be denoted by $h(\mathbf{x}) = 0$. Show that if

$$\Phi_{pq} = \left(\frac{\partial^2 \phi}{\partial x_p\, \partial x_q} - k\frac{\partial^2 h}{\partial x_p\, \partial x_q}\right)\bigg|_{\mathbf{x}=\mathbf{x}_0},$$

$$\nabla\phi\big|_{\mathbf{x}=\mathbf{x}_0} = k\nabla h\big|_{\mathbf{x}=\mathbf{x}_0}, \tag{8.6.14}$$

then the matrix L of Exercise 8.7(c) is defined by

$$L_{ij} = \frac{\partial^2 \psi(\boldsymbol{\sigma})}{\partial \sigma_i\, \partial \sigma_j}\bigg|_{\boldsymbol{\sigma}=0}, \qquad \psi(\boldsymbol{\sigma}) = \phi(\mathbf{x}(\boldsymbol{\sigma})).$$

Hence,

$$J = \prod_{i=1}^{n-1} \alpha_i, \tag{8.6.15}$$

where J is defined by (8.6.9) and the α_i's are as introduced in Exercise 8.7(d).

8.9. In Exercise 8.7 denote by Φ the matrix with elements Φ_{pq}.

(a) From (8.6.10) show that the numbers $\alpha_1, \ldots, \alpha_{n-1}$ are the roots of the polynomial equation

$$\mathbf{N}\,(\Phi - \alpha I)^{-1}\,\mathbf{N}^T = 0, \qquad \mathbf{N} = (N_1, \ldots, N_n). \tag{8.6.16}$$

(b) Show that the product of the roots in (8.6.16) is given by

$$\prod_{p=1}^{n-1} \alpha_p = \mathbf{N}\,\Phi^{-1}\,\mathbf{N}^T \det\Phi = \mathbf{N}\,\mathrm{cof}(\Phi)\,\mathbf{N}^T.$$

Here cof(A) denotes the matrix whose elements are the cofactors of the elements of the matrix A.

8.10. With Φ defined as in Exercise 8.8 use the results of Exercises 8.6 to 8.9 to show that

$$J = \frac{\partial\phi}{\partial x_p}\frac{\partial\phi}{\partial x_q}\left[\mathrm{cof}(\phi_{x_p x_q} - k h_{x_p x_q})\right]|\nabla\phi|^{-2} \tag{8.6.17}$$

and thus verify (8.3.64).

8.11. Specialize the result (8.6.17) to two dimensions and show that the result agrees with (8.2.21).

8.12. In two dimensions let ρ be the radius of curvature of the curve $\phi(\mathbf{x}) = \phi(\mathbf{x}_0)$ at $\mathbf{x} = \mathbf{x}_0$ and let ρ' be the radius of curvature of the boundary curve $h(\mathbf{x}) = 0$ at $\mathbf{x} = \mathbf{x}_0$. Show that in (8.6.17)

$$J = \left|\frac{1}{\rho} - \frac{1}{\rho'}\right| |\nabla\phi|, \qquad n = 2. \tag{8.6.18}$$

8.13. Let $b_{\alpha\beta}^{(\phi)}$ be the elements of the second fundamental form of differential geometry at $\mathbf{x} = \mathbf{x}_0$ for the surface $\phi(\mathbf{x}) = \phi(\mathbf{x}_0)$. Similarly, let $b_{\alpha\beta}^{(h)}$ be the elements of this same form at $\mathbf{x} = \mathbf{x}_0$ for the surface $h(\mathbf{x}) = 0$. Then show that in (8.6.17)

$$J = |\nabla\phi|^{n-1} \left|\det(b_{\alpha\beta}^{(\phi)} - b_{\alpha\beta}^{(h)}\right|. \tag{8.6.19}$$

8.14. Consider

$$I(\lambda) = \int_{\mathscr{D}} \left(x^2 + \frac{y^2}{2}\right) \exp\left\{i\lambda\left[y \sin\frac{\pi x}{2} - \frac{y^2}{2}\right]\right\} dx\, dy \tag{8.6.20}$$

as $\lambda \to \infty$. Here $\mathscr{D}$ is the domain $x^2 + y^2 < 4$.

(a) Identify the critical points for this integral.

(b) Let $v(\mathbf{x};\mathbf{x}_0)$, $\mathbf{x} = (x, y)$, be the neutralizer defined by (8.4.6) in two dimensions with $\delta_1 = \frac{1}{3}$, $\delta_2 = \frac{2}{3}$. Employ this neutralizer to write $I(\lambda)$ as a sum of integrals thereby isolating the boundary of $\mathscr{D}$ from the interior critical point.

8.15. Let $I(\lambda)$ be given by (8.6.20). For each of the domains listed below, calculate the asymptotic expansion of I to order λ^{-2}.

(a) $|x + y| < 1, \qquad |x - y| < 1.$

(b) $x^2 + y^2 < 4.$

(c) $x^2 + y^2 < 2.$

8.16. Let

$$I(\lambda;\alpha) = \int_{\mathscr{D}} \exp\{i\lambda[\sin x_1 \cos x_2 \sin\alpha + \cos x_1 \cos\alpha]\}\, dx_1\, dx_2. \tag{8.6.21}$$

Here $\mathscr{D}$ is the domain $0 \le x_1 \le \pi/4$, $0 \le x_2 \le 2\pi$. Calculate the asymptotic expansion to order $\lambda^{-3/2}$ of each of the following integrals:

(a) $I(\lambda;0)$.

(b) $I\left(\lambda;\frac{\pi}{8}\right)$.

(c) $I\left(\lambda;\frac{\pi}{4}\right)$.

(d) $I\left(\lambda; \frac{3\pi}{8}\right)$.

(e) $I(-\lambda;0)$.

8.17. Suppose that the function $\phi(\mathbf{x})$ has a simple stationary point at $\mathbf{x} = \mathbf{x}_0$ and is C^∞ in a neighborhood of $\mathbf{x} = \mathbf{x}_0$. Consider the transformation (8.4.16)–(8.4.18) from $\mathbf{z}$ to $\boldsymbol{\xi}$. Use the Taylor series expansion for $f(\mathbf{z})$ to write down the Taylor series for each ξ_i^2, $i = 1, 2, \ldots, n$. Explain, in terms of these Taylor series, the nonuniqueness of the transformation from $\mathbf{z}$ to $\boldsymbol{\xi}$.

8.18. Consider the integral $K(\lambda)$ defined by (8.4.33). Introduce the domain S_1, which is a hypercube centered at $\boldsymbol{\xi} = \mathbf{0}$, and completely contained in D. Introduce a second hypercube S_2 completely contained in S_1 and centered at $\boldsymbol{\xi} = \mathbf{0}$.

(a) Construct an n-dimensional neutralizer $v_s(\boldsymbol{\xi};\mathbf{0})$ having the properties that (i) it is the product of one-dimensional neutralizers and (ii) it is identically one inside S_2 and identically zero outside S_1.

(b) Show that we can replace D by S_1 in (8.4.33) with an error $o(\lambda^{-R})$ for any R.

(c) Write the integral over the new domain S_1 as a product of one-dimensional integrals and verify (8.4.34).

8.19. Prove Lemma 8.4.5.

8.20. Suppose that

$$I(\lambda) = \frac{1}{i\lambda} \int_{\mathscr{P}} \frac{g_0 \, \mathbf{N} \cdot \nabla\phi}{|\nabla\phi|^2} W \, v(\boldsymbol{\sigma};\mathbf{0}) \exp\{i\lambda\psi\} \, d\boldsymbol{\sigma}. \tag{8.6.22}$$

Here $\mathscr{P}$, ψ and W are defined in the discussion below (8.4.2). The neutralizer $v(\boldsymbol{\sigma};\mathbf{0})$ isolates a simple stationary point of $\psi(\boldsymbol{\sigma})$ at $\boldsymbol{\sigma} = \mathbf{0}$.

(a) Assume the results of Exercises 8.7 to 8.10 and show that

$$I(\lambda) \sim \frac{1}{\lambda}\left(\frac{2\pi}{\lambda}\right)^{(n-1)/2} \frac{\operatorname{sgn}\,[\mathbf{N} \cdot \nabla\phi(\mathbf{x}_0)]}{i} \frac{g(\mathbf{x}_0)}{|\nabla\phi|} \frac{\exp\left\{i\lambda\phi(\mathbf{x}_0) + \dfrac{i\mu\pi}{4}\right\}}{|J|^{1/2}}. \tag{8.6.23}$$

Here $\mathbf{x} = \mathbf{x}_0$ for $\boldsymbol{\sigma} = \mathbf{0}$, J is given by (8.6.17), and

$$\mu = \operatorname{sig}\left(\frac{\partial^2 \psi}{\partial\sigma_i \, \partial\sigma_j}\right). \tag{8.6.24}$$

(b) Show that

$$\mu = \operatorname{sig}\left\{(\phi_{pq} - kh_{pq}) \frac{\partial x_p}{\partial\sigma_i} \frac{\partial x_q}{\partial\sigma_j}\right\}. \tag{8.6.25}$$

(c) Show that

$$\mu = \text{sig}[b^{(\phi)}_{\alpha\beta} - b^{(h)}_{\alpha\beta}] \tag{8.6.26}$$

with the relevant quantities defined in Exercise 8.13.

8.21. (a) Suppose ϕ satisfies (8.4.64) and (8.4.65). Show that

$$\phi(\mathbf{x}) - \phi(\mathbf{x}_0) = \frac{1}{2}\mu\left[|\phi_{11}|^{1/2}\eta_1 + \frac{\mu\,\phi_{12}}{|\phi_{11}|^{1/2}}\eta_2\right]^2 + \frac{1}{3!}[\phi_{ijk}\eta_i\eta_j\eta_k] + \cdots.$$

Here, we sum over repeated indices from 1 to 2 and

$$\eta_i = x_i - x_{i0}, \qquad \phi_i = \frac{\partial\phi}{\partial x_i}, \qquad i = 1, 2, \qquad \mu = \text{sign}\,\phi_{11}.$$

(b) Thus show that to obtain (8.4.70), we first set

$$z_1 = |\phi_{11}|^{1/2}\eta_1 + \frac{\mu\,\phi_{12}}{|\phi_{11}|^{1/2}}\eta_2$$

and determine the remaining constants of the matrix B by requiring that

$$\phi_{ijk}\,B_{i1}\,B_{j2}\,B_{k2} = 0$$

and

$$\phi_{ijk}\,B_{i2}\,B_{j2}\,B_{k2} = 2.$$

Here again we sum over repeated indices from 1 to 2.

REFERENCES

N. CHAKO. Asymptotic expansions of double and multiple integrals arising in diffraction theory. *J. Inst. Math. Applic.* **1,** 372–422, 1965.

In this paper asymptotic expansions of multiple integrals of Fourier type are obtained. Contributions from various types of critical points are considered.

A. ERDÉLYI. On the principle of stationary phase. *Proceedings of the Fourth Canadian Mathematical Congress.* University of Toronto Press, Toronto, 1959.

In this paper the author uses an integration by parts technique similar to the one used here, as a means of identifying critical points of two-dimensional integrals of Fourier type.

J. FOCKE. Asymptotische Entwicklungen mittels der Methode der stationären Phase. *Ber. Verh. Sächs. Akad. Wiss. Leipzig Math.* Nat. Kl. **101,** 3, 1954.

This is a basic paper underlying the development of asymptotic expansions of multiple integrals. The author treats single and double integrals of Fourier type and uses neutralizers to isolate critical points.

L. C. HSU. On the asymptotic evaluation of a class of multiple integrals involving a parameter. *Duke Math. J.* **15,** 625–634, 1948.

In this paper, the author extends Laplace's method to multidimensional integrals.

D. S. JONES. *Generalized Functions.* McGraw-Hill, London, 1966.

The author presents an extensive discussion of the asymptotic expansion of multidimensional Fourier integrals.

D. S. JONES and M. KLINE. Asymptotic expansion of multiple integrals and the method of stationary phase. *J. Math. Phys.* **37,** 1–28, 1958.

A method for the asymptotic analysis of multidimensional Fourier integrals is presented in which the problem is reduced to the analysis of a one-dimensional Fourier transform.

R. M. LEWIS. Asymptotic methods for the solution of dispersive hyperbolic equations. *Asymptotic Solutions of Differential Equations and their Applications.* C. Wilcox (ed.). *Univ. of Wisc. Symposium Proceedings.* John Wiley, New York, 1964.

The author gives a heuristic derivation of the leading term of the asymptotic expansion of a multidimensional integral of Fourier type having a simple interior stationary point.

J. MILNOR. *Morse Theory.* University Press, Princeton, 1963.

On pages 6–8 the author shows that there exists a transformation which diagonalizes the exponent function (as we need in Sections 3 and 4) and is regular in some neighborhood of the stationary point.

9 Uniform Asymptotic Expansions

9.1. Introduction

As we have seen in earlier chapters, it often occurs that an asymptotic expansion obtained with respect to a parameter λ depends on a second parameter, say θ. Furthermore, it can occur that when this second parameter takes on a *critical value* θ_c, the asymptotic expansion becomes invalid. In that event we say that the asymptotic expansion is *nonuniform* with respect to θ.

The Hankel functions $H_{ka}^{(j)}(kr)$, considered in Examples 7.2.2 to 7.2.4 serve to illustrate what can happen. Here $\lambda = kr$ is a large parameter and $\theta = \beta = a/r$ is the second parameter. The expansions (7.2.35) and (7.2.45), valid for $\beta < 1$ and $\beta > 1$, respectively, are seen to fail when $\beta = \beta_c = 1$. For $\beta = 1$, the correct expansion is given by (7.2.42). As we observed in Section 7.2, this latter result is not the limit of either of the former results as $\beta \to 1$.

Returning to the general situation, it is often possible and very desirable to obtain an asymptotic expansion which remains valid as θ varies over a domain containing a critical value θ_c. Indeed, this entire chapter shall be devoted to the development of techniques for the determination of such *uniform asymptotic expansions.*

It seems reasonable to expect that, in order to develop a technique for obtaining a given uniform expansion, we must first ascertain just what is causing the nonuniformity or anomaly. Thus let us again consider the Hankel functions and, in particular, the integral representation (7.2.22)–(7.2.23). We know that for $\beta \neq 1$ the exponent $w(z;\beta)$ has two simple saddle points $z = z_{\pm}$ defined by (7.2.26). As $\beta \to 1$, these saddle points approach each other and, in the limit $\beta = 1$, coalesce, producing a single saddle point of order 2 located at $z = \pi/2$. Thus, in the case of the Hankel functions we find that the observed nonuniformities in the expansions (7.2.35) and (7.2.45) are due to the coalescence of two simple saddle points.

Of course, there are many other anomalies that can arise. The coalescence of a saddle point on an endpoint of integration and of a saddle point on a singularity of the integrand are two further examples.

We might anticipate that the techniques for obtaining the various uniform expansions will be widely diverse. Fortunately, this is not the case. Indeed, for the class of problems to be considered in this chapter, namely uniform asymptotic expansions of integrals of the form

$$I(\lambda;\theta) = \int_C g(z;\theta) \exp\{\lambda w(z;\theta)\}\, dz, \tag{9.1.1}$$

we shall find that there are certain fundamental underlying principles that can be universally applied.

In the following section, we shall motivate these principles by considering in detail the problem of two coalescing simple saddle points. We then formulate and discuss the principles in Section 9.3. Finally in Sections 9.4 to 9.6 we apply them in the analyses of other important and interesting anomalies.

To conclude this section, we wish to introduce the concept of a *useful* uniform asymptotic expansion. Technically, the only criterion for the validity of a uniform expansion is that the error estimate be independent of the second parameter as that parameter varies throughout some fixed domain. This criterion is trivially satisfied, however, by the original integral representation itself. Clearly, such a result is useless.

In general, an integral representation of a function can be viewed as a simply derived uniform asymptotic expansion with respect to a very complicated asymptotic sequence. When we seek an asymptotic expansion, the major purpose is to represent the given function, in terms of functions much simpler than the original integral representation. It is when we endeavor to employ too simple an asymptotic sequence, however, that nonuniformities arise.

Thus when dealing with uniform expansions, we must continuously trade off gains in uniformity against increases in the complexity of the underlying asymptotic sequence. In practice, we seek a compromise appropriate for the particular problem at hand. The one qualitative criterion that should always be kept in mind is that for the final asymptotic expansion to be useful, the terms in the underlying asymptotic sequence must be significantly simpler than the original integral representation, at least in the limit of interest.

9.2. Asymptotic Expansions of Integrals with Two Nearby Saddle Points

Let us consider the integral

$$I(\lambda;\boldsymbol{\alpha}) = \int_C g(z) \exp\{\lambda\, w(z;\boldsymbol{\alpha})\}\, dz, \qquad \boldsymbol{\alpha} = (\alpha_+, \alpha_-). \tag{9.2.1}$$

Here $g(z)$ and $w(z;\boldsymbol{\alpha})$ are analytic functions of z in some simply connected domain containing the contour C and the points $z = \alpha_\pm$.

The exponent $w(z;\boldsymbol{\alpha})$ is assumed to have simple saddle points at $z = \alpha_+$ and $z = \alpha_-$ when $\alpha_+ \neq \alpha_-$. Thus,

$$w_z(\alpha_+;\boldsymbol{\alpha}) = w_z(\alpha_-;\boldsymbol{\alpha}) = 0; \tag{9.2.2}$$

$$w_{zz}(\alpha_+;\boldsymbol{\alpha}) \neq 0, \qquad w_{zz}(\alpha_-;\boldsymbol{\alpha}) \neq 0; \qquad \alpha_+ \neq \alpha_-. \tag{9.2.3}$$

We shall assume that these saddle points are free to move in some simply connected domain which we denote by D_1. In particular, therefore, they are permitted to coalesce in D_1. When this occurs we suppose that a single saddle point of order 2 is produced. This implies that

$$w_z(\alpha_+;\boldsymbol{\alpha}) = w_{zz}(\alpha_+;\boldsymbol{\alpha}) = 0, \qquad w_{zzz}(\alpha_+;\boldsymbol{\alpha}) \neq 0; \qquad \alpha_+ = \alpha_-. \tag{9.2.4}$$

We further suppose that, for each choice of $\alpha_\pm$ in D_1, there exists a domain D_2 containing D_1, outside of which all other saddle points of w lie. Finally, we assume that $z = \alpha_\pm$ are the dominant critical points for (9.2.1). In other words, all other saddle points of w are assumed to lie in the valleys of w with respect to both $z = \alpha_\pm$.

We seek an asymptotic expansion of $I(\lambda;\boldsymbol{\alpha})$, as $\lambda \to \infty$, that remains valid for α_+ nearby α_- and, in particular, for $\alpha_+ = \alpha_-$. Before proceeding, we wish to point out that, for this problem, we can consider the distance between the saddle points $z = \alpha_\pm$ as the second parameter θ introduced in Section 9.1. Thus,

$$\theta = |\alpha_+ - \alpha_-| \tag{9.2.5}$$

and the critical value θ_c is zero. We shall not need to make explicit use of (9.2.5) however.

As in our discussion of the ordinary method of steepest descent, we anticipate introducing in (9.2.6) a new variable of integration, say t, which in some sense will simplify our analysis. To help us arrive at the desired change of variable $z = z(t)$, let us list some reasonable criteria:

(1) $z = z(t)$ should yield a conformal map of some disc $D_{\boldsymbol{\alpha}} \subset D_2$, containing $z = \alpha_\pm$, onto a domain $\hat{D}_{\boldsymbol{\alpha}}$ in the complex t plane.
(2) The new exponent $\phi(t;\boldsymbol{\alpha}) = w(z(t);\boldsymbol{\alpha})$ should have in $\hat{D}_{\boldsymbol{\alpha}}$ two simple saddle points when $\alpha_+ \neq \alpha_-$ that coalesce to a single saddle point of order 2 when $\alpha_+ = \alpha_-$.[1]

[1] Criteria (1) and (2) are not independent as we shall see.

(3) $\phi(t;\alpha)$ should be significantly "simpler" than $w(z;\alpha)$.

The simplest form for ϕ would be a polynomial in t. For (2) to be satisfied, $\dot{\phi} = d\phi/dt$ must be a polynomial of at least degree 2. Let us therefore define $z(t)$ by the equation

$$w(z;\alpha) = -\left(\frac{t^3}{3} - \gamma^2 t\right) + \rho = \phi(t;\alpha). \tag{9.2.6}$$

Here $\gamma = \gamma(\alpha)$ and $\rho = \rho(\alpha)$ are to be determined.

By differentiating (9.2.6) with respect to t we obtain

$$\dot{z} = \frac{dz}{dt} = \frac{\gamma^2 - t^2}{w_z(z;\alpha)}. \tag{9.2.7}$$

To satisfy (1) we must require that $\dot{z}$ be finite and nonzero for all t in $\hat{D}_\alpha$ (and all z in D_α). From (9.2.7) we see that difficulties can only arise when $z = \alpha_\pm$ and when $t = \pm\gamma$. It should be clear that our change of variable must be such that these points correspond, that is, we must have

$$t = \pm\gamma \quad \text{when} \quad z = \alpha_\pm. \tag{9.2.8}$$

If we make the correspondence (9.2.8) in (9.2.6), then we obtain the following expressions for γ and ρ.

$$\frac{4\gamma^3}{3} = w(\alpha_+;\alpha) - w(\alpha_-;\alpha), \tag{9.2.9}$$

$$\rho = \tfrac{1}{2}\{w(\alpha_+;\alpha) + w(\alpha_-;\alpha)\}. \tag{9.2.10}$$

We note that γ is not uniquely determined by (9.2.9). Indeed, when $\alpha_+ \neq \alpha_-$, (9.2.9) defines three values of γ. We shall discuss this ambiguity further below. For now, we merely assume that γ and ρ satisfy (9.2.8) and (9.2.9) respectively and wish to consider the behavior of z at the saddle points $t = \pm\gamma$.

If $\alpha_+ \neq \alpha_-$ so that $\gamma \neq 0$, then by applying L'Hospital's rule in (9.2.7) we obtain

$$\dot{z}^2\Big|_{\substack{t=\pm\gamma\\ z=\alpha_\pm}} = \frac{\mp 2\gamma}{w_{zz}(\alpha_\pm;\alpha)}, \qquad \alpha_+ \neq \alpha_- \tag{9.2.11}$$

which is finite and nonzero. If, however, $\alpha_+ = \alpha_-$ so that $\gamma = 0$, then we must apply L'Hospital's rule twice. This yields

$$\dot{z}^3\Big|_{\substack{t=0\\ z=\alpha_+}} = \frac{-2}{w_{zzz}(\alpha_+;\alpha)} \qquad \alpha_+ = \alpha_- \tag{9.2.12}$$

which is also finite and nonzero.

We have seen that for $z = z(t)$ to define a conformal map of D_α it is necessary that γ and ρ be defined by (9.2.9) and (9.2.10). It is by no means clear that these conditions are sufficient. Indeed, we see that, for each value of z, (9.2.6) defines three possible values of t, that is, there are three branches of the inverse transformation.

It turns out that there is one branch of the transformation (9.2.6) that defines for each $\alpha_\pm$ in D_1 a conformal map of D_α. The proof of this assertion is quite involved and we will merely state the relevant theorem which is due to Chester, Friedman, and Ursell.

THEOREM 9.2.1. For each $\alpha_\pm$ in D_1, the transformation (9.2.6) has just one branch which defines a conformal map of some disc D_α containing $\alpha_\pm$.[2] On this branch the points $z = \alpha_+$, $z = \alpha_-$ correspond respectively to $t = \gamma$ and $t = -\gamma$.

Let us now reconsider the ambiguity in the determination of γ. We can show that for each of the three possible choices of γ, the corresponding regular branch of (9.2.6) referred to in Theorem 9.2.1 maps the restriction of C to the domain D_α onto a contour asymptotically equivalent to the restriction of one of the contours C_1, C_2, C_3 of Figure 2.5 to the image domain $\hat{D}_\alpha$. In what follows, we shall always select that determination of γ which leads to an image contour asymptotically equivalent to $C_1 \cap \hat{D}_\alpha$.

To help clarify this last point, let us consider the mapping $z = z(t)$ near the origin in the t plane. We define $z = z_0$ to be the preimage of $t = 0$. If $\Delta z = z - z_0$ is an increment directed from $z = z_0$ to the contour C, then its image Δt is approximately a directed increment from the origin to the contour C_1. (See Figure 9.2.1.) Then we have

$$\frac{2\pi}{3} < \arg(\Delta t) < \frac{4\pi}{3}, \qquad \text{mod } 2\pi. \tag{9.2.13}$$

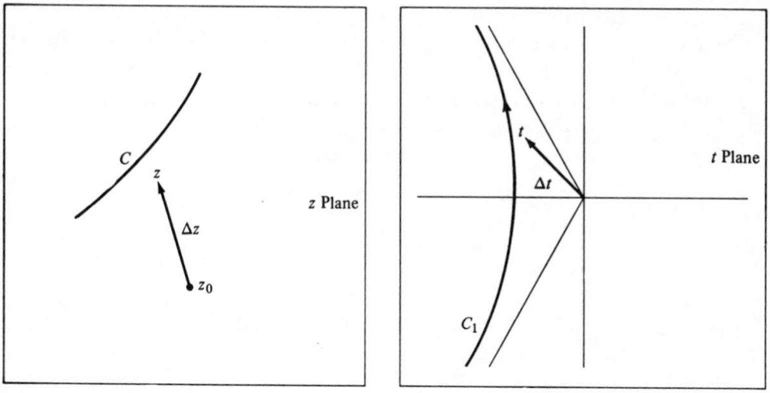

Figure 9.2.1.

[2] For definiteness we can think of D_α as the largest disc for which Theorem 9.2.1 holds.

We now use the approximation

$$\Delta z = \dot{z}\big|_{t=0} \, \Delta t \tag{9.2.14}$$

which, together with (9.2.7) yields

$$\Delta z = \frac{\gamma^2 \, \Delta t}{w_z(z_0;\alpha)}. \tag{9.2.15}$$

Hence

$$\arg(\Delta t) = \arg\left[\frac{w_z(z_0;\alpha)\,\Delta z}{\gamma^2}\right] \tag{9.2.16}$$

so that it follows from (9.2.13) that

$$\frac{1}{2}\arg \Delta z + \frac{1}{2}\arg(w_z(z_0;\alpha)) - \frac{2\pi}{3} < \arg \gamma$$

$$< \frac{1}{2}\arg \Delta z + \frac{1}{2}\arg(w_z(z_0;\alpha)) - \frac{\pi}{3}, \qquad \text{mod } \pi. \tag{9.2.17}$$

This restricts γ to be in one of two supplementary sectors of the complex plane, each having angle $\pi/3$. This serves to uniquely determine the cube root in (9.2.9).

Under the transformation (9.2.6) we can write (9.2.1) as

$$I(\lambda;\alpha) = \int_{C_1 \cap \hat{D}_\alpha} G_0(t;\alpha) \exp\{\lambda\phi(t;\gamma)\}\, dt + \mathscr{E}. \tag{9.2.18}$$

Here

$$G_0(t;\alpha) = g(z(t))\frac{dz}{dt} \tag{9.2.19}$$

which is regular in $\hat{D}_\alpha$ and $\mathscr{E}$ is asymptotically negligible being by assumption exponentially smaller than I itself.

The next step in our procedure is to expand the amplitude G_0 in a manner that will allow for the derivation of the uniform expansion. We have previously introduced the notion that when the integrand vanishes at a critical point of an integral, the contribution to the asymptotic expansion corresponding to that point is diminished. To exploit this we set

$$G_0(t;\alpha) = a_0 + a_1 t + (t^2 - \gamma^2)\, H_0(t;\alpha) \tag{9.2.20}$$

with a_0, a_1, and H_0 to be determined.

We note that if $H_0(t;\alpha)$ is regular in $\hat{D}_\alpha$, then the last term in (9.2.20) vanishes at the two saddle points $t = \pm\gamma$. Indeed, this last term is proportional to $\dot{\phi}$. To determine a_0 and a_1, we assume H_0 regular and set $t = \pm\gamma$ in (9.2.20). This yields

$$a_0 = \frac{G_0(\gamma;\alpha) + G_0(-\gamma;\alpha)}{2},$$

$$a_1 = \frac{G_0(\gamma;\alpha) - G_0(-\gamma;\alpha)}{2\gamma}. \tag{9.2.21}$$

The regularity of G_0 assures us that a_1 has a removable singularity at $\gamma = 0$. Indeed, we have

$$\lim_{\gamma \to 0} a_0 = G_0(0;\alpha), \qquad \lim_{\gamma \to 0} a_1 = \dot{G}_0(0;\alpha). \tag{9.2.22}$$

With a_0 and a_1 so determined, it is easy to see that

$$H_0 = \frac{G_0(t;\alpha) - a_0 - a_1 t}{t^2 - \gamma^2} \tag{9.2.23}$$

is regular in $\hat{D}_\alpha$ and has removable singularities at $t = \pm\gamma$. In fact, we have

$$\lim_{t \to \pm\gamma} H_0(t;\alpha) = \pm \frac{\dot{G}_0(\pm\gamma;\alpha) - a_1}{2\gamma}. \tag{9.2.24}$$

Upon inserting (9.2.20) into (9.2.18) we obtain

$$I(\lambda;\alpha) \sim \exp\{\lambda\rho\} \int_{C_1 \cap \hat{D}_\alpha} \exp\left\{ -\lambda\left(\frac{t^3}{3} - \gamma^2 t\right) \right\} (a_0 + a_1 t)\, dt + R_0(\lambda;\alpha). \tag{9.2.25}$$

Here

$$R_0(\lambda;\alpha) = \exp\{\lambda\rho\} \int_{C_1 \cap \hat{D}_\alpha} (t^2 - \gamma^2) H_0(t;\alpha) \exp\left\{ -\lambda\left(\frac{t^3}{3} - \gamma^2 t\right) \right\} dt. \tag{9.2.26}$$

In the first integral on the right of (9.2.25) we can replace $C_1 \cap \hat{D}_\alpha$ by C_1 itself, introducing thereby an asymptotically negligible error. From (2.5.10) we see that the resulting integral can be expressed in terms of the Airy function $\mathrm{Ai}(x)$ and its derivative.

In R_0, we integrate by parts. The boundary terms can be ignored again to within an asymptotically negligible error. Thus, we have

$$I(\lambda;\alpha) \sim 2\pi i \exp\{\lambda\rho\} \left[\frac{a_0}{\lambda^{1/3}} \mathrm{Ai}(\lambda^{2/3}\gamma^2) + \frac{a_1}{\lambda^{2/3}} \mathrm{Ai}'(\lambda^{2/3}\gamma^2) \right] + \frac{\exp\{\lambda\rho\}}{\lambda} \int_{C_1 \cap \hat{D}_\alpha} G_1(t;\alpha) \exp\left\{ -\lambda\left(\frac{t^3}{3} - \gamma^2 t\right) \right\} dt. \tag{9.2.27}$$

Here

$$G_1(t;\alpha) = \frac{d}{dt} H_0(t;\alpha). \tag{9.2.28}$$

We note that the last term on the right of (9.2.27) is an integral of the form (9.2.18) multiplied by λ^{-1}. Hence, we can apply the above process repeatedly. In this manner we obtain after $N + 1$ applications

$$I(\lambda;\alpha) \sim 2\pi i \exp\{\lambda\rho\} \left[\frac{\mathrm{Ai}(\lambda^{2/3}\gamma^2)}{\lambda^{1/3}} \sum_{n=0}^{N} \frac{a_{2n}}{\lambda^n} + \frac{\mathrm{Ai}'(\lambda^{2/3}\gamma^2)}{\lambda^{2/3}} \sum_{n=0}^{N} \frac{a_{2n+1}}{\lambda^n} \right] + R_N(\lambda;\alpha), \tag{9.2.29}$$

$$R_N(\lambda;\alpha) = \lambda^{-(N+1)} \exp\{\lambda\rho\} \int_{C_1 \cap \hat{D}_\alpha} G_{N+1}(t;\alpha) \exp\left\{ -\lambda\left(\frac{t^3}{3} - \gamma^2 t\right) \right\} dt. \tag{9.2.30}$$

The coefficients a_j are given recursively by

$$a_{2n} = \frac{\{G_n(\gamma;\alpha) + G_n(-\gamma;\alpha)\}}{2},$$

$$a_{2n+1} = \frac{\{G_n(\gamma;\alpha) - G_n(-\gamma;\alpha)\}}{2\gamma},$$

$$G_n(t;\alpha) = a_{2n} + a_{2n+1}\, t + (t^2 - \gamma^2)\, H_n(t;\alpha), \tag{9.2.31}$$

$$G_{n+1}(t;\alpha) = \frac{d}{dt}\, H_n(t;\alpha),$$

$n = 0, 1, 2, \ldots$.

Our claim is that the above formal procedure yields an asymptotic expansion of $I(\lambda;\alpha)$ that is uniformly valid for $\theta = |\alpha_+ - \alpha_-|$ small. This, of course, has to be proven. Before offering a proof, we wish to emphasize that the expansion we have obtained involves a single special function, namely the Airy function. Indeed, the Airy function and its derivative effect a smooth transition in the *algebraic* order in λ of I, as $\gamma \to 0$. We know that for separated simple saddle points, this algebraic order is $\lambda^{-1/2}$, while for a single saddle point of order 2, the order is $\lambda^{-1/3}$. Our expansion mirrors this because both $\lambda^{-1/3}\,\mathrm{Ai}(\lambda^{2/3}\gamma^2)$ and $\lambda^{-2/3}\,\mathrm{Ai}'(\lambda^{2/3}\gamma^2)$ have algebraic order $\lambda^{-1/2}$ when $|\gamma|^2 \geq |\gamma_0|^2 > 0$ while both Ai(0) and Ai′(0) are $O(1)$ in λ.

The asymptotic nature of our result is established in the following.

THEOREM 9.2.2. Under the assumptions leading to (9.2.18), the recursive system (9.2.29)–(9.2.31) yields an asymptotic expansion of $I(\lambda;\alpha)$ as $\lambda \to \infty$ with respect to the auxiliary asymptotic sequence $\{\Phi_n(\lambda;\alpha)\}$. Here

$$\Phi_n(\lambda;\alpha) = \exp\{\mathrm{Re}(\lambda\rho)\}\,[\lambda^{-n-1/3}|\mathrm{Ai}(\lambda^{2/3}\gamma^2)| + \lambda^{-n-2/3}|\mathrm{Ai}'(\lambda^{2/3}\gamma^2)|]. \tag{9.2.32}$$

Moreover, this expansion remains uniformly valid for all α_+, α_- in D_1.

PROOF.[3] We first use the fact that (9.2.29) is an exact representation except for an error which, as $\lambda \to \infty$, is exponentially smaller than I itself. Then, because each term in the sums of that equation has the factor λ^{-n} and because G_n, $n = 0, 1, 2, \ldots$ is analytic in $\hat{D}_\alpha$, we need only consider I and show that for λ sufficiently large

$$|I(\lambda;\alpha)| < \exp\{\mathrm{Re}(\lambda\rho)\}\left[\frac{d_0}{\lambda^{1/3}}|\mathrm{Ai}(\lambda^{2/3}\gamma^2)| + \frac{d_1}{\lambda^{2/3}}|\mathrm{Ai}'(\lambda^{2/3}\gamma^2)|\right] \tag{9.2.33}$$

for some positive constants d_0, d_1 independent of λ.

Let us then consider (9.2.27). Clearly, we have that

[3] Some of the details of this proof will be left to the exercises.

$$|\exp\{\lambda\rho\}| \left|\frac{a_0}{\lambda^{1/3}}\operatorname{Ai}(\lambda^{2/3}\gamma^2) + \frac{a_1}{\lambda^{2/3}}\operatorname{Ai}'(\lambda^{2/3}\gamma^2)\right|$$

$$\leq \exp\{\operatorname{Re}(\lambda\rho)\}\left[\frac{|a_0|}{\lambda^{1/3}}|\operatorname{Ai}(\lambda^{2/3}\gamma^2)| + \frac{|a_1|}{\lambda^{2/3}}|\operatorname{Ai}'(\lambda^{2/3}\gamma^2)|\right] \quad (9.2.34)$$

so that we need only obtain an analogous estimate for

$$R_0(\lambda;\alpha) = \frac{\exp\{\lambda\rho\}}{\lambda}\int_{C_1\cap\hat{D}_\alpha} G_1(t;\alpha)\exp\left\{-\lambda\left(\frac{t^3}{3} - \gamma^2 t\right)\right\} dt. \quad (9.2.35)$$

It is convenient to treat two distinct cases.

Case I. $|\gamma| \leq \delta\,\lambda^{-1/3}$. Here δ is independent of λ and will be determined in Case II below. For λ sufficiently large, the saddle points in (9.2.35) are confined to a small circle about the origin. To estimate R_0 in this case, we first introduce the variable of integration $t/\gamma = \tau$. This yields

$$R_0 = \frac{\gamma}{\lambda}\exp\{\lambda\rho\}\int_{\bar{C}} \exp\left\{\lambda\gamma^3\left(\tau - \frac{\tau^3}{3}\right)\right\} G_1(\gamma\tau;\alpha)\, d\tau \quad (9.2.36)$$

with $\bar{C}$ the image contour. From this representation it is readily seen that the estimate

$$|R_0| \leq \frac{M}{\lambda}\exp\{K\delta^3\}\exp\{\operatorname{Re}(\lambda\rho)\} \quad (9.2.37)$$

holds, where M and K are independent of both λ and γ.

Because the arguments of the Airy functions that appear in (9.2.33) are bounded in magnitude by δ^2 and, from the theory of ordinary differential equations, because $\operatorname{Ai}(t)$ and $\operatorname{Ai}'(t)$ are never simultaneously zero, we can find constants b_0 and b_1 independent of λ such that

$$|R_0| \leq \exp\{\operatorname{Re}(\lambda\rho)\}\left[\frac{b_0\,|\operatorname{Ai}(\lambda^{2/3}\gamma^2)|}{\lambda^{1/3}} + \frac{b_1\,|\operatorname{Ai}'(\lambda^{2/3}\gamma^2)|}{\lambda^{2/3}}\right] \quad (9.2.38)$$

for λ sufficiently large.

Case II. $|\gamma| > \lambda^{-1/3}\,\delta$. Here we introduce the new variable of integration $\tau = t/\delta$ in (9.2.35) to obtain

$$R_0 = \delta\,\lambda^{-1}\exp\{\lambda\rho\}\int_C G_1(\delta\tau;\alpha)\exp\left\{-\lambda\delta^3\left(\frac{\tau^3}{3} - \frac{\gamma^2}{\delta^2}\tau\right)\right\} d\tau. \quad (9.2.39)$$

To estimate R_0, we shall apply the method of steepest descent with $\lambda\delta^3$ considered the large parameter.

Although we shall leave the details of this analysis to the exercises, we note that the exponential factor that results in this estimate is the same as that which results when we replace each Airy function in (9.2.32) by the leading term of its asymptotic expansion for large argument. For this to be a valid procedure, in this case, δ^2 must be sufficiently large. More precisely, δ^2 must be large enough so that for $\lambda^{2/3}\gamma^2 > \delta^2$ we can bound the right side of (9.2.33) by replacing $\mathrm{Ai}(\lambda^{2/3}\delta^2)$ and $\mathrm{Ai}'(\lambda^{2/3}\delta^2)$ by the leading terms of their asymptotic expansions.

With δ so determined we can use the stated result concerning the asymptotic analysis of (9.2.39) to conclude that the estimate (9.2.38) holds in this case also with possibly different constants b_0, b_1.

Upon combining the results of the two cases above with the estimate (9.2.34) and upon recalling that (9.2.27) is exact to within an error exponentially smaller than I, we find that there exist constants d_0 and d_1 such that (9.2.33) holds. This completes the proof.

EXAMPLE 9.2.1. The Hankel function of nearly equal order and argument. The Hankel function $H_{ka}^{(1)}(kr)$ has the integral representation

$$H_{ka}^{(1)}(kr) = \frac{1}{\pi}\int_{C_1} \exp\left\{ik\left[r\cos z + a\left(z - \frac{\pi}{2}\right)\right]\right\} dz \tag{9.2.40}$$

with the contour C_1 as depicted in Figure 7.2.1. As in Example 7.2.2 we set $\lambda = kr$, $\beta = a/r$. Then (9.2.40) becomes

$$H_{\lambda\beta}^{(1)}(\lambda) = \frac{1}{\pi}\int_{C_1} \exp\{\lambda w(z;\beta)\}\, dz, \tag{9.2.41}$$

$$w(z;\beta) = i\left[\cos z + \beta\left(z - \frac{\pi}{2}\right)\right]. \tag{9.2.42}$$

In the fundamental strip $-\pi < \mathrm{Re}(z) < \pi$, w has two simple saddle points $z = z_\pm$ when $\beta \neq 1$ and a single saddle point of order 2 located at $z = \pi/2$ when $\beta = 1$. We wish to obtain an asymptotic expansion of $H_{\lambda\beta}^{(1)}(\lambda)$ as $\lambda \to \infty$ that remains uniformly valid as $\beta \to 1$.

We first consider the case $\beta > 1$, that is, the case where the order is larger than the argument. The saddle points $z = z_\pm$ are now located as shown in Figure 7.2.4. We readily find that

$$w(z_\pm;\beta) = \mp[\beta\cosh^{-1}\beta - \sqrt{\beta^2 - 1}], \tag{9.2.43}$$

where the expression in brackets is positive. It thus follows from (9.2.9), (9.2.10), and (9.2.43) that the constants γ and ρ in the transformation (9.2.6) are given by

$$\frac{2\gamma^3}{3} = -[\beta\cosh^{-1}\beta - \sqrt{\beta^2 - 1}]; \qquad \rho = 0. \tag{9.2.44}$$

From (9.2.44) we see that the three possible choices for $\arg(\gamma)$ are $\pm\pi/3$ and π. To determine the correct one we use (9.2.17). Here the preimage of $t = 0$ is $z = z_0 = \pi/2$ and we have $\arg(w_z(\pi/2)) = \pi/2$. Furthermore, we can take $\arg(\Delta z) = \pi$ in (9.2.17) which then yields

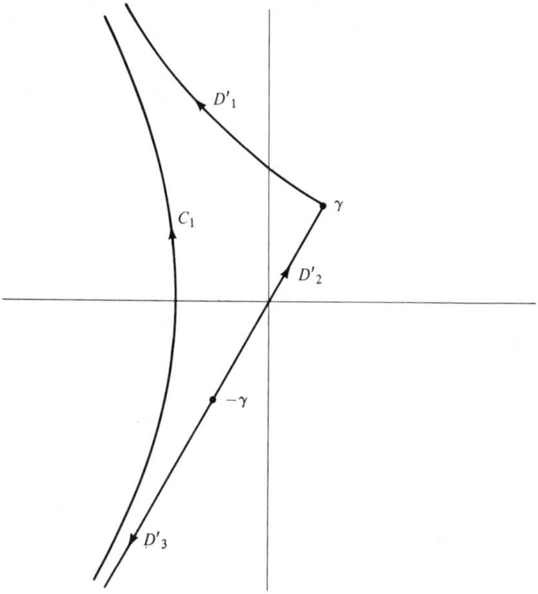

Figure 9.2.2. The Contours D_1', D_2', D_3' Are the Images of D_1, D_2, D_3, Respectively, in Figure 7.2.4.

$$\frac{\pi}{12} < \arg(\gamma) < \frac{5\pi}{12}, \qquad \text{mod } \pi. \tag{9.2.45}$$

Thus $\arg(\gamma)$ must be $\pi/3$ and we have

$$\gamma = e^{\pi i/3} \left\{\tfrac{3}{2} \left[\beta \cosh^{-1} \beta - \sqrt{\beta^2 - 1}\right]\right\}^{1/3}, \qquad \beta > 1. \tag{9.2.46}$$

The local configuration of the image contour and the saddle points in the t plane is shown in Figure 9.2.2. We remark that, except for the different orientation of the contours, Figure 9.2.2 looks locally like a rotation of Figure 7.2.4. This is as it must be for a conformal mapping. We should check that this would not be the case if any of the other determinations of $\arg(\gamma)$ is used.

For $\beta < 1$, we use (7.2.28) in (9.2.9) and (9.2.10) to obtain

$$\tfrac{2}{3} \gamma^3 = i\left[\sqrt{1 - \beta^2} - \beta \cos^{-1} \beta\right]; \qquad \rho = 0. \tag{9.2.47}$$

Again the term within brackets is positive. Now $\arg(\gamma)$ is either $\pi/6 \pm 2\pi/3$ or $\pi/6$. If we again take $\arg(\Delta z) = \pi$ and use the fact that now $\arg(w_z(\pi/2)) = -\pi/2$, then we find from (9.2.17) that

$$-\frac{5\pi}{12} < \arg(\gamma) < -\frac{\pi}{12}, \qquad \text{mod } \pi. \tag{9.2.48}$$

Hence $\arg(\gamma) = 5\pi/6$ and we have

$$\gamma = e^{5\pi i/6}\{\tfrac{3}{2}[\sqrt{1-\beta^2} - \beta\cos^{-1}\beta]\}^{1/3}, \qquad \beta < 1. \tag{9.2.49}$$

With γ determined as above we are now prepared to calculate the coefficients a_j in the uniform expansion (9.2.29). We shall only determine a_0 and a_1 here however. Because $g(z) \equiv 1/\pi$ in this case, we have

$$G_0(t;\alpha) = -\frac{1}{\pi}\frac{dz}{dt}. \tag{9.2.50}$$

The minus sign comes from the fact that the image contour in the t plane is oppositely directed to the contour C_1 in Figure 2.5.

To find a_0 and a_1 we must determine dz/dt at the saddle points. It follows from (9.2.11), the above determination of γ, and the easily calculated expressions for $w_{zz}(z_\pm;\beta)$ that

$$\dot{z}^2\Big|_{t=\pm\gamma} = \frac{2|\gamma|}{\sqrt{|\beta^2-1|}}\,e^{\pi i/3}. \tag{9.2.51}$$

We, of course, must choose the correct square root. To do this we observe that for γ small

$$\dot{z}\Big|_{t=\pm\gamma} \approx \frac{z_+ - z_-}{2\gamma}; \qquad \arg(\dot{z})\Big|_{t=\pm\gamma} \approx \arg(z_+ - z_-) - \arg(\gamma). \tag{9.2.52}$$

By reading the angles in (9.2.52) from (9.2.46) and Figure 7.2.4 or from (9.2.49) and Figure 7.2.2, and by using the explicit expressions for γ and $z_\pm$, we find that

$$\dot{z}\Big|_{t=\pm\gamma} = \frac{e^{\pi i/6}\sqrt{2|\gamma|}}{|\beta^2-1|^{1/4}}, \qquad \beta \neq 1. \tag{9.2.53}$$

By taking the limit either as $\beta \uparrow 1$ or as $\beta \downarrow 1$ we find (see Exercise 9.5)

$$\dot{z}\big|_{t=0} = 2^{1/3}e^{\pi i/6}, \qquad \beta = 1. \tag{9.2.54}$$

We now insert (9.2.53) and (9.2.54) into (9.2.21) to obtain

$$a_0 = \begin{cases} -\dfrac{\sqrt{2|\gamma|}}{\pi|\beta^2-1|^{1/4}}\,e^{\pi i/6}, & \beta \neq 1, \\[2ex] -\dfrac{2^{1/3}e^{\pi i/6}}{\pi}, & \beta = 1, \end{cases} \tag{9.2.55}$$

$$a_1 = 0.$$

The uniform asymptotic expansion is given to leading order by

$$H^{(1)}_{\lambda\beta}(\lambda) \sim \frac{2^{3/2}e^{-\pi i/3}}{\lambda^{1/3}}\left|\frac{\sigma}{\beta^2-1}\right|^{1/4}\mathrm{Ai}(\lambda^{2/3}e^{2\pi i/3}\sigma). \tag{9.2.56}$$

Here

$$\sigma = \begin{cases} \left(\frac{3}{2}\left[\beta \cosh^{-1} \beta - \sqrt{\beta^2 - 1}\right]\right)^{2/3}; & \beta \geq 1, \\ -\left(\frac{3}{2}\left[\sqrt{1 - \beta^2} - \beta \cos^{-1} \beta\right]\right)^{2/3}; & \beta \leq 1. \end{cases} \tag{9.2.57}$$

Finally, we point out that to leading order, the limiting nonuniform expansions, valid when either β is bounded away from 1 or when β is equal to 1, can be readily recovered from (9.2.56). Indeed, suppose that β is bounded away from 1. Then we can replace the Airy function in (9.2.56) by the leading term of its asymptotic expansion for complex argument (7.3.16) to recover (7.2.45).

9.3. Underlying Principles

The procedure developed in Section 9.2 for obtaining a uniform asymptotic expansion, in the case of two nearby saddle points, can be applied with minor modifications to obtain a variety of uniform expansions. Indeed, the basic steps can actually be formulated as underlying principles. In this section we shall briefly discuss these principles.

The first step in the asymptotic analysis of (9.1.1) is always to introduce a change of variable $z = z(t)$ that replaces the original exponent w by a new exponent ϕ. This transformation should be one-to-one in some domain containing the critical points of interest while being as simple as possible. In most instances ϕ will be a polynomial in t of degree $n + 1$, where n is the number of the saddle points of w under consideration.[4] Because a whole class of problems can be considered by reducing w to a particular ϕ, we usually call ϕ the *canonical* exponent for the class. Thus, we found in Section 9.2 that, when considering two nearly simple saddle points, the canonical exponent is a polynomial of degree three.

As a result of the change of variable just discussed a new amplitude function is obtained. The next step in the analysis is to make a finite expansion of this amplitude with the following criteria in mind:

(1) The remainder should vanish at all of the critical points that are to be involved in the uniform expansion. In particular, therefore, the remainder should be proportional to $\dot{\phi}$.
(2) For all permissible locations of the critical points, the remainder should have the same smoothness properties as the transformed amplitude.

Upon replacing the transformed amplitude by a finite expansion satisfying (1) and (2), the integral involving the remainder can be uniformly integrated by parts. This introduces a remainder integral that is of the same form as I multiplied by λ^{-1}. The boundary terms are either zero or asymptotically

[4] Here a saddle point of order m contributes m to the total n.

negligible when compared to I. Thus, the leading term of the uniform expansion involves a finite sum of *canonical integrals*.

In many instances each canonical integral is asymptotically equivalent to a well-studied special function. In very complicated problems this need not be so. Nevertheless, the canonical integrals still define the special function appropriate for the particular anomaly being studied.

If smoothness permits, then further terms in the uniform expansion can be obtained by applying the above process to the remainder integral. Indeed, as in Section 9.2, in most cases an infinite expansion can be obtained by applying the process repeatedly.

To conclude, we wish to discuss a technical point. In transforming w into a polynomial of degree $n+1$, there are $n+2$ constants to be determined. One of these is simply a scaling factor. If we map the n saddle points of w into the n saddle points of ϕ, then there remains one free constant. In general, we try to select this free constant so as to facilitate the identification of the canonical integrals with known special functions. Thus, in Section 9.2, we arbitrarily selected the coefficient of t^2 in the cubic transformation (9.2.6) to be zero. This enabled us to directly express the canonical integrals in (9.2.25) in terms of the Airy function and its derivatives.

9.4. Saddle Point near an Amplitude Critical Point

In this section we shall consider integrals of the form

$$I(\lambda;\alpha) = \int_C z^r\, g(z) \exp\{\lambda\, w(z;\alpha)\}\, dz. \tag{9.4.1}$$

Here g and w are analytic functions in some domain containing the contour of integration C, the origin, and the point $z=\alpha$. Furthermore, w has a simple saddle point at $z=\alpha$, so that

$$w_z(\alpha;\alpha) = 0, \qquad w_{zz}(\alpha;\alpha) \neq 0. \tag{9.4.2}$$

For r not an integer, the factor z^r in (9.4.1) introduces a branch point of the integrand at the origin. For r a negative integer, the origin is a pole, while for r a positive integer, the origin is a zero.

We assume that the contour C is such that the origin and the saddle point $z=\alpha$ are the dominant critical points for (9.4.1) as $\lambda \to \infty$ and we seek an asymptotic expansion of I in this limit which remains valid for α near zero, that is, for a saddle point near an amplitude critical point. All other saddle points of w are, of course, bounded away from $z=\alpha$ and $z=0$.

If we formally differentiate (9.4.1) with respect to r, then we obtain

$$J(\lambda;\alpha) = \frac{d}{dr} I(\lambda;\alpha) = \int_C (\log z)\, z^r\, g(z) \exp\{\lambda w(z;\alpha)\}\, dz. \tag{9.4.3}$$

Thus, the case where a saddle point is near a logarithmic branch point can be recovered from our asymptotic analysis of I.

Before proceeding with the analysis, let us make some remarks concerning the contour C. There are several possibilities that can arise. Indeed, C might be a contour with endpoints that lie in different valleys of w with respect to $z = \alpha$. Alternatively, C may loop around the origin and have both endpoints in the same valley. Also, for $r > -1$, one endpoint of C can be the origin. We shall not exclude any of these possibilities in our analysis.

Following the discussion of Section 9.3, we first introduce in (9.4.1) the variable of integration t defined by

$$w(z;\alpha) = -\frac{t^2}{2} - \gamma t + \rho = \phi(t;\alpha). \tag{9.4.4}$$

Here we wish to choose γ and ρ so that $t = -\gamma$ is the image of the saddle point $z = \alpha$, and the origin is preserved. These conditions yield the relations

$$\gamma^2 = 2\{w(\alpha;\alpha) - w(0;\alpha)\}, \qquad \rho = w(0;\alpha). \tag{9.4.5}$$

As in the problem considered in Section 9.2, we find that for each of the two possible choices of γ, one branch of the inverse transformation $t = t(z;\alpha)$ remains conformal in some domain containing the critical points even as $\alpha \to 0$. The point is more readily established here than in Section 9.2. Indeed, upon solving (9.4.4) for t, we obtain

$$t + \gamma = \sqrt{\gamma^2 + 2[\rho - w(z;\alpha)]} = \sqrt{2[w(\alpha;\alpha) - w(z;\alpha)]}. \tag{9.4.6}$$

Because $t = 0$ corresponds to $z = 0$ we have that the branch of $t = t(z;\alpha)$ must be chosen to that

$$\gamma = \sqrt{\gamma^2}. \tag{9.4.7}$$

Then, for the two choices of γ, the images differ only by a rotation through π. Also, for z near α, we have from (9.4.6)

$$t + \gamma \approx \sqrt{-w_{zz}(\alpha;\alpha)}\,(z - \alpha)\,[1 + 0(z - \alpha)]. \tag{9.4.8}$$

Hence $t(z;\alpha)$ is analytic near $z = \alpha$ even when $\alpha = 0$.

When we use (9.4.4) in (9.4.1) we obtain

$$I(\lambda;\alpha) \sim \int_{\tilde{C}} t^r\, G_0(t;\alpha) \exp\{\lambda\phi(t;\alpha)\}\, dt, \tag{9.4.9}$$

$$G_0(t;\alpha) = \left(\frac{z}{t}\right)^r g(z)\,\frac{dz}{dt}. \tag{9.4.10}$$

Here $\tilde{C}$ is a truncation of the image of C under (9.4.4). It is such that G_0 and ϕ are analytic functions of t in some domain containing $\tilde{C}$, $t = 0$, and $t = -\gamma$. It is readily seen that the error caused by this truncation is asymptotically negligible.

The function

$$\frac{dz}{dt} = -\frac{(t + \gamma)}{w_z(z;\alpha)} \tag{9.4.11}$$

is finite and nonzero near the critical points. Indeed we have

$$\left(\frac{dz}{dt}\right)^2\bigg|_{\substack{z=\alpha\\ t=-\gamma}} = \frac{-1}{w_{zz}(\alpha;\alpha)}. \tag{9.4.12}$$

Also $(z/t)^r$ has a removable singularity at the origin even when $\alpha = 0$. We find that

$$\lim_{\alpha\to 0}\lim_{t\to 0}\left(\frac{z}{t}\right)^r = (-w_{zz}(0;0))^{-r/2}. \tag{9.4.13}$$

Our next step is to expand G_0 in the appropriate manner. Thus, we set

$$G_0(t;\alpha) = a_0 + a_1 t + t(t+\gamma)\, H_0(t;\alpha) \tag{9.4.14}$$

and observe that the last term vanishes at both of the critical points for (9.4.9). To determine a_0 and a_1 we evaluate (9.4.14) at $t = 0$ and $t = -\gamma$. In this manner we obtain

$$a_0 = G_0(0;\alpha), \qquad a_1 = \frac{G_0(0;\alpha) - G_0(-\gamma;\alpha)}{\gamma}. \tag{9.4.15}$$

Hence

$$H_0(t;\alpha) = \frac{G_0(t;\alpha) - a_0 - a_1 t}{t(t+\gamma)} \tag{9.4.16}$$

has removable singularities at $t = 0$ and $t = -\gamma$.

Upon inserting (9.4.14) into (9.4.9) we obtain

$$I(\lambda;\alpha) \sim \exp\{\lambda\rho\}\left[\frac{a_0}{\lambda^{(r+1)/2}}\, W_r(\sqrt{\lambda}\,\gamma) + \frac{a_1}{\lambda^{(r+2)/2}}\, W_{r+1}(\sqrt{\lambda}\,\gamma)\right] + R_0(\lambda;\alpha). \tag{9.4.17}$$

Here

$$W_r(z) = \int_{\hat{C}} t^r \exp\left\{-\frac{t^2}{2} - zt\right\} dt \tag{9.4.18}$$

with $\hat{C}$ one of the contours depicted in Figure 9.4. Just which contour appears in (9.4.18) depends on the form of the original contour C. Thus if C is a loop around the origin, then so is $\hat{C}$, and so on. The error made in replacing $\tilde{C}$ by $\hat{C}$ is again asymptotically negligible. We remark that W_r is related to a solution of Weber's differential equation which we briefly discussed in Section 2.5. We shall say more on this below.

The remainder integral

$$R_0(\lambda;\alpha) = \int_{\tilde{C}} t^{r+1}\,(t+\gamma)\, H_0(t;\alpha) \exp\{\lambda\phi(t;\alpha)\}\, dt \tag{9.4.19}$$

becomes after an integration by parts,

$$R_0(\lambda;\alpha) \sim \lambda^{-1} \int_{\tilde{C}} t^r\, G_1(t;\alpha) \exp\{\lambda\phi(t;\alpha)\}\, dt. \tag{9.4.20}$$

Here

$$G_1 = \left\{(r+1)\, H_0(t;\alpha) + t\frac{dH_0}{dt}\right\} \tag{9.4.21}$$

and we have neglected the boundary contributions because they too are asymptotically negligible.

Because the integral in (9.4.20) is of the same form as (9.4.9) multiplied by λ^{-1} we can apply the above process repeatedly. Indeed after $N+1$ steps we obtain

$$I(\lambda;\alpha) \sim \exp\{\lambda\rho\}\left[\frac{W_r(\sqrt{\lambda}\,\gamma)}{\lambda^{(r+1)/2}} \sum_{n=0}^{N} \frac{a_{2n}}{\lambda^n} + \frac{W_{r+1}(\sqrt{\lambda}\,\gamma)}{\lambda^{(r+2)/2}} \sum_{n=0}^{N} \frac{a_{2n+1}}{\lambda^n}\right] + R_N(\lambda;\alpha). \tag{9.4.22}$$

The coefficients a_j are defined recursively by

$$a_{2n} = G_n(0;\alpha), \qquad a_{2n+1} = \frac{G_n(0;\alpha) - G_n(-\gamma;\alpha)}{\gamma},$$

$$G_n(t;\alpha) = \left\{(r+1)\, H_{n-1}(t;\alpha) + t\,\frac{dH_{n-1}(t;\alpha)}{dt}\right\}, \tag{9.4.23}$$

$$H_{n-1}(t;\alpha) = \frac{G_{n-1}(t;\alpha) - a_{2(n-1)} - a_{2n-1}\, t}{t(t+\gamma)}, \qquad n = 1, 2, \ldots.$$

The remainder $R_N(\lambda;\alpha)$ is given asymptotically by

$$R_N(\lambda;\alpha) \sim \lambda^{-(N+1)} \int_{\tilde{C}} t^r\, G_{N+1}(t;\alpha) \exp\{\lambda\phi\}\, dt. \tag{9.4.24}$$

The proof of the asymptotic nature of (9.4.22), (9.4.23) follows the same lines as the corresponding proof in Section 9.2 and will be left to the exercises.

Let us now consider the function $W_r(z)$ in more detail. We can readily show that

$$\Psi(z) = e^{-z^2/4}\, W_r(z) \tag{9.4.25}$$

is a solution to Weber's differential equation of order $-r-1$:

$$\Psi'' - \left(r + \frac{1}{2} + \frac{z^2}{4}\right)\Psi = 0. \tag{9.4.26}$$

Two linearly independent solutions to this equation often referred to in the literature are $D_{-r-1}(z)$ and $D_r(-iz)$. These functions are defined by

$$W_r^{(1)}(z) = 2i \exp\left[i\pi r + \frac{z^2}{4}\right] \sin \pi r\, \Gamma(r+1)\, D_{-r-1}(z) \tag{9.4.27}$$

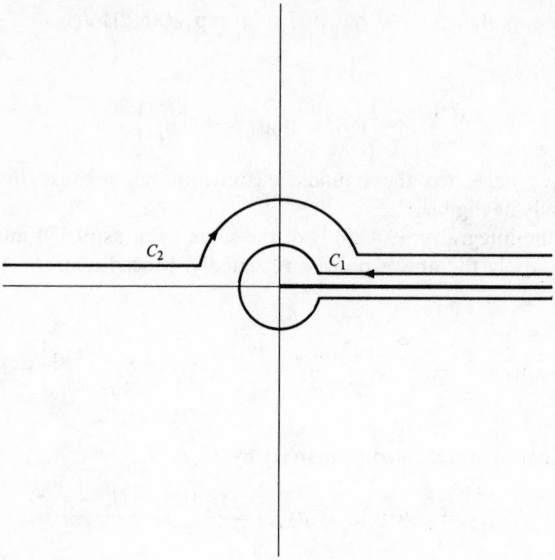

Figure 9.4. Two Choices of the Contour $\hat{C}$ in (9.4.18).

and

$$W_r^{(2)}(z) = \sqrt{2\pi}\exp\left\{\frac{i\pi r}{2} + \frac{z^2}{4}\right\} D_r(iz). \tag{9.4.28}$$

Here

$$W_r^{(j)}(z) = \int_{C_j} t^r \exp\left\{-\frac{t^2}{2} - zt\right\} dt \tag{9.4.29}$$

with the contours C_1, C_2 as shown in Figure 9.4.

There are certain special cases that are of particular interest. First, suppose that $r > -1$. Then

$$W_r^{(1)}(z) = [2i \sin \pi r\, e^{\pi ir}] \int_0^\infty t^r \exp\left\{-\frac{t^2}{2} - zt\right\} dt \tag{9.4.30}$$

which is the appropriate special function for the case where the origin is not only a branch point but an endpoint of integration as well.

When r is a nonnegative integer

$$W_n^{(1)}(z) \equiv 0, \qquad W_n^{(2)}(z) = \sqrt{2\pi}\left(-\frac{d}{dz}\right)^n e^{z^2/2}, \qquad n = 0, 1, 2, \ldots$$

$$= \sqrt{2\pi}\, 2^{-n/2} \exp\left\{\frac{z^2}{2} + \frac{n\pi i}{2}\right\} \mathscr{H}_n\left\{\frac{iz}{\sqrt{2}}\right\}. \tag{9.4.31}$$

Here $\mathscr{H}_n(t)$ is the Hermite polynomial of degree n. A solution, linearly independent of $W_n^{(2)}(z)$ in this case is given by

$$W_n^{(3)}(z) = \sqrt{2}\,(-1)^n \left(\frac{d}{dz}\right)^n \left\{\exp\left(\frac{z^2}{2}\right) \operatorname{erfc}\left(\frac{z}{\sqrt{2}}\right)\right\}. \tag{9.4.32}$$

Here

$$\operatorname{erfc}(x) = \int_x^\infty \exp(-\xi^2)\, d\xi \tag{9.4.33}$$

is the complementary error function. We see then that the complimentary error function is the canonical function for the problem with a fixed saddle point at the origin and a variable endpoint x that approaches the origin.

Finally, when r is a negative integer

$$W_{-n}^{(1)}(z) = \frac{2\pi i}{(n-1)!}\, 2^{-(n-1)/2}\,(-1)^{n-1}\,\mathscr{H}_{n-1}\left(\frac{z}{\sqrt{2}}\right),$$

$$W_{-n}^{(2)}(z) = \frac{(-i)^n\sqrt{2\pi}}{(n-1)!} \int_{-iz}^{\infty} (\xi + iz)^{n-1}\, e^{-\xi^2/2}\, d\xi, \qquad n = 1, 2, \ldots. \tag{9.4.34}$$

We note that the second of these is simply an $(n-1)$-fold iterated integral of $\operatorname{erfc}(-iz/\sqrt{2})$.

EXAMPLE 9.4.1. In Section 7.5 we considered a source problem for the Klein-Gordon equation. In dimensionless variables, the solution to that problem is given by (7.5.8)

$$U(\lambda;\theta) = \frac{1}{4\pi bc} \int_{\Gamma'} \frac{\exp\{\lambda\psi(v;\theta)\}}{(v - v_0)(v^2 - 1)^{1/2}}\, dv. \tag{9.4.35}$$

Here

$$\psi(v;\theta) = i\{(v^2 - 1)^{1/2}\,\theta - v\}, \tag{9.4.36}$$

$\theta = |x|/ct$, $\lambda = bt$, and Γ' is the contour as depicted in Figure 7.5.1. The constant b is some characteristic frequency of the problem, x is a spatial coordinate, and t is time.

As our analysis in Section 7.5 showed, the dominant critical points for (9.4.35) as $\lambda \to \infty$ are the two saddle points $\hat{v} = \pm\hat{v}(\theta)$ defined by (7.5.13) and the pole $v = v_0$. Our objective here is to obtain an asymptotic expansion of $U(\lambda;\theta)$ as $\lambda \to \infty$ that remains valid for the saddle point $v = \hat{v}(\theta)$ near and at the fixed pole $v = v_0$. In terms of θ, we wish our expansion to remain valid for θ near and at

$$\theta_0 = \frac{(v_0^2 - 1)^{1/2}}{v_0}. \tag{9.4.37}$$

To begin, we suppose that $\theta > \theta_0$ and use the integral representation (7.5.18) for $U(\lambda;\theta)$. The descent contours D_1, D_2, D_3, D_4 along with the relative positions of the saddle points and the pole are shown in Figure 7.5.3. Because the saddle point at $-\hat{v}(\theta)$ is not near the pole, the integral over the contour D_3, D_4 can be asymptotically evaluated by the method of steepest descent. Indeed (7.5.21) is uniformly valid in θ for θ near θ_0.

Thus, we shall focus our attention here on

$$I(\lambda;\theta) = \int_{D_1 - D_2} \frac{\exp\{\lambda\psi(v;\theta)\}}{(v - v_0)(v^2 - 1)^{1/2}}\, dv. \tag{9.4.38}$$

The first step in our analysis should be a translation of coordinates to make v_0 the new origin. Then $I(\lambda;\theta)$ will be of the form (9.4.1). Clearly, such a translation can be incorporated into the general transformation (9.4.6). We shall therefore think of the variable v in (9.4.38) and z in (9.4.1) as being the same, except for a translation, and shall treat them interchangeably for the remainder of this example.

The contour $D_1 - D_2$ passes from one valley of the saddle point $\hat{v}(\theta)$ to the other. Thus we should anticipate that the final contour in the t plane will be the contour C_2 of Figure 9.4.

Let us now think of θ near θ_0 and consider a small neighborhood of v_0 and $\hat{v}(\theta)$ containing a segment of $D_1 - D_2$ near $\hat{v}(\theta)$. We note that in the t plane all steepest descent paths are horizontal through the saddle points. If we examine Figure 7.5.3, we see that the transformation from v to t must be such that the mapping is locally a rotation through angle $\pi/4$.

For this choice of mapping, we find that in (9.4.6)

$$\gamma = -\sqrt{2}\, e^{i\pi/4}\left[-\sqrt{1-\theta^2} - \theta\sqrt{v_0^2 - 1} + v_0\right]^{1/2} \tag{9.4.39}$$

and

$$\rho = \psi(v_0;\theta). \tag{9.4.40}$$

We can check that the bracketed expression in (9.4.39) is positive. The saddle point at $t = -\gamma$ will now be located on the ray through the origin with angle $\pi/4$.

To determine the coefficients a_0 and a_1 that appear in the uniform expansion (9.4.17) we first note that

$$G_0(t) = \frac{t}{(v - v_0)}(v^2 - 1)^{-1/2}\frac{dv}{dt}, \qquad v = v(t). \tag{9.4.41}$$

Furthermore, we find that

$$a_0 = G_0(0) = (v_0^2 - 1)^{-1/2} \tag{9.4.42}$$

and

$$G_0(-\gamma) = \frac{-e^{-i\pi/4}\,\gamma\,\theta\,[\hat{v}^2(\theta) - 1]^{-1/2}}{(\hat{v}(\theta) - v_0)(1 - \theta^2)^{3/4}}. \tag{9.4.43}$$

Hence,

$$a_1 = \frac{(\nu_0^2 - 1)^{-1/2}(\hat{\nu}(\theta) - \nu_0)(1 - \theta^2)^{3/4} + e^{-\pi i/4}\gamma\,\theta\,[\hat{\nu}^2(\theta) - 1]^{-1/2}}{\gamma(\hat{\nu}(\theta) - \nu_0)(1 - \theta^2)^{3/4}}. \tag{9.4.44}$$

The uniform expansion is now given by

$$I(\lambda;\theta) \sim \exp\{\lambda\psi(\nu_0;\theta)\}\left[a_0\,W_{-1}^{(2)}(\sqrt{\lambda}\gamma) + \frac{a_1}{\sqrt{\lambda}}\,W_0^{(2)}(\sqrt{\lambda}\gamma)\right]. \tag{9.4.45}$$

Here the special functions $W_{-1}^{(2)}$ and $W_0^{(2)}$ are defined by (9.4.29).

To extend the result to the range $\theta < \theta_0$, we remark that it is only necessary to define γ so that the saddle point is on the ray through the origin with angle $5\pi/4$. We find that this can be achieved by simply taking γ to be the negative of (9.4.39). Thus, in (9.4.45), we have

$$\gamma = \mp\sqrt{2}\,e^{i\pi/4}\left\{-\sqrt{1-\theta^2} - \theta\sqrt{\nu_0^2 - 1} + \nu_0\right\}^{1/2}, \qquad \pm(\theta - \theta_0) \geq 0. \tag{9.4.46}$$

We leave showing that the nonuniform results (7.5.22)–(7.5.24) can be recovered from the uniform expansion (9.4.45) to the exercises.

9.5. A Class of Integrals That Arise in the Analysis of Precursors

When we studied the Klein-Gordon equation in Section 7.5, we obtained an asymptotic expansion, for $\lambda = bt \to \infty$, of that portion of the solution called the precursor wave. Indeed, that expansion is given by (7.5.22) and is valid for $(\nu_0 - 1)^{1/2}/\nu_0 < \theta = |x|/ct < 1$. Here ν_0 is a dimensionless source frequency.

An examination of (7.5.22) shows that the leading term in the expansion goes to zero as θ goes to 1, that is, as the front of propagation is approached. If we examine subsequent terms, however, then we find that they all become infinite as $\theta \to 1$. Indeed, the second term is proportional to $(1 - \theta^2)^{-1/4}$ Thus, (7.5.22) must be suspect for θ near and at 1.

Let us attempt to describe the nature of the anomaly that arises in the expansion (7.5.22) in the limit $\theta = 1$. We have that, for $0 < \theta < 1$, there are two simple saddle points of

$$\psi = i[(\nu^2 - 1)^{1/2}\,\theta - \nu]$$

located at $\nu = \pm\nu(\theta) = \pm(1 - \theta^2)^{-1/2}$. As $\theta \to 1-$, these saddle points coalesce at infinity producing a saddle point of *infinite* order. This then is the mathematical nature of the anomaly, while the underlying physical problem, we repeat, is to study a signal near its front of propagation.

To motivate further the general problem to be treated below, let us consider a typical integral representation of a wave or signal traveling in a one-dimensional dispersive medium. As we have seen in Section 2.6, such a representation has the form

$$I(x,t) = \frac{1}{2\pi} \int_{\Gamma} g(k) \exp\{i[kx - \omega(k)t]\}\, dk. \tag{9.5.1}$$

Here x is a spatial coordinate, t is time, and $\omega = \omega(k)$ is called the dispersion relation for the problem.

In most physical problems, the group speed (see Section 8.5) $C(k) = |d\omega/dk|$ is bounded above by a characteristic speed c and $\lim_{|k|\to\infty} C(k) = c$. This behavior is obtained when the dispersion relation is such that

$$\omega = \omega(k) = c\left[k + \frac{\text{const}}{k} + o(k^{-1})\right], \qquad |k| \to \infty. \tag{9.5.2}$$

Indeed, (9.5.2) will serve to motivate the form of the general problem to be considered in this section.

Let us now consider the integral

$$I(\lambda;\theta) = \frac{1}{2\pi} \int_{\Gamma} g(k) \exp\{i\lambda\, \Phi(k;\theta)\}\, dk, \qquad 0 < \theta \le 1. \tag{9.5.3}$$

Here g and Φ are analytic functions of k in some neighborhood of the contour Γ. Γ itself is a contour that typically arises in the Fourier transform analysis of a partial differential equation. In particular, we shall take Γ to be infinite in extent and along which $\text{Re}(k)$ varies from $-\infty$ to $+\infty$ and $|\exp\{i\lambda\Phi\}|$ remains bounded. Furthermore, we shall assume that Γ passes above all singularities of the functions Φ and g whose branch cuts are drawn vertically downward in the k plane.

We further suppose that there exist constants R_1 and θ_1 such that, for $|k| \ge R_1$ and all θ in the interval $[\theta_1, 1]$, Φ has a convergent Laurent expansion

$$\Phi(k;\theta) = -\left\{k(1-\theta) + \sum_{n=0}^{\infty} \alpha_n(\theta)\, k^{-n}\right\}. \tag{9.5.4}$$

Here $\alpha_n(\theta)$, $n = 1, 2, 3, \ldots$ are known bounded functions of θ with $\alpha_1(\theta) > 0$. [The form of (9.5.4) has, of course, been motivated by (9.5.2).]

We readily find that, for θ near 1, Φ has two simple saddle points located at

$$k = k_{\pm}(\theta) = \pm\sqrt{\frac{\alpha_1(\theta)}{1-\theta}} + O(1). \tag{9.5.5}$$

Moreover, because

$$\lim_{\theta\to 1-} \Phi^{(n)}(k_{\pm}(\theta);\theta) = 0, \qquad n = 1, 2, \ldots \tag{9.5.6}$$

we have that, in this limit, these saddle points coalesce at infinity producing a saddle point of infinite order.

Let us assume that R_1 and θ_1 have been chosen so that, for $\theta \,\varepsilon\, [\theta_1, 1]$, $k_{\pm}(\theta)$ lie in the region $|k| > R_1$. We shall allow Φ to have a finite number of additional saddle points but shall insist that they be confined for all θ in $[\theta_1, 1]$

to some bounded region $|k| \leq R_2 < R_1$.

Finally, to be assured that $\lim_{\theta \to 1-} I(\lambda;\theta)$ exists, we further assume that

$$g = k^{-(1+r)}\, \tilde{g}, \qquad r > 0. \tag{9.5.7}$$

Here $\tilde{g}$ has a convergent Laurent expansion for $|k| \geq R_1$ and $\lim_{|k|\to\infty} \tilde{g} \neq 0$.

Under the above assumptions, we can deform Γ upward in the k plane onto a new contour Γ' along which the Laurent expansions of Φ and $\tilde{g}$ about infinity are convergent. Indeed, we may take Γ' to consist of the real k axis for $|k| \geq \hat{R} > R_1$ and the upper half of the circle $|k| = \hat{R}$, as shown in Figure 9.5.

Our objective is to obtain an asymptotic expansion of $I(\lambda;\theta)$, as $\lambda \to \infty$, that remains uniformly valid for $\theta \in [\theta_1, 1]$. We have been quite detailed in our formulation of the problem. We shall be somewhat brief, however, in our discussion of the solution because the analysis follows closely those of previous sections.

As usual, the first step is to make a change of variable that reduces the exponent to a canonical form. With this in mind we set

$$\Phi(k;\theta) = -\left[\gamma^2(\theta)\, t + \frac{1}{4t} + \rho(\theta)\right] = \phi(t;\theta). \tag{9.5.8}$$

Because

$$\frac{dk}{dt} = -\frac{\left(\gamma^2 - \dfrac{1}{4t^2}\right)}{\Phi_k(k;\theta)} \tag{9.5.9}$$

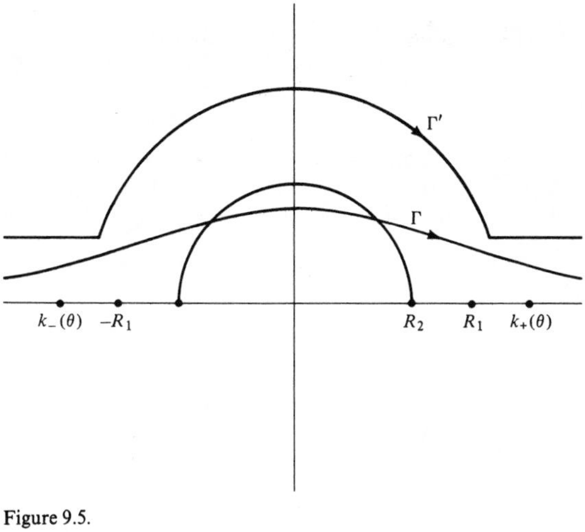

Figure 9.5.

we must require that the points $k = k_{\pm}(\theta)$ correspond to the points $t = \pm(2\gamma)^{-1}$, respectively. This serves to determine γ and ρ with the result

$$\gamma(\theta) = -\left\{\frac{\Phi(k_+(\theta);\theta) - \Phi(k_-(\theta);\theta)}{2}\right\},$$

$$\rho(\theta) = -\left\{\frac{\Phi(k_+(\theta);\theta) + \Phi(k_-(\theta);\theta)}{2}\right\} \tag{9.5.10}$$

By using (9.5.4) and (9.5.5), we find that, as $\theta \to 1-$,

$$\gamma(\theta) = 2\sqrt{(1-\theta)\,\alpha_1(\theta)} + O((1-\theta)^{3/2}),$$

$$\rho(\theta) = \alpha_0(\theta) + O(1-\theta). \tag{9.5.11}$$

The value of dk/dt at the saddle points is obtained by applying L' Hospital's rule in (9.5.9).

$$\left(\frac{dk}{dt}\right)\bigg|_{t=\pm(2\gamma)^{-1}} = \left\{\frac{\mp 4\gamma^3(\theta)}{\Phi_{kk}(k_\pm(\theta);\theta)}\right\}^{1/2}$$

$$= 4\alpha_1(\theta)\,[1 + O(\sqrt{1-\theta})], \qquad \theta \to 1-. \tag{9.5.12}$$

If we use (9.5.8) in (9.5.3), then we obtain

$$I(\lambda;\theta) = \frac{1}{2\pi}\int_{\bar\Gamma} t^{-(1+r)}\,G_0(t;\theta)\exp\{i\lambda\phi(t;\theta)\}\,dt. \tag{9.5.13}$$

Here

$$t^{-(1+r)}\,G_0(t;\theta) = g(k(t;\theta))\,\frac{dk(t;\theta)}{dt} \tag{9.5.14}$$

and we note that $G_0(t;\theta)$ has a Laurent expansion that converges all along $\bar\Gamma$ which is the image of Γ' under the change of variable to t. Furthermore, because the transformation (9.5.8) is essentially linear for $|k|$ large, the value of r in (9.5.14) is the same as in (9.5.7), $\lim_{|t|\to\infty} G_0 \neq 0$ and $\bar\Gamma$ is deformable onto a contour which is of the same form as Γ'. We shall assume that this deformation has been accomplished but shall continue to denote the contour by $\bar\Gamma$.

We now expand G_0 with the usual criteria in mind. Indeed, we set

$$G_0(t;\theta) = a_0(\theta) + a_1(\theta)\,t^{-1} + [\gamma^2 - (4t^2)^{-1}]\,H_0(t;\theta). \tag{9.5.15}$$

Here

$$a_0(\theta) = \left\{\frac{G_0\left(\frac{1}{2\gamma};\theta\right) + G_0\left(-\frac{1}{2\gamma};\theta\right)}{2}\right\}, \tag{9.5.16}$$

$$a_1(\theta) = \left\{\frac{G_0\left(\frac{1}{2\gamma};\theta\right) - G_0\left(-\frac{1}{2\gamma};\theta\right)}{4\gamma}\right\},$$

and we observe that

$$H_0(t;\theta) = \frac{G_0(t;\theta) - a_0 - a_1 t^{-1}}{\gamma^2 - (4t^2)^{-1}} \tag{9.5.17}$$

is regular at infinity and has removable singularities at $t = \pm(2\gamma)^{-1}$. Moreover, the functions a_0 and a_1 are both $0(1)$ for θ in $[\theta_1;1]$.

We now have

$$I(\lambda;\theta) = a_0(\theta) I_0(\lambda;\theta) + a_1(\theta) I_1(\lambda;\theta) + R_0(\lambda;\theta). \tag{9.5.18}$$

Here

$$I_j(\lambda;\theta) = \frac{1}{2\pi}\int_{\bar{\Gamma}} t^{-(j+r+1)} \exp\{i\lambda\phi\}\, dt, \qquad j = 0, 1 \tag{9.5.19}$$

and, after an integration by parts,

$$R_0(\lambda;\theta) = -\frac{i}{2\pi\lambda}\int_{\bar{\Gamma}} t^{-(r+2)} G_1(t;\theta) \exp\{i\lambda\phi\}\, dt. \tag{9.5.20}$$

Here

$$G_1(t;\theta) = t^{2+r}\frac{d}{dt}[t^{-(r+1)} H_0(t;\theta)] \tag{9.5.21}$$

which has the same domain of regularity as G_0 and is $0(1)$ in t as $|t| \to \infty$.

The functions I_0 and I_1 are readily expressible in terms of Bessel functions of the first kind. Indeed, by using a standard integral representation for such Bessel functions, we obtain

$$I_j(\lambda;\theta) = -\exp\{-i\lambda\rho\}\, i(2\gamma\, e^{-\pi i/2})^{r+j} J_{r+j}(\lambda\gamma). \tag{9.5.22}$$

By applying the above process repeatedly we obtain the following asymptotic expansion:

$$\begin{aligned} I(\lambda;\theta) = -\exp\{-i\lambda\rho\}\, i \sum_{n=0}^{N} (2\gamma\, e^{-\pi i/2})^{r+n} (i\lambda)^{-n} \\ \times [a_{2n} J_{r+n}(\lambda\gamma) - 2i\gamma\, a_{2n+1} J_{r+n+1}(\lambda\gamma)] + R_N(\lambda;\theta). \end{aligned} \tag{9.5.23}$$

Here, the coefficients a_j are defined recursively by

$$a_{2n}(\theta) = \left\{\frac{G_n\left(\frac{1}{2\gamma};\theta\right) + G_n\left(-\frac{1}{2\gamma};\theta\right)}{2}\right\},$$

$$a_{2n+1}(\theta) = \left\{\frac{G_n\left(\frac{1}{2\gamma};\theta\right) - G_n\left(-\frac{1}{2\gamma};\theta\right)}{4\gamma}\right\}, \qquad n = 0, 1, 2, \ldots,$$

$$G_n(t;\theta) = t^{r+n+1} \frac{d}{dt} \{t^{-(r+n)} H_{n-1}(t;\theta)\},$$

$$H_{n-1}(t;\theta) = \frac{G_{n-1}(t;\theta) - a_{2n-2} - a_{2n-1} t^{-1}}{\gamma^2 - (4t^2)^{-1}}, \qquad n = 1, 2, \ldots . \tag{9.5.24}$$

We find that all of the a_j's are $O(1)$ for θ in $[\theta_1, 1]$.

The remainder R_N is given by

$$R_N(\lambda;\theta) = \frac{(i\lambda)^{-(N+1)}}{2\pi} \int_{\bar{\Gamma}} t^{-(2+r+N)} G_{N+1}(t;\theta) \exp\{i\lambda\phi\}\, dt. \tag{9.5.25}$$

The asymptotic nature of our result can be rigorously established. Indeed, we have the following theorem which we state without proof.

THEOREM 9.5. The formal series (9.5.23) yields an asymptotic expansion of $I(\lambda;\theta)$, as $\lambda \to \infty$, with respect to the asymptotic sequence $\{\Psi_n(\lambda;\theta)\}$. Here

$$\Psi_n(\lambda;\theta) = \left(\frac{2\gamma}{\lambda}\right)^n (2\gamma)^r \left[|J_{r+n}(\lambda\gamma)| + |J_{r+n+1}(\lambda\gamma)|\right]. \tag{9.5.26}$$

Moreover, this expansion is uniformly valid for all θ in $[\theta_1, 1]$.

We wish to point out that not only is $\{\Psi_n(\lambda;\theta)\}$ an asymptotic sequence for $\lambda \to \infty$ and all θ in $[\theta_1, 1]$, but it is also such a sequence for fixed λ and γ near zero. Thus, when considering the underlying physical problem, the result (9.5.23) is useful in describing the precursor region for finite time near the front of propagation.

EXAMPLE 9.5. We wish here to reconsider the problem treated in Section 7.5. In particular let us investigate the integral

$$U(\lambda;\theta) = \frac{1}{4\pi bc} \int_\Gamma \frac{\exp\{i\lambda\Phi\}\, dv}{(v - v_0)(v^2 - 1)^{1/2}}, \tag{9.5.27}$$

$$\Phi = (v^2 - 1)^{1/2}\, \theta - v \sim -\left\{v(1-\theta) + \frac{\theta}{2v} + o(v^{-1})\right\}, \qquad |v| \to \infty, \tag{9.5.28}$$

in the limit $\lambda \to \infty$. We desire an expansion valid for θ near 1.

We first note that all of the conditions imposed on the general problem treated in this section are satisfied here. Upon proceeding as in the general case with k replaced by v, we find that in (9.5.7) $r = 1$ and

$$\gamma(\theta) = \sqrt{1 - \theta^2}, \qquad \rho(\theta) = 0. \tag{9.5.29}$$

We have

$$G_0(t;\theta) = \frac{t^2 \dfrac{dv}{dt}}{(v - v_0)(v^2 - 1)^{1/2}\, 2bc} \tag{9.5.30}$$

from which it is a simple matter to calculate a_0 and a_1. Indeed, we find that

$$a_0(\theta) = \frac{[1 - v_0^2(1-\theta^2)]^{-1}}{4bc}, \qquad (9.5.31)$$

$$a_1(\theta) = \frac{v_0[1 - v_0^2(1-\theta^2)]^{-1}}{8bc}.$$

Thus to leading order with respect to the sequence $\{\Psi_n\}$ defined by (9.5.26) we have

$$U(\lambda;\theta) \sim -[2bc]^{-1}[1 - v_0^2(1-\theta^2)]^{-1}\sqrt{1-\theta^2}$$

$$\times [J_1(\lambda\sqrt{1-\theta^2}) - i\sqrt{1-\theta^2}\,\gamma_0 J_2(\lambda\sqrt{1-\theta^2})]. \qquad (9.5.32)$$

We note that here we must choose $\theta_1 > \theta_0 = (v_0^2 - 1)^{1/2}/v_0$.

9.6. Double Integrals of Fourier Type

The principles of Section 9.3 can also be applied to multiple integrals. To illustrate this, let us consider the double integral

$$I(\lambda;\mathbf{x}_0) = \int_{\mathscr{D}} g(\mathbf{x}) \exp\{i\lambda\,\phi(\mathbf{x};\mathbf{x}_0)\}\, d\mathbf{x}, \qquad \mathbf{x} = (x_1, x_2). \qquad (9.6.1)$$

Here $\mathbf{x}_0 = (x_{10}, x_{20})$ represents the position of the only stationary point of ϕ in a domain $\mathscr{D}^*$ containing the bounded simply connected domain of integration $\mathscr{D}$. Moreover, we assume that this stationary point is simple, that is,

$$\nabla\phi(\mathbf{x}_0;\mathbf{x}_0) = 0, \qquad \det(\phi_{x_ix_j}(\mathbf{x}_0;\mathbf{x}_0)) \neq 0. \qquad (9.6.2)$$

Finally, we assume that both g and ϕ are infinitely differentiable with respect to x_1 and x_2 in $\mathscr{D}$.

We have seen in Section 8.4 that the nature of the asymptotic expansion of $I(\lambda;\mathbf{x}_0)$, as $\lambda \to \infty$, is different depending on whether $\mathbf{x} = \mathbf{x}_0$ is interior to $\mathscr{D}$, lies on Γ, the boundary of $\mathscr{D}$, or is exterior to $\mathscr{D}$. We seek here an asymptotic expansion that remains uniformly valid as $\mathbf{x}_0$ varies throughout the as yet undefined domain $\mathscr{D}^*$. We shall define $\mathscr{D}^*$ more precisely below. We remark now, however, that the assumption $\mathscr{D}^* \supset \mathscr{D}$ allows the stationary point to "pass through" the boundary Γ.

We begin by applying the transformations used in Section 8.4 to reduce ϕ to either a sum or difference of squares. For convenience we repeat the relevant equations here but refer the reader to Section 8.4 for a more detailed discussion. We denote by r the number of positive eigenvalues of the matrix $(\phi_{x_ix_j}(\mathbf{x}_0;\mathbf{x}_0))$. If $r = 0$ or $r = 2$, then $\mathbf{x} = \mathbf{x}_0$ is called a *center* of ϕ while if $r = 1$, then $\mathbf{x} = \mathbf{x}_0$ is called a *saddle* of ϕ.

We set

$$(\mathbf{x}-\mathbf{x}_0)^T = QR(\mathbf{z}-\mathbf{x}_0)^T, \qquad R = \begin{pmatrix} |\lambda_1|^{-1/2} & 0 \\ 0 & |\lambda_2|^{-1/2} \end{pmatrix},^{5} \tag{9.6.3}$$

$$\begin{aligned} f(\mathbf{z}) &= \phi(\mathbf{x}(\mathbf{z});\mathbf{x}_0) - \phi(\mathbf{x}_0;\mathbf{x}_0) \\ &\sim \frac{1}{2}\sum_{i=1}^{r}(z_i - x_{i0})^2 - \frac{1}{2}\sum_{i=r+1}^{2}(z_i - x_{i0})^2; \qquad |\mathbf{z}-\mathbf{x}_0| \to 0, \end{aligned} \tag{9.6.4}$$

$$\begin{aligned} (\xi_i - x_{i0}) = h_i(\mathbf{z}) &= (z_i - x_{i0}) + o(|\mathbf{z}-\mathbf{x}_0|), \\ &|\mathbf{z}-\mathbf{x}_0| \to 0, \qquad i = 1, 2, \end{aligned} \tag{9.6.5}$$

$$\sum_{i=1}^{r} h_i^2 - \sum_{i=r+1}^{2} h_i^2 = \sum_{i=1}^{r}(\xi_i - x_{i0})^2 - \sum_{i=r+1}^{2}(\xi_i - x_{i0})^2 = 2f. \tag{9.6.6}$$

Under the above transformations the domain $\mathscr{D}$ is mapped on to some domain $D(\mathbf{x}_0)$ in the ξ plane. We require that this mapping be one-to-one and remark that this will be the case whenever the level curves of ϕ along with their orthogonal trajectories form a simple covering of $\bar{\mathscr{D}} - \mathbf{x}_0$. We now define $\mathscr{D}^*$ to be the largest domain containing $\mathscr{D}$ for which the covering property remains true for all $\mathbf{x}_0$ in $\mathscr{D}^*$. We denote the image of $\mathscr{D}^*$ in the ξ plane by D^*.

If we introduce ξ_1, ξ_2 as new variables of integration in (9.6.1), then we obtain

$$I(\lambda;\mathbf{x}_0) = \exp\{i\lambda\phi(\mathbf{x}_0;\mathbf{x}_0)\}\int_D G_0(\xi;\mathbf{x}_0)\exp\left\{\frac{i\lambda}{2}\boldsymbol{\rho}\cdot(\xi-\mathbf{x}_0)\right\} d\xi. \tag{9.6.7}$$

Here

$$\begin{aligned} \boldsymbol{\rho} &= \begin{cases} -(\xi_1 - x_{10}, \xi_2 - x_{20}), & r = 0, \\ (\xi_1 - x_{10}, x_{20} - \xi_2), & r = 1, \\ (\xi_1 - x_{10}, \xi_2 - x_{20}), & r = 2, \end{cases} \\ &= (\operatorname{sgn}\lambda_1\,(\xi_1 - x_{10}), \operatorname{sgn}\lambda_2\,(\xi_2 - x_{20})), \end{aligned} \tag{9.6.8}$$

and

$$G_0(\xi;\mathbf{x}_0) = g(\mathbf{x}(\xi))\,J(\xi), \qquad J(\xi) = \left|\frac{\partial(x_1, x_2)}{\partial(\xi_1, \xi_2)}\right|. \tag{9.6.9}$$

We now set

$$G_0(\xi) = G_0(\mathbf{x}_0) + \boldsymbol{\rho}\cdot\mathbf{H}_0.^{6} \tag{9.6.10}$$

[5] Here Q is the orthogonal matrix that diagonalizes $(\phi_{x_i x_j}(\mathbf{x}_0;\mathbf{x}_0))$ such that (9.6.4) holds.

[6] For ease of notation we suppress the explicit dependence of G_0 on $\mathbf{x}_0$.

Although $\mathbf{H}_0$ is not uniquely determined by (9.6.10) we may choose for its components

$$H_{01} = \frac{\mu_1}{2}\left\{\frac{G_0(\xi_1,\xi_2) - G_0(x_{10},\xi_2) + G_0(\xi_1,x_{20}) - G_0(x_{10},x_{20})}{\xi_1 - x_{10}}\right\},$$

$$H_{02} = \frac{\mu_2}{2}\left\{\frac{G_0(\xi_1,\xi_2) - G_0(\xi_1,x_{20}) + G_0(x_{10},\xi_2) - G_0(x_{10},x_{20})}{\xi_2 - x_{20}}\right\}. \tag{9.6.11}$$

Here

$$\mu_1 = \mu_2 = -1, \quad r = 0; \qquad \mu_1 = 1, \quad \mu_2 = -1, \quad r = 1;$$

$$\mu_1 = \mu_2 = 1, \quad r = 2. \tag{9.6.12}$$

The assumed smoothness properties of ϕ and g imply that the functions H_{0i}, $i = 1, 2$ are infinitely differentiable in D.

Upon inserting (9.6.10) into (9.6.7) we obtain

$$I(\lambda;\mathbf{x}_0) = \exp\{i\lambda\phi(\mathbf{x}_0;\mathbf{x}_0)\}\,\{G_0(\mathbf{x}_0;\mathbf{x}_0)\mathscr{F}(\lambda;\mathbf{x}_0) + I_0(\lambda;\mathbf{x}_0)\}, \tag{9.6.13}$$

where

$$\mathscr{F}(\lambda;\mathbf{x}_0) = \int_D \exp\left\{\frac{i\lambda}{2}\,\boldsymbol{\rho}\cdot(\boldsymbol{\xi} - \mathbf{x}_0)\right\} d\boldsymbol{\xi} \tag{9.6.14}$$

and

$$I_0(\lambda;\mathbf{x}_0) = \int_D (\boldsymbol{\rho}\cdot\mathbf{H}_0)\exp\left\{\frac{i\lambda\boldsymbol{\rho}}{2}\cdot(\boldsymbol{\xi} - \mathbf{x}_0)\right\} a_{\varsigma}. \tag{9.6.15}$$

We now integrate by parts in (9.6.15). This yields

$$I_0(\lambda;\mathbf{x}_0) = (-i\lambda)^{-1}\left[-\int_{\hat{\Gamma}} (\mathbf{H}_0\cdot\mathbf{N})\exp\left\{\frac{i\lambda}{2}\,\boldsymbol{\rho}\cdot(\boldsymbol{\xi} - \mathbf{x}_0)\right\} d\sigma\right.$$
$$\left. + \int_D G_1(\boldsymbol{\xi};\mathbf{x}_0)\exp\left\{\frac{i\lambda}{2}\,\boldsymbol{\rho}\cdot(\boldsymbol{\xi} - \mathbf{x}_0)\right\} d\boldsymbol{\xi}\right]. \tag{9.6.16}$$

Here $\hat{\Gamma}$ is the boundary of D, σ is arc-length along $\hat{\Gamma}$, $\mathbf{N}$ is the unit outward normal to $\hat{\Gamma}$, and

$$G_1(\boldsymbol{\xi};\mathbf{x}_0) = \nabla_{\xi}\cdot\mathbf{H}_0. \tag{9.6.17}$$

We observe that the last integral on the right side of (9.6.16) is of the same form as $I(\lambda;\mathbf{x}_0)$. The multiplicative factor $(i\lambda)^{-1}$ then suggests that

$$I(\lambda;\mathbf{x}_0) \sim \exp\{i\lambda\phi(\mathbf{x}_0;\mathbf{x}_0)\}\left[G_0(\mathbf{x}_0;\mathbf{x}_0)\mathscr{F}(\lambda;\mathbf{x}_0)\right.$$
$$\left. - (-i\lambda)^{-1}\int_{\hat{\Gamma}} (\mathbf{H}_0\cdot\mathbf{N})\exp\left\{\frac{i\lambda}{2}\,\boldsymbol{\rho}\cdot(\boldsymbol{\xi} - \mathbf{x}_0)\right\} d\sigma\right] \tag{9.6.18}$$

represents, to leading order, a uniformly valid asymptotic expansion of I as $\lambda \to \infty$. Moreover,

$$R_1(\lambda;\mathbf{x}_0) = (-i\lambda)^{-1} \exp\{i\lambda\phi(\mathbf{x}_0;\mathbf{x}_0)\} \int_D G_1(\boldsymbol{\xi};\mathbf{x}_0) \exp\left\{\frac{i\lambda}{2}\boldsymbol{\rho}\cdot(\boldsymbol{\xi}-\mathbf{x}_0)\right\} d\boldsymbol{\xi} \tag{9.6.19}$$

is the remainder or error integral.

The above process can be applied to (9.6.19). Indeed, it can be repeatedly applied yielding after p steps

$$\begin{aligned} I(\lambda;\mathbf{x}_0) = \exp\{i\lambda\phi(\mathbf{x}_0;\mathbf{x}_0)\} & \left[\mathscr{F}(\lambda;\mathbf{x}_0) \sum_{j=0}^{p-1} G_j(\mathbf{x}_0;\mathbf{x}_0)(-i\lambda)^{-j}\right. \\ & \left. - \sum_{j=0}^{p-1} (-i\lambda)^{-j-1} \int_{\hat{\Gamma}} (\mathbf{H}_j\cdot\mathbf{N}) \exp\left\{\frac{i\lambda}{2}\boldsymbol{\rho}\cdot(\boldsymbol{\xi}-\mathbf{x}_0)\right\} d\sigma\right] \\ & + (-i\lambda)^{-p} R_p(\lambda;\mathbf{x}_0). \end{aligned} \tag{9.6.20}$$

The functions G_j and $\mathbf{H}_j$ are defined recursively by

$$H_{j1} = \frac{\mu_1}{2}\left[\frac{G_j(\xi_1,\xi_2) - G_j(x_{10},\xi_2) + G_j(\xi_1,x_{20}) - G_j(x_{10},x_{20})}{\xi_1 - x_{10}}\right],$$

$$H_{j2} = \frac{\mu_2}{2}\left[\frac{G_j(\xi_1,\xi_2) - G_j(\xi_1,x_{20}) + G_j(x_{10},\xi_2) - G_j(x_{10},x_{20})}{\xi_2 - x_{20}}\right],$$

$$j = 0, 1, 2, \ldots . \tag{9.6.21}$$

$$G_j(\xi) = \nabla_\xi \cdot \mathbf{H}_{j-1}, \qquad j = 1, 2, \ldots . \tag{9.6.22}$$

In (9.6.21) the quantities μ_1 and μ_2 are defined by (9.6.12). The error integral R_p is given by

$$R_p(\lambda;\mathbf{x}_0) = \exp\{i\lambda\phi(\mathbf{x}_0;\mathbf{x}_0)\} \int_D G_p(\boldsymbol{\xi};\mathbf{x}_0) \exp\left\{\frac{i\lambda}{2}\boldsymbol{\rho}\cdot(\boldsymbol{\xi}-\mathbf{x}_0)\right\} d\boldsymbol{\xi}. \tag{9.6.23}$$

An examination of (9.6.20) shows that the uniform asymptotic expansion involves only one double integral $\mathscr{F}(\lambda;\mathbf{x}_0)$, which can be considered the canonical special function for the problem, and a sequence of one-dimensional integrals of Fourier type.

The uniform expansion we have derived suffers from some defects. Indeed, although (9.6.20) appears to involve functions simpler than I itself, these functions are by no means simple. The canonical function $\mathscr{F}(\lambda;\mathbf{x}_0)$ is not a well-studied special function and, in particular, it has not been tabulated. To complicate matters, the domain of integration in (9.6.14) depends not only on the original domain $\mathscr{D}$ but on the location of the stationary point $\mathbf{x}_0$ as well.

We must therefore conclude that the uniform expansion is of limited utility. We might obtain a more useful result, however, if we give up total uniformity. We wish now to investigate certain aspects of this idea.

Let us suppose that the stationary point $\boldsymbol{\xi} = \mathbf{x}_0$ approaches a smooth convex segment of the boundary curve $\hat{\Gamma}$.[7] We shall denote this segment by $\hat{\Gamma}$. Let us

[7] By a smooth segment we mean one with a continuously turning tangent and a well-defined curvature.

further suppose that $\mathbf{x} = \mathbf{x}_0$ is a center of ϕ so that either $r = 0$ or $r = 2$.

Our objective here is to obtain what we might term a *semiuniform* asymptotic expansion of $I(\lambda; \mathbf{x}_0)$, as $\lambda \to \infty$. It is to remain valid for all locations of the stationary point $\boldsymbol{\xi} = \mathbf{x}_0$ in a sufficiently small neighborhood of $\hat{\Gamma}$.

We start with the uniform expansion (9.6.18). The boundary integral in that expansion is a one-dimensional integral of Fourier type. If $\mathbf{x}_0$ is sufficiently close to $\hat{\Gamma}$ and, in particular, bounded away from the evolute[8] of Γ, then no special uniform method is needed for the asymptotic analysis of this integral. In fact, its asymptotic expansion can be obtained via the method described in Chapter 6. Therefore, in the present discussion we shall not consider the boundary integral further.

Our main concern will be with the asymptotic expansion of the canonical function $\mathscr{F}(\lambda; \mathbf{x}_0)$. It is, of course, defined by (9.6.14). We first isolate the segment $\hat{\Gamma}$ and the stationary point $\boldsymbol{\xi} = \mathbf{x}_0$ by introducing the two-dimensional neutralizer $v(\xi_1, \xi_2)$. We require that

$$v = \begin{cases} 1, & \text{on } D_1, \\ 0, & \text{on } D_2^c,^{9} \end{cases}$$

where the domains D_1 and D_2 are as depicted in Figure 9.6. On $D_2 - D_1$, the value of v varies from 0 to 1 and v is infinitely differentiable everywhere. Finally, we require that the evolute of $\hat{\Gamma}$ lies in D_2^c.

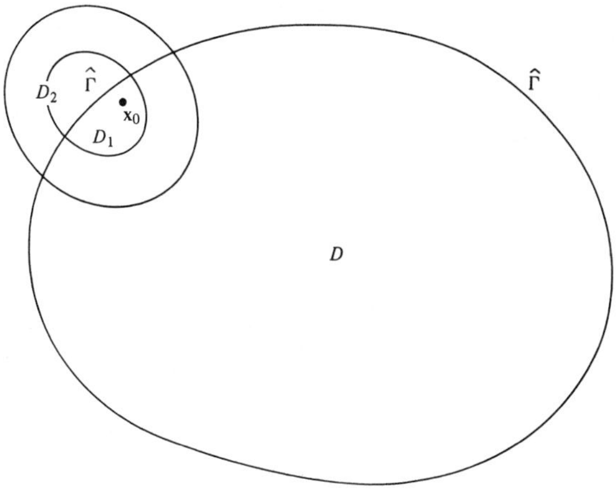

Figure 9.6.

[8] The evolute of $\hat{\Gamma}$ is the envelope of the normals to $\hat{\Gamma}$ all drawn in the ξ_1, ξ_2 plane.

[9] D_2^c denotes the complement of the domain D_2.

We now write

$$\mathscr{F}(\lambda;\mathbf{x}_0) = \int_D v(\boldsymbol{\xi}) \exp\left\{\frac{i\lambda}{2}\,\boldsymbol{\rho}\cdot(\boldsymbol{\xi}-\mathbf{x}_0)\right\}\,d\boldsymbol{\xi}$$

$$+ \int_D (1-v(\boldsymbol{\xi}))\exp\left\{\frac{i\lambda}{2}\,\boldsymbol{\rho}\cdot(\boldsymbol{\xi}-\mathbf{x}_0)\right\}\,d\boldsymbol{\xi}$$

$$= \hat{\mathscr{F}}(\lambda;\mathbf{x}_0) + \tilde{\mathscr{F}}(\lambda;\mathbf{x}_0). \qquad (9.6.24)$$

We shall insist that the stationary point $\boldsymbol{\xi}=\mathbf{x}_0$ remain in the domain D_1 and hence bounded away from the effective domain of integration in $\tilde{\mathscr{F}}(\lambda;\mathbf{x}_0)$. As a result, the asymptotic expansion of $\tilde{\mathscr{F}}$ can be obtained by the method of Section 8.4 and it too need not be discussed further.

What we have found therefore is that only $\hat{\mathscr{F}}(\lambda;\mathbf{x}_0)$ requires any special uniform analysis. We first introduce in $\hat{\mathscr{F}}(\lambda;\mathbf{x}_0)$ new variables of integration consisting of σ, arc-length along $\hat{\Gamma}$ and $\tilde{\sigma}$, arc-length along the normal to $\hat{\Gamma}$. We assume that σ increases as $\hat{\Gamma}$ is traversed in a counterclockwise manner and $\tilde{\sigma}$ increases in the direction of the inward normal to $\hat{\Gamma}$. We also assume that $\hat{\Gamma}$ is convex.

If $\boldsymbol{\xi}=\boldsymbol{\xi}(\sigma)$ is the equation of $\hat{\Gamma}$, then the change of variables is defined by

$$\boldsymbol{\xi} = \boldsymbol{\xi}(\sigma) - \tilde{\sigma}\,\mathbf{N}(\sigma), \qquad (9.6.25)$$

where $\mathbf{N}(\sigma)$ is the unit outward normal to $\hat{\Gamma}$. We have

$$\frac{1}{2}\,\boldsymbol{\rho}\cdot(\boldsymbol{\xi}-\mathbf{x}_0) = \frac{\mu}{2}\,|\boldsymbol{\xi}(\sigma)-\tilde{\sigma}\,\mathbf{N}(\sigma)-\mathbf{x}_0|^2$$

$$= s(\sigma;\tilde{\sigma}); \qquad \mu = (-1)^{r/2+1} = \pm 1. \qquad (9.6.26)$$

Here μ is positive when ϕ has a minimum at $\mathbf{x}_0$ and μ is negative when ϕ has a maximum at $\mathbf{x}_0$. Now we write

$$\hat{\mathscr{F}}(\lambda;\mathbf{x}_0) = \int\int_{\hat{D}} v(\boldsymbol{\xi}(\sigma,\tilde{\sigma}))\left[1-\kappa(\sigma)\tilde{\sigma}\right]\exp\{i\lambda s(\sigma;\tilde{\sigma})\}\,d\sigma\,d\tilde{\sigma}. \qquad (9.6.27)$$

Here $\kappa(\sigma) = |d^2\boldsymbol{\xi}(\sigma)/d\sigma^2|$ is the curvature of the boundary $\hat{\Gamma}$, $\hat{D}$ is the image of $D \cap D_2$ under the transformation (9.6.25), and $1-\kappa(\sigma)\tilde{\sigma}$ is the Jacobian of this transformation. We note that under the assumptions made, in particular the one concerning the location of the evolute of $\hat{\Gamma}$, this Jacobian is positive in D.

To asymptotically evaluate $\hat{\mathscr{F}}(\lambda;\mathbf{x}_0)$ we seek stationary points of s with respect to σ for fixed $\tilde{\sigma}$. From (9.6.26) we obtain

$$\dot{s} = \mu\left[\dot{\boldsymbol{\xi}}-\tilde{\sigma}\dot{\mathbf{N}}\right]\cdot\left[\boldsymbol{\xi}-\tilde{\sigma}\mathbf{N}-\mathbf{x}_0\right], \qquad (\cdot)=\frac{d}{d\sigma}. \qquad (9.6.28)$$

Because both $\dot{\boldsymbol{\xi}}(\sigma)$ and $\dot{\mathbf{N}}(\sigma)$ are tangent to $\hat{\Gamma}$, it follows that $\dot{s}=0$ at that point

$\sigma = \sigma_0$ for which

$$\dot{\xi}(\sigma_0)\cdot[\boldsymbol{\beta} - \mathbf{x}_0] = 0, \qquad \boldsymbol{\beta} = \xi(\sigma_0). \tag{9.6.29}$$

Moreover, we have

$$\ddot{s}\big|_{\sigma=\sigma_0} = \mu[1 - \kappa(\sigma_0)\tilde{\sigma}]\,[1 - \eta\kappa(\sigma_0)\,|\boldsymbol{\beta} - \mathbf{x}_0|]. \tag{9.6.30}$$

Here $\eta = 1$ when $\mathbf{x}_0$ lies in $\bar{D}$ and $\eta = -1$ when $\mathbf{x}_0$ is exterior to D. In obtaining (9.6.30) we have used the Frênet formulas[10] of differential geometry. We leave the details of its derivation to the exercises.

We now apply the stationary phase formula (6.1.5) to obtain

$$\hat{\mathscr{F}}(\lambda;\mathbf{x}_0) \sim \left(\frac{2\pi}{\lambda}\right)^{1/2} e^{i\mu\pi/4} \int_0^\infty \left[\frac{1 - \kappa(\sigma_0)\,\tilde{\sigma}}{1 - \eta\kappa(\sigma_0)\,|\boldsymbol{\beta} - \mathbf{x}_0|}\right]^{1/2} \nu(\xi(\sigma_0, \tilde{\sigma}))$$
$$\times \exp\left\{\frac{i\mu}{2}\,\lambda\,[\tilde{\sigma} - \eta\,|\boldsymbol{\beta} - \mathbf{x}_0|]^2\right\} d\tilde{\sigma}. \tag{9.6.31}$$

Thus, we have reduced our problem to the analysis of a one-dimensional integral.

As $\mathbf{x}_0$ approaches the boundary $\hat{\Gamma}$, $|\boldsymbol{\beta} - \mathbf{x}_0|$ approaches zero. But for $\eta = 1$, that is, for $\mathbf{x}_0$ in D, the stationary point of the exponent in (9.6.31) is located at $\tilde{\sigma} = |\boldsymbol{\beta} - \mathbf{x}_0|$. Therefore, our problem is to obtain a uniform expansion of the integral (9.6.31) as the stationary point approaches the endpoint of integration $\tilde{\sigma} = 0$. This is a special case of the general anomaly treated in Section 9.4. We find that

$$\hat{\mathscr{F}}(\lambda;\mathbf{x}_0) \sim \left(\frac{2\pi}{\lambda}\right)^{1/2} e^{i\mu\pi/4} \left[\int_0^\infty \exp\left\{\frac{i\mu\lambda}{2}\,[\tilde{\sigma} - \eta\,|\boldsymbol{\beta} - \mathbf{x}_0|]^2\right\} d\tilde{\sigma}\right.$$
$$\left. + \frac{\exp\left\{\frac{i\mu\lambda}{2}\,|\boldsymbol{\beta} - \mathbf{x}_0|^2\right\}}{i\mu\lambda\eta}\; \frac{[1 - (1 - \eta\kappa(\sigma_0)\,|\boldsymbol{\beta} - \mathbf{x}_0|)^{1/2}]}{|\boldsymbol{\beta} - \mathbf{x}_0|\,(1 - \eta\kappa(\sigma_0)\,|\boldsymbol{\beta} - \mathbf{x}_0|)^{1/2}}\right], \tag{9.6.32}$$

which holds uniformly for all $\mathbf{x}_0$ in D_1.

We remark that the first term on the right side of (9.6.32) is a multiple of the special function known as the *Fresnel integral.* We readily find by using standard asymptotic techniques that $\hat{\mathscr{F}}(\lambda;\mathbf{x}_0) = 0(\lambda^{-1})$ when $\mathbf{x}_0$ lies in $D_1 \cap \bar{D}$ and $\hat{\mathscr{F}}(\lambda;\mathbf{x}_0) = 0(\lambda^{-3/2})$ when $\mathbf{x}_0$ lies in $D_1 \cap D^c$ and is bounded away from $\hat{\Gamma}$. The uniform expansion (9.6.32) effects a smooth transition as $\mathbf{x}_0$ passes through the boundary segment $\hat{\Gamma}$.

To obtain the semiuniform expansion of I to leading order we must obtain an asymptotic expansion of $\hat{\mathscr{F}}(\lambda;\mathbf{x}_0)$ correct to $0(\lambda^{-3/2})$ and an asymptotic expansion of the boundary integral in (9.6.18) correct to $0(\lambda^{-1/2})$. These results

[10] For plane curves these formulas are $\ddot{\xi} = -\kappa\,\mathbf{N}$ and $\dot{\mathbf{N}} = \kappa\dot{\xi}$ with $\mathbf{N}$ the outward normal and $\hat{\Gamma}$ convex.

when combined with (9.6.32) will then yield the desired expansion. We emphasize that the result is indeed semiuniform in that the stationary point $\boldsymbol{\xi} = \mathbf{x}_0$ is confined to the domain D_1.

We wish to comment briefly on one further semiuniform expansion. Let us again assume that $\mathbf{x} = \mathbf{x}_0$ is a center of ϕ. We can write the exponent of the boundary integral in (9.6.18) as

$$s(\sigma) = \frac{\mu}{2} |\boldsymbol{\xi}(\sigma) - \mathbf{x}_0|^2. \tag{9.6.33}$$

Here, as above, $\boldsymbol{\xi} = \boldsymbol{\xi}(\sigma)$ is the equation of the boundary $\hat{\Gamma}$ and μ is defined in (9.6.26).

Upon differentiating (9.6.33) we obtain

$$\dot{s}(\sigma) = \mu\ \mathbf{T}(\sigma)\cdot[\boldsymbol{\xi}(\sigma) - \mathbf{x}_0], \tag{9.6.34}$$

$$\ddot{s}(\sigma) = -\mu\kappa(\sigma)\ \mathbf{N}(\sigma)\cdot[\boldsymbol{\xi}(\sigma) - \mathbf{x}_0] + \mu, \tag{9.6.35}$$

where $\mathbf{T}$ is the unit tangent vector to $\hat{\Gamma}$ and we have again used the Frênet formulas.

It follows from (9.6.34) that $\sigma = \sigma_0$ is a stationary point of s if $\boldsymbol{\xi}(\sigma_0) - \mathbf{x}_0$ is normal to $\hat{\Gamma}$ at $\boldsymbol{\xi} = \boldsymbol{\xi}(\sigma_0)$. Moreover,

$$\ddot{s}(\sigma_0) = \mu[1 - \eta\kappa(\sigma_0)\,|\boldsymbol{\xi}(\sigma_0) - \mathbf{x}_0|]. \tag{9.6.36}$$

If $|\boldsymbol{\xi}(\sigma_0) - \mathbf{x}_0| = \kappa^{-1}(\sigma_0)$ and $\eta = 1$, that is, if $\mathbf{x}_0$ lies on the evolute of $\hat{\Gamma}$, then $\ddot{s}(\sigma_0) = 0$ and $\sigma = \sigma_0$ is not a simple stationary point. Indeed, for $\mathbf{x}_0$ near the evolute, $s(\sigma)$ must have at least two nearby stationary points. In particular, if $\dddot{s}(\sigma_0) \neq 0$, then $s(\sigma)$ has two stationary points that coalesce as the stationary point $\mathbf{x}_0$ of the original integral I approaches the evolute of the boundary. Let us suppose that, in fact, $\mathbf{x}_0$ is confined to a neighborhood of the evolute and that this neighborhood is small enough that $\mathbf{x}_0$ is bounded away from $\hat{\Gamma}$. Then we can rewrite $\mathcal{F}(\lambda;\mathbf{x}_0)$ as a term arising from an interior stationary point plus a boundary integral plus lower order terms. We leave the details to the exercises and simply state the result here:

$$I(\lambda) \sim \exp\{i\lambda\phi(\mathbf{x}_0;\mathbf{x}_0)\}\left[\frac{2\pi i\mu}{\lambda}\,G_0(\mathbf{x}_0;\mathbf{x}_0)\right.$$
$$\left. + \frac{1}{i\lambda}\int_{\Gamma}\left\{\frac{\mu G_0(\mathbf{x}_0;\mathbf{x}_0)(\boldsymbol{\xi}(\sigma) - \mathbf{x}_0)}{|\boldsymbol{\xi}(\sigma) - \mathbf{x}_0|^2} + \mathbf{H}_0(\boldsymbol{\xi}(\sigma))\right\}\cdot\mathbf{N}\exp[i\lambda s(\sigma)]\,d\sigma\right]. \tag{9.6.37}$$

The error in this result is an integral of the form of $I(\lambda)$ itself multiplied by $(i\lambda)^{-1}$. The line integral here has two stationary points which coalesce at σ_0 when $\mathbf{x}_0$ approaches the evolute and it has other simple stationary points as well. These latter may be treated by the method of stationary phase as developed in Section 6.1. The former may be treated by the method of Section 9.2. This leads to an asymptotic expansion involving an Airy function and its derivative. See Exercise 9.18.

9.7. Exercises

9.1. Consider

$$I(\lambda) = \frac{1}{2\pi} \int_{-\pi/2}^{\pi/2} g(z) \exp\{i\lambda w(z;\theta)\}\, dz. \tag{9.7.1}$$

Here

$$w(z;\theta) = (z \tan z)^{1/2} - \theta z. \tag{9.7.2}$$

(a) Show that for $\theta \gtrsim 1$, w has two simple stationary points

$$z_{\pm} = \pm\sqrt{2(1-\theta)}\,[1 + O(\theta - 1)], \tag{9.7.3}$$

which, as $\theta \to 1$, coalesce to a stationary point of order 2.

(b) Assume that $z_{\pm}$ are the dominant critical points for (9.7.1) and show formally that

$$\begin{aligned} I(\lambda) &= \lambda^{-1/3}\, 2^{-2/3} \operatorname{Ai}(\lambda^{2/3}\, 2^{1/3} (\theta - 1)) \Big[\sum_{\pm} g(\pm\sqrt{2(1-\theta)}) \Big] \\ &\quad \times \left[1 + O(\lambda^{1/3}(1-\theta)) + O\left(\frac{1}{\lambda}\right) \right] - i\, \lambda^{-2/3} 2^{-2/3} \operatorname{Ai}'(\lambda^{2/3}\, 2^{1/3} (\theta - 1)) \\ &\quad \times \left[\sum_{\pm} \pm \frac{g(\pm\sqrt{2(1-\theta)})}{2^{1/6}\sqrt{1-\theta}} \right] \left[1 + O(\lambda^{1/3}(1-\theta)) + O\left(\frac{1}{\lambda}\right) \right]. \end{aligned} \tag{9.7.4}$$

Here it is helpful to use the integral representation

$$\operatorname{Ai}(x) = \frac{1}{2\pi} \int_{-\infty}^{\infty} \exp\left\{ i \left[\frac{t^3}{3} + xt \right] \right\} dt \tag{9.7.5}$$

for the Airy function of real argument.

9.2. (a) For the integrals of Exercise 7.21, verify that the asymptotic expansions listed below are valid for t near 1. The functions appearing in these expansions are defined at the end of the exercise.

$$\begin{aligned} u_2 &\sim 2\pi g \lambda^{-1/3} \exp\left(-\frac{i\pi}{6}\right) \operatorname{Ai}[\lambda^{2/3} f^2(t)\, e^{-2\pi i/3}] \\ &\quad - a \exp\left\{ \lambda\phi_2 - i\left(\frac{\pi}{3} + \theta\right) \right\}, \end{aligned} \tag{9.7.6}$$

$$\begin{aligned} u_3 &\sim -2\pi g \lambda^{-1/3} \exp\left(-\frac{i\pi}{6}\right) \operatorname{Ai}(\lambda^{2/3} f^2(t)\, e^{-2\pi i/3}) \\ &\quad + a \exp\left\{ \lambda\phi_3 - i\left(\frac{\pi}{6} - \theta\right) \right\}, \end{aligned} \tag{9.7.7}$$

$$u_4 \sim 2\pi i g \lambda^{-1/3} \operatorname{Ai}(\lambda^{2/3} f^2(t)) + 2i \operatorname{Im}\left(\exp\left[-\lambda\phi_2 + i\left(\frac{\pi}{6} - \theta\right)\right]\right), \tag{9.7.8}$$

$$u_5 \sim -2\pi g \lambda^{-1/3} \exp\left(\frac{i\pi}{6}\right) \operatorname{Ai}(\lambda^{2/3} f^2(t)\, e^{-2\pi i/3})$$
$$- a \exp\left\{-\lambda\phi_2 + i\left(\frac{\pi}{6} - \theta\right)\right\}, \tag{9.7.9}$$

$$u_6 \sim 2\pi g \lambda^{-1/3} \exp\left(\frac{i\pi}{6}\right) \operatorname{Ai}(\lambda^{2/3} f^2(t)\, e^{2\pi i/3})$$
$$- a \exp\left\{-\lambda\phi_3 + i\left(\frac{\pi}{3} + \theta\right)\right\}. \tag{9.7.10}$$

Here

$$f(t) = \begin{cases} (\frac{3}{2}\phi_1)^{1/3}, & 0 < t \le 1, \\ i|\frac{3}{2}\phi_1|^{1/3}, & 1 \le t, \end{cases} \tag{9.7.11}$$

$$\phi_1(t) = \frac{(1 - t^{1/3})^{3/2}}{35}(30t^{2/3} + 24t^{1/3} + 16), \tag{9.7.12}$$

$$\phi_2(t) = \phi_1(te^{4\pi i}), \tag{9.7.13}$$

$$\phi_3(t) = \phi_1(te^{2\pi i}), \tag{9.7.14}$$

$$a = \sqrt{\frac{\pi}{3\lambda}}\, t^{-1/3}(t^{2/3} + t^{1/3} + 1)^{-1/8}, \tag{9.7.15}$$

$$\theta = \frac{1}{4}\arctan\left(\frac{\sqrt{3}\,t^{1/3}}{t^{1/3} + 2}\right), \qquad 0 < \theta < \frac{\pi}{2}, \tag{9.7.16}$$

$$g = \left|\frac{2f(t)}{6t^{2/3}(1 - t^{1/3})^{1/2}}\right|^{1/2}. \tag{9.7.17}$$

(b) Show that

$$\lim_{t \to 1} g(t) = 3^{-1/3}.$$

9.3. Consider

$$I(\lambda) = \int_{C_1} G(\delta\tau) \exp\{-\lambda\delta^3 \psi(\tau)\}\, d\tau. \tag{9.7.18}$$

Here

$$\psi(\tau) = \frac{\tau^3}{3} - \frac{\gamma^2}{\delta^2}\tau. \tag{9.7.19}$$

Suppose that C_1 is such that the dominant critical point for (9.7.19) is the saddle point at $\tau = \gamma/\delta$.

(a) Show that

$$\lambda\delta^3 \left[\psi(\tau) - \psi\left(\frac{\gamma}{\delta}\right)\right] = e^{i\theta}\,\sigma y^2 \left[1 + \frac{e^{-i\theta}}{3}\frac{y}{\sigma}\right] \tag{9.7.20}$$

when we set

$$\sigma = \lambda^{1/3}\,|\gamma| = \frac{\lambda^{1/3}\,\delta\left|\psi''\left(\frac{\gamma}{\delta}\right)\right|}{2}, \qquad \theta = \arg(\gamma),$$

$$y = \lambda^{1/3}\,\delta\left(\tau - \frac{\gamma}{\delta}\right). \tag{9.7.21}$$

(b) Introduce a new variable of integration ζ by the equation

$$\sigma\,\zeta^2 = \lambda\delta^3 \left[\psi(\tau) - \psi\left(\frac{\gamma}{\delta}\right)\right] \tag{9.7.22}$$

and then show that

$$\frac{\zeta}{\sigma} = e^{i\theta/2}\,\frac{y}{\sigma}\left[1 + f_1\left(\frac{y}{\sigma}\right)\right]. \tag{9.7.23}$$

Also show that, in a sufficiently small neighborhood of $y = 0$,

$$\left|\tau - \frac{\gamma}{\delta}\right| < 2\,\frac{|\gamma|}{\delta} \qquad \text{or} \qquad |y| < 2\lambda^{+1/3}\,|\gamma|,$$

we may invert (9.7.23) to obtain

$$y = e^{-i\theta/2}\,\zeta\left[1 + f_2\left(\frac{\zeta}{\sigma}\right)\right] \tag{9.7.24}$$

and

$$\frac{d\tau}{d\zeta} = \delta^{-1}\,\lambda^{-1/3}\,e^{-i\theta/2}\left[1 + f_3\left(\frac{\zeta}{\sigma}\right)\right]. \tag{9.7.25}$$

Here the power series f_3 involves nonnegative powers of ζ/σ only.

(c) Show that we may write

$$G(\delta\tau) = \tilde{G}\left[\gamma + \frac{\zeta}{\sigma} f_4\left(\frac{\zeta}{\sigma}\right)\right], \tag{9.7.26}$$

where the power series expansion of $\tilde{G}$ near $\zeta = 0$ involves nonnegative powers of γ only.

(d) Recast $I(\lambda)$ in the form

$$I(\lambda) = \lambda^{-1/3}\int_{\tilde{C}_1} H\left(\frac{\zeta}{\sigma}\right)\exp(-\sigma\zeta^2)\,d\zeta \tag{9.7.27}$$

and conclude that, in fact, the asymptotic expansion of this integral is an expansion in inverse powers of σ. In particular, to leading order,

$$|I(\lambda)| \sim \lambda^{-1/3}\,\sigma^{-1/2}\,|G(\gamma)|\,\exp\{\tfrac{2}{3}\,\lambda\,\mathrm{Re}(\gamma^3)\}. \tag{9.7.28}$$

(e) Use (9.7.28) to estimate the integral (9.2.39) and to obtain analogous estimates for $\mathrm{Ai}(\lambda^{2/3}\,\gamma^2)$ and $\mathrm{Ai}'(\lambda^{2/3}\,\gamma^2)$.

(f) Verify the estimate (9.2.33) for the integral (9.2.39).

9.4. Consider

$$I(\lambda) = \int_C g(z)\exp\{\lambda\, w(z)\}\, dz \tag{9.7.29}$$

and let C be such that a simple saddle point at $z = z_0$ is the dominant critical point.

(a) Follow the lines of Exercise 9.1 to conclude that, in fact, $I(\lambda)$ has an asymptotic expansion in inverse powers of $\sigma = \lambda|w''(z_0)|/2$.

(b) Suppose that $w(z) = w(z;\theta)$ has two simple saddle points which coalesce to yield a saddle point of order 2 when $\theta = 0$. If $z_0 = z_0(\theta)$ is one of these saddle points, then what conjecture can be made about the "nonuniform" asymptotic expansion of (9.7.29) for θ "small"?

9.5. (a) Set

$$-\cos^{-1}\beta = \int_1^\beta \frac{d\beta'}{\sqrt{1-\beta'^2}} \tag{9.7.30}$$

and use integration by parts to show that

$$-\cos^{-1}\beta = -\frac{(1-\beta^2)^{1/2}}{\beta} + \frac{1}{3}\frac{(1-\beta^2)^{3/2}}{\beta^3} + \int_1^\beta \frac{(1-\beta'^2)^{5/2}}{\beta'^4}\, d\beta'. \tag{9.7.31}$$

(b) Use (9.7.31) to show that for γ given by (9.2.49)

$$\gamma \approx \frac{e^{5\pi i/6}}{2^{1/3}}\frac{(1-\beta^2)^{1/2}}{\beta} \qquad \beta \to 1. \tag{9.7.32}$$

(c) Use (9.7.32) to verify (9.2.54).

9.6. (a) In (9.2.56) suppose that $\beta \le \beta_0 < 1$. Replace the Airy function by its asymptotic expansion for complex argument (7.3.16) and show that the resulting nonuniform expansion agrees with (7.2.22), (7.2.34).

(b) Repeat (a) for $\beta \ge \beta_0 > 1$ and compare with (7.2.44), (7.2.22).

(c) Set $\beta = 1$ in (9.2.56) and show that the result agrees with (7.2.42).

9.7. For the integrals of Exercise 7.21 discuss the nature of the uniformly valid asymptotic expansions for t near zero. In particular discuss

(i) the forms of the canonical exponents,

(ii) the types of special functions that result,
(iii) the general expansion procedure.

9.8. Let

$$I(\lambda; x, y) = \int_{-1}^{1} g(\xi) \exp\left\{i\lambda\left[\xi(\xi - x) + y\sqrt{1-\xi^2} - \frac{\xi^2}{2}\right]\right\} d\xi. \qquad (9.7.33)$$

(a) Discuss the nature of the asymptotic expansion of I for $y > 0$, x bounded away from zero, and

$$x^{2/3} + y^{2/3} = 1.$$

(b) Discuss the nature of the asymptotic expansion of I in the neighborhood of $(x, y) = (0, 1)$.

9.9. Suppose that

$$I(\lambda; \boldsymbol{\alpha}) = \int_C g(z) \exp\{\lambda w(z; \boldsymbol{\alpha})\}\, dz, \qquad \boldsymbol{\alpha} = (\alpha_1, \ldots, \alpha_m).$$

Let $w(z; \boldsymbol{\alpha})$ have n saddle points, $n \geq m$, one or more of which coalesce for certain critical values of $\boldsymbol{\alpha}$. In particular, all of the saddle points coalesce for $\boldsymbol{\alpha} = \mathbf{0}$ yielding a single saddle point of order n. Formulate a conjecture about the nature of an asymptotic expansion of I as $\lambda \to \infty$, uniformly valid for $\boldsymbol{\alpha}$ near zero.

9.10. (a) In (9.4.45), suppose that $\theta - \theta_0 \geq \alpha > 0$. Use (9.4.31) and (9.4.34) to obtain asymptotic expansions of $W_{-1}^{(2)}$ and $W_0^{(2)}$ and verify that to leading order, the asymptotic expansion of $I(\lambda; \theta)$ given by (9.4.45) agrees with (7.5.22).
(b) Repeat for $\theta - \theta_0 \leq -\alpha < 0$ and (7.5.23).
(c) For $\theta = \theta_0$ verify that (9.4.45) agrees to leading order with (7.5.24).

9.11. Let

$$I(\lambda) = \int_\Gamma \frac{(z^2 - 1)^{1/2}}{(z^2 + 1)^{3/2}} \exp\left\{i\lambda\left[z\theta - \sqrt{z^2 + 1}\right]\right\} dz. \qquad (9.7.34)$$

Here Γ is a contour which passes above all singularities of the integrand and along which $\mathrm{Re}(z)$ goes from $-\infty$ to ∞.
(a) Show that the exponent has two saddle points

$$z_{\pm} = \pm \frac{\theta}{\sqrt{1 - \theta^2}}, \qquad (9.7.35)$$

which coalesce on the branch points at $z = \pm 1$ as $\theta \to 1/\sqrt{2}$.
(b) Show that for θ near $1/\sqrt{2}$

$$I(\lambda) \sim 2\,\mathrm{Re}\left\{\exp\left[i\lambda\,(\theta - \sqrt{2}) - \frac{3\pi i}{8}\right]\right.$$

$$\times \left[\frac{a_0}{\lambda^{3/4}} W^{(2)}_{1/2}(\sqrt{\lambda}\, a e^{i\pi/4}) + \frac{a_1}{\lambda^{5/4}} W^{(2)}_{3/2}(\sqrt{\lambda}\, a e^{i\pi/4})\right]\Bigg\}. \tag{9.7.36}$$

Here

$$a = -\operatorname{sgn}\left(\theta - \frac{1}{\sqrt{2}}\right)\left[2(\sqrt{2} - \theta - \sqrt{1-\theta^2})\right]^{1/2},$$

$$W^{(2)}_r(z) = \sqrt{2\pi}\, e^{r\pi i/2 + z^2/4} D_r(iz),$$

$$a_0 = \sqrt{2}(\sqrt{2} - \theta + \sqrt{1-\theta^2})^{-3/4}$$

$$a_1 = a^{-1} e^{-\pi i/4}\left[a_0 - \frac{|2\theta^2 - 1|^{1/2}(1-\theta^2)^{1/4}}{[2(\sqrt{2} - \theta - \sqrt{1-\theta^2})]^{1/4}}\right].$$

(c) Verify that as $\theta \to 1/\sqrt{2}$, $a_0 \to 2^{1/8}$ and $a_1 \to -2^{-1/8} e^{-i\pi/4}$

9.12. Use the integral representation

$$J_n(z) = \frac{1}{2\pi i}\left(\frac{z}{2}\right)^n \int_{0+} t^{-n-1} \exp\left\{t - \frac{z^2}{4t}\right\} dt \tag{9.7.37}$$

to verify (9.5.22).

9.13. In (9.5.32) assume that $0 \le \theta < 1$, replace the Bessel functions by the leading terms of their asymptotic expansions, and verify that, to leading order, this result agrees with (7.5.22).

9.14. In his analysis of precursors, Sommerfeld treats the integral

$$U(x,t) = -\frac{1}{\tau}\int_\Gamma \exp\{i[kx - \omega t]\}\left[\omega^2 - \left(\frac{2\pi}{\tau}\right)^2\right]^{-1} d\omega. \tag{9.7.38}$$

Here

$$k = \frac{\omega}{c}\sqrt{\frac{\omega^2 - \omega_1^2}{\omega^2 - \omega_0^2}}, \qquad \omega_1^2 = \omega_0^2(1 + a^2), \qquad \omega_0 > \frac{2\pi}{\tau}, \tag{9.7.39}$$

and Γ is the usual contour for Fourier integrals passing above all singularities of the integrand.

(a) Introduce dimensionless variables

$$\nu = \frac{\omega}{\omega_0}, \qquad b = \frac{2\pi}{\tau\omega_0}, \qquad \lambda = \omega_0 t, \qquad \theta = \frac{x}{ct}, \tag{9.7.40}$$

and then show that

$$U(x,t) = I(\lambda;\theta) = -\frac{b}{2\pi}\int_\Gamma \frac{\exp\{i\lambda\,\phi(\nu;\theta)\}}{\nu^2 - b^2}\, d\nu \tag{9.7.41}$$

with

$$\phi(v;\theta) = v\left\{\theta\sqrt{\frac{v^2-(1+a^2)}{v^2-1}} - 1\right\}. \tag{9.7.42}$$

Also, show that ϕ, as a function of v, is of the form (9.5.4) with $\alpha_0 = 0, \alpha_1 = \theta a^2/2$, and that $(v^2 - b^2)^{-1}$ is of the form (9.5.7) with $r = 1$ and $R_1 > b$.

(b) Show that, for this example $\gamma(\theta)$ and $\rho(\theta)$ defined by (9.8.11) have the estimates

$$\gamma(\theta) = a\sqrt{2(1-\theta)} + O((1-\theta)^{3/2}),$$

$$\rho(\theta) = O(1-\theta),$$

and that $a_0(\theta)$ and $a_1(\theta)$ defined by (9.5.16) have the estimates

$$a_0(\theta) = -\frac{b}{2a^2} + 0(1-\theta),$$

$$a_1(\theta) = O(\sqrt{1-\theta}).$$

(c) Show that

$$I(\lambda;\theta) = \frac{b}{a}\sqrt{2(1-\theta)}\, J_1(\lambda a\sqrt{2(1-\theta)})$$

$$\times\left[1 + O(1-\theta) + O(\lambda(1-\theta))\right]. \tag{9.7.43}$$

(d) From the error estimates in (9.7.43) conclude that this result is a good approximation only for $1 - x/ct$ and $\omega_0 t(1 - x/ct)$ both small.

9.15. Consider

$$I(\lambda;\alpha) = \int_D v(x_1,x_2)\exp\{i\lambda\left[\sin x_1 \cos x_2 \sin\alpha + \cos x_1 \cos\alpha\right]\}\, dx_1\, dx_2. \tag{9.7.44}$$

Here D is the domain $0 \le x_1 \le \pi/4$, $-\pi/4 \le x_2 \le \pi/4$, and $v(x_1,x_2)$ is a neutralizer function equal to 1 inside the circle of radius $r < \pi/8$ centered at $(\pi/4, 0)$ and equal to zero outside of the circle of radius $\pi/8$.

(a) Show that for α near $\pi/4$

$$I \sim -\frac{i}{\lambda}\sqrt{\frac{2\pi}{\sin\alpha}}\exp\left\{i\lambda\cos\left(\alpha - \frac{\pi}{4}\right)\right\}\left[b_0\, V_0(\sqrt{\lambda}\, a e^{i\pi/4}) + \frac{a_1}{\lambda^{1/2}}\right]. \tag{9.7.45}$$

Here

$$b_0 = \frac{1}{\sqrt{\sin\alpha}}, \quad a_1 = a^{-1}e^{-i\pi/4}\left\{\frac{2^{1/4}}{\cos\left(\frac{\pi}{8}-\frac{\alpha}{2}\right)} - \frac{1}{\sqrt{\sin\alpha}}\right\},$$

$$a = -2\sin\left(\frac{\pi}{8} - \frac{\alpha}{2}\right),$$

$$V_0(\sqrt{\lambda}\, ae^{i\pi/4}) = \exp\left\{\frac{i\lambda a^2}{2} + \frac{i\pi}{4}\right\} \int_{\sqrt{\lambda}a}^{\infty} e^{-it^2/2}\, dt.$$

(b) Explain why a similar technique cannot be applied as $\alpha \to 0$.

9.16. (a) Use (9.4.29) to calculate $W_{-1}^{(2)}(0)$. Use this result to calculate $I(\lambda;\theta_0)$ defined by (9.4.45) to leading order in λ. Here θ_0 is defined by (9.4.37), a_0 by (9.4.42), and γ by (9.4.46).

(b) With this result and $I(\lambda;\theta)$ defined by (9.4.38), verify that (9.4.45) agrees with the nonuniform result (7.5.24).

(c) Use the definitions (9.4.28) for $W_{-1}^{(2)}$ and $W_0^{(2)}$ and the asymptotic expansions of $D_r(iz)$ in Exercise 4.6 to recover the nonuniform expansions (7.5.22), (7.5.23) for $\pm(\theta - \theta_0) > 0$.

9.17. (a) For $s(\sigma;\tilde{\sigma})$ defined by (9.6.26) show that

$$\dot{s} = \mu[1 - \kappa\tilde{\sigma}]\, \dot{\boldsymbol{\xi}}\cdot[\boldsymbol{\xi} - \mathbf{x}_0] \tag{9.7.46}$$

and thus verify that the stationary points of the integral (9.6.27) are given by those σ_0 for which (9.6.29) is true.

(b) Show that

$$\begin{aligned} \ddot{s}(\sigma) &= \mu[1 - \tilde{\sigma}\dot{\kappa}]\, \dot{\boldsymbol{\xi}}\cdot[\boldsymbol{\xi} - \mathbf{x}_0] \\ &\quad + \mu[1 - \tilde{\sigma}\kappa][-\kappa\mathbf{N}\cdot[\boldsymbol{\xi} - \mathbf{x}_0] + 1] \end{aligned} \tag{9.7.47}$$

and verify (9.6.30).

9.18. (a) In (9.6.34) and (9.6.35), show that both $\dot{s}(\sigma) = 0$ and $\ddot{s}(\sigma) = 0$ for $\sigma = \sigma_0$, if (i) $\boldsymbol{\xi}(\sigma_0) - \mathbf{x}_0$ lies along the normal to the boundary at $\boldsymbol{\xi}(\sigma_0)$ and (ii)

$$|\boldsymbol{\xi}(\sigma_0) - \mathbf{x}_0| = \rho(\sigma_0) \tag{9.7.48}$$

with $\rho(\sigma_0) = \kappa^{-1}(\sigma_0)$ the radius of curvature at σ_0; that is, $\mathbf{x}_0$ lies on the evolute of the boundary curve $\hat{\Gamma}$.

(b) Show that at such a point

$$\dddot{s}(\sigma_0) = -\mu\dot{\kappa}(\sigma_0)\, \rho(\sigma_0) \tag{9.7.49}$$

and thus conclude that at such a point, two simple stationary points coalesce to yield a stationary point of order 2 unless $\rho(\sigma_0) = 0$ or $\dot{\kappa}(\sigma_0) = 0$.

(c) For $\dddot{s}(\sigma_0) \neq 0$ discuss the uniform asymptotic expansion of (9.6.37) for $\mathbf{x}_0$ near the evolute under the assumption that the integrand is "neutralized" away from the two nearby stationary points.

9.19. Suppose in (9.6.13) that $\mathbf{x}_0$ is an interior center bounded away from $\hat{\Gamma}$, the boundary of D.

(a) In (9.6.14) justify the following:

$$\mathscr{F}(\lambda;x_0) = +\frac{2\pi i\mu}{\lambda} - \frac{i\mu}{\lambda}\int_{\hat{\Gamma}} \frac{(\xi(\sigma)-\mathbf{x}_0)\cdot\mathbf{N}}{|\xi(\sigma)-\mathbf{x}_0|^2} \exp\{+i\lambda\mu(\xi(\sigma)-\mathbf{x}_0)^2/2\}\,d\sigma \tag{9.7.50}$$
$$\cdot[1+O(\lambda^{-1})]$$

Here $\mu = -1$ if $\mathbf{x}_0$ is a maximum of ϕ and $\mu = +1$ if $\mathbf{x}_0$ is a minimum of ϕ.

(b) Explain why (9.6.37) is a two-term approximation in this case when $I(\lambda)$ is multiplied by $\exp\{-i\lambda\,\phi(\mathbf{x}_0;\mathbf{x}_0)\}$.

9.20. (a) For the integral (9.1.1) under the conditions (9.4.2) suppose that

$$w(z;\alpha) = i\Phi(z;\alpha) \tag{9.7.51}$$

with Φ real. Show that in this case

$$\arg\gamma = \arg\alpha - \frac{\mu\pi}{4} + 0 \text{ or } \pi, \qquad \mu = \operatorname{sgn}\Phi''(\alpha;\alpha). \tag{9.7.52}$$

(b) In particular, suppose that the contour C in (9.4.1) is the real axis from 0 to ∞. In this case, show that the appropriate canonical function is

$$W_r^{(4)}(z) = \int_0^\infty t^r \exp\left\{-\frac{t^2}{2} - zt\right\} dt. \tag{9.7.53}$$

(c) Show that

$$W_0^{(4)}(\zeta e^{\pm i\pi/4}) = \sqrt{2}\, e^{\pm i\zeta^2/2}\,\mathrm{Erfc}\left[\frac{\zeta e^{\pm i\pi/4}}{\sqrt{2}}\right] \tag{9.7.54}$$

or

$$W_0^{(4)}(\zeta e^{\pm i\pi/4}) = \frac{e^{\pm i\zeta^2/2 \pm i\pi/4}}{\sqrt{2}} \int_{\pm\zeta^2/2}^{\infty} \frac{e^{\mp iu}\,du}{u^{1/2}}. \tag{9.7.55}$$

Here Erfc$[x]$ is the complementary error function defined by (9.4.33) while the second form is related to the Fresnel integrals

$$C(x) = (2\pi)^{-1/2}\int_0^x t^{-1/2}\cos t\,dt,$$
$$S(x) = (2\pi)^{-1/2}\int_0^x t^{-1/2}\sin t\,dt. \tag{9.7.56}$$

(d) Show that

$$\mathrm{Erfc}[\zeta e^{\pm i\pi/4}] - \frac{1}{2}\sqrt{\pi} = -\sqrt{\frac{\pi}{2}}\, e^{\pm i\pi/4}\,[C(\zeta^2) \mp i\,S(\zeta^2)]. \tag{9.7.57}$$

9.21. Obtain a bound on the error R_N defined by (9.4.24) in terms of the functions W_r, W_{r+1} defined by (9.4.18). Follow the line of proof in Section 9.2 but replace $\lambda^{-1/3}$ in that discussion by $\lambda^{1/2}$. Use (9.4.25), (9.4.26) to determine a second-order ordinary differential equation for W_r and verify that W_r and W_{r+1} cannot simultaneously be zero, thus assuring that Case I follows as in the proof of Section 9.2. For Case II, follow the line of proof of Exercise 9.3 to establish an expansion for the contribution from the origin in powers of $[\lambda^{1/2} w_z(0;\alpha)]^{-1}$.

9.22. For the integral $S(\lambda;\phi)$ defined by (7.7.31) obtain a uniformly valid asymptotic expansion for ϕ near $\pi/2$ in terms of the functions $W_{-1}^{(2)}$ and $W_0^{(2)}$ defined by (9.4.34) and (9.4.31), respectively.

REFERENCES

N. BLEISTEIN. Uniform asymptotic expansions of integrals with stationary point near algebraic singularity. *Comm. Pure Appl. Math.* **XIX,** 353–370, 1966.
Section 9.4 is based on this paper.

N. BLEISTEIN. Uniform asymptotic expansions of integrals with many nearby stationary points and algebraic singularities. *J. Math. Mech.* **17,** 533–560, 1967.
This paper applies the general principles described in Section 9.3 to generalize the results of Sections 9.1 and 9.4.

N. BLEISTEIN and R. A. HANDELSMAN. Uniform asymptotic expansions of double integrals. *J. Math. Anal. Appl.* **27,** 434–453, 1969.
Section 9.6 is based on this paper.

C. CHESTER, B. FRIEDMAN, and F. URSELL. An extension of the method of steepest descents. *Proc. Camb. Phil. Soc.* **53,** 599–611, 1957.
Section 9.2 is based on this paper, which is the fundamental work in the area of uniform asymptotic expansions.

A. ERDÉLYI. Uniform asymptotic expansions of integrals. *Analytical Methods in Mathematical Physics.* Gordon and Breach, New York, 1968.
This is a survey of results on uniform expansions through 1967.

L. B. FELSEN. Radiation from a uniaxially anisotropic plasma half-space. *IEEE Trans. Ant. Prop.* **AP–11,** 469–484, 1963.

L. B. FELSEN and N. MARCUVITZ. Modal analysis and synthesis of electromagnetic fields. Polytechnic Inst. of Brooklyn, Microwave Res. Inst. Rep., 1959.
In these two papers the authors develop uniformly valid asymptotic expansions for a saddle point near a pole and for a saddle point near a branch point of order $\pm\frac{1}{2}$

B. FRIEDMAN. Stationary phase with neighboring critical points. *SIAM J. Appl. Math.* **7,** 280–289, 1959.
This paper offers a clear presentation of the material treated in Section 9.2.

R. A. HANDELSMAN and N. BLEISTEIN. Uniform asymptotic expansions of integrals that arise in the analysis of precursors. *Arch. Rat. Mech. Anal.* **35,** 267–283, 1969.
Section 9.5 is based on this paper.

L. LEVEY and L. B. FELSEN. On incomplete Airy functions and their application to diffraction problems. *Radio Science* **4,** 959–969, 1969.
The authors develop a uniform expansion in terms of functions which are related to the Airy function in the same way as the incomplete gamma functions are related to the gamma function.

R. M. LEWIS. Asymptotic theory of transients in electromagnetic wave theory. *URSI Symposium, Proceedings,* Delft, Pergamon Press, New York, 1967.
The author develops a uniform expansion of an integral with a stationary point near an endpoint.

D. LUDWIG. An extension of the validity of the stationary phase formula. *SIAM J. Appl. Math.* **15,** 915–923, 1967.
The author discusses the question of how close one stationary point may come to another and still have the stationary phase formula valid. Exercises 9.3 and 9.4 are based on this paper.

D. LUDWIG. Uniform asymptotic expansions for wave propagation and diffraction problems. *SIAM Review* **12,** 325–331, 1970.

The relationship between uniform asymptotic expansions of integrals and wave propagation problems is brought out in this paper.

E. L. REISS. The impact problem for the Klein-Gordon equation. *SIAM J. Appl. Math.* **17,** 526–542, 1969.

The author analyzes precursors directly from the differential equation and obtains a Bessel series expansion.

S. O. RICE. Uniform asymptotic expansions for saddle point integrals—application to a probability distribution occurring in noise theory. *Bell Syst. Tech. J.* **47,** 1971–2013, 1968.

This is an expository paper in which the problem of actually computing coefficients in a uniform expansion is discussed.

F. URSELL. Integrals with a large parameter; The continuation of uniformly valid asymptotic expansions. *Proc. Camb. Phil. Soc.* **61,** 113–128, 1965.

The region of validity of the Airy function expansion is extended here.

F. URSELL. Integrals with a large parameter; Paths of descent and conformal mapping. *Proc. Camb. Phil. Soc.* **67,** 371–381, 1970.

The author presents an alternative proof of the analyticity of the change of variables introduced in Section 9.2.

F. URSELL. Integrals with a large parameter; Several nearly coincident saddle-points, *Proc. Camb. Phil. Soc.* **72,** 49–65, 1972.

This paper discusses the problem of many nearly stationary points and improves on the results of the Bleistein paper listed above.

B. L. VAN DER WAERDEN. On the method of saddle points. *Appl. Sci. Res.* **B2,** 33–45, 1951.

The author develops a uniformly valid asymptotic expansion for the case in which a saddle point is near a pole.

Appendix

In this appendix we list, for convenience, the Mellin transforms of several functions.[1] In each case, the original domain of analyticity of the Mellin transform is given. We remark that in determining many of the Mellin transforms quoted in the text, relationships (2)–(5) below are used.

	Function	*Mellin Transform*	*Strip of Analyticity*
1	$f(t)$	$M[f;z] = \int_0^\infty t^{z-1} f(t)\,dt$	$\alpha < \mathrm{Re}(z) < \beta$
2	$t^\nu f(t)$	$M[f;z+\nu]$	$\alpha - \mathrm{Re}(\nu) < \mathrm{Re}(z) < \beta - \mathrm{Re}(\nu)$
3	$f(t^\rho);\quad \rho > 0$	$\frac{1}{\rho} M\left[f;\frac{z}{\rho}\right]$	$\alpha\rho < \mathrm{Re}(z) < \beta\rho$
4	$f(\mu t);\quad \mu > 0$	$\mu^{-z} M[f;z]$	$\alpha < \mathrm{Re}(z) < \beta$
5	$t^\nu f(\mu t^\rho);\quad \rho, \mu > 0$	$\frac{1}{\rho} \mu^{-(z+\nu)/\rho} M\left[f;\frac{z+\nu}{\rho}\right]$	$\rho\alpha - \mathrm{Re}(\nu) < \mathrm{Re}(z) < \rho\beta - \mathrm{Re}(\nu)$
6	$t^\nu,\quad 0 < t < 1$ $0,\quad 1 < t$	$\frac{1}{z+\nu}$	$\mathrm{Re}(z) > -\mathrm{Re}(\nu)$
7	$0,\quad 0 < t < 1$ $t^\nu,\quad 1 < t$	$-\frac{1}{z+\nu}$	$\mathrm{Re}(z) < -\mathrm{Re}(\nu)$
8	$(1-t)^{\nu-1},\quad 0 < t < 1$ $0,\quad 1 \le t;\quad \mathrm{Re}(\nu) > 0$	$B(\nu, z) = \frac{\Gamma(\nu)\,\Gamma(z)}{\Gamma(\nu+z)}$	$\mathrm{Re}(z) > 0$
9	$0,\quad 0 < t < 1$ $(t-1)^{\nu-1},\quad 1 \le t,$ $\mathrm{Re}(\nu) > 0$	$B(1-\nu-z, \nu)$	$\mathrm{Re}(z) < 1 - \mathrm{Re}(\nu)$
10	$(1+t)^{-\nu},\quad \mathrm{Re}(\nu) > 0$	$B(z, \nu - z)$	$0 < \mathrm{Re}(z) < \mathrm{Re}(\nu)$
11	$(t^2 + 2t\cos\psi + 1)^{-1};$ $-\pi < \psi < \pi$	$-\frac{\pi \sin\{(z-1)\psi\}}{\sin\psi \sin\pi z}$	$0 < \mathrm{Re}(z) < 2$

[1] For a more extensive list, the reader is referred to A. Erdélyi, Ed., *Tables of Integral Transforms*, vol. I, McGraw-Hill, New York, 1954. The results of this volume are used here with permission of the publisher.

	Function	*Mellin Transform*	*Strip of Analyticity*
12	$\log t,\ 0<t<1$; $0,\ 1<t$	$-\dfrac{1}{z^2}$	$0<\mathrm{Re}(z)$
13	$\log(1+t)$	$\dfrac{\pi}{z\sin\pi z}$	$-1<\mathrm{Re}(z)<0$
14	e^{-t}	$\Gamma(z)$	$\mathrm{Re}(z)>0$
15	$e^{\pm it}$	$e^{\pm\pi iz/2}\,\Gamma(z)$	$0<\mathrm{Re}(z)<1$
16	$\sin t$	$\Gamma(z)\sin\left(\dfrac{\pi z}{2}\right)$	$-1<\mathrm{Re}(z)<1$
17	$\cos t$	$\Gamma(z)\cos\left(\dfrac{\pi z}{2}\right)$	$0<\mathrm{Re}(z)<1$
18	$J_\mu(t)$	$\dfrac{2^{z-1}\,\Gamma\left(\dfrac{z+\mu}{2}\right)}{\Gamma\left(\dfrac{\mu-z+2}{2}\right)}$	$-\mathrm{Re}(\mu)<\mathrm{Re}(z)<\frac{3}{2}$
19	$K_\mu(t)$	$2^{z-2}\,\Gamma\left(\dfrac{z-\mu}{2}\right)\Gamma\left(\dfrac{z+\mu}{2}\right)$	$\mathrm{Re}(z)>\lvert\mathrm{Re}(\mu)\rvert$
20	$e^{-rt}\,J_\mu(t)$; $\mathrm{Re}(r)>0$	$\dfrac{\Gamma(z+\mu)}{2^\mu\,r^{\mu+z}\,\Gamma(1+\mu)} \times F\left(\dfrac{z+\mu}{2},\dfrac{z+\mu+1}{2};\mu+1;-\dfrac{1}{r^2}\right)$	$\mathrm{Re}(z)>-\mathrm{Re}(\mu)$
21	$H_\mu^{(1)}(t)$	$\dfrac{(1-i)\,2^{z-1}\,\Gamma\left(\dfrac{z+\mu}{2}\right)}{\Gamma\left(\dfrac{\mu+2-z}{2}\right)}$	$\lvert\mathrm{Re}(\mu)\rvert<\mathrm{Re}(z)<\frac{3}{2}$
22	$H_\mu^{(2)}(t)$	$\dfrac{(1+i)\,2^{z-1}\,\Gamma\left(\dfrac{z+\mu}{2}\right)}{\Gamma\left(\dfrac{\mu+2-z}{2}\right)}$	$\lvert\mathrm{Re}(\mu)\rvert<\mathrm{Re}(z)<\frac{3}{2}$
23	$e^{-t}\,I_\mu(t)$	$\dfrac{\Gamma(\frac{1}{2}-z)\,\Gamma(z+\mu)}{2^z\sqrt{\pi}\,\Gamma(1+\mu-z)}$	$-\mathrm{Re}(\mu)<\mathrm{Re}(z)<\frac{1}{2}$
24	$J_\mu(tR\sin\theta)e^{-tR\cos\theta}$; $0<\theta<\dfrac{\pi}{2}$	$R^{-z}\Gamma(z+\mu)\,P_{z-1}^{-\mu}(\cos\theta)$	$\mathrm{Re}(z)>-\mathrm{Re}(\mu)$

	Function	*Mellin Transform*	*Strip of Analyticity*
25	$J_\mu(rt)\,J_\nu(t), \quad r>0$	$\dfrac{\Gamma[\frac{1}{2}(\mu+\nu+z)]\,r^\mu}{2^{1-z}\,\Gamma(\mu+1)} \times \dfrac{F\left(\frac{\mu+\nu+z}{2},\frac{\mu+z-\nu}{2};\mu+1;r^2\right)}{\Gamma[\frac{1}{2}(\nu-\mu+2-z)]},$ $0\le r<1;$ $\dfrac{2^{z-1}\,B(1-z,\frac{1}{2}(\mu+\nu+z))}{\Gamma\left(\frac{\nu-\mu-z+2}{2}\right)\Gamma\left(\frac{\mu-\nu-z+2}{2}\right)},$ $r=1;$ $\dfrac{2^{z-1}\,\Gamma[\frac{1}{2}(\mu+\nu+z)]}{r^{\nu+z}\,\Gamma(\nu+1)} \times \dfrac{F\left(\frac{\mu+\nu+z}{2},\frac{\mu+z-\nu}{2};\mu+1;r^2\right)}{\Gamma[\frac{1}{2}(\mu-\nu-z+2)]},$ $r>1$	$-\mathrm{Re}(\mu+\nu)<\mathrm{Re}(z)<1$
26	$\mathrm{Ai}(t)$	$\dfrac{3^{2z/3-7/6}}{2\pi}\Gamma\left(\frac{z}{3}\right)\Gamma\left(\frac{z+1}{3}\right)$	$\mathrm{Re}(z)>0$
27	$\mathrm{Ai}(-t)$	$\dfrac{3^{2z/3-7/6}}{\pi}\Gamma\left(\frac{z}{3}\right)\Gamma\left(\frac{z+1}{3}\right) \times \sin\left(\frac{\pi z}{3}+\frac{\pi}{6}\right)$	$0<\mathrm{Re}(z)<\frac{7}{4}$
28	$D_\mu(t)$	$\dfrac{\sqrt{\pi}\,\Gamma(z)\,2^{(1+\mu)/2}}{\Gamma\left(\frac{z+1-\mu}{2}\right)} \times F\left(\frac{z+1}{2},\frac{1-\mu}{2};\frac{z+1-\mu}{2};-1\right)$	$\mathrm{Re}(z)>0$

General References

Books

L. BERG. *Asymptotische Darstellungen und Entwicklungen.* VEB Deutscher Verlag der Wissenschaften, Berlin, 1968.

E. T. COPSON. *Asymptotic Expansions.* University Press, Cambridge, 1965.

N. G. DE BRUIJN. *Asymptotic Methods in Analysis.* North-Holland Publishing, Amsterdam, 1958.

A. ERDÉLYI. *Asymptotic Expansions.* Dover, New York, 1956.

K. O. FRIEDRICHS. *Special Topics in Analysis* (Part B). Lecture Notes, Courant Institute of Mathematical Sciences, N.Y.U., New York, 1953–1954.

H. JEFFREYS. *Asymptotic Approximations.* Clarendon Press, Cambridge, 1965.

F. W. J. OLVER. *Asymptotics and Special Functions.* Academic Press, New York, 1974.

L. SIROVICH. *Techniques of Asymptotic Analysis.* Springer-Verlag, New York, 1971.

Survey Papers

A. ERDÉLYI. Uniform asymptotic expansions of integrals. *Analytical Methods in Mathematical Physics.* Gordon and Breach, New York, 1968.

D. S. JONES. Asymptotic behavior of integrals. *SIAM Rev.* **14,** 286–317, 1972.

E. JA. RIEKSTYŅŠ and T. T. CĪRULIS. Methods that can be used for the asymptotic representation of functions defined by integrals when the parameters take large values, Latviiskii Matematiceskii Ezegodnik (Latvian Mathematical Yearbook) **7,** 193–253, Riga, 1970.

Index

Index

Abramowitz, M., 68, 178
Airy functions
  asymptotic expansion, 166, 268, 285
    complex argument, 281ff
  first kind, 52
  second kind, 52
Airy's differential equation, 50
Ai(x), Airy function of the first kind, 52
Ascent, direction of, 254
  path of, 254
Asymptotic expansion
  Poincaré type, 15
    in the complex plane, 22
    finite expansion, 16
    uniqueness, 16
  with respect to auxiliary sequence, 19
Asymptotic power series,
  definitions, 10, 11
  relation to Taylor's theorem, 12
Asymptotic sequence
  in the complex plane, 22
  definition and examples, 15
  multiplicative, 26
Asymptotically equivalent, 19
Auxiliary sequence, 19

Baker, B. B., 68
Barcilon, V., 319
Berg, L., 417
Bessel functions
  $H_0^{(1)}(t), t \to \infty$, 98
    as kernel of an integral transform, 98
  $H_{ka}^{(1)}(kr)$, $H_{ka}^{(2)}(kr)$, asymptotic expansions, $k \to \infty$, 268ff
    $r > a$, 271
    $r = a$, 273
    $r < a$, 276
  $H_{ka}^{(1)}(kr)$, $k \to \infty$, uniform for $r \approx a$, 376ff
    formula, 378
  $J_0(x), x \to 0$, 37
  $J_\nu(\lambda), \lambda \to \infty$
    via integration by parts, 97
    via stationary phase, 223
    via steepest descents, 309, 310
    $\lambda = \nu$, 224
  $J_\nu(\lambda t)$, as kernel of Hankel transform, 124
  $J_\nu(\lambda\phi(t))$, as kernel of Hankel transform, 242
    $\phi = 0$ at critical point, 230ff

Bessel functions (*Cont.*)
$\phi \neq 0$ at critical point, 243ff
$K_\nu(\lambda), \lambda \to \infty$
via Laplace's method, 193
via Watson's lemma, 165
$K_\nu(a), \nu \to \infty$, 213
$K_\mu(\lambda t)$
as kernel of an integral transform, 144
with nonmonotonic argument, 193ff
Bessel's differential equation, 64
integral representation of solutions to, 64
Beyer, W. A., 178
$\mathrm{Bi}(x)$, Airy function of the second kind, 52
Bleistein, N., 251, 319, 411
Boundary
in the context of the generalization of the method of steepest descents to h-transforms, 280
in the context of the method of steepest descents, 254
Bouwkamp, C. J., 178
Bruijn, N. G. de., 319, 417

Carleman, T., 40
Carrier, G. F., 319
Central limit theorem, 298ff
Cesàro sum, 62
Chako, N., 366
Characteristic function, 45
Chester, C., 411
Cirulis, T. T., 417
Coddington, E. A., 68
Compact support, 55
Copson, E. T., 40, 68, 101, 218, 319, 417
Corput, J. G. van der, 86, 101
Critical points, 84
Crook, M., 319

$D_\nu(\lambda)$, Weber function, 63, 383
Daniels, H. E., 178
Davis, P., 40
Descent, direction of, 254
path of, 254
Digamma function, 96
Dirichlet kernel, 61
Dispersion relation
form for precursor analysis, 388
for Klein-Gordon equation, 54
for three-dimensional waves, 354
Doetsch, G., 178

$E_1(x)$, exponential integral, 2
Elliptic integrals, 172
Erdélyi, A., 40, 68, 90, 101, 218, 251, 411, 417
Euler-Mascheroni constant, 2
Exponential integral
asymptotic expansion for large argument, 3
asymptotic expansion for small argument, 2
numerical values, 5

Far-field approximations, 55
Féjer kernel, 62
Feller, W., 68, 178
Felsen, L. B., 411
Feshbach, H., 68
Focke, J., 366
Fourier cosine transform, 77
Fourier integral, asymptotic expansion formulae, 80, 91, 92
Fourier integral, multidimensional, 340ff
corner contribution, two dimensions, 349ff
correction to stationary phase formula, 347
stationary phase formula, 347
nonsimple stationary point, 351
stationary point of the boundary integral, 360ff
stationary point of the boundary, 348
two dimensions, center, 360
Fourier sine transform, 77
Fourier transform, 77
Fox, C., 178
Fractional integrals, asymptotic expansion, large argument, 155ff
Friedman, B., 411
Friedrichs, K. O., 319, 417
Fulks, W., 218

Gamma function, 42ff
  asymptotic expansion
    for complex argument, via integration by parts, 81ff
    for complex argument, via steepest descents, 286ff
    formulae, 83, 291
    via Laplace's method, 185
  incomplete, 43
Granoff, B., 44, 62, 319

$H_\nu^{(j)}(r)$, Hankel function, 98
Handelsman, R. A., 178, 218, 251, 319, 411
Hankel transform, 77
  asymptotic expansion, 124ff
  of nonmonotonic argument, 233
    nonzero interior stationary point, 244
Heller, L., 178
Hill, in the context of the generalization of the method of steepest descents, 280
  in the context of the method of steepest descents, 254
Holomorphic function, 106
h-Transforms of monotonic argument, $\lambda \to \infty$, 117ff
  formulae, 121, 170
  for generalized transforms, 127
h-Transforms of monotonic argument, $\lambda \to 0$, 130ff
h-Transforms of nonmontonic argument, $\lambda \to \infty$
  algebraic kernel, 199ff
    nonzero argument at critical point, 200
    zero argument at critical point, 201ff
  exponential kernel, 187ff
    interior critical point, 197
    left endpoint critical, 189ff
    right endpoint critical, 195ff
  oscillatory kernel
    left endpoint critical, 230ff
    right endpoint critical, 234ff
    specialized to $J_\nu(\lambda\phi)$, 233, 242, 244
    stationary point of order ν_0, 231

Incomplete gamma function, 43, 62, 135
Integral transforms, 76ff
Integration by parts, 69ff
  for kernel $h(t;\lambda)$, 71
    for $(1-t^2)^\lambda$, 73
    for t^λ, 72
  for kernel $h(\lambda t)$; h-transforms, 76ff
    for $h(\lambda\phi(t))$, $\phi'(t)$ nonzero, 79
    for $\exp(i\lambda t)$, 80
    for $\exp(-\lambda t)$, 81
  for kernel $(t-a)^{\alpha-1}\exp(i\lambda t)$, 89ff

$J_\nu(t)$, Bessel function, 37, 64
Jeffreys, H., 68, 319, 417
Joint probability distribution function, 45
Jones, D. S., 178, 218, 251, 319, 366, 417

$K_\nu(t)$, modified Bessel function, 144, 165
K_ν transform, 144, 193
Kaplun, S., 40
Kelvin, Lord W. T., 251
Kernel of a transform, 76
Kirchhoff approximation, 55ff, 247, 357
Klein-Gordon equation, 54, 292ff
Kline, M., 366

Lanczos, C., 68
Laplace, P. S., 218
Laplace integral, 181
  and integration by parts, 81
  multidimensional, 331ff
    with simple boundary maximum, $\nabla\phi = 0$, 339
    with simple boundary maximum, $\nabla\phi \neq 0$, 340
    with simple interior maximum, 337, 338, 339
  in two dimensions, 322ff
    with simple boundary maximum, $\nabla\phi = 0$, 331
    with simple boundary maximum, $\nabla\phi \neq 0$, 325
    with simple interior maximum, 329, 360
  and Watson's lemma, 102ff

Laplace transform, 46, 77
  asymptotic expansion, $\lambda \to \infty$, 122
  asymptotic expansion, $\lambda \to 0$, 134
  bilateral, 106
  generalized, 50ff
  inversion formula, 77
Laplace's formula, 183
Laplace's method, 180ff
  for interior maximum, 183, 196
  for interior maximum of order $2n$, 197, 210
  for the kernel $[\phi(t)]^\lambda$
    interior maximum, 212
    left endpoint maximum, 211
    maximum of order β, 212
    right endpoint maximum, 212
  for left endpoint maximum, 184
  for left endpoint maximum of order $2n$, 210
  for left endpoint maximum of order β, 211
  for right endpoint critical, 184
Legendre functions
  associated, second kind, 63
  asymptotic expansions, 213
Levey, L., 411
Lew, J. S., 178, 218, 319
Lewis, R. M., 366, 411
"Locally integrable" defined, 103
Ludwig, D., 411

Marcuvitz, N., 411
Maximum modulus theorem, 257, 302
Mellin transform, 77, 106ff
  analytic continuation, 110ff
  generalized, 115
  Parseval formula, 108
    generalized, 126
Millar, R. F., 179
Monkey saddle, 258, 259
Morse, P. M., 68

Neutralized function, 88
Neutralizer, 86ff
Nussenzveig, H. M., 319

Olmstead, W. E., 177
Olver, F. W. J., 179, 218, 319, 417
Order of vanishing, 8
Order relations, 6ff
  large "O" defined, 6
  small "O" defined, 6

Parseval formula for Mellin transform, 108, 126
Pearson, C. E., 319
Poincaré, H., 40
Poincaré type, *see* Asymptotic expansion of
Precursor, 294, 297, 387
Primary field, 56
Probability density function, 44
  of a quotient of random variables, 46
  of a sum of random variables, 44
Probability distribution function, 44
  of a product of random variables, 45

Radiation condition, 56
Random variable, 44
Reiss, E. L., 412
Relaxation time, 48
Rice, S. O., 412
Riekstyn'š, E. Ja., 417
Riemann, B., 320
Rosser, J. B., 101

Saddle point, 258
Saddle point method, 265
Scattered field, 56
Sirovich, L., 320, 417
Stationary phase, 219ff
  formula, 220
    with correction terms, 248
    left stationary point of order ν, 222
    right stationary point of order ρ, 222
  generalized to h-transforms, *see* h-Transforms of nonmonotonic argument, oscillatory kernels
  method of, 222
Stationary phase, multidimensional, 340ff
  formula, 347
    correction term, 347

two-dimensional center, 360
see also Fourier integral, multidimensional
Stationary point, 220
in n dimensions, 341
of order ν, 221
simple, 221
Steepest ascent, directions of, paths of, 254
Steepest descent, directions of, paths of, 254
Steepest descents, 262ff
formula, 265
generalized to h-transforms, 280, 316ff
for loop integrals, 305
Stenger, F., 179
Stieltjes transform, 77
asymptotic expansion for large argument, 123ff
generalized, 140
Stirling's formula, 83, 186
Stokes, G. G., 40, 251
Stokes line, 23
for the Airy function, 284
Stokes' phenomenon, 23

Titchmarsh, E. C., 68, 101, 179, 320

Uniformly valid asymptotic expansions, 367ff
double integrals with variable position stationary point, 393ff
formula, 395
semiuniform, stationary point near boundary, 396ff
formula, 399
precursor integral, saddle point attaining infinite order, 387ff
formula, 391
saddle point near amplitude critical point, 380ff
formula, 383ff
two nearby saddle points, 369ff
formula, 373
generalized to n nearby saddle points, 405
underlying principles, 379ff
Ursell, F., 411, 412

Valley, in the method of steepest descents, 254

Waerden, B. van der, 320
Watson, G. N., 40, 179, 251, 320
Watson's lemma, 103
for loop integrals, 162, 167ff
Weber function, $D_\nu(\lambda)$, $\lambda \to \infty$, 167
complex λ, 307
Weber's differential equation, 53
solutions represented as integrals, 63
Whittaker, E. T., 68
Widder, D. V., 68, 101, 179, 218
Woolcock, W. S., 179

Zimmering, S., 179

A CATALOG OF SELECTED DOVER BOOKS IN ALL FIELDS OF INTEREST

THE MURDER BOOK OF J.G. REEDER, Edgar Wallace. Eight suspenseful stories by bestselling mystery writer of 20s and 30s. Features the donnish Mr. J.G. Reeder of Public Prosecutor's Office. 128pp. 5⅜ × 8½. (Available in U.S. only)
24374-5 Pa. $3.50

ANNE ORR'S CHARTED DESIGNS, Anne Orr. Best designs by premier needlework designer, all on charts: flowers, borders, birds, children, alphabets, etc. Over 100 charts, 10 in color. Total of 40pp. 8¼ × 11. 23704-4 Pa. $2.25

BASIC CONSTRUCTION TECHNIQUES FOR HOUSES AND SMALL BUILDINGS SIMPLY EXPLAINED, U.S. Bureau of Naval Personnel. Grading, masonry, woodworking, floor and wall framing, roof framing, plastering, tile setting, much more. Over 675 illustrations. 568pp. 6½ × 9¼. 20242-9 Pa. $8.95

MATISSE LINE DRAWINGS AND PRINTS, Henri Matisse. Representative collection of female nudes, faces, still lifes, experimental works, etc., from 1898 to 1948. 50 illustrations. 48pp. 8⅜ × 11¼. 23877-6 Pa. $2.50

HOW TO PLAY THE CHESS OPENINGS, Eugene Znosko-Borovsky. Clear, profound examinations of just what each opening is intended to do and how opponent can counter. Many sample games. 147pp. 5⅜ × 8½. 22795-2 Pa. $2.95

DUPLICATE BRIDGE, Alfred Sheinwold. Clear, thorough, easily followed account: rules, etiquette, scoring, strategy, bidding; Goren's point-count system, Blackwood and Gerber conventions, etc. 158pp. 5⅜ × 8½. 22741-3 Pa. $3.00

SARGENT PORTRAIT DRAWINGS, J.S. Sargent. Collection of 42 portraits reveals technical skill and intuitive eye of noted American portrait painter, John Singer Sargent. 48pp. 8¼ × 11⅛. 24524-1 Pa. $2.95

ENTERTAINING SCIENCE EXPERIMENTS WITH EVERYDAY OBJECTS, Martin Gardner. Over 100 experiments for youngsters. Will amuse, astonish, teach, and entertain. Over 100 illustrations. 127pp. 5⅜ × 8½. 24201-3 Pa. $2.50

TEDDY BEAR PAPER DOLLS IN FULL COLOR: A Family of Four Bears and Their Costumes, Crystal Collins. A family of four Teddy Bear paper dolls and nearly 60 cut-out costumes. Full color, printed one side only. 32pp. 9¼ × 12¼.
24550-0 Pa. $3.50

NEW CALLIGRAPHIC ORNAMENTS AND FLOURISHES, Arthur Baker. Unusual, multi-useable material: arrows, pointing hands, brackets and frames, ovals, swirls, birds, etc. Nearly 700 illustrations. 80pp. 8⅜ × 11¼.
24095-9 Pa. $3.75

DINOSAUR DIORAMAS TO CUT & ASSEMBLE, M. Kalmenoff. Two complete three-dimensional scenes in full color, with 31 cut-out animals and plants. Excellent educational toy for youngsters. Instructions; 2 assembly diagrams. 32pp. 9¼ × 12¼. 24541-1 Pa. $3.95

SILHOUETTES: A PICTORIAL ARCHIVE OF VARIED ILLUSTRATIONS, edited by Carol Belanger Grafton. Over 600 silhouettes from the 18th to 20th centuries. Profiles and full figures of men, women, children, birds, animals, groups and scenes, nature, ships, an alphabet. 144pp. 8⅜ × 11¼. 23781-8 Pa. $4.95

JAPANESE DESIGN MOTIFS, Matsuya Co. Mon, or heraldic designs. Over 4000 typical, beautiful designs: birds, animals, flowers, swords, fans, geometrics; all beautifully stylized. 213pp. 11⅜ × 8¼. 22874-6 Pa. $7.95

THE TALE OF BENJAMIN BUNNY, Beatrix Potter. Peter Rabbit's cousin coaxes him back into Mr. McGregor's garden for a whole new set of adventures. All 27 full-color illustrations. 59pp. 4¼ × 5½. (Available in U.S. only) 21102-9 Pa. $1.50

THE TALE OF PETER RABBIT AND OTHER FAVORITE STORIES BOXED SET, Beatrix Potter. Seven of Beatrix Potter's best-loved tales including Peter Rabbit in a specially designed, durable boxed set. 4¼ × 5½. Total of 447pp. 158 color illustrations. (Available in U.S. only) 23903-9 Pa. $10.80

PRACTICAL MENTAL MAGIC, Theodore Annemann. Nearly 200 astonishing feats of mental magic revealed in step-by-step detail. Complete advice on staging, patter, etc. Illustrated. 320pp. 5⅜ × 8½. 24426-1 Pa. $5.95

CELEBRATED CASES OF JUDGE DEE (DEE GOONG AN), translated by Robert Van Gulik. Authentic 18th-century Chinese detective novel; Dee and associates solve three interlocked cases. Led to van Gulik's own stories with same characters. Extensive introduction. 9 illustrations. 237pp. 5⅜ × 8½. 23337-5 Pa. $4.50

CUT & FOLD EXTRATERRESTRIAL INVADERS THAT FLY, M. Grater. Stage your own lilliputian space battles. By following the step-by-step instructions and explanatory diagrams you can launch 22 full-color fliers into space. 36pp. 8¼ × 11. 24478-4 Pa. $2.95

CUT & ASSEMBLE VICTORIAN HOUSES, Edmund V. Gillon, Jr. Printed in full color on heavy cardboard stock, 4 authentic Victorian houses in H-O scale: Italian-style Villa, Octagon, Second Empire, Stick Style. 48pp. 9¼ × 12¼. 23849-0 Pa. $3.95

BEST SCIENCE FICTION STORIES OF H.G. WELLS, H.G. Wells. Full novel *The Invisible Man*, plus 17 short stories: "The Crystal Egg," "Aepyornis Island," "The Strange Orchid," etc. 303pp. 5⅜ × 8½. (Available in U.S. only) 21531-8 Pa. $4.95

TRADEMARK DESIGNS OF THE WORLD, Yusaku Kamekura. A lavish collection of nearly 700 trademarks, the work of Wright, Loewy, Klee, Binder, hundreds of others. 160pp. 8¾ × 8. (Available in U.S. only) 24191-2 Pa. $5.00

THE ARTIST'S AND CRAFTSMAN'S GUIDE TO REDUCING, ENLARGING AND TRANSFERRING DESIGNS, Rita Weiss. Discover, reduce, enlarge, transfer designs from any objects to any craft project. 12pp. plus 16 sheets special graph paper. 8¼ × 11. 24142-4 Pa. $3.25

TREASURY OF JAPANESE DESIGNS AND MOTIFS FOR ARTISTS AND CRAFTSMEN, edited by Carol Belanger Grafton. Indispensable collection of 360 traditional Japanese designs and motifs redrawn in clean, crisp black-and-white, copyright-free illustrations. 96pp. 8¼ × 11. 24435-0 Pa. $3.95

KEYBOARD WORKS FOR SOLO INSTRUMENTS, G.F. Handel. 35 neglected works from Handel's vast oeuvre, originally jotted down as improvisations. Includes Eight Great Suites, others. New sequence. 174pp. 9⅜ × 12¼.
24338-9 Pa. $7.50

AMERICAN LEAGUE BASEBALL CARD CLASSICS, Bert Randolph Sugar. 82 stars from 1900s to 60s on facsimile cards. Ruth, Cobb, Mantle, Williams, plus advertising, info, no duplications. Perforated, detachable. 16pp. 8¼ × 11.
24286-2 Pa. $2.95

A TREASURY OF CHARTED DESIGNS FOR NEEDLEWORKERS, Georgia Gorham and Jeanne Warth. 141 charted designs: owl, cat with yarn, tulips, piano, spinning wheel, covered bridge, Victorian house and many others. 48pp. 8¼ × 11.
23558-0 Pa. $1.95

DANISH FLORAL CHARTED DESIGNS, Gerda Bengtsson. Exquisite collection of over 40 different florals: anemone, Iceland poppy, wild fruit, pansies, many others. 45 illustrations. 48pp. 8¼ × 11. 23957-8 Pa. $1.75

OLD PHILADELPHIA IN EARLY PHOTOGRAPHS 1839-1914, Robert F. Looney. 215 photographs: panoramas, street scenes, landmarks, President-elect Lincoln's visit, 1876 Centennial Exposition, much more. 230pp. 8⅜ × 11¾.
23345-6 Pa. $9.95

PRELUDE TO MATHEMATICS, W.W. Sawyer. Noted mathematician's lively, stimulating account of non-Euclidean geometry, matrices, determinants, group theory, other topics. Emphasis on novel, striking aspects. 224pp. 5⅜ × 8½.
24401-6 Pa. $4.50

ADVENTURES WITH A MICROSCOPE, Richard Headstrom. 59 adventures with clothing fibers, protozoa, ferns and lichens, roots and leaves, much more. 142 illustrations. 232pp. 5⅜ × 8½. 23471-1 Pa. $3.95

IDENTIFYING ANIMAL TRACKS: MAMMALS, BIRDS, AND OTHER ANIMALS OF THE EASTERN UNITED STATES, Richard Headstrom. For hunters, naturalists, scouts, nature-lovers. Diagrams of tracks, tips on identification. 128pp. 5⅜ × 8. 24442-3 Pa. $3.50

VICTORIAN FASHIONS AND COSTUMES FROM HARPER'S BAZAR, 1867-1898, edited by Stella Blum. Day costumes, evening wear, sports clothes, shoes, hats, other accessories in over 1,000 detailed engravings. 320pp. 9⅜ × 12¼.
22990-4 Pa. $9.95

EVERYDAY FASHIONS OF THE TWENTIES AS PICTURED IN SEARS AND OTHER CATALOGS, edited by Stella Blum. Actual dress of the Roaring Twenties, with text by Stella Blum. Over 750 illustrations, captions. 156pp. 9 × 12.
24134-3 Pa. $8.50

HALL OF FAME BASEBALL CARDS, edited by Bert Randolph Sugar. Cy Young, Ted Williams, Lou Gehrig, and many other Hall of Fame greats on 92 full-color, detachable reprints of early baseball cards. No duplication of cards with *Classic Baseball Cards.* 16pp. 8¼ × 11. 23624-2 Pa. $3.50

THE ART OF HAND LETTERING, Helm Wotzkow. Course in hand lettering, Roman, Gothic, Italic, Block, Script. Tools, proportions, optical aspects, individual variation. Very quality conscious. Hundreds of specimens. 320pp. 5⅜ × 8½.
21797-3 Pa. $4.95

THE RIME OF THE ANCIENT MARINER, Gustave Doré, S.T. Coleridge. Doré's finest work, 34 plates capture moods, subtleties of poem. Full text. 77pp. 9¼ × 12. 22305-1 Pa. $4.95

SONGS OF INNOCENCE, William Blake. The first and most popular of Blake's famous "Illuminated Books," in a facsimile edition reproducing all 31 brightly colored plates. Additional printed text of each poem. 64pp. 5¼ × 7. 22764-2 Pa. $3.00

AN INTRODUCTION TO INFORMATION THEORY, J.R. Pierce. Second (1980) edition of most impressive non-technical account available. Encoding, entropy, noisy channel, related areas, etc. 320pp. 5⅜ × 8½. 24061-4 Pa. $4.95

THE DIVINE PROPORTION: A STUDY IN MATHEMATICAL BEAUTY, H.E. Huntley. "Divine proportion" or "golden ratio" in poetry, Pascal's triangle, philosophy, psychology, music, mathematical figures, etc. Excellent bridge between science and art. 58 figures. 185pp. 5⅜ × 8½. 22254-3 Pa. $3.95

THE DOVER NEW YORK WALKING GUIDE: From the Battery to Wall Street, Mary J. Shapiro. Superb inexpensive guide to historic buildings and locales in lower Manhattan: Trinity Church, Bowling Green, more. Complete Text; maps. 36 illustrations. 48pp. 3⅞ × 9¼. 24225-0 Pa. $2.50

NEW YORK THEN AND NOW, Edward B. Watson, Edmund V. Gillon, Jr. 83 important Manhattan sites: on facing pages early photographs (1875-1925) and 1976 photos by Gillon. 172 illustrations. 171pp. 9¼ × 10. 23361-8 Pa. $7.95

HISTORIC COSTUME IN PICTURES, Braun & Schneider. Over 1450 costumed figures from dawn of civilization to end of 19th century. English captions. 125 plates. 256pp. 8⅜ × 11¼. 23150-X Pa. $7.50

VICTORIAN AND EDWARDIAN FASHION: A Photographic Survey, Alison Gernsheim. First fashion history completely illustrated by contemporary photographs. Full text plus 235 photos, 1840-1914, in which many celebrities appear. 240pp. 6½ × 9¼. 24205-6 Pa. $6.00

CHARTED CHRISTMAS DESIGNS FOR COUNTED CROSS-STITCH AND OTHER NEEDLECRAFTS, Lindberg Press. Charted designs for 45 beautiful needlecraft projects with many yuletide and wintertime motifs. 48pp. 8¼ × 11. 24356-7 Pa. $1.95

101 FOLK DESIGNS FOR COUNTED CROSS-STITCH AND OTHER NEEDLE-CRAFTS, Carter Houck. 101 authentic charted folk designs in a wide array of lovely representations with many suggestions for effective use. 48pp. 8¼ × 11. 24369-9 Pa. $2.25

FIVE ACRES AND INDEPENDENCE, Maurice G. Kains. Great back-to-the-land classic explains basics of self-sufficient farming. The one book to get. 95 illustrations. 397pp. 5⅜ × 8½. 20974-1 Pa. $4.95

A MODERN HERBAL, Margaret Grieve. Much the fullest, most exact, most useful compilation of herbal material. Gigantic alphabetical encyclopedia, from aconite to zedoary, gives botanical information, medical properties, folklore, economic uses, and much else. Indispensable to serious reader. 161 illustrations. 888pp. 6½ × 9¼. (Available in U.S. only) 22798-7, 22799-5 Pa., Two-vol. set $16.45

REASON IN ART, George Santayana. Renowned philosopher's provocative, seminal treatment of basis of art in instinct and experience. Volume Four of *The Life of Reason*. 230pp. 5⅜ × 8. 24358-3 Pa. $4.50

LANGUAGE, TRUTH AND LOGIC, Alfred J. Ayer. Famous, clear introduction to Vienna, Cambridge schools of Logical Positivism. Role of philosophy, elimination of metaphysics, nature of analysis, etc. 160pp. 5⅜ × 8½. (USCO) 20010-8 Pa. $2.75

BASIC ELECTRONICS, U.S. Bureau of Naval Personnel. Electron tubes, circuits, antennas, AM, FM, and CW transmission and receiving, etc. 560 illustrations. 567pp. 6½ × 9¼. 21076-6 Pa. $8.95

THE ART DECO STYLE, edited by Theodore Menten. Furniture, jewelry, metalwork, ceramics, fabrics, lighting fixtures, interior decors, exteriors, graphics from pure French sources. Over 400 photographs. 183pp. 8⅜ × 11¼. 22824-X Pa. $6.95

THE FOUR BOOKS OF ARCHITECTURE, Andrea Palladio. 16th-century classic covers classical architectural remains, Renaissance revivals, classical orders, etc. 1738 Ware English edition. 216 plates. 110pp. of text. 9½ × 12¾. 21308-0 Pa. $11.50

THE WIT AND HUMOR OF OSCAR WILDE, edited by Alvin Redman. More than 1000 ripostes, paradoxes, wisecracks: Work is the curse of the drinking classes, I can resist everything except temptations, etc. 258pp. 5⅜ × 8½. (USCO) 20602-5 Pa. $3.50

THE DEVIL'S DICTIONARY, Ambrose Bierce. Barbed, bitter, brilliant witticisms in the form of a dictionary. Best, most ferocious satire America has produced. 145pp. 5⅜ × 8½. 20487-1 Pa. $2.50

ERTÉ'S FASHION DESIGNS, Erté. 210 black-and-white inventions from *Harper's Bazar*, 1918-32, plus 8pp. full-color covers. Captions. 88pp. 9 × 12. 24203-X Pa. $6.50

ERTÉ GRAPHICS, Erté. Collection of striking color graphics: *Seasons, Alphabet, Numerals, Aces* and *Precious Stones*. 50 plates, including 4 on covers. 48pp. 9⅜ × 12¼. 23580-7 Pa. $6.95

PAPER FOLDING FOR BEGINNERS, William D. Murray and Francis J. Rigney. Clearest book for making origami sail boats, roosters, frogs that move legs, etc. 40 projects. More than 275 illustrations. 94pp. 5⅜ × 8½. 20713-7 Pa. $2.25

ORIGAMI FOR THE ENTHUSIAST, John Montroll. Fish, ostrich, peacock, squirrel, rhinoceros, Pegasus, 19 other intricate subjects. Instructions. Diagrams. 128pp. 9 × 12. 23799-0 Pa. $4.95

CROCHETING NOVELTY POT HOLDERS, edited by Linda Macho. 64 useful, whimsical pot holders feature kitchen themes, animals, flowers, other novelties. Surprisingly easy to crochet. Complete instructions. 48pp. 8¼ × 11. 24296-X Pa. $1.95

CROCHETING DOILIES, edited by Rita Weiss. Irish Crochet, Jewel, Star Wheel, Vanity Fair and more. Also luncheon and console sets, runners and centerpieces. 51 illustrations. 48pp. 8¼ × 11. 23424-X Pa. $2.00

YUCATAN BEFORE AND AFTER THE CONQUEST, Diego de Landa. Only significant account of Yucatan written in the early post-Conquest era. Translated by William Gates. Over 120 illustrations. 162pp. 5⅜ × 8½. 23622-6 Pa. $3.50

ORNATE PICTORIAL CALLIGRAPHY, E.A. Lupfer. Complete instructions, over 150 examples help you create magnificent "flourishes" from which beautiful animals and objects gracefully emerge. 8⅛ × 11. 21957-7 Pa. $2.95

DOLLY DINGLE PAPER DOLLS, Grace Drayton. Cute chubby children by same artist who did Campbell Kids. Rare plates from 1910s. 30 paper dolls and over 100 outfits reproduced in full color. 32pp. 9¼ × 12¼. 23711-7 Pa. $3.50

CURIOUS GEORGE PAPER DOLLS IN FULL COLOR, H. A. Rey, Kathy Allert. Naughty little monkey-hero of children's books in two doll figures, plus 48 full-color costumes: pirate, Indian chief, fireman, more. 32pp. 9¼ × 12¼. 24386-9 Pa. $3.50

GERMAN: HOW TO SPEAK AND WRITE IT, Joseph Rosenberg. Like *French, How to Speak and Write It.* Very rich modern course, with a wealth of pictorial material. 330 illustrations. 384pp. 5⅜ × 8½. (USUKO) 20271-2 Pa. $4.75

CATS AND KITTENS: 24 Ready-to-Mail Color Photo Postcards, D. Holby. Handsome collection; feline in a variety of adorable poses. Identifications. 12pp. on postcard stock. 8¼ × 11. 24469-5 Pa. $2.95

MARILYN MONROE PAPER DOLLS, Tom Tierney. 31 full-color designs on heavy stock, from *The Asphalt Jungle, Gentlemen Prefer Blondes,* 22 others. 1 doll. 16 plates. 32pp. 9⅜ × 12¼. 23769-9 Pa. $3.50

FUNDAMENTALS OF LAYOUT, F.H. Wills. All phases of layout design discussed and illustrated in 121 illustrations. Indispensable as student's text or handbook for professional. 124pp. 8⅛ × 11. 21279-3 Pa. $4.50

FANTASTIC SUPER STICKERS, Ed Sibbett, Jr. 75 colorful pressure-sensitive stickers. Peel off and place for a touch of pizzazz: clowns, penguins, teddy bears, etc. Full color. 16pp. 8¼ × 11. 24471-7 Pa. $2.95

LABELS FOR ALL OCCASIONS, Ed Sibbett, Jr. 6 labels each of 16 different designs—baroque, art nouveau, art deco, Pennsylvania Dutch, etc.—in full color. 24pp. 8¼ × 11. 23688-9 Pa. $2.95

HOW TO CALCULATE QUICKLY: RAPID METHODS IN BASIC MATHEMATICS, Henry Sticker. Addition, subtraction, multiplication, division, checks, etc. More than 8000 problems, solutions. 185pp. 5 × 7¼. 20295-X Pa. $2.95

THE CAT COLORING BOOK, Karen Baldauski. Handsome, realistic renderings of 40 splendid felines, from American shorthair to exotic types. 44 plates. Captions. 48pp. 8¼ × 11. 24011-8 Pa. $2.25

THE TALE OF PETER RABBIT, Beatrix Potter. The inimitable Peter's terrifying adventure in Mr. McGregor's garden, with all 27 wonderful, full-color Potter illustrations. 55pp. 4¼ × 5½. (Available in U.S. only) 22827-4 Pa. $1.60

BASIC ELECTRICITY, U.S. Bureau of Naval Personnel. Batteries, circuits, conductors, AC and DC, inductance and capacitance, generators, motors, transformers, amplifiers, etc. 349 illustrations. 448pp. 6½ × 9¼. 20973-3 Pa. $7.95